W9-ACX-200

WRITING WELL FOR THE TECHNICAL PROFESSIONS

CARL A. RUDISILL LIBRARY
LENOIR-RHYNE COLLEGE

WRITING WELL FOR THE TECHNICAL PROFESSIONS

Anne Eisenberg
Polytechnic University

CARL A. RUDISILL LIBRARY
LENOIR-RHYNE COLLEGE

1817 HARPER & ROW, PUBLISHERS

New York Cambridge Philadelphia
San Francisco Washington London Mexico City
São Paulo Singapore Sydney

T
11
.E38
1989
154257
Nov. 1991

Sponsoring Editor: Lucy Rosendahl

Project Coordination, Text Design, Text Art: Publishing Synthesis Ltd.

Cover Design: Patrick Deffenbaugh

Compositor: Ruttle, Shaw, & Wetherill, Inc.

Printer and Binder: R. R. Donnelley & Sons Company

Writing Well for the Technical Professions

Copyright © 1989 by Harper & Row, Publishers, Inc.

All rights reserved. Printed in the United States of America. No part of this book may be used or reproduced in any manner whatsoever without written permission, except in the case of brief quotations embodied in critical articles and reviews. For information address Harper & Row, Publishers, Inc., 10 East 53d Street, New York, NY 10022.

Library of Congress Cataloging in Publication Data

Eisenberg, Anne, 1942–
Writing well for the technical professions / Anne Eisenberg.
p. cm.
Includes index.
ISBN 0-06-041892-3
1. Technical writing. I. Title.
T11.E38 1989
808′.0666—dc19 88-23510
CIP

90 91 9 8 7 6 5 4 3 2

Contents

PART VI REPORTS, PROPOSALS, MANUALS, ORAL PRESENTATIONS

Preface

- A memo to a client explaining a provision of the 1986 Tax Reform Act
- A set of safety instructions for laboratory workers
- A report on advances in the laying of deep-water pipelines

These are all examples of technical communication, communication that plays a fundamental part in the work of technical professionals, whether they are chemists or computer scientists, accountants or engineers.

This book is an introduction to technical communication for students who are beginning to explore the writing of their professional fields: instructions, process explanations, descriptions, abstracts, letters, notebooks, memos, manuals, proposals, and reports of many types.

DOING THE JOB WELL

What is good professional communication? It is writing or speaking that is accurate, complete, and understandable to its audience—that tells the truth about the data directly and clearly. Doing this takes **research, analysis of the audience,** and the mastering of the three interrelated elements of **organization, language,** and **design and illustration.**

- **research.** Lots of it—understanding the technical content of the material by reading, interviewing, listening, observing, ordering, classifying, inferring.
- **analysis of the audience.** One of the most important ideas students practice in a writing course for the professions is audience analysis. In many forms of academic writing—term papers in Vic-

torian Literature, for instance, or in Management—the audience or readership is one person: an expert, the instructor who assigns and then reads the paper. In the writing you will do as a technical professional, however, the audience is rarely a single academic reader. Instead, you will often write for people whose technical backgrounds are varied, or who are not specialists in your field. You will often find yourself writing for people who are paying you to solve a problem, and to tell the solution in a brief, accessible, well-supported argument. You will find yourself writing reports for people who are very busy, and who therefore need a salient abstract or summary first so they will know the "What?" and the "So What?" *before* the details. In all these instances, the writer needs to adjust the technical information so that it is appropriate and effective for the intended audience.

- **organization.** Strong organizational patterns to guide the reader or listener.
- **language.** Direct, accessible, and helpful to your audience.
- **design and illustration** that give the text its strongest visual impact.

This textbook is about ways to handle these imperatives of *research, audience analysis, organization, language,* and *visual impact.*

THE WRITING PROCESS

As a student learning to write about technical materials, you may sometimes feel you have three roles to juggle:

- **the writer** deciding how to order and word the argument
- **the editor/critic** thinking about how to revise the text to increase its clarity
- **the researcher** handling comments from interviews, data from tables, or page after page of details from, for instance, a tax code or regulatory document

Indeed, writing in the technical professions does require all three of these intricate, overlapping roles, and the examples and exercises in this textbook are designed to develop your ability to handle them as part of the writing process.

If you look at examples of technical writing that you admire—a clearly presented set of instructions, an understandable article explaining a complicated procedure—you may think that such excellent text sprang full-blown from the writer's mind onto the page.

Actually, such facility is rare. True, there are the famous few writers who sit down and rattle out word-perfect prose in first drafts, but most writers describe a process of writing and rewriting to get that final, readable version.

The process they describe is rarely an arrow-straight, linear progression of outline, draft, revision, and submission. Instead, they describe a shuttling process that often loops back on itself. They move back and forth, from a draft of the evolving report to their notes to the report, adjusting the logic, the details, and the language so that the final version will be accessible to the reader. Writers describe a progression in which they collect and assimilate material, and then move through several drafts, revising as they go, mindful throughout of the demands of the audience.

Many of the examples in this book show how technical writers adjust text during the writing process, the sorts of changes they make as they take the measure of their audience and then set out to write for them, revising their drafts as they head for a final product. Audience informs most of the revisions, from choice of language to decisions about which data to illustrate. The more sophisticated the audience's background knowledge of the subject material, the denser the technical detail, including detail in illustration.

STRUCTURE OF THIS BOOK

This book is divided into six parts:

- Part I. The tradition of technical reporting, and the importance of audience analysis.
- Part II. The techniques that underlie readability: logic and organizational patterns, direct language, effective visual display.
- Part III. Gathering the data—interview, observation, library search.
- Part IV. Applications—definitions and glossaries, descriptions, process explanations, instructions.
- Part V. Correspondence—letters and memos.
- Part VI. Reports, proposals, manuals, and oral presentations.

ACKNOWLEDGMENTS

I'd like to thank the following people who reviewed early drafts of the manuscript: Deborah C. Andrews, University of Delaware; Stuart

C. Brown, University of Arizona; David H. Covington, North Carolina State University; R. S. Krishnan, North Dakota State University; John A. Muller, Air Force Institute of Technology—Wright Patterson Air Force Base; Mary Beth Raven, Rensselaer Polytechnic Institute; Thomas L. Warren, Oklahoma State University; Don D. Wilson, Hartford State Technical College; Muriel Zimmerman, University of California at Santa Barbara.

Anne Eisenberg

PART I

INTRODUCING A LEGACY OF GOOD WRITING

CHAPTER 1

A Legacy of Good Writing

"I never developed my writing skills; then I went to work and discovered they are at the heart of my job."

This is the comment of a chemist who as an undergraduate scrupulously avoided "any course that had a lot of writing." Then she went to work, discovered that her work entailed communication skills, and returned to the university to sign up for a report writing class.

She wasn't the only undergraduate, of course, who shared the pervasive attitude that verbal skills are irrelevant to the true work of technical professionals.

But whether they are civil engineers or physicists, tax accountants or economists, working professionals are indeed expected to be articulate beings. They spend a good deal of their time on communication skills, that is, on

- interviewing
- writing
- editing
- speaking

How much is "a good deal of time"? Studies of technical professionals show they typically spend anywhere from a quarter to a third of their workweek using communication skills.

Throughout their varied work lives, which are full of changing imperatives, one imperative remains constant—it is vital that they be able to talk and write adeptly.

THE TRADITION OF TECHNICAL REPORTING

There is a negative stereotype of the engineer, scientist, or technical professional as a person who cannot write, who comes from a tradition that is essentially mute. Neither situation is the case.

A fine tradition of writing in technical professions exists, extending from instructions in an ancient Egyptian surgery, through Frederick II's *The Art of Falconry* (see exercises at end of chapter), to the present. The technical report, for instance, that orderly progression of objective, procedure, results, and conclusions, has a form as venerable and elegant as the sonnet; in fact, both the literary form of the sonnet and the technical form of the report flowered during the English Renaissance. While English poets such as Spenser took on the Petrarchian sonnet form, scientists such as Francis Bacon adopted an ancient form of the forensic report when they sought a way to cast the results of their investigations of nature. Like the Elizabethan poets, Bacon and other Renaissance scientists also borrowed from the Greeks for an extensive new vocabulary to express their rapidly expanding knowledge.

There is a legacy of good writing in technology and in the many overlapping fields of applied science. A sampling is included in this textbook to introduce a tradition that is neither mute nor inglorious.

The stereotype of the poor technical writer is refuted by a history of writing that is both true and engaging, logically rigorous and illuminating.

THE PURPOSE OF TECHNICAL COMMUNICATION

Perhaps you've read the Marianne Moore poem "Granite and Steel," where she describes the Brooklyn Bridge as

> Enfranchising cable, silvered by the sea,
> . . . "O catenary curve" from tower to pier . . .
> first seen by the eye of the mind,
> then by the eye. . . .
> Climactic ornament, double rainbow,
> as if inverted by French perspicacity,
> John Roebling's monument,
> German tenacity's also;
> composite span—an actuality.

Moore uses technical terms ("catenary curve"), but her lovely poem is not what one means by technical writing. She says her writing shares the purpose articulated by William Faulkner: "It should help a man endure by lifting up his heart."

The purpose of technical writing is quite different from that. It does not deal with what Yeats called "the rag and bone shop of the heart." Instead, its purpose grows out of its content, content rooted in the word *technique*, defined in the American Heritage Dictionary as "the systematic procedure by which a complex or scientific task is accomplished . . . hence technical." Describing scientific results is technical; describing a complex task is technical, whether the task is dry-column chromatography, the tuning of a harpsichord, or the wording of a consumer loan application.

As the content of technical communication is different, so is its goal: It seeks to inform, to instruct, and often to persuade that its interpretation of data is the correct one.

Many people distinguish science and technology by their general ends and say that the end of science is truth, that of technology utility—the putting into practice of that which did not previously exist. In this view, technology and applied science run together after the Renaissance in the "same powerful channels . . . with only the smallest distinction between applied science—the application of the principles of pure science—and technology." (James K. Feibleman, "Pure Science, Applied Science, and Technology: An Attempt at Definitions," *Technology and Culture,* 11, No. 4, Fall 1961.)

If the most general end of technology and applied science is utility, certainly that end is shared by their prose. But good technical writing—and there is a wealth of it—combines practicality with aesthetics. It tells us what we would know with such graceful logic and language that it achieves a form of literary elegance.

TECHNICAL WRITING MAY BE PERSUASIVE

Note that the purpose of technical writing is usually to inform, but that it may also seek to persuade.

"How can technical writing be persuasive?" students often ask. "Isn't technical writing objective? And doesn't 'objective' mean that you don't take a side?"

Technical writing does deal with the objective—with that which can be categorized, measured, analyzed, disproved. But "objective" and "persuasive" are not mutually exclusive terms. For instance, the writer of a feasibility study may present five or six methods, then argue for one of them. A report writer may have a series of spectra, and argue for one interpretation of the series. Data don't stand alone in solitary splendor; often they are only as good as the persuasive argument that supports and interprets them.

There is still another, more subtle sense in which technical communication may be persuasive. In this book there are many examples of daily technical writing: brief trip reports, technical memos of an investigation, summaries of an instrument search. Many were submitted in a context where the writing stood for the writer. How can this be? In a large organization where matters are discussed more by memo or report than face to face, writing is a key way that people get to know one another. Well-written technical documents stand as representatives of the writer, small flags of correctness and accuracy. They have a role in proving that the writer is a competent, professional person.

CLASSIFYING TECHNICAL COMMUNICATION

There is probably no single way all the various types of technical communication can fit into one, all-purpose model.

Here is a listing that categorizes technical communication by task, by type, and by audience.

By Task	**Example**
PROPOSING What you want to do	Proposals for new methods, new materials, new equipment feasibility studies, research proposals

By Task	Example
REPORTING	
What you did	Trip and conference reports, research reports, quality control reports, troubleshooting reports, progress reports
DOCUMENTING	
The way it is done	Procedures
What we know now	Specifications
What exists now	Interviews
	Descriptions
	Definitions
	State-of-the-art report
	Laboratory notebooks
INSTRUCTING	
How to do it	Training and orientation talks
	Safety procedures
	User manuals
EDITING	
	Revising a report for publication
	Writing summaries and overviews to accompany text
	Writing headings and subheadings

By Type

Instructions	Manuals	Proposals
Procedures	Abstracts and briefs	Newsletters
Descriptions	Notebooks	Articles
Documentation	Reports	Speeches

By Audience	Example
To peers	Research report in professional journal
To a distribution up and down the corporate ladder	Executive abstract Progress report
To the public	Letters Instructional booklets

Who does most of this writing? There are a handful of specialists in technical or professional communication who have taken courses and even degrees in specialized writing. But most technical writing is done by people as part of their profession. It is simply part of the job—and a very important part at that.

SUMMING UP

- Verbal skills are at the heart of technical jobs. Studies of technical professions show they typically spend a quarter to a third of their workweek using communication skills—writing, editing, interviewing, speaking. Contrary to stock images, technical professionals are expected to write and to write well.
- Fine technical writing abounds, from accounts by Babylonian astrologers, through the medieval chronicles of Frederick II, to the present.
- Technical writing seeks to inform, and often to persuade. "Objective" and "persuasive" are not mutually exclusive terms.
- Technical writing may be characterized by task, by type, or by audience. Tasks include proposing, reporting, documenting, instructing, and editing. Types include procedures, instructions, reports, proposals, articles, newsletters, and manuals. Audiences range from a peer group to distribution up and down the corporate ladder.

EXERCISES

1. Find an example of technical writing for group discussion. Make photocopies for the group.

Some questions to answer:

What type of writing is it? (A report? A proposal? Instructions? An explanation of a process? A physical description?)

To what audience or audiences is it addressed?

What does it seek to do?

Is it an effective piece of writing? Explain.

2. Two Descriptions of Birds Example 1.1 is a short excerpt from a medieval work, *The Art of Falconry*. In it the author, Emperor Frederick II, distinguishes between two kinds of species. What char-

Of the Division of Birds into Raptorial and Nonraptorial Species

Birds may be classified in still another manner—as raptorial and nonraptorial species. We call raptorial all those birds who, employing their powerful flight and the special fitness of their members, prey upon any other bird or beast they are able to hold and whose sole sustenance is the flesh of such animals. These are the eagles, hawks, owls, falcons, and other similar genera. They feed only on their prey—never upon dead flesh or carrion (*carnibus cadaverum neque residuis*)—and are therefore called rapacious birds.

• • •

Among the characteristic forms of their organs may be mentioned: the beak, which in birds of prey is generally curved, strong, hard, and sharp; claws that are bent inward and are hard and needle-pointed; retracted eyes; a short neck, short legs, and the posterior toe of each foot very strong. The female is larger than the male. Not all of the foregoing is true of nonraptorial birds.

• • •

As to plumage, it varies among raptores; the first year after hatching (when they are called sorehawks [*sàure*]) they moult only once, while other birds (generally) shed their feathers twice. The large quill feathers of the wings and tail are limited to a definite number; this is not true of other birds.

In numbers also the two classes differ, for there are fewer rapacious birds than nonrapacious; and there are no raptores among aquatic and neutral birds, but only among land birds, and even here they are few in number; so that all water and neutral birds and the greater part of land birds are nonrapacious.

Rapacious birds (which are universally warmer and drier than aquatics and neutrals) dislike water for two reasons, one active and the other passive. Since they have not members and plumage of a suitable form, they do not live in the water, nor can they do so, because they cannot continue to stand in deep water, lacking long legs like those of herons and cranes, nor can they swim about with ease, as their feet and toes are not webbed like those of geese, ducks, coots, and nearly all aquatic birds. Were a raptorial bird overturned, or submerged, in water her feathers and quills would be more inclined than those of aquatic birds to become soaked, so that she could hardly fly, and her claws would become so softened that she would be unable to wound or hold her prey. For these reasons, birds of prey dread remaining in the water, since they are extremely feeble in that element. There are certain birds, however, similar to eagles but smaller, that perch above bodies of water (or on high banks) and, when they perceive fish in the water, suddenly drop on them, draw them out alive, and feed on them. They are, therefore, called fish eagles. Their members and plumage are better adapted for this purpose than are those of other raptores. . . .

SOURCE: The Art of Falconry, by Emperor Frederick II of Hohenstaufen

Example 1.1

The Eagle
(Fragment)

He clasps the crag with crooked hands;
Close to the sun in lonely lands,
Ring'd with the azure world, he stands.

The wrinkled sea beneath him crawls;
He watches from his mountain walls,
And like a thunderbolt he falls.

Example 1.2

acteristics make this piece a technical document rather than, say, a literary one? Who do you think is the intended audience? What do you think was the author's objective in writing the document? For purposes of comparison, a poem by Alfred, Lord Tennyson, "The Eagle," follows (Example 1.2). Are both works descriptions? How do they differ?

3. Interview someone who does technical writing as part of his or her job. You may talk with someone who writes technical textbooks and is on the school faculty. You may talk with a person who writes press releases with technical content on behalf of a company, university public relations office, or government agency. You may interview a technical writer at a local company or someone whose job is as an engineer, scientist, or business person, but who spends a certain amount of time writing within a specialized field. Ask the person about the kinds of writing done, and the main problems the writer has to solve. Summarize the interview in a brief account that includes the problems the writer faces, and the writer's solutions.

4. What value do you think is to be found in clear writing? In a letter to the editors of *The New York Times,* Joseph Weizenbaum, Professor of Computer Science at MIT, writes,

> On the Education Watch page of the Dec. 8 Week in Review, Dr. George W. Tressel of the National Science Foundation complains that "most people are essentially unequipped to read and understand" any of the many articles concerning, for example, toxic chemicals and nuclear safety that appear with great frequency in the press. He says this "functional illiteracy" is "terrible" because "our lives are controlled by chemicals and computers, not Gothic cathedrals, and science is the 'humanities' of our time." He goes on to suggest that "considerably more time should be devoted to mathematics and science throughout the elementary and secondary grades."
>
> In over 20 years spent at MIT teaching students who had had an abundance of mathematics and science instruction before coming

> to college, I found, and still find, distressingly many of them essentially unable to read discursive texts critically. Nor does it make much difference what the text deals with or whether it appears in a newspaper or book.
>
> Many more students are hardly able to write a single paragraph of grammatically correct English that says what they intend it to say. That such students get through MIT at all does not imply that the traditional humanities, as opposed to Dr. Tressel's new ones, are irrelevant to science and engineering. It merely shows that MIT is not, at least in one respect, sufficiently demanding of its students.
>
> The disaster induced by America's failure to assign as high a priority to the education of its children as it does to its military might is that generations of children are condemned to grow to adulthood without having mastered their own language. The tragedy isn't, as Dr. Tressel has it, that masses of people are functionally illiterate with respect to science; it is the steady, if not rising, tide of *general* functional illiteracy among our youth.
>
> If our lives are not to be controlled by chemicals and computers, our schools had better get on with what is their overwhelmingly most important task: teaching their charges to express themselves clearly and with precision in both speech and writing; in other words, leading them toward mastery of their own language. Failing that, all their instruction in mathematics and science is a waste of time.

Is instruction in mathematics and science a waste of time if people do not have mastery of their own language? What use and value do you think writing has to the profession you are studying? Give examples in your discussion.

CHAPTER 2

Writing for the Audience: Bridging the Gap

My primary concern in writing is to render my work as clear and intelligible as it is true. . . . An author should not expect that people will understand his little hints as he understands them himself, since he is filled with what he knows. He must realize that if he doesn't arrange his subject matter in an appropriate order and if he doesn't explain it simply, people will reap very little from the fruit of his labor.

Principles of the Harpsichord by Monsieur de Saint Lambert, 17th century French musician and writer.

There are striking differences between the two pieces of technical writing represented by Figures 2.1a and 2.1b. The difference lies not in what they say—both documents have the same content—but in how they say it.

PERSONAL FINANCE DEPARTMENT - NEW YORK
APPLICATION
NUMBER ____________

ANNUAL PERCENTAGE RATE ____________ %

$ ____________
TOTAL OF PAYMENTS (4) + (7)

PROCEEDS TO BORROWER
PROPERTY INS. PREMIUM
FILING FEE
AMOUNT FINANCED (1) + (2) + (3)
PREPAID FINANCE CHARGE
GROUP CREDIT LIFE INS. PREMIUM
FINANCE CHARGE (5) + (6)

____________ ($ ________) (TOTAL OF PAYMENTS) ( ) IN ________ EQUAL CONSECUTIVE MONTHLY INSTALMENTS OF $ ________ EACH ON THE SAME DAY OF EACH MONTH, COMMENCING ________ DAYS FROM THE DATE THE LOAN IS MADE; OR ( ) IN ________ EQUAL CONSECUTIVE WEEKLY INSTALMENTS OF $ ________ EACH ON THE SAME DAY OF EACH WEEK, COMMENCING NOT EARLIER THAN 5 DAYS NOR LATER THAN 45 DAYS FROM THE DATE THE LOAN IS MADE; OR ( ) IN ________ EQUAL CONSECUTIVE BI-WEEKLY INSTALMENTS OF $ ________ EACH, COMMENCING NOT EARLIER THAN 10 DAYS NOR LATER THAN 45 DAYS FROM THE DATE THE LOAN IS MADE, AND ON THE SAME DAY OF EACH SECOND WEEK THEREAFTER; OR ( ) IN ________ EQUAL CONSECUTIVE SEMI-MONTHLY INSTALMENTS OF $ ________ EACH, COMMENCING NOT EARLIER THAN 10 DAYS NOR LATER THAN 45 DAYS FROM THE DATE THE LOAN IS MADE, AND ON THE SAME DAY OF EACH SEMI-MONTHLY PERIOD THEREAFTER, (ii) A FINE COMPUTED AT THE RATE OF 5¢ PER $1 ON ANY INSTALMENT WHICH HAS BECOME DUE AND REMAINED UNPAID FOR A PERIOD IN EXCESS OF 10 DAYS, PROVIDED (A) IF THE PROCEEDS TO THE BORROWER ARE $10,000 OR LESS, NO SUCH FINE SHALL EXCEED $5 AND THE AGGREGATE OF ALL SUCH FINES SHALL NOT EXCEED THE LESSER OF 2% OF THE AMOUNT OF THIS NOTE OR $25, OR (B) IF THE ANNUAL PERCENTAGE RATE STATED ABOVE IS 7.50% OR LESS, THE LIMITATIONS PROVIDED IN (A) SHALL NOT APPLY AND NO SUCH FINE SHALL EXCEED $25 AND THE AGGREGATE OF ALL SUCH FINES SHALL NOT EXCEED 2% OF THE AMOUNT OF THIS NOTE, AND SUCH FINE(S) SHALL BE DEEMED LIQUIDATED DAMAGES OCCASIONED BY THE LATE PAYMENT(S); (iii) IN THE EVENT OF THIS NOTE MATURING, SUBJECT TO AN ALLOWANCE FOR UNEARNED INTEREST ATTRIBUTABLE TO THE MATURED AMOUNT, INTEREST AT A RATE EQUAL TO 1% PER MONTH AND (iv) IF THIS NOTE IS REFERRED TO AN ATTORNEY FOR COLLECTION, A SUM EQUAL TO ALL COSTS AND EXPENSES THEREOF, INCLUDING AN ATTORNEY'S FEE EQUAL TO 15% OF THE AMOUNT OWING ON THIS NOTE AT THE TIME OF SUCH REFERENCE, FOR NECESSARY COURT COSTS. THE ACCEPTANCE BY THE BANK OF ANY PAYMENT(S) EVEN IF MARKED PAYMENT IN FULL OR SIMILAR WORDING, OR IF MADE AFTER ANY DEFAULT HEREUNDER, SHALL NOT OPERATE TO EXTEND THE TIME OF PAYMENT OF OR TO WAIVE ANY AMOUNT(S) THEN REMAINING UNPAID OR CONSTITUTE A WAIVER OF ANY RIGHTS OF THE BANK HEREUNDER.

IN THE EVENT THIS NOTE IS PREPAID IN FULL OR REFINANCED, THE BORROWER SHALL RECEIVE A REFUND OF THE UNEARNED PORTION OF THE PREPAID FINANCE CHARGE COMPUTED IN ACCORDANCE WITH THE RULE OF 78 (THE "SUM OF THE DIGITS" METHOD), PROVIDED THAT THE BANK MAY RETAIN A MINIMUM FINANCE CHARGE OF $10, WHETHER OR NOT EARNED, AND, EXCEPT IN THE CASE OF A REFINANCING, NO REFUND SHALL BE MADE IF IT AMOUNTS TO LESS THAN $1. IN ADDITION, UPON ANY SUCH PREPAYMENT OR REFINANCING, THE BORROWER SHALL RECEIVE A REFUND OF THE CHARGE, IF ANY, FOR GROUP CREDIT LIFE INSURANCE INCLUDED IN THE LOAN EQUAL TO THE UNEARNED PORTION OF THE PREMIUM PAID OR PAYABLE BY THE HOLDER OF THE OBLIGATION (COMPUTED IN ACCORDANCE WITH THE RULE OF 78), PROVIDED THAT NO REFUND SHALL BE MADE OF AMOUNTS LESS THAN $1.

AS COLLATERAL SECURITY FOR THE PAYMENT OF THE INDEBTEDNESS OF THE UNDERSIGNED HEREUNDER AND ALL OTHER INDEBTEDNESS OR LIABILITIES OF THE UNDERSIGNED TO THE BANK, WHETHER JOINT, SEVERAL, ABSOLUTE, CONTINGENT, SECURED, UNSECURED, MATURED OR UNMATURED, UNDER ANY PRESENT OR FUTURE NOTE OR CONTRACT OR AGREEMENT WITH THE BANK (ALL SUCH INDEBTEDNESS AND LIABILITIES BEING HEREINAFTER COLLECTIVELY CALLED THE "OBLIGATIONS"), THE BANK SHALL HAVE, AND IS HEREBY GRANTED, A SECURITY INTEREST AND/OR RIGHT OF SET-OFF IN AND TO (a) ALL MONIES, SECURITIES AND OTHER PROPERTY OF THE UNDERSIGNED NOW OR HEREAFTER ON DEPOSIT WITH OR OTHERWISE HELD BY OR COMING TO THE POSSESSION OR UNDER THE CONTROL OF THE BANK, WHETHER HELD FOR SAFEKEEPING, COLLECTION, TRANSMISSION OR OTHERWISE OR AS CUSTODIAN, INCLUDING THE PROCEEDS THEREOF, AND ANY AND ALL CLAIMS OF THE UNDERSIGNED AGAINST THE BANK, WHETHER NOW OR HEREAFTER EXISTING, AND (b) THE FOLLOWING DESCRIBED PERSONAL PROPERTY (ALL SUCH MONIES, SECURITIES, PROPERTY, PROCEEDS, CLAIMS AND PERSONAL PROPERTY BEING HEREINAFTER COLLECTIVELY CALLED THE "COLLATERAL"): ( ) Motor Vehicle ( ) Boat ( ) Stocks, ( ) Bonds, ( ) Savings, and/or ____________

SEE CUSTOMER'S COPY OF SECURITY AGREEMENT(S) OR COLLATERAL RECEIPT(S) RELATIVE TO THIS LOAN FOR FULL DESCRIPTION.

IF THIS NOTE IS SECURED BY A MOTOR VEHICLE, BOAT OR AIRCRAFT, PROPERTY INSURANCE ON THE COLLATERAL IS REQUIRED, AND THE BORROWER MAY OBTAIN THE SAME THROUGH A PERSON OF HIS OWN CHOICE.

IF THIS NOTE IS NOT FULLY SECURED BY THE COLLATERAL SPECIFIED ABOVE, AS FURTHER SECURITY FOR THE PAYMENT OF THIS NOTE, THE BANK HAS TAKEN AN ASSIGNMENT OF 10% OF THE UNDERSIGNED BORROWER'S WAGES IN ACCORDANCE WITH THE WAGE ASSIGNMENT ATTACHED TO THIS NOTE.

Figure 2.1a

Figure 2.1a (Version A) is a technical document written by lawyers, for lawyers. It is a standard installment loan form of the sort typical in traditional consumer contracts. So long as lawyers conversant in its terms are the audience, no problem arises. It is when the readership is wider than a circle of experts that a document's "readability" becomes critical.

For Figure 2.1b (Version B), the original document was revised so that it is more readable for those who need to understand it: the people who are actually applying for loans.

Consumer Loan Note Date________________, 19____

(In this note, the words **I, me, mine** and **my** mean each and all of those who signed it.

Terms of Repayment To repay my loan, I promise to pay you ____________________ Dollars ($__________). I'll pay this sum at one of your branches in __________ uninterrupted __________ installments of $__________ each. Payments will be due ____________________, starting from the date the loan is made.

Here's the breakdown of my payments:

1. **Amount of the Loan** $__________
2. **Property Insurance Premium** $__________
3. **Filing Fee for Security Interest** $__________
4. **Amount Financed** (1+2+3) $__________
5. **Finance Charge** $__________
6. Total of Payments (4+5) $__________

Annual Percentage Rate______%

Prepayment of Whole Note Even though I needn't pay more than the fixed installments, I have the right to prepay the whole outstanding amount of this note at any time. If I do, or if this loan is refinanced—that is, replaced by a new note—you will refund the unearned **finance charge,** figured by the rule of 78—a commonly used formula for figuring rebates on installment loans. However, you can charge a minimum **finance charge** of $10.

Late Charge If I fall more than 10 days behind in paying an installment, I promise to pay a late charge of 5% of the overdue installment, but no more than $5. However, the sum total of late charges on all installments can't be more than 2% of the total of payments or $25, whichever is less.

Security To protect you if I default on this or any other debt to you, I give you what is known as a security interest in my O Motor Vehicle and/or ____________________ (see the Security Agreement I have given you for a full description of this property), O Stocks, O Bonds, O Savings Account (more fully described in the receipt you gave me today) **and** any account or other property of mine coming into your possession.

Insurance I understand I must maintain property insurance on the property covered by the Security Agreement for its full insurable value, but I can buy this insurance through a person of my own choosing.

Default I'll be in default:

1. If I don't pay an installment on time; or
2. If any other creditor tries by legal process to take any money of mine in your possession.

You can then demand immediate payment of the balance of this note, minus the part of the **finance charge** which hasn't been earned figured by the rule of 78. You will also have other legal rights, for instance, the right to repossess, sell and apply security to the payments under this note and any other debts I may then owe you.

Irregular Payments You can accept late payments or partial payments, even though marked "payment in full", without losing any of your rights under this note.

Delay in Enforcement You can delay enforcing any of your rights under this note without losing them.

Collection Costs If I'm in default under this note and you demand full payment, I agree to pay you interest on the unpaid balance at the rate of 1% per month, after an allowance for the unearned **finance charge.** If you have to sue me, I also agree to pay your attorney's fees equal to 15% of the amount due, and court costs. But if I defend and the court decides I am right, I understand that you will pay my reasonable attorney's fees and the court costs.

Comakers If I'm signing this note as a comaker, I agree to be equally responsible with the borrower. You don't have to notify me that this note hasn't been paid. You can change the terms of payment and release any security without notifying or releasing me from responsibility on this note.

Copy Received The borrower acknowledges receipt of a completely filled-in copy of this note.

Signatures Addresses

Borrower: ____________________ ____________________

Figure 2.1b

To do a good job in technical communication—whether it is a consumer contract or a set of instructions for assembling a tricycle—you have to think about your readers, or audience. The writer must develop what E. B. White calls a "deep sympathy for the reader," a person White compares to a man floundering in a swamp. The writer has a duty "to drain this swamp quickly and get his man up on dry ground, or at least throw him a rope."

It's the main job of technical writing to throw the reader the rope of a clear explanation.

The specific ways to make an explanation clear are taught through techniques such as organizational patterns, direct language, and effective illustration. But there is one central idea that underlies all the techniques writers use to make text more readable—the idea of audience analysis. Because this idea is so central, it is treated separately within this chapter.

ANALYZING THE AUDIENCE

The goal of technical writing is to convey information; the audience reads for understanding, not entertainment.

The problem is that different audiences come to the text with different levels of understanding. The writer who wants to be effective has to adjust terminology, level of detail, organizational pattern, language—in short, the entire piece of writing—according to the expectations and abilities of the audience.

Such an adjustment is not easy to make. For the beginning writer who is grappling with the complexities of the subject matter, the manner in which the information is cast seems secondary when the imperatives of the content are so great.

Simply *stating* the information seems a sufficiently demanding chore to the novice. "I have my hands full just writing down an explanation," a student commented. "Thinking about who's actually going to have to understand it seems a remote, academic exercise."

Yet audience analysis is not an academic exercise; it is the first, and probably most important, step to clarity in technical communication. Should terms be defined? How much supporting detail should be introduced? How much time will be available for reading? What does the reader want to know? These and other similar questions are part of the first, essential steps in any technical writing job as the author tries to imagine the audience.

Audience analysis takes place before the first word is written, before the first table is drawn up.

Audience analysis makes itself felt at every level of communication. For instance, many people use terms without defining them, or define them only as an afterthought when the user is 20 pages into the document. Examples of this abound in poorly written instructions and particularly in computer software documentation, much of which is notorious for its cavalier use of undefined terms.

Another problem lies in learning how to adjust the use of detail and example. At one extreme there are accounts with detail after detail, where the reader has to search in annoyance for the main points. At the other end, there are accounts so stark, so stripped of example, that a newcomer to the project finds the account incomprehensible. In *Communicating Technical Information,* Robert Rathbone, MIT professor and author, comments: "There are a few who believe that terseness is next to godliness. They have a special talent for brevity, but in their desire to be economical they frequently sacrifice both clarity and readability. They mean well; they're earnest, hard-working individuals—liked by everyone except their readers—but they have the strange notion that 'objectivity' and 'economy of speech' are one."

TALKING TO A PEER GROUP

To write about a technical subject is by definition to deal with a highly specialized vocabulary and highly specialized concepts. If there is only one kind of reader to consider, and that reader has a background equal in sophistication to the writer, the task of audience analysis is mitigated.

The ideal reader in this instance is completely at home with the terminology of the field, saturated in the related literature, and therefore highly capable of reading between the lines whenever a nuance or inference is needed.

You may sometimes have this informed reader; he or she is called a "peer," or equal. One place you are likely to find a peer group is within the distribution of a research journal. "There are maybe forty or fifty people in the world who understand what I'm doing," a laser chemist comments. "Some of them are the referees of my papers. I write for this group when I publish. We all know one another's work. This is the one audience I know how to please. It's homogeneous; it shares my background."

"The problem for me comes when I have to explain what I am doing to everyone else in the world besides those forty people—namely my unit manager and the people he talks to when we need funds for research. I'm not writing clearly unless I write for these people. And 'clear' for them is different from 'clear' for my referees."

Example 2.1 is taken from a publication in which the authors can reasonably assume their readers are peers. The piece is designed not for electrical and electronics engineers in general, but for spe-

Example 2.1

***Abstract*—A fully self-consistent computer model of the steady-state behavior of the zero-order lateral optical field of a GaAs twin-stripe injection laser is presented which takes into account current spreading in the p-type confining layer, the effect of lateral diffusion of carriers in the active layer, and bimolecular and stimulated radiative recombination. The results predict the lateral movement of the near field of the optical signal under asymmetric drive conditions, as observed in practice. Also calculated are the corresponding carrier and current density distributions. It is shown that the near-field zero order lateral optical field can be beam steered across the facet by only 2μm, typically. However, the initial position of the beam can be controlled by the two-stripe currents and also the geometry of the device. For the case where $I_{s1} \simeq I_{s2}$ the beam movement is seen to be proportional to either I_{s1} or I_{s2}. The results show that beam steering is not accompanied by a negative slope to the *I–L* characteristics. The effect of geometry and diffusion coefficient on the value of maximum current allowed before modal instability occurs is also given.**

I. Introduction

There has been much interest [1]–[7] shown in twin-stripe injection lasers following the observation of current controllable near-field and far-field beam steering by Scifres *et al.* [8]. Fig. 1 shows a typical twin-stripe geometry DH laser. The two laser stripes are electrically coupled by: 1) current spreading in the P-GaAlAs confining layer, and 2) lateral carrier diffusion in the active layer beneath the electrodes.

Optically, the near-field modes supported depend upon the local gain profile and the refractive index profile, which are both dependent upon the carrier density distribution. Consequently, the beam position can be controlled by changing the current in either electrode. However, above threshold, stimulated recombination has a major influence upon the carrier density distribution. This paper includes the interaction of the optical field with the carriers in the active region, via the stimulated recombination coefficient in the continuity equation. This requires that the one-dimensional wave equation in the lateral direction is solved simultaneously with the continuity equation, describing the carrier density distribution in the active layer, and Laplace's equation, defining current spreading in the passive layers.

Several self-consistent models of single-stripe lasers now exist [9]–[14]. In this paper we present a self-consistent computer model of the lateral behavior of a twin-stripe laser. The principal objectives of the paper are: 1) to examine the effect of diffusion coefficient and geometry on both the threshold condition and the stability of the zero order mode, 2) to predict the steady-state beam steering effect of the near-field of an asymmetrically pumped device, and the range of currents for which it is stable, and 3) to show that gain guiding is largely responsible for near-field beam steering.

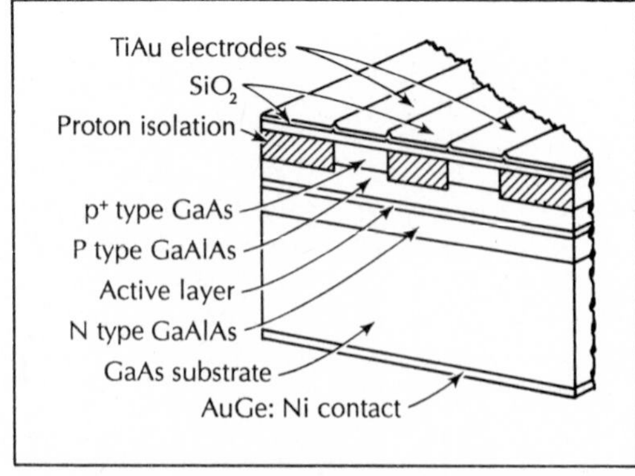

Figure 1

cialists in quantum electronics. Because the writer is aiming the argument at colleagues, the language is shorn of example, elaboration, definition. Such elaborations are unnecessary, given the shared background of the audience.

Example 2.2 is an article from the *New England Journal of Medicine*. This journal has a slightly more diversified audience than that of Example 2.1; readers include many types of specialists, from pediatricians to anesthesiologists, as well as public health professionals, science writers, and others who scan its refereed pages for reliable medical information. The journal makes small adjustments for this slight diversity, but it is still very much a publication for peers. An understanding of the text depends on a mastery of the specialized vocabulary and concepts. *The writers make no attempt to bridge the gap.* For instance, "azoospermic" and "nulliparous" aren't defined; there's no need. Only terms used in special senses, such as "fecundity," are delineated.

Version A of the consumer loan contract (Figure 2.1a), at the beginning of this chapter, is a classic example of writing that is perfect for a homogeneous audience that understands its conventions, but it is inappropriate for the very people who need to read and understand it.

Example 2.2

Female Fecundity as a Function of Age

Results of Artificial Insemination in 2193 Nulliparous Women with Azoospermic Husbands

FÉDÉRATION CECOS,* D. SCHWARTZ, PH.D., AND M. J. MAYAUX, B.A.

The decrease in the fecundity of women who have passed a certain age is generally acknowledged, but supporting data on natural reproduction are scarce. (We use the term "fecundity" in the sense of "capacity for procreation"; "fertility" denotes actual procreation.) In an analysis of data from three large studies, Leridon has suggested that fecundity is decreased in women over 30 years of age.[1] However, this group includes women with a wide range of ages. Furthermore, it was not possible to determine from the data whether the decrease in fecundity was biologic or simply the consequence of diminished sexual activity.

Artificial insemination with donor semen (AID) seems to present an opportunity to control certain variables in the study of female fecundity over time, but the few studies published to date have been carried out in small populations. Moreover, the husbands of most

Example 2.2 (*continues*)

women who have received AID cannot be considered to have been totally sterile. The probability that these men will procreate increases as the fecundity of their wives increases.[2] Accordingly, among women treated by AID, there are very probably some with reduced fecundity; the degree of reduction increases as the time without conception ("exposure time") increases. Therefore an observed decrease in fecundity with age in these women could simply be due to a sampling bias.

We studied 2193 women who were receiving AID and whose husbands were totally sterile. The curve of the cumulative success rate for the women 25 years of age or younger was similar to that for women 26 to 30 years old. However, this curve showed a significant decrease in the cumulative success rate for women 31 to 35 years of age. This decrease was even greater for those over 35. Similar decreases with age were observed for the mean conception rate per cycle. Our data therefore provide evidence of reduced fecundity with age, which begins at some point after the age of 30 years.

Methods

In France, most AID procedures are performed with frozen semen furnished by the Centres d'Etude et de Conservation du Sperme Humain (CECOS), according to a protocol developed with the Unité de Recherches Statistiques of the Institut National de la Santé et de la Recherche Médicale (INSERM). This protocol is followed as closely as possible by each center and includes the definition of patients, the selection of donors, the conditions of semen supply, the freezing technique, and the insemination procedure. All women entering the program are presumed to be normally fecund on the basis of physical examination and hysterosalpingography. Before treatment begins, temperature charts are used to obtain evidence of ovulation. The appropriate time for insemination is estimated on the basis of records of basal body temperature and an examination of cervical mucus. The gynecologist is at liberty to decide when and how often insemination will be carried out and to prescribe supplemental treatment to improve fecundity, such as induction of ovulation. The couples are given no specific instructions concerning their pattern of coitus during insemination cycles.

Our data were obtained from 11 CECOS centers. All 2193 women who had azoospermic husbands and were treated by AID from the inception of CECOS in 1973 until February 1980 were studied; those returning for treatment after a first pregnancy did not reenter the study. The women were divided into four age groups: 25 years old or younger, 371 women; 26 to 30, 1079; 31 to 35, 599; and 35 or older, 144 (16 of these women were over 40). The ages and characteristics of the donors were similar within each age group; the age groups were alike in the number of insemination procedures per cycle, the interval between insemination cycles, and the frequency of treatments prescribed to assist fertilization.

At the end of the study period, each woman was classified as belonging to one of four categories: success, lost to follow-up, open case, or dropout. Success was defined as conception that was confirmed by the presence of a hyperthermic plateau for at least 21 days and by an affirmative clinical examination or a positive immunologic pregnancy test. Thus, all pregnancies were included in the study, regardless of outcome. A woman was considered to be lost to follow-up if the result of the last AID cycle was not known. In an open case the result of the last AID cycle was known, but the next insemination procedure had not yet taken place. A woman was regarded as a dropout if she stated that she would discontinue treatment.

The cumulative success rates were calculated after 12 cycles with the life-table technique adapted to AID,[3] as though there were no dropouts (theoretical cumulative rates). The curves obtained from the cumulative rate as a function of the number of treatment cycles were compared for different age groups with use of the Mantel-Haenszel test.[4]

Results

The cumulative success rates for the four age groups are given in Figure 1. The curves for those 25 or under and for those 26 to 30 were very similar, and they are represented by a single curve. The four curves differed significantly (chi-square = 15.72, with three degrees of freedom; P<0.01). The curves for the two younger groups did not differ significantly from each other, but both differed from the curves for the two older groups (P<0.001, in comparison to the group >35; and P<0.03 in comparison to the group 31 to 35).

Table 1 shows the cumulative success rate after 12 cycles, as well as the mean success rate per cycle, the dropout rate, and the rate for loss to follow-up. The results for a subgroup 36 to 40 years old are also included. The mean success rates per cycle for the women 35 or under and for those over 35 were also calculated separately for four centers with large numbers of subjects and for the other seven centers. In all cases, the older women had lower success rates (Table 2).

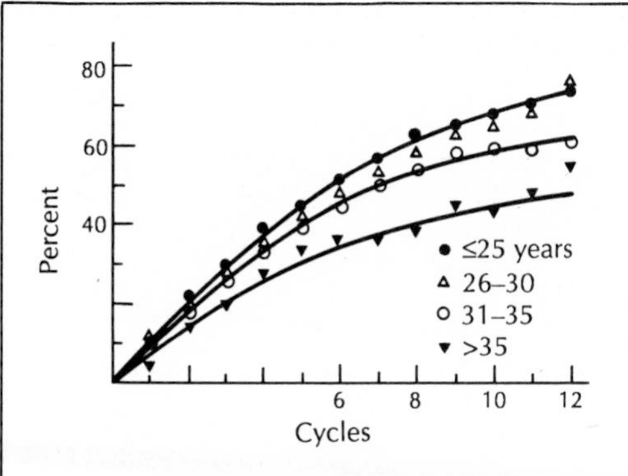

Figure 1. Theoretical Cumulative Success Rates in the Age Groups. The four curves differ significantly (P<0.01). Because the curves of the two younger groups were similar, they are represented by a single tracing. These curves differ significantly from those of the two older groups (P<0.03 for those 30 to 35, and P<0.001 for those over 35). There were 371 women in the <25 group, 1079 in the 26–30 group, 599 in the 31–35 group, and 144 in the >35 group.

Table 1. Rates for Success, Loss to Follow-up, and Dropping Out, According to Age Group.

Rate	Percentage			
	<25 yr	26–30 yr	31–35 yr	>35 yr (36–40)
Mean rate per cycle				
Successes	11.0	10.5	9.1	6.5 (6.5)
Losses to follow-up	2.8	2.5	2.4	2.4
Dropouts	4.0	4.0	4.7	4.9
Cumulative success rate after 12 cycles	73.0	74.1	61.5	53.6 (55.8)

Discussion

This study has shown that a decrease in fecundability (conception rate per cycle) as a function of a woman's age is slight but significant after 30 years of age and marked after 35 years. The probability of success of AID for 12 cycles, which was 73 per cent and 74 per cent for the two groups of women under 31, dropped to 61 per cent for those 31 to 35 and to 54 per cent for those over 35. The large decrease in this last group was not simply due to the inclusion of women over 40, since the subgroup that was 36 to 40 had the same low probability of success. The decrease in fecundability with age is consistent across the CECOS centers (Table 2) and supports the reliability of the findings.

Table 2. Mean Success Rates at CECOS Centers.

Age Group	Per Cent of Successes per Cycle*				
	Center 1 (2701)	Center 2 (1854)	Center 3 (1789)	Center 4 (779)	Center 5–11 (2877)
<35 yr	10.9	12.6	7.7	12.6	8.9
>35 yr	6.5	7.9	5.6	6.8	5.4

* Figures in parentheses denote number of cycles.

Example 2.2 (*continues*)

In any attempt to study variations in fecundity as a function of a woman's age, two major problems are encountered. The first is the need to separate the influence of the age of the woman from associated variables such as the pattern of coitus and the age of the husband. In our study it was possible to control these variables through the use of artificial insemination with frozen semen. Furthermore, the characteristics of the donors and the insemination cycles were similar in all age groups.

The second problem is that the variable under study—the age of the woman—can itself result in bias, since time introduces a type of selection. In AID this possibility is especially high if a husband has reduced fecundity but is not sterile; if his wife is very fecund, she may be precluded from study because she has previously conceived. This bias becomes more pronounced with the age of the women studied. This is the reason for our choosing to study only women with azoospermic husbands. It is indeed possible that the choice of population could have introduced bias because of factors such as previous marriages, previous attempts at conception by AID, or adultery. In our study, the proportions of remarried women and of those who had already been inseminated were very low (<1 per cent). It is possible to evaluate the incidence of adultery, but it is unlikely that adultery accounted for the findings of this study.

The many variables encountered in attempts to study the effect of age on female fecundity make it necessary to find various approaches to this problem. One such approach is through artificial insemination with frozen semen, as in this study. It is difficut to know to what extent our results approximate those of natural reproduction, but artificial insemination with frozen donor semen now appears to provide the best means of reducing the influences of associated variables and sources of bias to a minimum.

We are indebted to Dr. W. S. Price for help in translating and reviewing the manuscript.

References

1. Leridon H. Human fertility: the basic components. Chicago: University of Chicago Press, 1977.
2. Emperaire JC, Gauzere E, Audebert A. Female fertility and donor insemination. Lancet. 1980; 1:1423–4.
3. Schwartz D, Mayaux MJ. Mode of evaluation of results in artificial insemination. In: David G, Price WS, eds. Human artificial insemination and semen preservation. New York: Plenum Press, 1980:197–210.
4. Mantel N. Chi-square tests with one degree of freedom: extensions of the Mantel-Haenszel procedure. J Am Stat Assoc. 1963; 58:690–700.

WRITING FOR A DIVERSIFIED READERSHIP: TALKING TO TECHNICAL NONSPECIALISTS

The audience for a good part of technical writing is diversified, from people trained in related technical specialties to those in business with little or no technical background.

In such instances, it's the writer's job to bridge the gap. That gap can be considerable, whether the document is a consumer loan application, a trip report, or software documentation. For instance, the audience for a user guide to a graphics package for a personal computer might include

- management
- engineers
- students
- professors
- clerical workers
- programmers
- business and professional staff
- dealers and distribution networks

Talking to the audience directly and clearly—throwing the man in the swamp a rope—is particularly difficult in technical communication, where there is likely to be a variety of readers, each in need of different information, conveyed differently.

If you write reports, many of your readers will have a technical background, but it is not likely to be as close to the material as yours; for your purposes, they will be technical nonspecialists—educated, but not in your line of work. You are also likely to have non-technical readers from management, sales, and marketing as readers.

An example of writing for a diversified group of technical nonspecialists appears in Example 2.3, the opening pages of an article from *IEEE Spectrum*. While the publication is written for electrical and electronics engineers, it does not assume that subsets within the group are intimately acquainted with one another's research. Terms are defined; essential concepts are established by example and elaboration.

Example 2.3

Zero-defect Software: The Elusive Goal

It is theoretically possible but difficult to achieve; logic and interface errors are most common, but errors in user intent may also occur

MARGARET H. HAMILTON Hamilton Technologies Inc.

In October 1960, shortly after a new radar network to warn the United States of missile attacks had become operational, a radar station in Greenland reported the appearance of a massive attack—a large number of radar returns coming over the eastern horizon. The

Example 2.3 (*continues*)

real cause of the alarm: the moon was rising.

In late 1985, as activity in financial markets escalated, the operations of one financial services company were brought to a halt as its computers reported error after error. The designers of a bond-tracking program had built room for only 32,767 bond issues into their tables, and the 32,768th had just appeared.

In 25 years, although the speed, memory capacity, and reliability of computer hardware have increased manyfold, the reliability of computer software has not. Certainly software has become more reliable, but bugs still crop up in programs of all kinds, from the smallest game on a micro to the largest operating system on a mainframe.

As the Government proposes to build immense real-time systems like antimissile shields, which require enormous amounts of trouble-free software, critics question whether such systems can ever be made to function reliably. Although software development methods have improved measurably in the last several years, error-free software is, in the opinion of most software engineers, an impossible goal.

But some software developers believe complex software can be developed that approaches zero defects by using formal specification techniques and computer-based tools. These tools first check the consistency and logical completeness of a set of formal specifications and then generate program code that matches them. Checking the specifications for completeness and consistency eliminates errors of logic that arise from oversights, while tools that produce code directly from the specifications eliminate errors that might arise in implementing the specifications by hand. Software developed by such techniques may not always end up doing what the user wants, but it will do what the user asks it to do.

Most software development techniques proceed from requirements to specifications to designs to program code, using people to carry out the transformations from one level to another. Only the final step in development—generating machine code—is usually done by machine. In the other steps, two kinds of errors arise: those in which the user's intent is recorded incorrectly—like a misplaced comma in one National Aeronautics and Space Administration (NASA) program that sent a Voyager spacecraft toward Mars instead of Venus—and errors in which the wrong intent, considered in some larger context, is set down in logically complete and consistent fashion. An example of the latter might be billing programs that send threatening letters to customers who owe $0.00.

Defining terms

Interface error: an error that occurs because of improper use of a program module; for example, a module might be given too many arguments for input, or the arguments might be passed in the wrong order.

Primitive operation: a procedure that cannot be broken down into other operations; depending on the application, a primitive operation in a specification may translate to only a few machine instructions or to an entire software subsystem.

Specification: a formal description of what a program will do, phrased in terms of its inputs, its output, and the relationships between them, rather than in procedural form.

Strong typing: a characteristic of some programming languages that enforces constraints on the use of variables to reduce mistakes; for example, a strongly typed language would not allow a variable of the type *apple* to be added to one of the type *orange*, even if both types were represented as integers.

User-intent error: an error that occurs because the user did not properly think through a program before committing it to software.

Many techniques have been developed to deal with the first kind of error—incorrect statements in software—especially at the program-code level. Compilers can check the syntax of statements submitted to them, and "strongly typed" languages can enforce consistency between different uses of the same variable. A variable defined as an integer in one place, for example, cannot be used for character operations somewhere else. These remedies are static methods for software verification, which work by examining program source code rather than by testing a program's execution.

Static methods can also be used to check program specifications, or any other formal representation of a program, provided those specifications have been written in machine-readable form. Recently tools have begun to appear that can check program specifications for inconsistencies, ambiguities, and incompleteness in the same way that compilers check the syntax of program code. But static methods cannot eliminate all errors from either code or specifications; in particular, they cannot deal with errors of user intent.

Errors in user intent are the hardest to catch, because a program containing them can be consistent and complete but still give the wrong results. Some errors of intent arise from oversights—the equivalent of typographical errors in program code—while others come from a genuine confusion on the part of the user as to what the program should do.

An additional complication is that software is almost always part of a larger system that also includes hardware—and humans. The software can be reliable and free of defects, but the system that contains it can still fail, and in ways that resemble software errors.

ADJUSTING FORMAT TO AUDIENCE

Beginning writers often imagine there is one absolute format for each type of technical writing. For instance, it is common to think that the format for writing abstracts suits all occasions. In the manner of a person following a cookbook, the user simply takes abstract format and fits in the ingredients—procedure, results, conclusions—and produces the all-purpose abstract, as appropriate to water chemistry specialists as civil engineers.

This is not so. The abstract for the water chemistry specialists will be different from one written for engineers. True, the two abstracts will share common ingredients—but the language, order and emphasis will vary according to the readers' needs and expectations.

The same may be said for any other form of technical communication, from instructions to software documentation. The content is adjusted according to the audience. That's why *IEEE Spectrum* and *IEEE Journal of Quantum Electronics* are written differently, although both are technical publications with technically trained readerships.

A dramatic example of audience adjustment appears in the following article from *The New York Times* (Example 2.4). The article is based on the paper that appeared in *The New England Journal of Medicine* (reprinted in Example 2.2).

Example 2.4

Study Shows Female Fertility Drops Sharply After Age of 30

By BAYARD WEBSTER

An unusually large and rigorous study of female fertility appears to demonstrate that the ability of women to become pregnant decreases sharply from the age of 31 to that of 35. The study shows the decline coming earlier and more precipitously than had generally been thought.

The common wisdom had been that the chances for pregnancy were high from the age of 25 to that of 35 and declined rapidly thereafter. But in the new research, conducted by French scientists, the most rapid decline was in the 31- to 35-year-old range, with a lesser decline thereafter in the childbearing years.

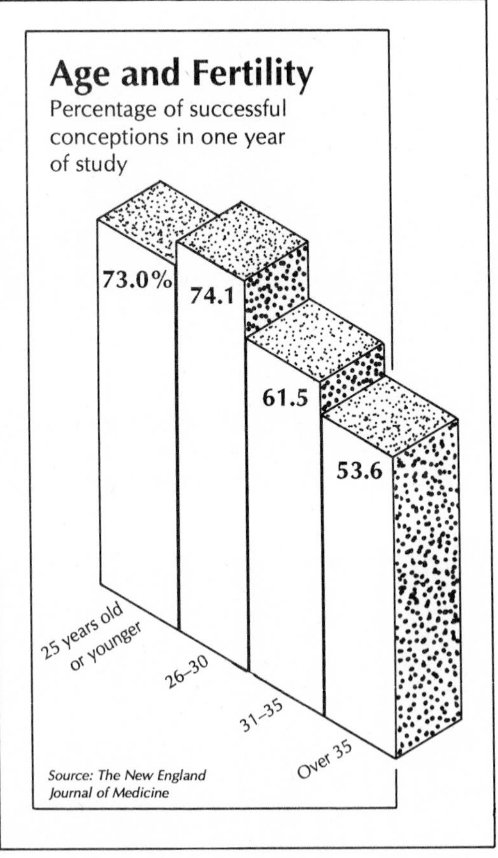

Health and Social Implications

An editorial accompanying the report, which appeared in the Feb. 18 issue of The New England Journal of Medicine, said, "If the decline in fecundity after 30 is as great as the French investigation indicates, new guidelines for counseling on reproduction may have to be formulated."

The research has health and social implications for women who, in increasing numbers, are delaying child-rearing to pursue professional careers or for other reasons.

The research was conducted among 2,193 women who were married to sterile men and who were trying to conceive through artificial insemination with donor semen. Over a period of about one year women were inseminated in 12 ovulation cycles and were given pregnancy tests at frequent intervals.

The physicians and scientists who conducted the study assert that this was a more reliable fertility research than had previously been done. The few previous studies had been complicated by such unknown factors as the degree of the husband's fertility, the frequency of the couple's sexual relations, and latent diseases.

The results showed that women 25 and under had a 73 percent rate of successful conceptions, those from 26 to 30 had a 74 percent rate, those from 31 to 35 had a rate of 61 percent, a significant drop of 13 percent from the lower age group. A relatively small group, 144 women, of those older than 35 had a 53 percent rate of success. In the latter group, there were 16 women over the age of 40.

The studies were conducted at 11 artificial insemination centers across France by physicians working with a federation of organizations for the study of human reproduction.

Although the subjects in the study were a special group with special problems, experts in the field said they believed the success in conceiving in this experiment would be roughly similar to that of women studied under different circumstances.

In an editorial in the medical journal, Dr. Alan H. DeCherney and Dr. Gertrud S. Berkowitz of the Yale University School of Medicine cited the importance of the study, noting the changing patterns of childbearing in the United States. They said national statistics showed that, while women 30 or over had contributed 6.8 percent of the country's first live births in 1960, they contributed 8 percent in 1979.

"Although these figures might not suggest a dramatic shift in childbearing patterns, the increase in first as well as second and third births among women in their 30's has been particularly pronounced in recent years, and there is no evidence this trend is shifting," they said.

The editorial also notes that current age-and-reproductive counseling is generally limited to the increased risk of Down's syndrome and other genetic abnormalities with advanced maternal age. And the two scientists wondered if, because of the study, women would reverse the current trend and return to having children in their 20's and concentrating on professional careers in their 30's. If so, they said, it might entail restructuring educational and training programs, and the labor market.

Although there is a growing body of knowledge about woman's reproductive functions, the French report, limiting itself to statistical analysis, gave no explanation of possible biological underpinnings for the decline in fecundity they recorded.

The editorial writers pointed out that some loss of fertility can be the result of gynecological diseases such as endometriosis, the failure of the female tubal organs and hormonal disorders. But why from 25 to 47 percent of the women did not conceive is not known.

READER PROFILE SHEETS

Because an understanding of the reader is so fundamental to successful technical writing, many writers prepare a *Reader Profile Sheet* before beginning any complicated assignment.

The profile sorts readers by

- education
- subject literacy (for example, background knowledge in computers, finance, or management)
- need for information.

Once the list is made, the writer categorizes readers as *primary, secondary, and fringe,* and shapes the report according to the varying needs of the audience, starting with the primary readers.

Here is a brief example of how a reader profile sheet might work. George Pilton, a young scientist just hired by a corporate research lab, was asked by his manager to look into the possibilities of buying a new instrument, a spectrometer. The manager, Henry Silks, knew that spectrometers were George's specialty.

George, who was conscientious, launched himself on the assignment. He submitted a densely written, seven-page document bristling with technical detail. In many ways it resembled Figure 2.1a (Version A of the consumer loan contract) at the beginning of this chapter. His manager called him in and asked him several questions about the instrument. "I said all this in the report," George answered. "Maybe you did," responded his manager, "but I can't find it. And I don't have two or three hours to spend figuring out what's important," he added.

George went back to his desk and did some thinking. He resisted a strong desire to write a yet longer and more technical version of the same report. He knew something completely different was necessary. The second time around, he tried using a Reader Profile Sheet. Listing his readers and then thinking about their needs was a mechanical step, but it helped him reorganize his report so that it was more readable.

Every audience is different, and each person's solution will vary, but here is the way George worked out his problem.

Step One: He started with a list of who would be reading the report and what they would want to know. He sorted them into primary and secondary readers (Example 2.5).

Step Two: He decided which sections each person was likely to read, and then, starting with his primary audience, shaped each section to the readers' needs.

George reasoned that people in management may not necessarily read an entire report, but they could be counted on to read the sections that would help them make a decision. Lincoln was therefore highly likely to read an abstract or summary, the budget, and

Example 2.5

1. Henry Silks. Supervisor. PhD, physics. Section head, will forward budget request to Vice-President. Silks is conversant with the subject, he initiated the idea.

Silks will take pieces of the proposal and use them to support his argument to the V-P. A summary of technical arguments would be particularly important for him; if it were well written, he could adopt it.

2. George Lincoln. Manager. M.B.A. Vice-President, approves budget requests for section; Silks will present the request to Lincoln. Lincoln is interested in the benefits we'll receive vs. the costs. He has no background or interest in physical chemistry, except in relation to selling the product. The argument must be made in terms of increased revenue in the future. This must be in the summary so that Henry can use this argument in presenting the case.

3. Bruce Sinclair, Larry Knowell. Colleagues. B.A., Chemistry. Helped with references. They will read the descriptions carefully to see if I have the details down correctly.

Primary Audience: Manager, George Lincoln.
Secondary Audience: Supervisor, Henry Silks.
Fringe Audience: Colleagues Sinclair and Knowell.

the budget justification. Lincoln would appreciate nontechnical language, as would Silks—spectrometers weren't their subject. Lincoln and Silk would both need examples put as nontechnically as possible.

At this point, George looked back at his original report and saw to his amazement that the only people who could understand it were the fringe audience, *his colleagues* Sinclair and Knowell. He'd failed to emphasize the very things his primary and secondary reader would need:

- *There was no summary whatsoever.* Lincoln would have to read the entire report to find out what Henry thought.
- *There were no definitions.*
- *There were no examples of cost effectiveness.*

Analysis complete, George revised his paper successfully. He added an abstract, examples, definitions. He shifted a good deal of the technical material to the appendixes. The report had the same information, but was shaped according to the needs of his readers.

George's mistake is a common one—he lost sight of what his audience needed to know in his concentration on the marshalling of technical detail.

Enmeshed in the difficulties of simply writing down what he knew, he forgot to provide his readers with a clear explanation. That makes him neither a villain nor a fool—just a person who was beginning to learn about the importance of audience.

The techniques of technical writing all descend from this central idea: **Think about your readers.**

AUDIENCE AND THE WRITING PROCESS

It is tempting to look at a polished piece of technical writing and conclude that the writer simply had a knack for producing such smooth prose. Actually, writing that is readable—clear and understandable to its audience—usually owes more to thought and practice than to some inborn flair for composition.

Writing is a difficult job. T. S. Eliot once called it a "raid on the inarticulate, with shabby machinery always deteriorating." The job might be simpler if writing were only transcribing one's thoughts, taking them down word-for-word as a tape recorder captures a discussion during a meeting.

But that's not the way writing works. For most of us, writing crystallizes understanding. The process of setting down our thoughts shapes them. Not surprisingly, people learn how to do this difficult job in different ways. Research suggests that people take many paths on their way to becoming fluent writers. No one, linear path exists for the beginner to follow. In fact, most paths in writing fail to be linear at all. Instead, fluent writers describe a process in which, for instance, they may work first on the organization of a piece, then on the language, then back to organization.

Skilled writers tend to think more about the audience and about ways to meet audience expectations. They tend to do more planning before they write and to adjust their organizational patterns and language according to the readers. In time, they become more fluent. That doesn't mean that writing is easier for them—most say it isn't. But in the process, they become more skillful.

The final versions that you may admire got that way through much planning, adjustment, analysis, and practice. The text did not emerge word-perfect in its first draft.

WRITING FOR A PROFESSIONAL AUDIENCE

Much of the writing students do in college is aimed at just one person: an expert. In contrast, most professional writing is done for a variety of readers with different levels of expertise in different subjects. In college, when you write about the 19th century novel, the reader is a person thoroughly conversant in the field. At work, when you write about the tax code, it is often for someone who expects you to be the expert, and to use that expertise to solve a problem and then present the solution in an understandable way.

You will find yourself writing for people who are busy and who need the abstract, examples, and tightly structured report George provided in his rewrite—rather than the long, academic document he provided in the original draft.

Audience analysis will change the way you approach a writing assignment. While there isn't any "right" way to go through the writing process, there are some procedures we have learned from people who write professionally that may be useful to you. All of these procedures stem from analyzing the audience.

- Many people who write a great deal spend time on planning—on thinking about what they are going to write, on reading and digesting sources—before they begin. This is a different pattern from that of many beginning writers, who tend to plunge right into a first draft, and then, quite often, grind to a halt. If you are writing a report and you are inexperienced at organizing a writing job, you might simply sit down with a few reference notes and start writing. The more experienced writer thinks first about the audience. What do they want to know? Is there a central problem you are supposed to analyze? What is the solution? How good is the evidence? How can you write the document so that it can be read and understood quickly?

• Experienced writers collect material as they work on a project. The material may be interviews, first-person observation, reference notes. While thinking about the audience, they spend time assimilating the material they are collecting, reading and digesting background information. Usually this process ends with a scratch outline, one that is much revised as the writer proceeds through the document.

• Some professional writers use the outline to divide any writing job into smaller, manageable parts. They tackle one part at a time.

• Some people skip the opening—always a difficult part to do, as it must make the path for the reader—and start in the middle with the easier, procedural sections, doubling back to write the introduction at the end. Others feel that they must have the introduction first, for from it the whole piece flows.

• Most writers expect to do two or even three drafts before the final version.

SUMMING UP

• There is no such thing as a standard format, no cookbook recipe for the forms of technical writing. There are, however, standard ingredients. These are shaped and tempered by audience background and needs.

• If you are grappling with the complexities of the subject, the manner in which the information is cast may seem secondary to you. It is not. There are people out there who have to read what you've written.

• Audience analysis is the most important step to clarity. The writer who wants to be effective has to adjust terminology, level of detail, organizational pattern and language according to the background of the audience.

• A peer group knows the basic terminology, can quickly fit detail into a pattern, and will be capable of appreciating the subtleties of your work.

• A heterogeneous audience has a diversified background. It's the writer's job to bridge the gap between what these readers know already and what they need to learn.

• The Reader Profile Sheet sorts audience by education, subject literacy, and need for information. The writer uses the profile to categorize readers into primary, secondary, and fringe groups. Then the writer adjusts the organization, emphasis, and language of the report according to the audience. This is a mechanical step, but it may get you out of a quagmire when you've worked hard and produced a document that is correct, but unreadable for your audience.

EXERCISES

1. Locate two discussions on the same subject, each written for a different audience. You may use the two contrasting articles in this chapter, Example 2.2 from the *New England Journal of Medicine* and Example 2.4 from *The New York Times,* or any other two contrasting pieces on the same subject. In a brief (1- or 2-page) paper, analyze the differences in approach between the two articles.

2. Describe a technical term, item, process, or piece of equipment with which you are familiar—a table of commands for Lotus 1-2-3, for instance, or the Dow Jones Industrial Index. Assume your description is part of an orientation in which you are introducing new employees to the subject.

   When you are done, write the same description or definition in a *second* version, this time aimed at a group of specialists in the field.

3. Choose five or six journals in your field. If you are studying in a class, form a group with other majors to do this exercise.

   a. Locate the guidelines for authors in each journal.

   b. After looking at both the journals and the guidelines, prepare a summary of the style of each journal, discussing the differences in treatment.

   c. Rank the journals from the most homogeneous to the most diversified. What adjustments are made for audience by the authors in the more diversified publications?

PART II

MAKING TECHNICAL TEXT READABLE: LOGIC AND ORGANIZATIONAL PATTERNS, LANGUAGE, VISUAL DISPLAY

CHAPTER 3

Logic and Organizational Patterns

Elizabeth David's famed recipe for Roman stew starts with fundamentals: First, she advises, catch your hare.

Step one in technical writing—the equivalent of "Catch your hare"—is, "Analyze your audience."

What comes after audience analysis? What subsequent elements characterize good technical prose? To be sure, most of the devices literary writers use are off-limits for technical writers. Plot, characters, dialogue, psychological realism—the Harris cartoon (Fig. 3.1) gently pokes fun at the incongruity of using such literary techniques in a book on nuclear physics.

Yet there are certain paths that, when followed, make technical prose—text that is by definition informative, explicatory, objective—effective and even elegant. There are ways to structure and express

Chapter 7
THE STRUCTURE OF THE NUCLEUS
OF THE ATOM

"What?" exclaimed Roger, as Karen rolled over on the bed, and rested her warm body against his. "I know some nuclei are spherical and some are ellipsoidal, but where did you find out that some fluctuate in between?"

Karen pursed her lips. "They've been observed with a short-wavelength probe . . ."

Figure 3.1 © 1975 by Sidney Harris—*American Scientist Magazine*

technical text powerfully and coherently so that ideas make the quickest, most efficient journey from the page to the reader. To do this—to write so that text makes an efficient journey from the silence of print to information the reader comprehends—is to write readable prose.

In Chapter 2 you saw two versions of the same consumer loan application; Version B (Figure 2.1b) was more readable than Version A (Figure 2.1a). The techniques that make it more readable can be divided into three broad categories

- logic and organizational pattern
- accessible language
- visual display

This and the two succeeding chapters analyze these three interrelated areas, areas fundamental to the writing of instructions, procedures, letters, memos, descriptions, manuals, and reports of every sort.

IMPORTANCE OF ORGANIZATIONAL PATTERNS

Communication that is logical is reasoned in the proposition, order, interconnection, development, and disposition of its elements.

That this path is clear to the writer does not guarantee that it is clear to the listener or reader. **It is the first, most urgent job of the writer to make the logic apparent.**

When you write an academic paper in college, aimed at a reader who is an expert, you may safely assume that this reader knows thoroughly the roots and implications of the problem you are discussing, as well as possible solutions. Your reader, after all, is an expert. At work, you will find that these classic roles shift: You will become the expert, the person working closest to the problem. If, for instance, you are inspecting a beverage bottling plant for quality control, it is you who will go to the plant, inspect the surfaces for sanitation, do the analytical tests. You will be the person with first-hand, intricate knowledge of the problem. Your reader or listener, in contrast, will be farther from the details, less familiar with the problem. This person will need a guide, or at least a clearly marked path.

When you write as a technical professional, you will find that the reader or listener needs first and foremost to understand the structure or path of the argument, whether the writing is a problem-solution memo, an orientation talk introducing new equipment and procedures, or a formal investigative report that moves from hypotheses through procedure, methods, and materials, to the results and conclusions.

Instead of being forced to conquer the text line by line, readers or listeners will need to know in advance where they are going throughout a report, memo, paper, or talk. And just at the moment the audience wonders, "What next?" the next sequence in the logic should follow.

In this way, the report or memo becomes not a morass of everything the author knows about the problem, but a highly focused account in which the main ideas are first encapsulated and then developed systematically.

Yet many beginning writers don't realize the importance of logic and the organizational patterns used to develop argument. Instead of using a structure in which the author's path is well marked, they write anecdotal accounts where the main idea only appears on the last page. This murder-mystery approach is the wrong format for readers who want the gist of the problem served up first, followed by the orderly development of the supporting details.

In the following excerpt from *Crystals and Crystal Growing* (Example 3.1) part of the clarity of the text can be attributed to its logic and organizational pattern. Part of the appeal also lies in the directness of the language and effective use of illustration. Logic, language, and visual display are all interrelated in this example, as in most explicatory text that seeks to be readable.

(Text *continues* on page 46.)

Example 3.1

Two Methods for Growing Crystals

Two general procedures for growing large single crystals of salts can conveniently be used at home. In both methods you suspend a seed crystal by a thread in a Mason jar containing the solution. In one, the "sealed-jar method," you supersaturate the solution and seal the jar to keep water from evaporating. The seed will grow as excess salt in the solution slowly crystallizes on it. This is the quickest and most useful way of growing most of the substances mentioned later.

In the other method, the "evaporation method," you start with a saturated solution and permit it to evaporate slowly. You leave the jar unsealed, and cover the top with a piece of cloth, both to reduce the rate of evaporation and to keep dust out of the solution. As water evaporates, the solution becomes supersaturated and the seed grows.

In both methods fairly constant temperatures are quite important, because changes in temperature change the degree of supersaturation. Consequently, it is wise to keep the jar somewhere in the house, possibly the basement, where it will not be disturbed and where the temperature varies the least.

Preparing a Saturated Solution

In both methods of growing crystals the first step is to make a solution that is saturated at the temperature at which the crystals will be growing. In the evaporation method you will then let the solution evaporate slowly after you have hung a seed in it. In the sealed-jar method you will heat the saturated solution to a higher temperature, where it is unsaturated. Then you will dissolve a little more salt in it, hang a seed in it, and cool it to the original temperature, where it will find itself supersaturated.

To prepare the saturated solution, you could proceed either by dissolving solid in an unsaturated solution or by withdrawing solid from a supersaturated solution. Notice now why the latter is the better procedure.

A solid salt at the bottom of a jar of water will dissolve quickly at first, but it will soon be surrounded by a concentrated solution. Since the solution is denser than the water, it will tend to stay at the bottom. If you do not stir the solution, further progress toward saturation will depend on diffusion of the salt upward into the more unsaturated part of the

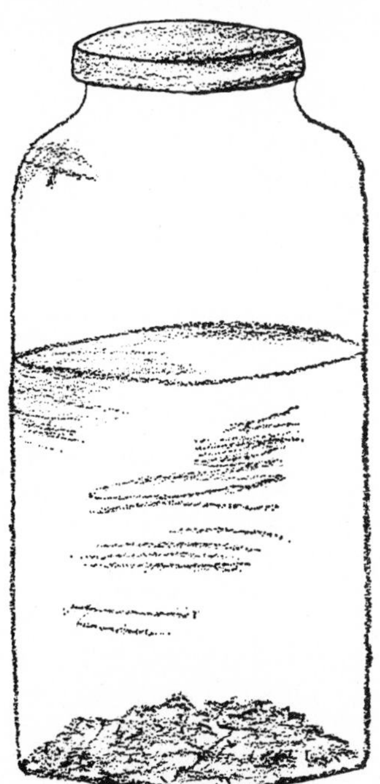

Fig. 35. TO MAKE A SATURATED SOLUTION, *seed a supersaturated solution and shake twice daily. It will put its excess on the seeds and become saturated in two or three days.*

solution, a very slow process. If you stir the solution, you soon meet another problem. As the solution comes closer to the saturation point, the solid dissolves more slowly. The procedure needs a lot of attention over a long time

A better procedure is to reach saturation by letting a supersaturated solution deposit its excess solid, as shown in Figure 35. The crystals at the bottom of the jar take solid out of the liquid, leaving the solution around them less dense than the rest of it. The less dense solution rises and a more concentrated solution replaces it in contact with the crystal surfaces. Thus the solution stirs itself, so to speak. Shaking the solution occasionally will speed the process; but in time the solution will reach saturation even without agitation.

The recipes in the next chapter give quantities of salt and of water that produce solutions supersaturated at temperatures below about 27° centigrade. These quantities have been worked out, whenever possible, to use a full jar of the salt as suppliers usually package it. Make the water measurements as accurately as you can, preferably in cubic centimeters. The appendix shows how to convert temperatures from the centigrade to the Fahrenheit scale.

By heating the mixture of the salt and water to about 50° centigrade, you will be able to dissolve the salt in the water quite rapidly if you stir the mixture occasionally. Do not use aluminum vessels; they may be attacked by some of the solutions. A suitable vessel is a stainless steel double boiler, or a Mason jar placed in a saucepan of hot water. Keep a lid on the vessel between stirrings to reduce loss of water by evaporation from the solution.

When the salt has dissolved, pour the hot solution into a one-quart Mason jar and seal it to prevent evaporation. Then cool the solution to the temperature at which you expect to grow crystals. Since the solution is now supersaturated, seed it with a pinch of the salt to provide a place for the excess salt in the solution to deposit. Suitable seeds for this purpose are the crystalline powder left in the supply jar, or the powder left after evaporating a drop of the solution to dryness.

Seal the jar again, shake it well, and keep the solution at your expected growing temperature for at least two days; shake it twice a day and give it time to become saturated. If the temperature varies much, place the sealed jar in a pail of water. Since the water rises and falls in temperature more slowly than the surrounding air, it will act as a "thermal ballast," reducing the temperature fluctuations of the immersed jar.

Example 3.1 (*continues*)

When the precipitate stops growing, the solution has reached its saturation point. Pour off the clear solution into another container, taking care that the solution carries with it the least possible amount of the salt at the bottom of the jar. Then scrape that deposited salt onto a saucer, and when it has dried, return it to the supply bottle. Wash and drain the Mason jar, pour the saturated solution back into it, and seal it, since further evaporation would make the solution supersaturated in a short time.

Preparing a Seed Crystal

Any fragment of the solid, no matter how tiny, is a potential seed. But in order to be conveniently suspended by a thread, a seed must be ⅛ to ¼ inch long. Furthermore, it must be a single crystal so that the crystal growing from it will also be single.

You can prepare such seeds by pouring an ounce of your saturated solution into a small glass and setting it in an undisturbed place. As the solution evaporates, a few crystals will usually begin to grow on the bottom of the glass (Figure 36). If it becomes supersaturated without depositing crystals, add a very small amount of crystalline powder from the supply bottle, or of the powder left after evaporating a drop of solution. Look at the glass and its contents once or twice a day; harvest the seeds when they have grown large enough for convenient handling, but before they grow so large that they touch and interfere with one another. Pick out the good seeds with tweezers, or pour off the solution and dump all the seeds on a paper tissue, where you can dry them well.

Save all good seeds. Your first crystal-growing efforts may fail, and when they succeed you will probably want to grow several crystals of the same substance. Furthermore, good seed crystals are excellent subjects for determining the angles between crystal faces with the "reflecting goniometer" you will read about further on.

You will notice that one face of each of these little crystals is slightly concave. It is the face that rested on the bottom of the glass. Not much solution is able to get under such a face to make it grow. But slight vibrations of the glass let a tiny amount of liquid under the edges of the crystal, and those edges slowly grow until finally the hollow left at the center of the face becomes deep enough to be noticeable.

Fig. 36. TO GROW SEED CRYSTALS, *allow an ounce of saturated solution in a glass to evaporate slowly, and remove and dry the crystals when they have reached the proper size.*

Preparing the Growing Solution for the Sealed-Jar (Supercooling) Method

The sealed-jar method requires the preliminary preparation of a supersaturated solution from the saturated solution described at the beginning of this chapter. You will prepare that solution by dissolving more salt in the saturated solution at a higher temperature, and then cooling the solution. The proper degree of supersaturation varies with the behavior of each salt: how fast it can order itself into a crystal without faults, and how highly its solution can be supersaturated without depositing seeds spontaneously. Of course, the crystal you grow at constant temperature cannot become larger than the amount of salt you add to a solution originally saturated at that temperature.

The recipes in the next chapter, specifying amounts of salt to add to a saturated solution in order to supersaturate it, take these considerations into account. They are suitable for growing crystals in a room whose temperature fluctuates between 23° and 25° centigrade. If your growing conditions differ from this, you may find after trial that you get better results by slightly increasing or reducing the amount of added salt.

Weigh the amount of salt needed, put it in the double boiler, and pour the saturated solution over it. Figure 37 shows how to make a balance accurate enough for this purpose. Heat the solution slowly, stirring it until the salt has dissolved. In this operation, any tiny crystals that remained in the solution when you decanted it after saturating it will also be dissolved. Wash the Mason jar, and let it drain dry enough so that the amount of wash-water you will be adding to the solution is negligible. Pour the solution into the jar, seal the jar, and let the solution cool slowly.

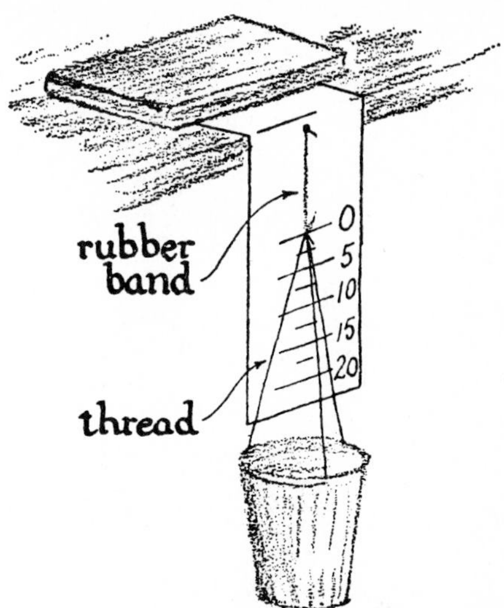

Fig. 37. **BALANCES FOR WEIGHING SALTS.** *In a rough spring balance, a thin rubber band can be used for the spring. Make a mark on the cardboard at the point where the rubber band and string meet when the paper cup is empty. Then mark for 2½ grams by putting a dime in the cup, and continue to about 20 grams. The rubber band will eventually show "fatigue"; the balance will not return to the zero mark when the cup is empty. Then you must replace the rubber band and recalibrate the scale. A more accurate and permanent balance is fairly easy to make of scraps of wood, with razor blade suspensions. Study a manufactured balance for a pattern.*

Seeding the Solution in the Sealed-Jar Method

While the solution is cooling, prepare a cardboard disc to hold the thread for suspending the seed, as shown in Figure 38. Lay the seed on a piece of paper in preparation to tying it. Then wash your hands, for by this time there will be many invisible seed crystals on them. With clean hands, make a slip knot in a piece of sewing thread and tighten the loop around the seed, as shown in Figure 39. Attach the thread to the cardboard disc, and leave such a length of thread between seed and disc that the seed will be suspended an inch or two above the bottom of the jar.

Bring the solution to a temperature about three degrees centigrade above the expected growing temperature. If the solution is still warmer than that, put the jar in a pail of cold water and stir the solution with a thermometer; if the solution is too cool, use hot water in the pail. In either circumstance stir the solution well to equalize the temperature throughout it. Now you are ready to seed the solution. To plant one seed, excluding all others and preserving that one, takes care; it is one of the most critical steps in growing the crystal. The seed you have prepared has other microscopic seeds on it; the air carries a dust of seeds. At the planting temperature, the solution is unsaturated. Soon after you have planted the pre-

Example 3.1 (continues)

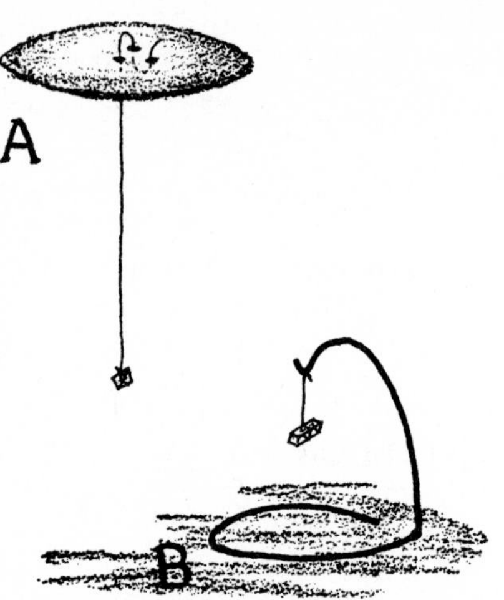

Fig. 38. SUSPENSIONS FOR SEEDS. *A—To make the disc for suspending the seed in a sealed jar, turn the jar upside down on a piece of cardboard, draw a circle around the mouth, and cut out the circular piece. This disc will fit over the top of the jar without falling in, yet permit the lid to be screwed down tightly. Make three small holes near the center of the disc, large enough for thread to go through. After the seed is tied, send the loose end of the thread through these holes, first up from the bottom, then down, then up again. The thread will shorten if you pull on the loose end. Adjust it so that the seed hangs about an inch above the bottom of the jar. B—A wire bent into a "cobra" is used to suspend the seed while crystals grow by evaporation. The important things to remember are that the top of the wire must be below the surface of the solution and that the base must be wide enough to prevent tipping.*

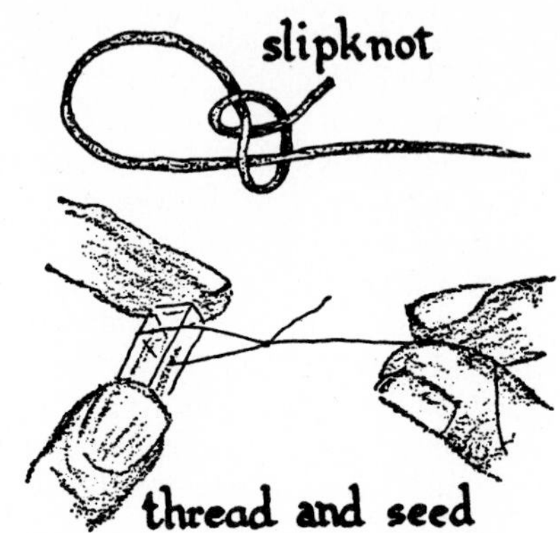

Fig. 39. TO TIE THE SEED, *use common sewing thread with a slip knot.*

pared seed, all the microscopic seeds, and part of the prepared seed, will dissolve. Since the prepared seed is the largest, some of it will be left when the other seeds have dissolved completely.

As the seeded solution cools to the growing temperature, it will become supersaturated, and the prepared seed will stop dissolving and begin to grow. By watching the density currents around the crystal, shown in Figure 40, you can tell whether it is growing or dissolving.

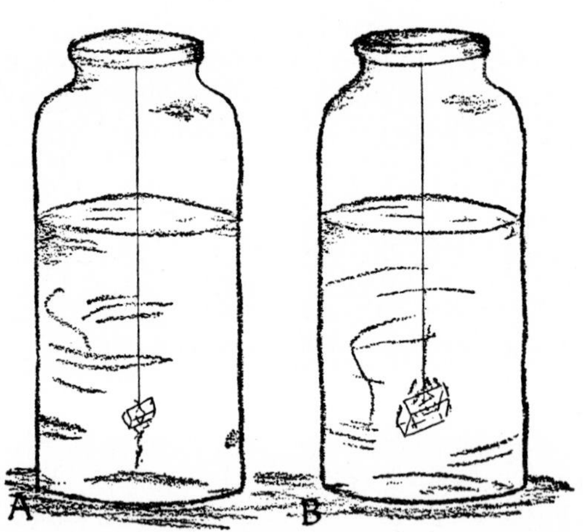

Fig. 40. A DENSITY CURRENT *will either rise or descend from a crystal hung in a solution if the solution is quiet. When the current descends (A) it is carrying extra salt dissolved from the crystal. Because the crystal is dissolving, the solution must be unsaturated. When the current ascends (B) it consists of a solution depleted of some of its salt. The lost salt has been deposited on the crystal; the crystal is growing, and the solution in the jar must be supersaturated.*

Changes in Procedure for Growing by Evaporation

In the method of growing by evaporation, you change the foregoing procedure in some details. Heat the saturated solution to dissolve all unwanted seeds, return it to a clean jar, and cool it to a temperature a degree or two above the expected growing temperature, stirring it with a thermometer to make sure that the temperature is uniform throughout.

Tie the seed thread to a wire "cobra" (Figure 38B) instead of a cardboard disc. Be sure that the cobra and the thread do not extend above the surface of the solution. A thread projecting from the solution will act as a wick. In this method, unlike the sealed-jar method, water will evaporate from such a wick, additional seeds will form on it, and there is danger that they will drop on the growing crystal. In any event, some unwanted seeds will drop from the surface of the solution as the water evaporates, but they cause trouble less often.

Place the cobra, with its suspended seed crystal, in the slightly unsaturated solution, and cover the top of the jar with a cloth held in place by a rubber band, as shown in Plate 12. Again growth will start when the solution cools to the growing temperature. The rate of growth thereafter depends on the rate at which water evaporates from the solution.

Growing and Harvesting a Crystal

Do not disturb the crystal during growth. Try to keep the temperature fairly constant, using a bucket of water for thermal ballast if necessary. By the sealed-jar method the crystal should grow to a good size in from three to six days. Some crystal "debris" may accumulate in the bottom of the jar, growing from unwanted seeds, but the debris will cause no damage so long as the desired crystal continues to grow without interference, as the crystal shown in Plate 13 is growing.

The crystals pictured later with the recipes have all been grown by these methods; the photographs will help you to decide when to harvest your crystals. When a crystal has attained full growth, pull it out and dry it immediately with a paper tissue or a soft cloth. Be careful how you handle it, especially if you intend to use it in the optical experiments described later, for it is soluble in water, and perspiration on your hands will damage its clear, plane faces. The best way to store it is to wrap it in a scrap of cloth and put it in a screw-topped jar to keep it from damage in either too dry or too humid air.

To grow a second crystal of the same substance, weigh the total amount of solid material you took out of the jar—both the crystal and the debris at the bottom. Then dissolve that weight of material in the old solution, warming and stirring to make the new growing solution.

An Example of a Recipe

At this point an example will help to make the foregoing procedures clear. The recipe is for growing a crystal of Rochelle Salt. You will find how to convert pounds into grams, and fluid ounces into cubic centimeters, in the Appendix.

I. Supersaturated solution:
   a. 1 pound Rochelle Salt in 349 cc. (11.5 oz.) water
   b. 130 grams Rochelle Salt per 100 cc. water.

II. Add to saturated solution:
   a. 31 grams Rochelle Salt
   b. 9 grams per original 100 cc. water.

The "a" quantities are based on your buying a certain quantity of the salt. They always specify the smallest amount of salt that will give satisfactory results, and they will spare you much weighing. The "b" quantities are ratios; from them you can calculate any desired quantity of solution.

To grow a Rochelle Salt crystal, dissolve

Example 3.1 *(continues)*

one pound of the salt in 349 cc. of water, measured by a graduated cylinder, or 11½ ounces of water, measured by a kitchen measuring cup. Heat the mixture to dissolve the salt, seal it in a Mason jar, and let it cool. Then add some grains of Rochelle Salt; the supersaturated solution will deposit its excess on the added grains, and in a couple of days the solution will become saturated. Pour off the solution, and grow some seeds from an ounce of it.

Now you are ready to make the growing solution. To grow by evaporation, warm the saturated solution to dissolve the unwanted seeds, then let it cool again. To grow by the sealed-jar method, warm the saturated solution, add the "a" quantity of salt—31 grams—given in Part II of the recipe, and dissolve it. This is the growing solution in which you plant the seed, using the cardboard disc to suspend it.

Comparison of the Two Methods

Each method—sealed-jar and evaporation—has advantages and disadvantages. Both, of course, provide the indispensable condition for a crystal to grow from solution, supersaturation. In the sealed-jar method, the supersaturation arises from supercooling the solution. In the other method, evaporation of some of the water provides a progressive supersaturation of the solution.

Growing a crystal by evaporation, you can, at least in principle, get back almost all the dissolved solid in the form of a single crystal. But the rate of evaporation is hard to control: it depends on how humid the environment is, and how effectively casual drafts remove the evaporated moisture. Since evaporation occurs at the surface of the solution, the degree of supersaturation tends to be greatest there; unwanted seeds often form at the surface and may drop on the desired crystal. Any droplets of solution splashed on the sides of the jar, at the time it was filled, will evaporate to dryness, and the residue of crystalline dust may drop into the solution, providing a host of nuclei for crystallization.

On the other hand, supersaturating the solution by cooling it below the saturation temperature provides a control of supersaturation as good as the control of temperature. Often you can cool the solution three or more degrees centigrade below the saturation temperature without causing additional seeds to form spontaneously. Then a crystal can be grown for as long as a week at constant temperature. As the crystal grows, the supersaturation declines, and thus automatically provides the slower growth rate usually desirable as a crystal becomes larger. But the amount of material that can be deposited from the solution is clearly limited, even if you reduce the temperature again after the initial supersaturation has been exhausted.

USE OF CONTRAST AND COMPARISON

Part of the clarity and readability of *Crystals and Crystal Growing* comes from its organization. The authors have taken a pattern—comparison and contrast—and used it as the scaffold for their information. The excerpt compares and contrasts two methods for growing large single crystals of salts: the sealed-jar method and the evaporation method. Comparison and contrast are used not only for the overall pattern, but for developing the ideas in the subsection "Preparing a Saturated Solution" and in the captions for Figures 38 and 40.

One of the devices that makes the contrast effective is the use of strong lead sentences. These provide the audience—novices, in this case—with a pattern, a scheme of the authors' main points.

Here is part of *Crystals and Crystal Growing* repeated, with some of the lead sentences shown in boldface. The divisions of main ideas are underlined. Lead sentences are found not only at the beginning of the chapter and the beginning of each subsection, but before any technical exposition.

Two Methods of Growing Crystals

Two general procedures for growing large single crystals of salts can conveniently be used at home. In both methods you suspend a seed crystal by a thread in a Mason jar containing the solution. In one, the "sealed-jar method," you supersaturate the solution and seal the jar to keep water from evaporating. . . .

In the other method, the "evaporation method," you start with a saturated solution and permit it to evaporate slowly.

Preparing a Saturated Solution

In both methods of growing crystals the first step is to make a solution that is saturated at the temperature at which the crystals will be growing. In the evaporation method. . . .

In the sealed-jar method. . . .

To prepare the saturated solution, you could proceed either by dissolving solid in an unsaturated solution or by withdrawing solid from a supersaturated solution. Notice now why the latter is the better procedure.

A solid salt at the bottom. . . .

A better procedure is to reach saturation by . . .

FINDING A SHAPE: Two overview formats

Technical prose usually follows the explicatory mode of telling or previewing main ideas first, then fitting in details. There are few mystery story approaches, although occasionally there are patterns that deliberately try to intrigue the reader—such as a speech which opens with a puzzle that the speaker then proceeds to solve.

In the excerpt from *Crystals and Crystal Growing*, the useful organizational pattern **comparison and contrast** is the skeleton that supports the technical argument. There are many other organizational devices that you can use in technical exposition to shape or frame an argument so the reader finds the content accessible and therefore more readable.

Most of these patterns are predicated on *giving an overview*—focusing and presenting main points before developing details.

Two formats that help provide this path are **problem-solution** and **main idea-significance.**

Problem-Solution

Consider the following memo (Example 3.2). Notice the order in which the information is presented.

The writer uses the first paragraph to state the technical problem. The second paragraph follows with the solution. In the third and fourth paragraphs, the author develops the point by providing necessary information, using definition, process description, and example.

What is the problem?

What is the solution?

What are the important supporting details?

Main Idea-Significance Format

If you are writing for one person, and that one person knows your area of expertise well, then your reports and memos may concentrate on the "what"—that is, on what you found out in the experiment or troubleshooting trip or product showing.

But if, for instance, you write for a range of readers whose background is as likely to be in marketing or management as in engineering, you'll need to include not only the "what," but the "So what?" Someone steeped in the problem will understand its signif-

To: Date:

From:

Re: Separating Isotopes with Laser Light

Problem

Isotopes of the same element have identical chemical properties; they cannot be separated chemically. Isotope separation has been a complicated and expensive undertaking, most notably in the World War II program to make an atomic bomb by separating fissionable uranium 235 from nonfissionable uranium 238.

Solution

But the different isotopes of an element have very slightly different patterns of spectral lines, which means that they emit and absorb slightly different wavelengths of light. The possibility of exploiting the slight differences in absorption spectra to separate the two isotopes of uranium was examined and rejected, because the available light sources were unsuitable. In recent years, however, the development of lasers has provided a source of light whose wavelength may be controlled precisely enough so that one isotope will absorb the energy while another isotope will not. In addition, laser light is intense enough to be suitable for efficient isotope separation.

Details
•definition
•process explanation
•examples

Laser light differs from ordinary light because it is ''coherent''; that is, its waves all have the same wavelength, frequency, and orientation. Laser light is produced when a large number of molecules are induced to emit radiation of the same wavelength simultaneously. If the light is of the appropriate wavelength, it can be used to add energy to only one isotope of one element but not other isotopes. The more energetic isotopes could then be separated from others.

Potentially, laser isotope separation of uranium is 1000 times more efficient than gaseous diffusion separation. It has been estimated that the use of laser light instead of gaseous diffusion to enrich uranium for nuclear generating plants could save $100 billion by the year 2000. However, there are a number of practical problems in achieving useful laser separation of isotopes.

Example 3.2

icance without underscoring, but someone who is a few steps removed will appreciate the explicit connection.

A main idea-significance format is useful not only for such nonexpert readers, but for the writer as well, who can use the pattern as an aid to build an orderly piece of writing. Example 3.3 is a section of a report, written for nonspecialists, on a new surgical technique. First it gives the central point or main idea; then it follows with a brief explanation of *why the central point matters*.

Main idea-significance is a useful format for technical information, whether the information is in science or engineering, or business and finance. Example 3.4 is a neatly organized financial memo in what/so what format.

Example 3.3
Extract from a written report using main idea-significance format.

A new microscope will allow scientists to do something that has never been possible before: to look directly into the interior of both living tissue and one-of-a-kind fossils in a totally nondestructive manner; and, in what has been heralded as one of the most important advances of the century in light microscopy, it will enable the high-resolution images obtained to be recorded stereoscopically.

The Tandem Scanning Reflected Light Microscope (TSRLM), developed by Czech scientists, was brought to the attention of the Western world by anatomist Alan Boyde of University College, London. Surprisingly, the device makes no use of lasers, modern-day electronics or image processing. And not only is it cheaper than the best optical microscopes, it produces far better results.

The optical system of the TSRLM was developed nearly 20 years ago by Mojmir Petran and Milan Hadravsky of Charles University in Plzeň, Czechoslovakia, in an attempt to find a way to look at living brain tissue with reflected light. If living tissue is examined with an ordinary light microscope, the plane in focus is distorted by the many layers of overlying and underlying cells. For this reason, tissues are generally fixed to preserve their structure, embedded in a solid material and sliced before being viewed. The relationship of cells in this thin slice to those in the rest of the specimen can only be approximated by looking at a sequence of slices. This procedure has many drawbacks: It is tedious, destroys the intact specimen, introduces artifacts and, obviously, cannot be performed on living material.

Petran and Hadravsky sidestepped slicing and sectioning by developing a microscope that allows a viewer to see, undistorted, one very thin plane at a time. By focusing up and down, the viewer can see other planes within the object and, in effect, optically section the material.

The invention works because only light from the region focused on is allowed to pass into the eyepiece. The key is a spinning disk with pairs of spiral patterns of pinholes, developed from a century-old device called the Nipkow disk. In the TSRLM, light from an ordinary microscope lamp is formed into beams as it passes through the pinholes on one side of the disk; the objective lens of the microscope focuses these beams in a very shallow plane. Only light reflected from that plane can travel back through the corresponding pinholes on the opposite side of the disk to get to the eyepiece. Light from other planes will be focused either above or below the holes and will be blocked by the disk. Because the TSRLM forms images in real time, it can be used to monitor small changes in biological systems.

Example 3.4
Memo in what/so what format

	To: Date:
	From:
	Re:
Main Idea	Yesterday the Securities and Exchange Commission (SEC) voted to let the Chicago Board trade options on Government National Mortgage Association securities (GNMAs or Ginnie Maes).
Significance	This is the first time exchange trading has been approved for nonstock options. In the wake of the decision, other exchanges are expected to ask for permission to trade options on Ginnie Maes and certain Treasury securities.
Definition Explanation	Ginnie Maes are shares in pools of mortgages backed by the Federal Housing Administration or Veterans Administration and guaranteed by the GNMA. Put options in this security give a buyer the right to sell Ginnie Mae certificates at a certain price during a specified future time. Call options give the buyer the right to buy Ginnie Mae certificates at a certain price during a similarly specified time. Under the Chicago proposal, Ginnie Mae options contracts will be standardized and subject to Chicago margin requirements to reduce the likelihood of overcommitments. The SEC will regulate the market, and investors will be protected by rules on sales and disclosure.

CHRONOLOGY AND LISTING

Problem-solution, main idea-significance, and comparison-contrast are three powerful organizing tools to structure and clarify long sections of text, to help make them readable.

Sometimes you may be able to use a simpler pattern to develop details, depending on the nature of the material. In *Crystals and Crystal Growing*, for example, the main pattern is comparison and contrast; however, the authors use the simpler patterns of chronology and listing to present supporting details.

The subsection "Preparing a Seed Crystal" has a problem-solution pattern, within which detail is developed chronologically. Similarly, "Preparing a Saturated Solution" follows an overall pattern of contrast, but within the subsection, the details of how to prepare a saturated solution are developed in a chronological pattern, signalled by many transitional words such as "when," "then," "afterwards," and "later."

Listing is another way to develop detail within a larger organizational pattern. In fact, buttressing an argument by enumerating or listing details is probably one of the most common patterns in technical prose. Writers who use this pattern include transitional words or phrases to guarantee that the reader sees the connections.

In Example 3.5, for instance, the overall pattern of the selection is problem-solution, but the subsidiary pattern is listing.

Example 3.5

To: Date:

From:

Re: Separating Isotopes with Laser Light

Problem

Isotopes of the same element have identical chemical properties; they cannot be separated chemically. Isotope separation has been a complicated and expensive undertaking, most notably in the World War II program to make an atomic bomb by separating fissionable uranium 235 from nonfissionable uranium 238.

Solution

But, the different isotopes of an element have very slightly different patterns of spectral lines, which means that they emit and absorb slightly different wavelengths of light. The possibility of exploiting the slight differences in absorbtion spectra to separate the two isotopes of uranium was examined and rejected, because the available light sources were unsuitable. In recent years, however, the development of lasers has provided a source of light whose wavelength may be controlled precisely enough so that one isotope will absorb the energy while another isotope will not. In additon, laser light is intense enough to be suitable for efficient isotope separation.

Details
- definition
- process explanation
- examples

Laser light differs from ordinary light because it is "coherent"; that is, its waves all have the same wavelength, frequency, and orientation. Laser light is produced when a large number of molecules are induced to emit radiation of the same wavelength simultaneously. If the light is of the appropriate wavelength, it can be used to add energy to only one isotope of an element but not other isotopes. The more energetic isotopes could then be separated from the others.

Potentially, laser isotope separation of uranium is 1000 times more efficient than gaseous diffusion separation. It has been estimated that the use of laser light instead of gaseous diffusion to enrich uranium for nuclear generating plants could save $100 billion by the year 2000. However, there are a number of practical problems in achieving useful laser separation of isotopes.

Listing
•problems

One problem is the difficulty of developing lasers that emit the desired wavelengths of light. *Another problem* is that the thermal motion of atoms in a gas lessens the spectral distinctions between isotopes. Because thermal motion affects the absorption and emission of radiation, the spectral lines of the atoms may be blurred, which makes it difficult to excite only the desired isotope. There is *also the problem of* separating the excited isotope from the other isotopes.

•solutions

Several approaches are being tried to solve these problems. Some laboratories are using "scavengers" that absorb all but a very narrow wavelength of light, the wavelength that will be absorbed by the desired isotope. Thermal motion can be lessened if the atoms are cooled to a temperature approaching absolute zero. The energy absorbed by the isotope can be used to promote a chemical reaction that will make separation easier to accomplish.

Several laboratories have already reported successful separation of several isotopes, including those of chlorine and sulfur. An intensive effort is being made in several countries to achieve large-scale laser separation of uranium isotopes.

A listing pattern is useful when you want to give items a visual identity. You may emphasize this visual identity by setting off the list either with numbers (1,2,3), bullets (•), or underlining or boldface of key words in the list. The visual effect can be striking when you set off items this way. In Example 3.6a, for instance, the writer has written a problem-solution memo in which he gives the two solutions in narrative text without emphasizing the listing pattern. In Example 3.6b the memo has been revised so that the list has its own visual identity. Key words are chosen ("Interim Measures: New Mounting Blocks," "New Design for Pneumatic Control Loop") and set off by numbers and underlining. Example 3.6c gives the final typed version of the revision.

Example 3.6a
First version

To: Date:
From:
Re:

I met with the Alco serviceman, Carl Jones, at the Dayton Plant on March 31, 1987. He reviewed the two modifications he was making to the filler on line 5. The first modification was to install new mounting blocks which position the no jar-no fill valves in such a way as to prevent fluids from sitting on top of the valves. Alco has claimed this to be only an interim step that will reduce the frequency of failures but not eliminate the problem. I agree with their evaluation of this change and do not expect a noticeable improvement.

Five stations were fitted with a new design which added a pneumatic control loop that eliminates the above valve from the system and insulates the main air supply from the new valve that senses the presence of a jar. This new design has the potential to improve the reliability of the fillers. The new valve consists of a spring that allows air to escape from the control loop when bent by the presence of a jar. The spring valve does not appear to be rugged enough to stand the rigors of its environment, as indicated by only two of the five being serviceable for production on April 1.

The filler started with no problems on the 1st. With time, station number 11 began not filling sporadically. By the end of the shift there were approximately 100 bottles rejected because of this. Station 11 had only the interim modification.

Since then I have been in touch with Alco and indicated to them that their new design has potential but the spring type valve may be a weak link. I suggested that a one way valve could be installed in the present location of the no jar-no fill valve and perform the same function as the spring valve, but more reliably and positively.

I will keep in touch with Alco on a weekly basis to show our concern while pushing them to expedite improvements to their system.

Example 3.6b
Revised version

To: Date:
From:
Re:

I met with the Alco serviceman, Carl Jones, at the Dayton Plant on March 31, 1987. He reviewed the two modifications he was making to the filler on line 5.

1. Interim Measure: New Mounting Blocks. The first modification was to install new mounting blocks which position the no jar-no fill valves in such a way as to prevent fluids from sitting on top of the valves. Alco has claimed this to be only an interim step that will reduce the frequency of failures but not eliminate the problem. I agree with their evaluation of this change and do not expect a noticeable improvement.

2. New Design for Pneumatic Control Loop. Five stations were fitted with a new design which added a pneumatic control loop that eliminates the above valve from the system and insulates the main air supply from the new valve that senses the presence of a jar. This new design has the potential to improve the reliability of the fillers. The new valve consists of a spring that allows air to escape from the control loop when bent by the presence of a jar. The spring valve does not appear to be rugged enough to stand the rigors of its environment, as indicated by only two of the five being serviceable for production on April 1.

The filler started with no problems on the 1st. With time, station number 11 began not filling sporadically. By the end of the shift there were approximately 100 bottles rejected because of this. Station 11 had only the interim modification.

Since then I have been in touch with Alco and indicated to them that their new design has potential but the spring type valve may be a weak link. I suggested that a one way valve could be installed in the present location of the no jar-no fill valve and perform the same function as the spring valve, but more reliably and positively.

I will keep in touch with Alco on a weekly basis to show our concern while pushing them to expedite improvements to their system.

Example 3.6c
Final version

To: Date:
From:
Re:

I met with the Alco serviceman, Carl Jones, at the Dayton Plant on March 31, 1987. He reviewed the two modifications he was making to the filler on line 5.

1. Interim Measure: New Mounting Blocks. The first modification was to install new mounting blocks that position the no jar-no fill valves in such a way as to prevent fluids from sitting on top of the valves. Alco has claimed this to be only an interim step that will reduce the frequency of failures but not eliminate the problem.

   Comment: I agree with their evaluation of this change and do not expect a noticeable improvement.

2. New Design for Pneumatic Control Loop. Five stations were fitted with a new design that added a pneumatic control loop that eliminates the no jar-no fill valve from the system and insulated the main air supply from the new valve that senses the presence of a jar. This new design has the potential to improve the reliability of the fillers. The new valve consists of a spring that allows air to escape from the control loop when bent by the presence of a jar.

   Comment: The spring valve does not appear to be rugged enough to stand the rigors of its environment, as indicated by only two of the five being serviceable for production on April 1.

The filler started with no problems on April 1. Gradually, station number 11 began to fail: sporadically bottles did not fill, and by the end of the shift about 100 bottles were rejected. Station 11 had only the interim modification.

Since then I have been in touch with Alco and indicated that the new design has potential but the spring valve may be a weak link. I suggested that a one-way valve could be installed in the present location of the no jar-no fill valve and perform the same function as the spring valve, but more reliably and positively.

I will keep in touch with Alco on a weekly basis to show our concern while pushing them to expedite improvements to their system.

OBSERVATION-COMMENT PATTERN

In Example 3.6c the author has set off her comments to give them a distinct visual identity. Setting off comments and recommendations is a common pattern in technical writing when the job is to review extensive data, and then comment on these data in such a way that the reader can quickly get the gist of the information. Readers appreciate this device. Remember, the reader in a professional situation is often different from the academic reader, the one-person expert audience. At work, it is usually the writer who is the expert, the person who has visited the plant or sifted through the pages of regulatory data. The writers who use a pattern that visually distinguishes comments and recommendations will help their readers to see more quickly the relationship between details and conclusions.

Examples 3.7a and 3.7b show a revision in which a pattern makes the text more readable. In the first version, the author has merged observation, comment and recommendation. In her revised version, she has used an **observation-comment** pattern.

DEVELOPING DETAIL: Use of example, restatement, and analogy

Sometimes the text the writer seeks to elaborate is rooted in abstractions far removed from daily experience. When you are writing for an audience of fellow experts, this situation poses fewer problems, for the group shares a common technical background. But writing about abstractions for a diversified group is a challenge. It can be met in part by using devices such as restatement, examples, comparison, and analogies—which help give technical prose immediacy.

If the group you are writing for is diversified:

- **Use strong lead sentences that state the main idea clearly.**
- **Think about your readers. Will they understand the point as you've stated it? If not, try an example, restatement, contrast, or comparison.**

Here's an example of these suggestions. In this case the writer is working on a report on electromagnetic radiation and as part of

Example 3.7a
First version

To: Date
From:
Re:

The Division of Motor Vehicle's plan provides an excellent framework for improving its program; however, the plan does not fully describe all the improvements or identify the specific resources necessary to carry out the needed improvements. There are several areas where we would like to recommend that specific actions be taken. Our review and recommendations are discussed below on an improvement-by-improvement basis.

Problem One: During August 1987 two procedural changes were made to the thrice-annual audit program performed at each inspection station. First, each routine station audit now includes a detailed examination of the paper emission test records that are generated for emission inspection. Second, a vehicle inspected the same day of the audit and still on inspection premises will be retested in the presence of an Automobile Inspector.

At present the Division of Motor Vehicles audits each inspection station three times a year. The time needed to conduct an audit will increase with the additional activities. To maintain the three audits a year, it appears that more inspectors will have to be added.

Given this situation, we recommend that the number of inspectors and their assignments to accommodate the new schedule be demonstrated. The plan should discuss whether each inspection station will continue to be audited three times a year, or if the number of audits will drop. If the latter is the case, a formal request must be made by the Division of Motor Vehicles. Such a request would have to demonstrate that the reduced frequency would not affect the quality of the program. Given the current operating problems, such a demonstration would be difficult to make.

(p. 2 of memo not shown)

the background is describing the nature of a wave. The audience is multidisciplinary.

1. **Use lead sentences that state the point clearly.** The writer produces, "A wave can be thought of as a disturbance moving in a medium. The medium itself is not carried along."

2. **Think about your readers. Will they understand the point as stated?** If not, try an example, a restatement, a contrast, or a comparison.

Example 3.7b
Revised version

To: Date:
From:
Re:

The Division of Motor Vehicle's plan provides an excellent framework for improving its program; however, the plan does not fully describe all the improvements or identify the specific resources necessary to carry out the needed improvements. There are several areas where we would like to recommend that specific actions be taken. Our review and recommendations are discussed below on an improvement-by-improvement basis.

1. Procedural Changes to Regularly Scheduled Audits of Inspection Stations

During August 1987 two procedural changes were made to the thrice-annual audit program performed at each inspection station.

- Paper Emission Test. Each routine station audit now includes a detailed examination of the paper emission test records generated for emission inspection.

- Retested Vehicles. Any vehicle inspected the same day of the audit and still on inspection premises is retested in the presence of an Automobile Inspector.

Comments:

At present the Division of Motor Vehicles audits each inspection station three times a year. The time needed to conduct an audit will increase with the two additional activities (examining the paper emission test records and retesting vehicles). To maintain the three audits a year, it appears that more inspectors will have to be added.

Recommendations:

1. The Division of Motor Vehicles should present a plan that includes the number of inspectors it has available, and their assignments to accommodate the new schedule.

2. The plan should discuss whether each inspection station will continue to be audited three times a year, or if the number of audits will drop.

3. If the Division of Motor Vehicles plans on reducing the number of audits, it must make a formal request to the State Board. Such a request would have to demonstrate that the reduced frequency would not affect the quality of the program. Given the current operating problems, such a demonstration would be difficult to make.

(p. 2 of memo not shown)

"A wave can be thought of as a disturbance moving in a medium" is clearly said, but it may not be vivid enough to give the definitive immediacy. Some of the readers who are new to the information may need to visualize the idea, understand it in terms of what they already know. One way to help them do this is through the use of examples. Thus the writer comes up with:

Lead Sentences	A wave can be thought of as a disturbance that travels through a medium in a given direction.	
	The medium itself is not carried along. For example, consider a swimmer floating in the ocean. The swimmer bobs up and down with each passing wave, but is not carried toward the shore.	*Example*

3. **In your judgment the audience needs more input to understand the point, try a *restatement* amplifying and extending the initial explanation, or a *contrast* of key ideas, or both:**

Lead Sentences	A wave can be thought of as a disturbance that travels through a medium in a given direction.	
	The medium itself is not carried along. For example, consider a swimmer floating in the ocean. The swimmer bobs up and down with each passing wave, but is not carried toward the shore.	*Example*
Restatement	If molecules of water in the ocean could be followed, it would be seen that they do not travel toward the shore, although the peaks of the waves do.	
	(Objects are washed up on a beach by currents or turbulence, not by waves.)	*Contrast*

The examples and contrast make the essentially abstract thesis easier to visualize and therefore to remember. In expanding the point, the writer might also add a **comparison** and **example:**

Lead Sentences	Similarly, sound waves do not produce any substantial net movement of the gas molecules that make up air.	
	This fact can be shown if we float a balloon in front of a rock band at a concert. The balloon will not move toward the back of the hall, no matter how loud the sound.	*Example*

By now the writer has fleshed out the essentially abstract statement "the medium itself is not carried along" with examples of swimmers on the ocean and balloons at a rock concert.

Here is the complete section (Example 3.8):

Example 3.8

The Nature of Waves

A wave can be thought of as a disturbance that travels through a medium in a given direction. The medium itself is not carried along. For example, consider a swimmer floating in the ocean. The swimmer bobs up and down with each passing wave, but is not carried toward the shore. If molecules of water in the ocean could be followed, it would be seen that they do not travel toward shore, although the peaks of the waves do. (Objects are washed up on a beach by currents or turbulence, not by waves.)

Similarly, sound waves do not produce any substantial net movement of the gas molecules that make up air. This fact can be shown if we float a balloon in front of a rock band at a concert. The balloon will not move toward the back of the hall, no matter how loud the sound.

Because a wave is a moving disturbance, its location cannot be specified exactly. The disturbance is a series of peaks and valleys, crests and troughs, whose position changes constantly as the wave travels. Because a wave is spread out over a region of space, we can describe it only by specifying a number of its features.

One feature of a wave is the distance between two adjacent peaks, the **wavelength.** It is commonly represented by λ, the Greek letter lambda. Another characteristic of a wave is the distance from a horizontal midline to either the peak or the trough. This distance is called the **wave amplitude.**

Wavelength and wave amplitude are not enough to describe a wave. Information is also needed on the velocity of the wave, which can be described as the rate of motion of the peak (or any other point) in the direction of propagation. The value of the velocity, c, for light waves and all other electromagnetic radiation is approximately 3.00×10^8 m/s.

The **frequency** of the wave, which is represented by ν, the Greek letter nu, is directly related to wavelength and velocity. Suppose there is an observer looking at a fixed point as a wave goes by. If the observer counts the number of peaks that pass the fixed point in a given time period, he can specify the frequency of the wave as so many units per second. If there is a long distance between peaks—that is, if the wavelength is long—fewer peaks will pass in a given time period. In other words, *the frequency of a wave is inversely proportional to the wavelength.* The number of peaks that pass by in a given time period also depends on the velocity of the wave. Therefore, frequency is a function of two characteristics: the velocity c of the wave and the wavelength λ. The relationship is

$$c = \lambda\nu \tag{5.3}$$

Since the velocity c of light is known, the wavelength of light can be calculated if the frequency is known, and vice versa.

ADDING LANGUAGE: When is shorter better?

Notice what happens when the author uses elaborative devices like restatement and example. The point may become clearer, but the text also becomes longer. Is this a disadvantage?

Technical and scientific prose tends to be compressed. It is traditional to pack as much information as possible into each sentence. In fact, one of the major criticisms of technical writing is that it is sometimes too compressed—that vital examples and elaborations have been sacrificed at the altar of brevity.

Is there a time when more is better, when adding to the traditionally lean body of technical text is an effective technique? The answer is, "It depends on your audience."

In adding examples, contrasts, and comparisons, explanations grow longer. But if your readership includes people who are unfamiliar with your ideas—a diversified, multidisciplinary group, for instance—examples, contrasts, and comparisons will be useful. Adding them is not padding the prose, weighing down a report with window dressing. Here the additions are not verbiage, but readers' aides.

Here, for instance, is a simple contrast which has been expanded to suit the audience.

1. **Simple contrast**

> The structure of any crystal can be described by lattice points and unit cells. However, one can also describe the crystal as an assembly of closely packed spheres.

To sharpen the readers' understanding of the difference in crystal structure, the author extends the contrast.

2. **Simple contrast with transitional sentence**

> The structure of any crystal, no matter how complex, can be described by lattice points and unit cells. <u>But there is a simpler, alternative method of picturing the crystal structure of many substances.</u> One can describe the crystal as an assembly of closely packed spheres.

Transitional Sentence

If you want to sharpen the contrast further, you can add examples.

3. Contrast with transitional sentence and examples

The structure of any crystal, no matter how complex, can be described by lattice points and unit cells. The method works for a crystal of a complicated organic molecule, such as a protein or a nucleic acid, as well as it does for a crystal of a simple monatomic substance, such as a metal. *(Examples)*

(Transition) There is a simpler, alternative method of picturing the crystal structure of many substances. We can describe the crystal as an assembly of closely packed spheres. This alternative method works very well when the basic structural unit is an atom. Some molecular substances can also be described in this way, but only if the molecules are simple enough to be roughly spherical. Many minerals also lend themselves to a description of this sort.

SUMMING UP

- It is the first, most urgent job of the writer to make the logic of the argument apparent. Readers should not have to conquer the text line by line. Instead, the path of the argument should be there to guide them.
- Don't hold the main idea for the last page. Reports are not detective stories—suspense is the wrong format for readers who want the gist of the problem served up first, followed by the development of the supporting points.
- Some organizational patterns provide a useful scaffold for information, usually by focusing and presenting main points before giving details. Three of these patterns are
- **Problem-Solution Format.** The problem occurs in the first paragraph. The second paragraph gives the solution in a nutshell. The remaining paragraphs give the background and details of the solution.

• **Main Idea-Significance Format.** The main idea is in the first paragraph. The second paragraph gives the significance of the idea. The remaining paragraphs give the background and details of the findings. This pattern is also known as What? So What?

• **Comparison-Contrast Format.** All three patterns rely on the use of strong lead sentences that encapsulate main points.

• For a diversified group, expand lead sentences with **a combination of restatement, example, contrast, comparison, and analogy.** The text becomes longer, but the additions are not verbiage if the readers need the information to understand your point.

EXERCISES

1. Locate a piece of technical writing that you think is either well or poorly written, whether from instructions, a manual, a textbook, a magazine, a handbook, a reference work, or a report. Photocopy several pages of this example of technical writing. Write a brief (1- or 2-page) analysis of its structure. Consider these questions in your analysis: Does the writing have an overview before developing details? Does the author state the main point and its significance before developing details? Does the author use patterns such as listing, or comparison and contrast? How are examples used? Do logic and organizational patterns contribute to the readability of the document?

2. Assume that a research report, "The Use of Hypnosis to Enhance Recall," might intrigue your classmates, if only the authors had written the report so that it were readable for a diversified group of technical people, each of them from a different specialty.

   You decide to write a 1-page memo to members of the class telling them about the report and its significance. The report, "The Use of Hypnosis to Enhance Recall" is reprinted below (Example 3.9). Using the models in this chapter, write a 1-page memo that summarizes the paper in problem-solution format. In the first paragraph, state the problem the report addresses. In the second paragraph, summarize the solution. Discuss details in the remaining paragraphs.

3. Using the same report, "The Use of Hypnosis to Enhance Recall," write a 1-page memo organized in main idea-significance format.

The Use of Hypnosis to Enhance Recall

The increased use of hypnosis in forensic investigation has become controversial (1). Although numerous case reports attest to the utility of hypnosis in enhancing the recall of the eye witness (2), controlled studies have produced conflicting results. Some studies have failed to demonstrate hypnotic hypermnesia, whereas those that have (3), have not reported errors in a systematic way nor controlled for the natural hypermnesic effects that can be achieved through repeated testing (4). Still others (5) have found that hypnotized subjects are susceptible to leading questions. Although scientists are wary of the reliability of forensic hypnosis, police investigators are lobbying to sanction its use in criminal investigation and the judiciary is seeking evidence on which to base legal decisions. The relation between hypnosis and memory enhancement needs to be clarified.

We now report that any pressure to enhance recall beyond the initial attempt may increase the number of items recalled but increase the number of errors as well. The use of hypnosis exaggerates this process, particularly for those with hypnotic ability. When hypnotized, the highly hypnotizable subjects recalled twice as many new items as controls but made three times as many new errors.

Fifty-four subjects were selected on the basis of their hypnotic ability as measured by a group adaptation of the Stanford C Scale of Hypnotic Susceptibility (SHSS:C) (6). Subjects with low susceptibility had SHSS:C scores from 1 to 6, and those with high susceptibility, from 7 to 12. All subjects were presented with a series of 60 slides of simple black-and-white line drawings of common objects (7), presented at a rate of 3½ seconds per slide. They were then given a recall sheet and requested to write the name of a line drawing in each of the 60 blank spaces provided for this purpose, indicating as well which items represented memories and which were just guesses. This forced recall procedure is standard in hypermnesia studies (8). Subjects were initially given three trials in the laboratory with 3-minute rest periods between trials.

Subjects were then instructed that during the next week they were to recall as many of the line drawings as they could once each day, and to write their recollections on the take-home recall sheets provided. They were asked to deposit each recall sheet in a convenient dropbox daily for 6 days. Altogether, subjects completed nine trials over a period of 7 days before their second laboratory session.

The mean number of items recalled on the first trial was 30. By trial 9 the cumulative mean had risen to 38 items—an increase of 27 percent. The number of errors increased as well, from an average of less than one error on the first trial to an average of four errors by the ninth. Most subjects approached asymptotic levels of output by about trial 7, 4 days after a single viewing of the stimuli.

The next step was to see whether hypnotic suggestions for increased recall would enable subjects to retrieve more information after asymptotic recall had been reached. During this second laboratory session, subjects were told to relax and focus all their attention on the slides they had seen the week before. They did so either while hypnotized (hypnosis condition) or without hypnosis (task-motivated condition). Before this session subjects did not know which condition they would be in, and the experimenters were unaware of subjects' hypnotic ability. Consistent with these precautions, independent sample *t*-tests indicated no

Example 3.9

Example 3.9 (*continues*)

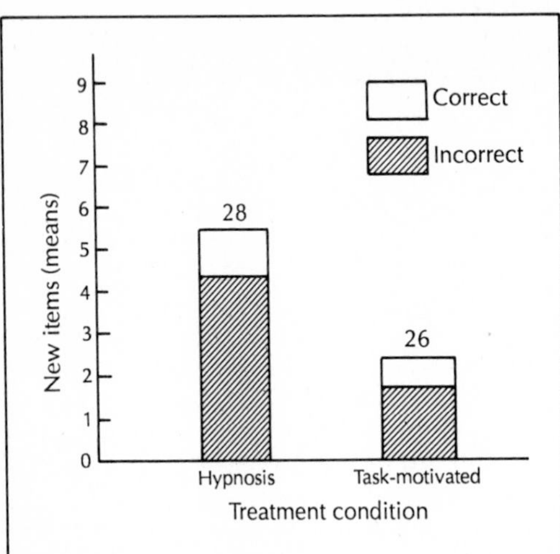

Figure 1. New items presented as memories by subjects after hypnotic or task-motivating suggestions to enhance recall. All items were designated by subjects as true memories. The number of subjects in each group is shown above each bar.

difference between high and low susceptible subjects in the cumulative number of correct items retrieved over the week before treatment [$t(26) = 0.49$] or for the cumulative errors retrieved prior to treatment [$t(26) = 0.14$].

Figure 1 illustrates the number of items reported on the treatment trials that had never been reported as memories before. Subjects in the hypnosis group reported over twice as many new items (both correct and incorrect) as subjects in the task-motivating condition did. The correct information retrieved by subjects in both conditions remained proportional to this shift in total output. Those higher in hypnotic ability in the hypnosis condition were primarily responsible for the increase in output, and hypnotic suggestion was no more potent than task motivating suggestion for those lower in hypnotic ability (Fig. 2).

A two-way analysis of variance based on the total increase in items indicates a significant main effect for condition [$F(1, 50) = 5.63$, $p < 0.03$] and a significant interaction of condition with hypnotic susceptibility [$F(1, 50) = 4.31$, $P < 0.05$]. When just the correct information was considered, the interaction between condition and hypnotic ability was significant as well [$F(1, 50) = 4.95$, $P < 0.05$]. Using new errors as the dependent measure yielded a significant main effect for condition [$F(1, 50) = 5.38$, $P = 0.03$], but the interaction in this case was not statistically significant [$F(1, 50) = 3.10$, $P = 0.08$]. Even though hypnotizable subjects in the hypnosis condition showed a satistically significant increase in accurate recall, this increase was small in absolute terms. No subject in even this most responsive group retrieved more than five new

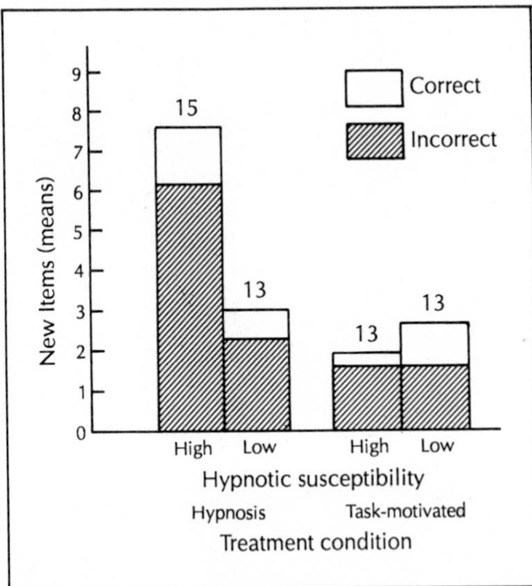

Figure 2. New items presented as memories by subjects of high and low susceptibility to hypnosis after hypnotic or task-motivating suggestions to enhance recall. The number of subjects in each group is shown above each bar.

correct items (mean = 1.40), and six of them failed to produce any new correct information at all. The cost of correctly recalling these few items was considerable, since it was accompanied by almost three times as many errors as were made by subjects in any other condition. We have replicated this pattern of results on a new sample of 56 subjects (9).

The probability of correctly recalling new items under hypnosis seems directly related to the number of items a subject is willing to report as memories, a finding that could be interpreted as being due to a shift in report criterion. That is, the increase in correct recall may not represent increased sensitivity to memory traces, but may instead result from less caution by subjects in what they are willing to report as memories. This criterion shift could be attributed to various demand characteristics, social cues, and expectations engendered by the hypnotic situation.

Another possible explanation for the effect of hypnosis on memory depends less on a shift in the report criterion than on the frequency with which the individuals' criterion for memorial judgment is subjectively met. Hypnosis may heighten the sense of recognition associated with even falsely recalled items, in effect "fooling" a central processor or editor responsible for memorial judgments (*10*). It may be that one of the criteria upon which this sense of recognition is based is the vividness with which the subject is able to envision those items generated as possible memories during recall attempts. If hypnosis enhances the vividness of mental imagery (*11*), perhaps the vividness with which the subject is able to envision these possibilities becomes compelling. Under these circumstances, the editor could mistake vividly imaged possibilities for memories of the stimuli; the enhanced vividness could lead to a false sense of recognition and hence the inflated output as well as the surprising certainty that subjects have about their hypnotically enhanced recall (*12*).

The role that affect may play in the relationship between hypnosis and memory has not been explored in this investigation, but may be relevant to the use of hypnosis in forensic settings. Nonetheless, our observations of hypnotically enhanced recall should give pause to those advocating the use of hypnosis in situations in which the veridicality of information is of prime concern.

JANE DYWAN
KENNETH BOWERS

*Department of Psychology,
University of Waterloo,
Waterloo, Ontario N2L 3G1*

References and Notes

1. C. Perry and J. Laurence, *Int. J. Clin. Exp. Hypn.* **30**, 443 (1982); B. L. Diamond, *Calif. Law Rev.* **68**, 313 (1980); H. Spiegel, *Ann. N.Y. Acad. Sci.* **347**, 73 (1980).
2. W. Kroger and R. G. Doucé, *Int. J. Clin. Exp. Hypn.* **27**, 358 (1979); T. S. Worthington, *ibid.*, p. 402; D. S. Carter, *Wash. Law Q.* **60**, 1059 (1982).
3. F. A. DePiano and H. C. Salzberg, *Int. J. Clin. Exp. Hypn.* **29**, 383 (1981); T. P. Dhanens and R. M. Lundy, *ibid.* **23**, 68 (1975).
4. The phenomenon was first described by W. Brown [*J. Exp. Psychol.* **6**, 337 (1923)] and brought into prominence again by M. H. Erdelyi and J. Becker [*Cognit. Psychol.* **6**, 159 (1974)].
5. M. Zelig and W. B. Beidleman, *Int. J. Clin. Exp. Hypn.* **29**, 401 (1981); W. H. Putnam, *ibid.* **27**, 437 (1979).
6. A. M. Weitzenhoffer and E. R. Hilgard, *Stanford Hypnotic Susceptibility Scale, Form C* (Consulting Psychologists' Press, Palo Alto, Calif., 1962). Test-retest reliability is 0.85 ($N = 307$) [E. R. Hilgard, *Hypnotic Susceptibility* (Harcourt, Brace & World, New York, 1965)].
7. The slides were selected to be high on image and name agreement from a set of pictures that had been normalized on a number of dimensions relevant to memory studies [J. B. Snodgrass and M. Vanderwart, *J. Exp. Psychol. Hum. Learn. Mem.* **6**, 174 (1980)]. This selection procedure ensured against ambiguity in scoring.

Example 3.9 (*continues*)

8. The hypermnesic function has been replicated in numerous studies that used repeated recall over the course of single laboratory session [for example, A. D. Yarmey, *Bull. Psychonom. Soc.* **8**, 115 (1976)] and over the course of a week [M. H. Erdelyi and J. Kleinbard, *J. Exp. Psychol.: Hum. Learn. Mem.* **4**, 275 (1978)].
9. All subjects were hypnotized, with those unresponsive to hypnotic suggestion serving as controls. The susceptible subjects reported significantly more information in response to hypnotic suggestion than the unsusceptible subjects [$t(27) = 2.68$, $P < 0.01$]. However, only 10 percent of the new information reported by either group was accurate. While hypnosis produced a small but significant increase in new information [$t(27) = 1.90$, $P < 0.05$], it also produced a significant increase in new errors [$t(27) = 2.67$, $P < 0.01$]. Since this replication examined specific hypotheses, one-tailed tests were used.
10. M. K. Johnson and C. L. Raye, *Psychol. Rev.* **88**, 67 (1981).
11. Experimental evidence on this point is controversial since most studies base their results solely on self-rating scales. H. J. Crawford, using self-rating scales and an objective performance criterion, reported a shift in the vividness of visual imagery during hypnosis for those in the upper range of hypnotic ability [paper presented at the National Meeting of the Society of Clinical and Experimental Hypnosis, Denver, Colo., October 1979].
12. Pseudomemories, originally developed in hypnosis, may come to be accepted by the subject as his actual recall of the original events; they are then remembered with great subjective certainty and reported with conviction [M. T. Orne, *Int. J. Clin. Exp. Hypn.* **27**, 331 (1979)].
13. We thank S. Vandermeulen and D. Chansonneuve for assistance in collecting and scoring the data; S. Segalowitz, J.-R. Laurence, E. Woody, and M. P. Bryden for their advice and editorial comments; and all members of the University of Waterloo Hypnosis Research Team for their suggestions and encouragement. Supported in part by Social Sciences and Humanities Research Council grant 037-8195.

21 March 1983; revised 21 July 1983.

Language

CHAPTER

4

Sir,

In your otherwise beautiful poem there is a verse which reads
Every moment dies a man,
Every moment one is born.

It must be manifest that if this were true, the population of the world would be at a standstill. In truth the rate of birth is slightly in excess of that of death. I would suggest that in the next edition of your poem you have it read
Every moment dies a man,
Every moment 1 1/16 is born.

Strictly speaking this is not correct, the actual figure is so long that I cannot give it into a line, but I believe the figure 1 1/16 will be sufficiently accurate for poetry.

Letter from Charles Babbage to Alfred, Lord Tennyson

Chapter 2 shows two strikingly different versions of a consumer loan application; the second version is far easier to read than the first. Part of this readability lies in improvement in the language. To illustrate the difference in language, here is a comparison of one section from the two documents:

Version A says

> IN THE EVENT THIS NOTE IS PREPAID IN FULL OR REFINANCED, THE BORROWER SHALL RECEIVE A REFUND OF THE UNEARNED PORTION OF THE PREPAID FINANCE CHARGE COMPUTED IN ACCORDANCE WITH THE RULE OF 78 (THE 'SUM OF THE DIGITS' METHOD) PROVIDED THAT THE BANK MAY RETAIN A MINIMUM FINANCE CHARGE OF $10

Figure 4.1 Cartoons poke fun at nominalizations. Courtesy *Trends in Analytical Chemistry*.

The same portion of the contract is expressed in Version B this way:

> . . . I have the right to prepay the whole outstanding amount of this note at any time. If I do, or if this loan is refinanced—that is, replaced by a new note—you will refund the unearned **finance charge,** figured by the rule of 78—a commonly used formula for figuring rebates on installment loans. However, you can charge a minimum **finance charge** of $10.

There are substantial differences in the language, not only in this portion, but in every part of the two documents, differences that make Version B easier to understand than Version A.

• **Sentence length.** If you look at the two examples above, you'll see that Version A is one long, sprawling sentence of 52 words.

The same information has been placed in three shorter sentences in Version B. Transitional words and phrases ("If" . . . "However,") link one sentence tightly to the next.

• **Verb forms.** Version A uses passive verbs; Version B, active verbs. In the first version we read, "The note is prepaid," in the second version, "I have the right to repay. . . ."

• **Definitions.** Version A does not define specialized terms; Version B does, both with "refinancing" and "the rule of 78."

Are the language problems in Version A attributable in part to the convoluted constructions and archaic words for which legal documents are famous? Probably not. Any professional field has convolutions in its language. Consider, for example, these two versions of a sentence from a computer manual.

> *Version A:* **"Providing documentation that is not inaccessible for configuration, user reference, and maintenance of the system has been proposed as the objective of this work."**
>
> *Version B:* **"We propose to provide readable instructions for how to configure, use, and maintain the computer system."**

Again the difference between the two examples lies not in what they say—both sentences have the same meaning—but in how they say it. The first sentence requires that readers concentrate their attention, lips moving perhaps, as they subvocalize their way to its end. After a moment's work, readers understand the sentence—but not without effort.

The second sentence does not exact the same toll on the reader. It is cast in a more direct way.

Is this language distinction crucial to good technical writing? Before you answer, multiply the wordiness of Version A in either the consumer loan application or the manual with the 20 or 30 similar sentences that would follow. Obviously, documents written in the style of Version A place far greater demands on the reader than do those of Version B.

Direct language is an integral part of readability in technical documents, from instructions like the consumer loan application to manuals, letters, reports, memos, and procedures. Such language is essential when you want as little static as possible between the message and the reader. Direct language helps a complex message make a swift journey from the page to the mind of the reader with a minimum of interference.

What are some of the ways that language in technical writing can be direct rather than indirect?

NOMINALIZATIONS VS. VERBS

English prose runs on verbs. If sentences are vehicles, verbs are the motors. If you want to write swift, readable prose, you have to provide the verbs that will give the sentences pace and readability.

Consider, for instance, a few simple examples of the difference between direct verbs and buried—or "nominalized"—ones. (To nominalize is to take a verb like "analyze" and make it a noun, "analysis.")

BURIED OR NOMINALIZED VERB	DIRECT VERB
An analysis was conducted of the problem	We analyzed the problem.
It is observed that acid has a corrosive effect on the substance.	Acid corrodes the substance.
There exists an urgent need for reevaluation of the proposal.	We need to reevaluate the proposal.

In "an analysis was conducted," the verb "analyze" has been buried or made into a noun. In the second example, the verb "corrodes" has been buried in "has a corrosive effect on." Because of the buried verbs, the sentence "There is an urgent need for a reevaluation of the proposal," takes 11 words to say what should be said in only five words.

Buried verbs are such annoying errors that many publications now send flyers to potential contributors telling them to remove buried verbs from their manuscripts. One such guideline has a picture of a waiter carrying a tray out the door. Sitting proudly on the tray is the word "measurement." The waiter is "carrying out a measurement." The guidelines advise, "Please don't carry out a measurement. Just measure." (For other examples of cartoons poking fun at nominalization, see Figure 4.1.)

If you are writing instructions, and you say "Make a measurement of the log," you can improve the sentence—make it shorter and more direct—by turning the nominalization into a verb.

Measure

~~Make a measurement of~~ the log.

ACTIVE VS. PASSIVE

The passive is a hallmark of scientific writing, a handy way to take first-person out of reports and substitute what many feel is a more objective tone. The passive (a combination of the verb "to be" and a past participle) creates an entirely different effect in writing than the active form of the verb.

PASSIVE	ACTIVE
It has been recommended by the Board that the measures be adopted.	The Board recommends we adopt the measures.
A measurement was made by the engineering staff of the apparatus.	The engineering staff measured the apparatus.

A comparison of the two forms of the verb shows that the passive form helps to soften or diffuse a point; in contrast, the active form is useful when you want to get to the point swiftly and directly.

ACTIVE	PASSIVE
The president unexpectedly rejected the candidate.	The candidate was unexpectedly rejected.

The passive is a perfect way to avoid naming the person who does the act. For instance, you may choose "The candidate was unexpectedly rejected" when you are sure that the president (the doer) wants no credit for the rejection. Here, it is your decision to prefer an indirect approach.

You may use the passive legitimately in many cases where you want to throw the spotlight on the action or the object instead of on the doer. For example, in Chapter 10, there is a description of rivets in the passive tense. This makes sense, for the authors are interested in describing rivets, not themselves:

> Rivets are used to join sheet metal.
>
> Rivets are made of soft copper, aluminum, or steel.

Similarly, an author describes floppy disks in the passive:

> The diskette is composed of a flexible material which has a surface that can be magnetized.

However, when the time comes for instructions, the author switches to active verbs:

> Don't touch the exposed areas of the disk.
>
> Store disks upright.

(For a discussion of verbs in descriptions, see Chapter 10; for a discussion of verbs in instructions, see Chapter 12.)

Whether you use active or passive verbs depends in part on your audience. Today many research journals ask their authors to write in the first person and to use active verbs, to prefer "we saw" to "it was seen"; "we selected fifty-four subjects" to "fifty-four subjects were selected." Many journals also ask contributors to avoid using passive clichés like "it has been concluded that," and "it has been found that." Instead, contributors are asked simply to conclude or to find. However, other journals still prefer the more traditional passive constructions.

Passives are appropriate, therefore, in some research journals that prefer them, and in the parts of technical writing where the writer wants the stress to fall on the object, not the person. But in most writing—particularly in memos, letters, instructions and informal reports—passives get in the way, lengthening, clogging, and obstructing the prose. They create the longer, indirect sentences of Version A of the loan application, sentences where we lose sight of the doer. Active verbs are usually your allies when you seek direct language.

SHORT SENTENCES VS. LONG ONES

The length of a sentence is no guarantor of clarity. Short, choppy sentences, so devoid of transitions that the reader must constantly provide the logical connections, are hard to follow. But it is longish, rambling sentences, rather than short, choppy ones, that are common in poor technical writing.

The average length of the sentences in the consumer loan application, Version A, is 50 words. It's easy to find similar long, sprawling sentences, such as the following one, in any technical area:

> As hydrocarbon mixtures derived from fossil fuel sources such as coal and petroleum are generally very complex, the determination of absolute structures is often impractical, and structural analysis of the materials consequently consists of evaluating average structural parameters by a suitable combination of analytical techniques, among them nuclear magnetic resonance (NMR) spectroscopy, the most useful and most commonly applied spectroscopic tool.

How long is too long? Critics agree that longer sentences (over 25 words) are harder for many readers to understand. If you are in doubt as to the readability of some longer sentences, you might find this example from the *Encyclopedia Britannica* (14th edition) both amusing and instructive:

> If the Roman government at the height of its power, and at a time when means of communication had been greatly improved, showed anxiety for the food supply of that Italy which was dominant in the Mediterranean world, it may be imagined that in the period preceding the great economic organization introduced by the Roman Principate the peoples of the Mediterranean region, peoples no one of which at the height of its power had controlled the visible food supply of the world so widely or so absolutely, had far graver cause for anxiety on the same subject, an anxiety such as would be, under ordinary circumstances, the main factor, or, even under the most favourable circumstances possible in those ages, a main factor, in moulding the life of the individual and the policy of the state. . . .

You are certainly safer with shorter sentences. A skilled writer, though, can do wonders with long sentences. In *The Great Bridge,* for instance, David McCullough describes the anchorages of the Brooklyn Bridge. The first sentence has 65 words, the second 71, the third, 64, the last 18.

> Below the underside of the anchor plate, through the nine eyes of each row, all matched in position as one, big steel pins were inserted and drawn up against the plate, fitting into semicylindrical grooves, and thereby forming the first link, a double-tiered link, of a gigantic double-tiered eyebar chain that extended up through the masonry in a gradual arc until it surfaced on top.
>
> The anchor bars were of slightly different sizes, depending on their position in the chain, but they averaged twelve and a half feet in length, and in the first three links—those nearest the anchor plates, where the pull from the cables would be felt the least—they were seven inches wide and three inches thick, swelling enough towards the ends to compensate for eyeholes five to six inches in diameter. The bars of the fourth, fifth, and sixth "links," however, were increased in thickness, to eight by three inches and from there to the top, as the bars became horizontal and so came directly in line with the pull of the cable, they measured nine by three, except for the last link, where the number of bars was doubled and their width was halved. The last link had in all thirty-eight bars, in four tiers, to catch hold of the cable wires.

NEGATIVES VS. POSITIVES

Negatives are hard to understand; the reader has to switch them mentally to positives. The bemused reader, for instance, who comes across the expression "a staunch anti-nuclear energy foe" has to pause for a moment before coming up with the translation "He is *for* nuclear energy."

Sometimes negatives are used to soften or conceal—as in the Rolls Royce that "fails to proceed," or the person who "does not tell the truth."

The lesson to learn from this? Write in the positive, and the reader will be free of on-the-spot translation. There's another advantage to the positive. Often a writer becomes tangled in negatives, and says the opposite of what was originally intended. "It's important that you don't fail to miss this seminar," one young accountant, swamped in negatives, inadvertently wrote in a memo. Another said, much to her subsequent embarrassment, "This delay shows a lack of disrespect for the food service profession."

Negatives obscure the message in technical writing, putting the burden on the reader to make sense of indirect prose. When you combine passives, nominalizations, and negatives, you achieve the effect of the sentence used in the original example:

> Providing documentation that is not inaccessible for configuration, user reference, and maintenance of the system has been proposed as the objective of this work.

PASSIVE	ACTIVE
Providing documentation has been proposed	We propose to provide documentation
NOMINALIZATIONS	**VERBS**
. . . for configuration, user reference, and maintenance of the system	for how to configure, use and maintain the computer system
NEGATIVE	**POSITIVE**
not inaccessible	(accessibility is implied)

Rewrite:

> We propose to provide readable instructions for how to configure, use and maintain the computer system.

HANDLING SPECIALIZED TERMS

One of the advantages of Version B in the consumer loan application was the prompt defining of specialized terms:

VERSION A	VERSION B
refinanced	refinanced—that is, replaced by a new note—

To write about technical matters is to use technical terms. If you are writing about loans, you may need to use the word "refinancing"; if you are writing about computers, the phrase "batch files." Specialized vocabulary is deeply woven into the fabric of technology.

Yet often, particularly when the audience mixes people from varying technical backgrounds, the writer cannot assume the reader understands all the technical terminology. One solution to the problem is to operate by the lowest common denominator—to eliminate all technical terms. Such a solution, though, is unrealistic most of the time. "I gotta use words when I talk to you," says a T. S. Eliot character. So, too, does the writer in a technical field, who will often need to use an integral technical term as part of an argument, whether it is "refinancing" in a consumer loan or "batch files" in a computer manual.

The solution? When the audience is diverse,

- use as few specialized terms as possible
- define them immediately

Prompt, clear definitions play a big part when you want to develop accessible language, when you want to turn Version A into Version B. (Specific ways to write definitions appear in Chapter 9. See also Chapters 10, 11, and 12 for ways to weave definitions into descriptions, process explanations, and instructions.)

WHEN SHORTER IS BETTER

"If you want your language to be direct, shorten and tighten what you have to say," is classic, well-founded advice for technical writers. For instance, sentences like these, taken from a set of instructions for a portable screen, are wordy. They are more effective shortened.

VERSION A	VERSION B
You should grasp the screen stem with one hand, and then depress the screen lock with your free hand. Once the latch is depressed, raise the screen stem by pushing and keeping the screen lock depressed.	Hold extension rod firmly with one hand. With the other hand, depress the screen latch and raise screen until it is fully extended and locks into place.

Certainly a sentence inflated by nominalizations and passives to eleven words is better off as six words:

There is an urgent need to reevaluate the proposal. →

We need to reevaluate the proposal.

Less is better when it comes to nominalizations, to inflated phrases, to redundant expressions, and to jargon:

has an impact upon us	→	affects us
rectangular in shape	→	rectangular
six hours time	→	six hours
at this point in time	→	now
consensus of opinion	→	consensus
local neighborhood	→	neighborhood
close proximity	→	close
combine together	→	combine

NOUN STRINGS: A Case Where Shorter Is Not Better

Noun strings are an interesting example of compression that usually fails to tighten and focus language; on the contrary, the device creates problems.

A noun string is a series of nouns placed shoulder to shoulder. The string often begins life as two nouns that are connected by a preposition: **institute for standards** is tightened to **standards institute.**

"Standards" in effect becomes an adjective explaining what kind of institute this is. You might at first glance think that tightening this phrase, making it shorter by eliminating the preposition, is a step in the right direction.

It certainly does no harm with "standards institute." But consider what happens when we add a third noun to the string:

industry standards institute

Technical people tend to be very fond of noun strings like "industry standards institute," probably because they are shorter. But noun strings are unclear. Their amgibuity becomes apparent when an adjective is added:

strong industry standards institute

Is it a strong industry we are talking about? Or a strong industry standard? Or a strong industry standard institute?

When you put more than two nouns side by side, you will have these problems with modification. The solution is to return to prepositional phrases. This is one case where long is better.

an institute for strong standards in industry

CORRECT USAGE

When you are involved in writing a document and struggling with the content, you may find it difficult to think about correct usage—about whether, for instance, verbs agree with their subjects. "I'm thinking about what I'm trying to say," a student working on a set of instructions commented. "Afterwards I check to see if people can follow the steps. I forget all about whether the verbs match the subjects. My eye skips errors like that, as though they were invisible."

Yet correct usage plays an important part in clear language. When you want readers to follow your instructions, usage errors become another form of static. A sentence like, "When depressed, the screen opens," written as part of a set of instructions, says that the screen is having mental problems. (The writer intended to say, "Depress the latch, and the screen will open.")

It is worth casting a cold eye over your writing for these common sentence-level trouble spots:

- subject-verb agreement
- pronoun antecedents
- modification
- parallel structure
- incomplete and fused sentences
- punctuation

To strengthen your command of these areas, see the explanations, examples, and exercises in the Usage Handbook at the end of this book.

Readers respond to the small, careful choices in language good writers make. Did you say *affect* when you meant *effect*? *Enormousness* when you meant *enormity*? These are "just picky little language things," as one student commented. That's true, but of such small distinctions as between *affect* and *effect* is language made. The reader who violates these distinctions introduces distortion that gets in the way of the message.

USE OF COMPARISON, METAPHOR, AND ANALOGY: Comparing the Known to the New

Charles Babbage, the great nineteenth-century scientist who virtually invented the modern computer, took issue with the poet Tennyson for figurative use of the expression "every moment dies a man" (see quotation at chapter head). Babbage is not the only one to tease poets about the distinction between literal and figurative language. But there are some types of figurative language both poets and technical writers agree are quite powerful. Comparison, metaphor, analogy—all are effective language tools a writer can use to link the new to the familiar, the abstract to the concrete.

Comparison

Here is J. S. Haldane, beginning a description:

> An insect going for a drink is in as great danger as a man leaning out over a precipice in search of food. If it once falls into the grip of the surface tension of the water, it is likely to remain so until it drowns. A few insects, such as waterbeetles, contrive to be unwettable. The majority keep well away from their drink by means of a long proboscis.

You can find illuminating comparison used throughout the history of technical writing.

Ada Augusta, Countess of Lovelace, includes in her description of the analytical engine, "The Analytical Engine weaves algebraical patterns, just as the Jacquard loom weaves flowers and leaves." The comparison is an apt one, for Babbage was inspired in his creation of the Analytical Engine by the punchcards used to generate Jacquard patterns.

Thomas Edison wrote in his notebook, "I am experimenting upon an instrument which does for the Eye what the phonograph does for the Ear, which is the recording and reproduction of things in motion."

Comparison is often used to bring numbers to life on a page. Here is Richard Feynman, explaining tests of the theory of quantum electrodynamics.

> I'll give you some recent numbers: experiments have Dirac's number at 1.00115965221 (with an uncertainty of about 4 in the last digit); the theory puts it at 1.00115965246 (with an uncertainty of about five times as much). To give you a feeling for the accuracy of these numbers, it comes out something like this: If you were to measure the distance from Los Angeles to New York to this accuracy, it would be exact to the thickness of a human hair. That's how delicately quantum electrodynamics has, in the past fifty years, been checked—both theoretically and experimentally.

There are some cases, though, where comparison is not the right technique. "Make measurements in centimeters, not in fruits, vegetables, or nuts," the medical textbook advises. "Pea-sized, lemon-sized and walnut-sized lesions vaguely convey an idea, but make accurate evaluations and future comparisons impossible. How big were the lemons or peas? Did the walnut have a shell?"

Walnuts come in different sizes; the comparison is not useful for the audience that needs not a flash of insight, but specific dimensions. On the other hand, describing an inorganic vapor detector as "badge-sized" is perfect when an image, not the exact dimensions, is what the writer seeks.

Metaphor

Simple and extended metaphors abound in good technical writing. A metaphor is an *implicit* comparison, a comparison that discards the signal words "like" or "as." Instead of saying A is like B, a metaphor says A is B.

> The operating system is a traffic cop, directing the passage of data through the computer's central processing unit.
>
> The microprocessor is the brain of the computer, manipulating data and performing calculations at the behest of the operating system.

Whether we say A is like B or A is B, we attempt to say something illuminating about the nature of A. Metaphors are extremely useful in technical writing; they are powerful teaching aids—research shows people use them spontaneously when learning about something new. The writer should, however, make an effort to make sure the metaphors do not mislead. Usually this is done by acknowledging the limits of the metaphor. For instance, researchers Carroll and Thomas recommend that those who use metaphors to teach about computer systems should explicitly point out when introducing a metaphor that it is not a perfect representation of the underlying system, and then indicate the boundaries of the specific metaphor (*IBM Research Report: Usability Specifications as a Tool in Iterative Development*).

Analogy

Analogy is a form of comparison, but an extended one. In logic, analogy is a form of inference in which it is reasoned that if two things agree with one another in one or more respects, they will probably agree in other respects.

Feynman, for instance, writes,

> When I say "light" in these lectures, I don't mean simply the light we can see, from red to blue. It turns out that visible light is just a part of a long scale that's analogous to a musical scale in which there are notes higher than you can hear and other notes lower than you can hear. The scale of light can be described by numbers—called the frequency—as the numbers get higher, the light goes from red to blue to violet to ultraviolet. . . .

As analogies go, this is a short one. Example 4.1 shows a longer one in which the author, Ed Edelson, begins with a comparison from everyday life, and then extends the comparison until it is an analogy. Analogy allows the writer to present the unfamiliar through that which is familiar to the reader. Note that in this case the author explicitly states the limits of the analogy, explains where it "breaks down."

Creating effective comparisons is a hard writing job; sometimes the author gets in trouble along the way. For instance, consider this excerpt:

> The way in which an atom absorbs and re-emits energy is governed by the energy level that represents the stable state of each type of atom. When an atom is struck by an electron of energy greater than that of its normal or ground state, it may absorb some of that energy (depending on the energy level and the characteristics of that type of atom) and reach one of several possible excited states.
>
> An atom in an excited state may be compared to a stepped, compressed spring in which energy is stored, whereas an atom in its ground state is like a spring at its natural length.* Whenever an atom jumps from a high energy level to a lower one (as when a stepped, compressed spring is partially released), the energy it loses is given up as a packet of radiation. This radiation consists

* The comparison shows that whereas the spring changes from one energy state to another continuously, the atom changes in discrete jumps.

We can understand some characteristics of crystals by examining a tile floor. Fig. 1 represents a regularly repeating two-dimensional pattern. To lay the tile floor shown in Fig. 2, we must repeat the basic unit of design, the tile that forms one complete pattern. We can describe the tile floor by describing the pattern and then stating that this pattern is repeated to cover the entire area of the floor.

The shape of the area covered by the tiles depends on how the tiles are laid. We can make a square floor by laying tiles equally on all four sides and a long, narrow floor by laying tiles to the two narrow ends of the floor. We can make an irregularly shaped floor by varying the way in which the tiles are laid. But no matter how irregular the floor might be, the angles between corresponding sides of the

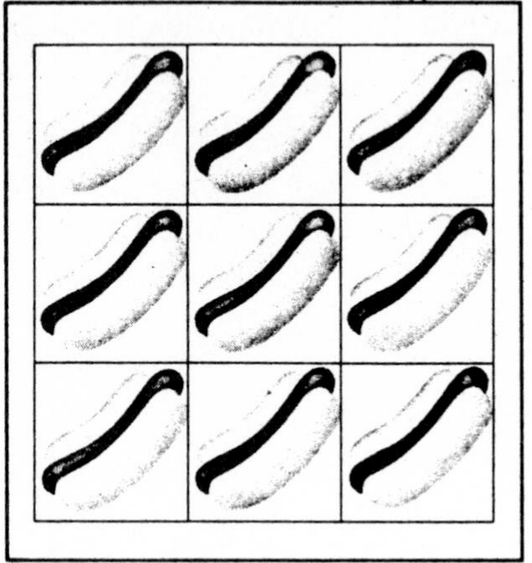

Figure 1 A tile floor with a regularly repeating two-dimensional pattern.

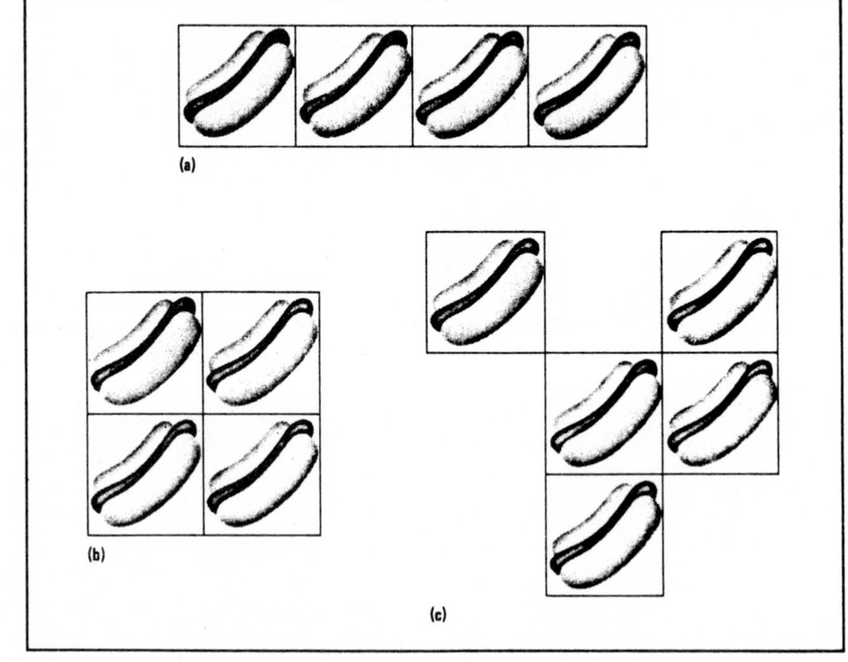

Figure 2 No matter how the pattern is shaped—as a rectangle (a), a square (b), or any irregular shape (c)—the angles between corresponding sides of the floor are still the same.

Example 4.1 A successful analogy. From Robert S. Boikess and Edward Edelson, *Chemical Principles,* Harper & Row, 1981.

floor will always be 90°, because the basic repeating unit is the same tile.

Another feature of tile floors is worth noting. Only a limited number of basic shapes will cover an area completely, with no gaps. An area can be covered entirely by square tiles or by tiles that are triangles, parallelograms, or regular hexagons. An area cannot be covered entirely by tiles that are regular pentagons or heptagons, or tiles of irregular shape. In fact, only a relatively small number of polygons will cover an area completely.

The floors are two-dimensional, but they have a number of features in common with three-dimensional crystals. **A crystal also has a characteristic pattern based on a relatively small number of basic structural units,** which can be atoms, ions, or molecules. This basic repeating three-dimensional pattern is called the **unit cell.** A single unit cell is analogous to the tiles that make up the basic repeating pattern of the floor. Just as the shape of the tiles determines the angles between corresponding sides of the floor, the shape of the unit cell can determine the angles between the corresponding faces on a crystal. And just as only tiles of certain shapes can cover a floor area completely, only unit cells of certain shapes can fill three-dimensional space completely.

But the analogy between floor tiles and crystals breaks down if it is carried too far. A floor tile is a clearly defined object with real sides. The unit cell of a crystal is an idealized shape. It is generated by imaginary lines connecting the points that form the regularly repeating pattern called the crystal lattice.

of electromagnetic waves with a frequency proportional to the difference between the two energy levels. A hydrogen atom emits ultraviolet radiation when it falls from its first excited state to the ground state. Visible light is produced when the atom descends from the second or higher excited state to the first, whereas transitions that end on the second excited state produce infrared radiation.

The author made a questionable choice when he decided to use compressed springs as a metaphor for excited states, for while the atom changes in discrete jumps, a spring can change continuously. The two objects are therefore difficult to compare. The author has tried to solve his difficulty by making the spring "stepped," and by adding a footnote.

It takes practice to write effective comparisons and analogies. Many experienced writers will tell you, though, that it is worth the effort. Bob Swofford, a physical chemist at BP America Research Laboratories, comments that "analogies are the best device I know for comparing the unknown—say a new research technique that we are developing—to something the audience already knows."

"The hardest trick in being a technical writer is to correctly gauge the audience," he comments. "People who write for a supervisor—who is probably a technical nonspecialist—have trouble shifting focus. They end up watering down what they have to say, rather than changing the whole approach.

"Without trivializing, I try to use analogies that are valid. For instance, we were doing some complex laser radar studies. I drew an analogy to a lighthouse. In the mind's eye the audience could see the lighthouse, scanning the horizon and picking up reflected signals. I took the analogy further: The lighthouse beam does not have to bounce off a reflective object to produce a signal. Light can be scattered off fog. Analogously, this laser beam did not have to bounce off a mirror to produce a signal. Instead, dust and haze were the distributed mirror bouncing signals back to the detector."

SUMMING UP

In most technical writing, passives create longer, indirect sentences where we lose sight of the doer.

Negatives and nominalizations make sentences harder to understand.

Noun strings are easier to understand when they are converted back into prepositional phrases; this is one case where longer is better.

If your audience is diversified, don't bombard them with specialized terms. Define any essential terms you use promptly and clearly.

Comparisons, metaphors, and analogies can help bring the abstract closer to the audience. These are difficult writing forms to master, but worth the effort, for they are natural teaching aids.

EXERCISES

1. What is "jargon"? Look up a definition in the dictionary. What are some examples of jargon from your field of study?

2. Find an example of technical writing that you either admire or dislike for its use of language. Write a brief (1-page) analysis, discussing either the strengths or weaknesses of the language. You might consider use of specialized terms, clarity of definitions, wording that is either especially clear or murky, use of comparisons or analogies, verb forms, involuted phrases, inaccessible or inflated expressions, or any of the other categories discussed in the chapter.

3. One metaphor that Carroll and Thomas (*IBM Research Report: Usability Specifications as a Tool in Iterative Development,*

RC10437) believe has caused difficulties is the "communication" metaphor. They point out that the phrase "man-computer communication" may mean one thing to an engineer who thinks of communication as the point-to-point transmission of information, and quite another thing to a lay person who thinks of communication as something that takes place between human beings and involves understanding of what is being communicated. They note that the use of the term invites the naive user to impute to the computer the ability to engage in communication in the same way human beings do. "This expectation can only result in frustration when the user begins to discover the limitations the computer really has in this regard."

Discuss this and other examples of metaphor used in your field, giving pluses and minuses.

4. Example 4.2 is an extended metaphor by Richard Feynman from *QED*. Comment on its strengths and weaknesses.

5. Writer Malcolm Browne, commenting on the language of science in the *New York Times*, cites these titles of research papers: "Long-Term Potentiation in Dentate Gyrus: Induction by Asynchronous Volleys in Separate Afferents," and "Surface Extended X-Ray-Absorption Fine-Structure Study of the O(2× 1)/Cu(110) System: Missing-Row Reconstruction, and Anisotropy in the Surface Mean Free Path and in the Surface Debye-Waller Factor." He says that with the right reference books, you might be able to translate the intimidating titles into plain language, but that does not imply that the crystallography paper would make sense to, say, a neurophysiologist. The reason? Times have changed since Lister wrote his paper. "One of the nice things about the frontiers of science a century ago was that they were more or less accessible even to ordinary people," Browne comments. Today, however, with information explosions in chemistry, microbiology, and many other disciplines, such accessibility is no longer the case. "As if this were not bad enough, there is a growing suspicion among many scientists that some of their colleagues deliberately try to impress their peers by confusing them." In support of this, Brown recounts an experiment where Dr. J. Scott Armstrong coached an actor in the delivery of a purportedly scientific lecture that actually consisted of double talk, meaningless words, false logic, contradictory statements, and meaningless references. No one who attended the lecture detected the hoax; in fact, people found the talk "clear and stimulating." From this and other experiments, Browne tells us, "Dr. Armstrong concluded that an unintelligible communication from a legitimate source in the recipient's area of expertise will increase the recipient's rating of the author's competence."

In a small group, discuss Browne's comments. Do you agree or

How am I going to explain to you the things I don't explain to my students until they are third-year graduate students? Let me explain it by analogy. The Maya Indians were interested in the rising and setting of Venus as a morning "star" and as an evening "star"—they were very interested in it when it would appear. After some years of observation, they noted that five cycles of Venus were very nearly equal to eight of their "nominal years" of 365 days (they were aware that the true year of seasons was different and they made calculations of that also). To make calculations, the Maya had invented a system of bars and dots to represent numbers (including zero), and had rules by which to calculate and predict not only the risings and settings of Venus, but other celestial phenomena, such as lunar eclipses.

In those days, only a few Maya priests could do such elaborate calculations. Now, suppose we were to ask one of them how to do just one step in the process of predicting when Venus will next rise as a morning star—subtracting two numbers. And let's assume that, unlike today, we had not gone to school and did not know how to subtract. How would the priest explain to us what subtraction is?

He could either teach us the numbers represented by the bars and dots and the rules for "subtracting" them, or he could tell us what he was really doing: "Suppose we want to subtract 236 from 584. First, count out 584 beans and put them in a pot. Then take out 236 beans and put them to one side. Finally, count the beans left in the pot. That number is the result of subtracting 236 from 584."

You might say, "My Quetzalcoatl! What *tedium*—counting beans, putting them in, taking them out—what a job!"

To which the priest would reply, "That's why we have the rules for the bars and dots. The rules are tricky, but they are a much more efficient way of getting the answer than by counting beans. The important thing is, it makes no difference as far as the *answer* is concerned: we can predict the appearance of Venus by counting beans (which is slow, but easy to understand) or by using the tricky rules (which is much faster, but you must spend years in school to learn them)."

To understand *how* subtraction works—as long as you don't have to actually carry it out—is really not so difficult. That's my position: I'm going to explain to you what the physicists are *doing* when they are predicting how Nature will behave, but I'm not going to teach you any tricks so you can do it *efficiently*. You will discover that in order to make any reasonable predictions with this new scheme of quantum electrodynamics, you would have to make an awful lot of little arrows on a piece of paper. It takes seven years—four undergraduate and three graduate—to train our physics students to do that in a tricky, efficient way. That's where we are going to skip seven years of education in physics: By explaining quantum electrodynamics to you in terms of what we are *really doing*, I hope you will be able to understand it better than do some of the students!

Taking the example of the Maya one step further, we could ask the priest *why* five cycles of Venus nearly equal 2,920 days, or eight years. There would be all kinds of theories about *why*, such as, "20 is an important number in our counting system, and if you divide 2,920 by 20, you get 146, which is one more than a number that can be represented by the sum of two squares in two different ways," and so forth. But the theory would have nothing to do with Venus, really. In modern times, we have found that theories of this kind are not useful. So again, we are not going to deal with *why* Nature behaves in the peculiar way that She does; there are no good theories to explain that.

Example 4.2 From Richard P. Feynman, *QED*, Princeton University Press, 1985.

disagree? Why? Are his points legitimate, and if so, what are the implications of his argument?

6. Vocabulary: When is specialized vocabulary appropriate? In *An Introduction to Scientific Research*, E. Bright Wilson quotes this description of an oak leaf approvingly, saying that "every word means something."

> The leaves are conduplicate in the bud, obovate-oblong and gradually narrowed and wedge-shaped at the base; they are divided into terminal lobes, and from three to nine but usually three pairs of lateral lobes by wide sinuses, which are rounded at the bottom, and are sometimes shallow and sometimes penetrate nearly to the midribs; the terminal lobe is short or elongated, obovate and three-lobed, or occasionally ovate, entire and acute or rounded; the lateral lobes are oblique, broad or narrow, entire or auriculate, and increase in size from the base to the apex of the leaf; or on vigorous shoots or small branches developed from the trunks of old trees, the leaves are often repand or slightly sinuately lobed or occasionally entire below and three-lobed at the broad apex; when they unfold they are bright red above, pale below, and coated with soft pubescence; the red colour fades at the end of a few days, and they become silvery white and very lustrous; their covering of tomentum then gradually disappears, and when fully grown the leaves are thin, firm and glabrous. . . .

When would this use of specialized vocabulary be appropriate? When might it fail?

7. Many a specialized term was deliberately developed by someone who wanted to avoid using a word in common currency that already had its full complement of meanings. For instance, an entirely new terminology was introduced by Faraday in 1833. Dampier tells us, in *A History of Science,*

> Instead of the word **pole,** which implied the old idea of attraction and repulsion, he used the word **electrode** (from the Greek for **"a way, path"**) and called the plate by which the current is usually said to enter the liquid, the **anode,** and that by which it leaves the liquid the **cathode.** The parts of the compound which travel in opposite directions through the solution he called **ions** (from the Greek for **"I go"**)—**cations** if they go towards the cathode, and **anions** if they go towards the anode. He also introduced the word **electrolysis** (from the Greek for **"I dissolve"**) to denote the whole process.

Select five of the key words in your field, and write a brief etymology for each.

Visual Display

CHAPTER 5

Part of the appeal of works like *Crystals and Crystal Growing* lies in their graphics, which supplement the job of language accurately and gracefully (Figure 5.1). The drawings are as helpful in their way as the text, for they

- orient the user

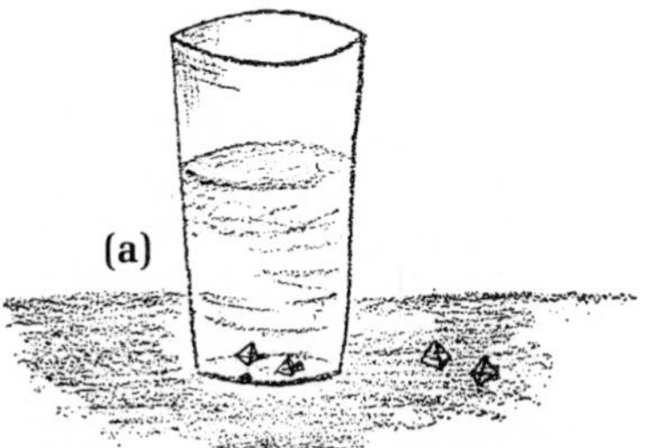

- show the parts of the whole

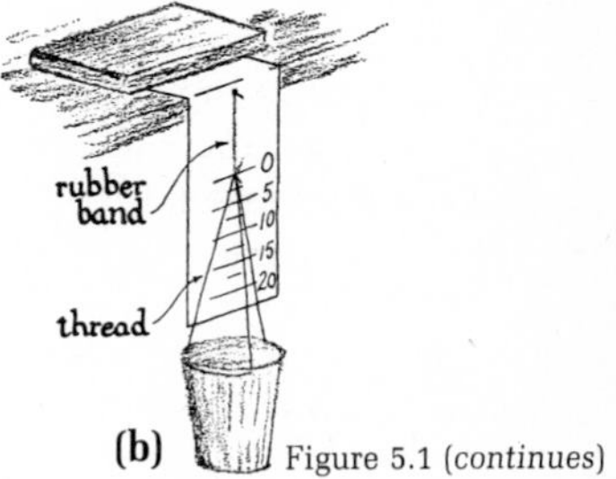

Figure 5.1 Writers use figures to orient the reader (a), show the parts of the whole (b), show contrast (c), provide close-ups (d). From Alan Holden and Phylis Morrison, *Crystals and Crystal Growing*, The MIT Press, 1982.

Figure 5.1 (*continues*)

- show contrast
- provide close-ups.

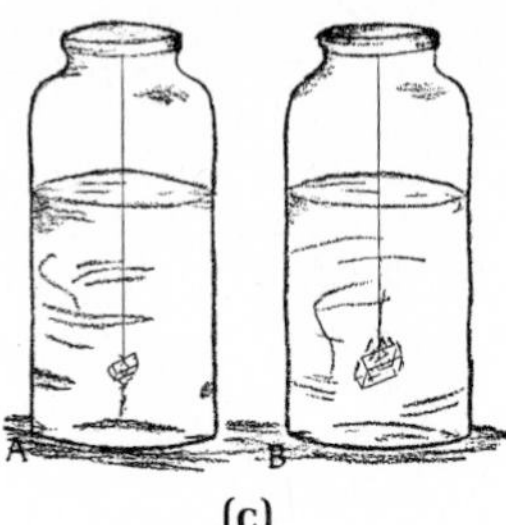

(c)

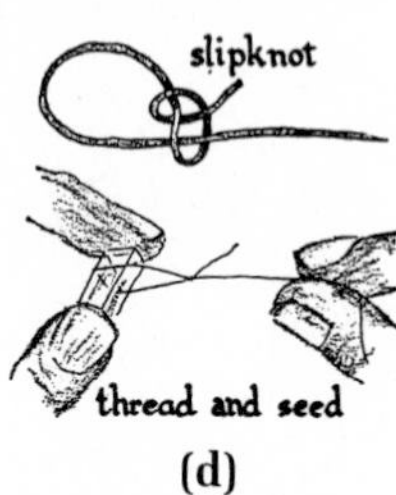

(d)

Figure 5.1 (*continued*)

Page design, too, plays a part in visual display when you seek to make technical text more readable for your audience. In the second version of the consumer loan application shown in Chapter 2, for example, the authors changed the type face, added headings and boldface emphasizers, and permitted more white space, or "air," on the page (see Figures 5.2a and b).

Figure 5.2a Page design plays a part in readability. This is the original version of an application for a consumer loan. Courtesy Alan Siegel, Siegel & Gale.

IN THE EVENT THIS NOTE IS PREPAID IN FULL OR REFINANCED, THE BORROWER SHALL RECEIVE A REFUND OF THE UNEARNED PORTION OF THE PREPAID FINANCE CHARGE COMPUTED IN ACCORDANCE WITH THE RULE OF 78 (THE "SUM OF THE DIGITS" METHOD), PROVIDED THAT THE BANK MAY RETAIN A MINIMUM FINANCE CHARGE OF $10, WHETHER OR NOT EARNED, AND, EXCEPT IN THE CASE OF A REFINANCING, NO REFUND SHALL BE MADE IF IT AMOUNTS TO LESS THAN $1. IN ADDITION, UPON ANY SUCH PREPAYMENT OR REFINANCING, THE BORROWER SHALL RECEIVE A REFUND OF THE CHARGE, IF ANY, FOR GROUP CREDIT LIFE INSURANCE INCLUDED IN THE LOAN EQUAL TO THE UNEARNED PORTION OF THE PREMIUM PAID OR PAYABLE BY THE HOLDER OF THE OBLIGATION (COMPUTED IN ACCORDANCE WITH THE RULE OF 78), PROVIDED THAT NO REFUND SHALL BE MADE OF AMOUNTS LESS THAN $1.

AS COLLATERAL SECURITY FOR THE PAYMENT OF THE INDEBTEDNESS OF THE UNDERSIGNED HEREUNDER AND ALL OTHER INDEBTEDNESS OR LIABILITIES OF THE UNDERSIGNED TO THE BANK, WHETHER JOINT, SEVERAL, ABSOLUTE, CONTINGENT, SECURED, UNSECURED, MATURED OR UNMATURED, UNDER ANY PRESENT OR FUTURE NOTE OR CONTRACT OR AGREEMENT WITH THE BANK (ALL SUCH INDEBTEDNESS AND LIABILITIES BEING HEREINAFTER COLLECTIVELY CALLED THE "OBLIGATIONS"), THE BANK SHALL HAVE, AND IS HEREBY GRANTED, A SECURITY INTEREST AND/OR RIGHT OF SET-OFF IN AND TO (a) ALL MONIES, SECURITIES AND OTHER PROPERTY OF THE UNDERSIGNED NOW OR HEREAFTER ON DEPOSIT WITH OR OTHERWISE HELD BY OR COMING TO THE POSSESSION OR UNDER THE CONTROL OF THE BANK, WHETHER HELD FOR SAFEKEEPING, COLLECTION, TRANSMISSION OR OTHERWISE OR AS CUSTODIAN, INCLUDING THE PROCEEDS THEREOF, AND ANY AND ALL CLAIMS OF THE UNDERSIGNED AGAINST THE BANK, WHETHER NOW OR HEREAFTER EXISTING, AND (b) THE FOLLOWING DESCRIBED PERSONAL PROPERTY (ALL SUCH MONIES, SECURITIES, PROPERTY, PROCEEDS, CLAIMS AND PERSONAL PROPERTY BEING HEREINAFTER COLLECTIVELY CALLED THE "COLLATERAL"): ( ) Motor Vehicle ( ) Boat ( ) Stocks, ( ) Bonds, ( ) Savings, and/or ______

Figure 5.2b This is part of the revised version of the application shown in Figure 5.2a. Note use of bold-faced headings, improved typeface, and increased white space. Courtesy Alan Siegel, Siegel & Gale.

Prepayment of Whole Note Even though I needn't pay more than the fixed installments, I have the right to prepay the whole outstanding amount of this note at any time. If I do, or if this loan is refinanced—that is, replaced by a new note—you will refund the unearned **finance charge,** figured by the rule of 78—a commonly used formula for figuring rebates on installment loans. However, you can charge a minimum **finance charge** of $10.

Late Charge If I fall more than 10 days behind in paying an installment, I promise to pay a late charge of 5% of the overdue installment, but no more than $5. However, the sum total of late charges on all installments can't be more than 2% of the total of payments or $25, whichever is less.

Security To protect you if I default on this or any other debt to you, I give you what is known as a security interest in my ○ Motor Vehicle and/or ______ (see the Security Agreement I have given you for a full description of this property), ○ Stocks, ○ Bonds, ○ Savings Account (more fully described in the receipt you gave me today) **and** any account or other property of mine coming into your possession.

Insurance I understand I must maintain property insurance on the property covered by the Security Agreement for its full insurable value, but I can buy this insurance through a person of my own choosing.

Default I'll be in default:

1. If I don't pay an installment on time; or
2. If any other creditor tries by legal process to take any money of mine in your possession.

You can then demand immediate payment of the balance of this note, minus the part of the **finance charge** which hasn't been earned figured by the rule of 78. You will also have other legal rights, for instance, the right to repossess, sell and apply security to the payments under this note and any other debts I may then owe you.

Irregular Payments You can accept late payments or partial payments, even though marked "payment in full", without losing any of your rights under this note.

Delay in Enforcement You can delay enforcing any of your rights under this note without losing them.

Collection Costs If I'm in default under this note and you demand full payment, I agree to pay you interest on the unpaid balance at the rate of 1% per month, after an allowance for the unearned **finance charge.** If you have to sue me, I also agree to pay your attorney's fees equal to 15% of the amount due, and court costs. But if I defend and the court decides I am right, I understand that you will pay my reasonable attorney's fees and the court costs.

Visual display, from illustrations to typography, matters a great deal in technical communication. Sam Howard, art director of *Scientific American,* has worked with illustration of scientific and technical concepts for many years, and comments on the power of technical illustration. "Visuals can make text readable. By readable I mean that graphics can make text understandable, make the information come out. That's their power."

TWO QUESTIONS FOR THE AUTHOR

The first question in illustration is, What do you want the reader to understand visually? The second is, What is the best technique to do this?

These questions must be answered in relation to the audience. Who will be reading the report or hearing the talk? What do they know already? What do they need to know?

You might, for instance, want to

• *orient readers.* In the final *Report of the Presidential Commission on the Space Shuttle Challenger Accident,* the detailed investigation of the O rings in the right Solid Rocket Booster begins with photographs of the Challenger that show the first, ominous puffs of smoke.

• *reveal what lies beneath the surface,* using such techniques as cutaways, exploded views, and cross sections (Figure 5.3).

• *select an aspect.* A schematic, or a sketch with one system marked, or a systems diagram—these are all devices that allow the reader to focus on one aspect of a complex design (Figure 5.4).

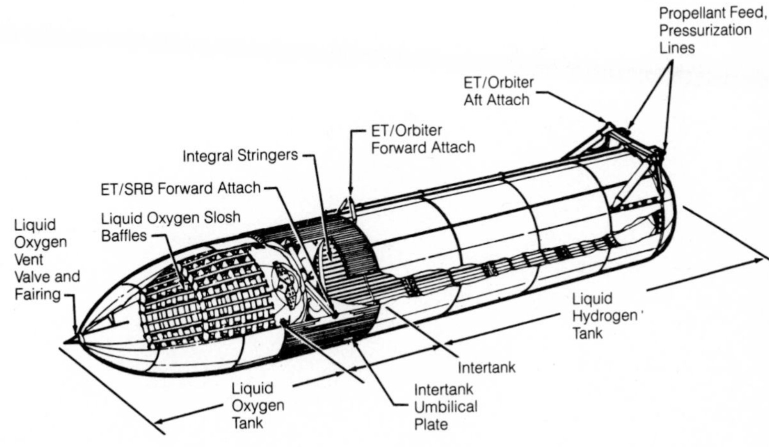

Figure 5.3 Example of a partial cutaway. From *Report of the Presidential Commission on the Space Shuttle Challenger Accident.*

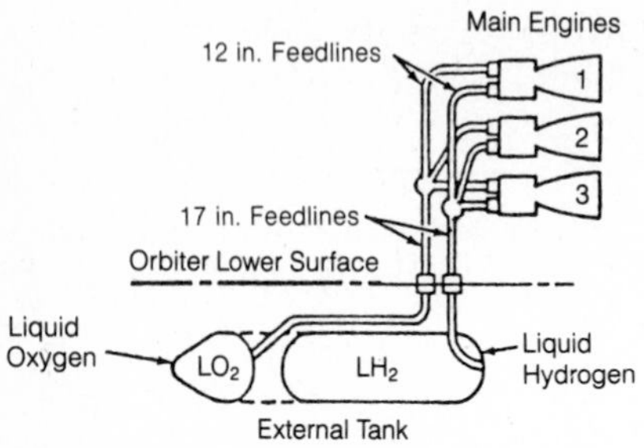

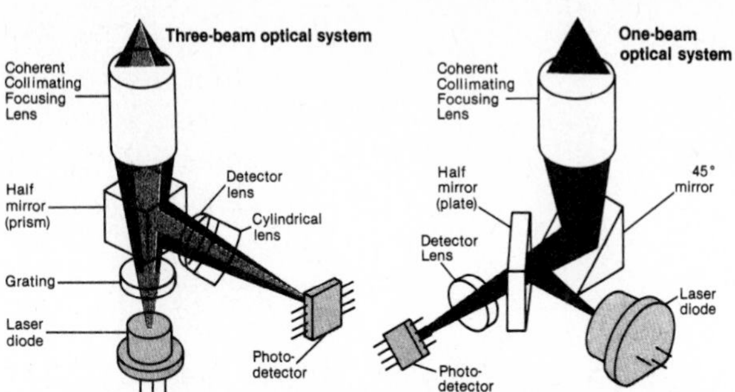

Designers of compact disk players have traditionally used a three-beam system (left) of reading the information from the disk. The three-beam system uses a grating to divide a laser beam into a major beam and two minor beams. The major beam reads the information off the disk, while the minor beams track the location to keep the beam in line. Matsushita is among the first CD manufacturers to switch to a one-beam system (right), which uses fewer parts and is thus cheaper and more reliable. In addition, because the beam is not split, it is stronger, making it capable of reading information on the disk through greasy fingerprints. To keep the beam on track, Matsushita's Technics CD player jitters the beam back and forth and digitally processes location information.

Figure 5.4 Example of a schematic drawing. From *Report of the Presidential Commission on the Space Shuttle Challenger Accident.*

Figure 5.5 Example of two figures used to show contrast. From IEEE *Spectrum*, January 1986, p. 72. © 1986 IEEE.

• *emphasize a difference.* Two illustrations or graphs can be juxtaposed to make a visual point that would have less effect if the two illustrations were separate (Figure 5.5).

• *show a trend or comparison.* Line graphs are powerful for showing and comparing trends. Tables can be used to convey the same information, but often the visual impact is less (Figure 5.6).

Figure 5.6 Data on world population displayed in table and in graph. Line graphs provide a powerful way to show trends. Note difference in visual impact when the same data are displayed as a table.

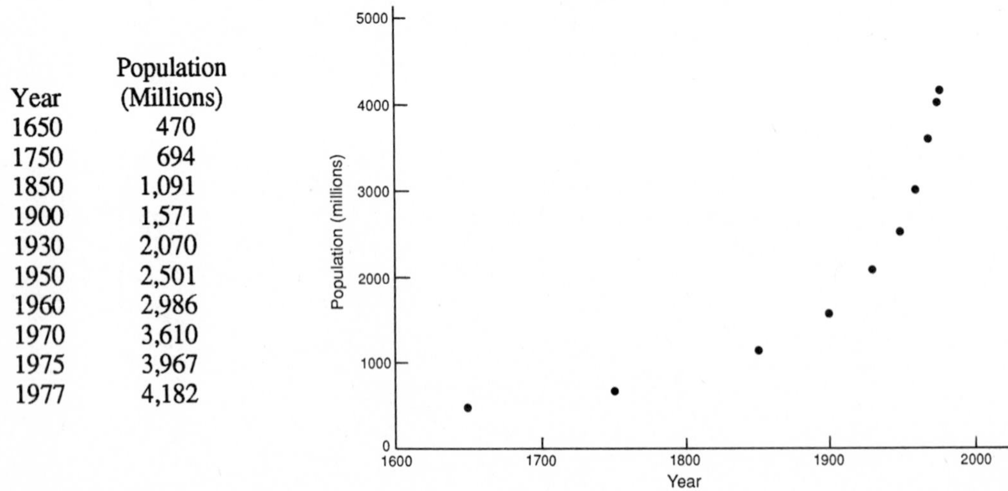

Year	Population (Millions)
1650	470
1750	694
1850	1,091
1900	1,571
1930	2,070
1950	2,501
1960	2,986
1970	3,610
1975	3,967
1977	4,182

Title →

Table II. Poly(methylphenylsiloxanes) Used as GC Liquid Phases

Stradddle rule separates levels of column heads →

Stub head →

Stub column →

GC grade**	Commercial grade	Percent phenyl***	Monomer(s) percent* w	x	y	z	Temperature range (C)
OV-73	-	5.5	94.5	-	5.5	-	20-350
-	SE-52	5	95	-	5	-	50-300
-	SE-54	5	93	-	5	2	100-300
OV-3	-	10	80	20	-	-	20-350
OV-7	-	20	60	40	-	-	20-350
-	OC-550	25	50	50	-	-	20-250
OV-61	-	33	67	-	33	-	20-350
OV11	-	35	30	70	-	-	0-350
OV17***	-	50	-	100	-	-	20-350
SP-2250	-	50	-	100	-	-	0-350
-	DC-710	50	-	100	-	-	20-250
OV-22	-	65	-	70	30	-	20-350
OV-25	-	75	-	50	50	-	20-350

Footnotes →

*See Figure 2 for monomer structures.
**Available from Ohio Valley Specialty Company., Marietta, OH and GC supply houses, unless otherwise noted.
***Recent work indicates that certain poly(methylphenylsiloxanes) actually have a different phenyl content than those reported here. See reference 23. For example, OV-17 has a phenyl content of 41 to 42%.
These polymers are either very viscous or are gums.
Available from Supelco,Inc. Bellefonte, PA.

Figure 5.7 Tables summarize detail, freeing the text for narrative. From Joel A. Yancy, "Liquid Phases Used in Packed Gas Chromatographic Columns," *Journal of Chromatographic Science*, Vol. 23, April 1985.

Simple graphs are used effectively at all levels of technical communication; some forms, such as logarithmic ones, are for specialized audiences.

• *show detail without congesting the text.* Both formal and informal tables are excellent devices for summarizing detail at the same time that they free the text for narrative (Figure 5.7).

• *show a process.* (Figure 5.8.)

Visual display in technical communication is not decoration; its aim is to reveal central ideas. Visuals are no mere handmaidens to language. Used correctly, they are powerful tools to tell the truth about the data.

Trudy E. Bell is a Senior Associate Editor at *IEEE Spectrum*, with an interest in ways to illustrate technical text effectively. "Illustrations are not window dressing," she says. "Illustrations are for the most important concepts. To think of them as decorative is to misunderstand their true worth. They should be used to lead people

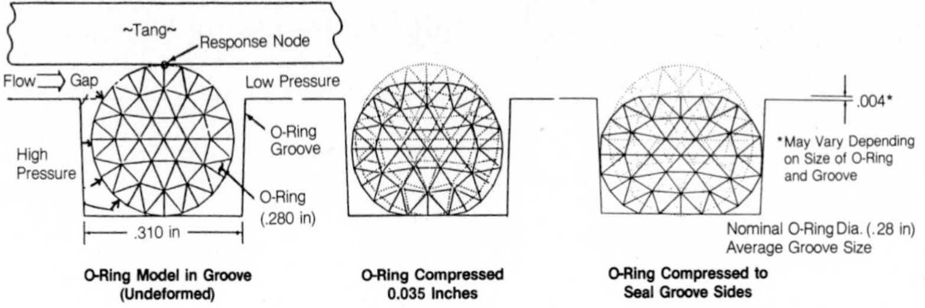

Figure 5.8 Example of figures used to show a process. From *Report of the Presidential Commission on the Space Shuttle Challenger Accident.*

to an immediate visual understanding of that which is crucial in the argument."

Many writers and speakers go through their first drafts and make a list of items they might want to illustrate, long before the appointment is made with the photographer or artist. This list goes hand in hand with the developing text.

Once the writer has decided which ideas are crucial for audience understanding, the question shifts to technique.

Not all visual display achieves what it sets out to do. There's an analogy to writing. Look over your writing, E. B. White advises, to see if you've said what you intended. The chances are good that you have not. The same might be said for graphics. It is as easy to create murky graphics as murky prose. The new graphics software packages are tremendously useful, but all the colors and instant pie-charts in the world won't help a display that is cluttered, inaccurate, or irrelevant.

AN EXAMPLE OF EFFECTIVE GRAPHICS: The Challenger Report

One powerful use of drawings is to display selective aspects of the same subject, using schematics, cutaways, close-ups, exploded views, or other techniques that may greatly aid the viewer's understanding.

An example is found in the *Report of the Presidential Commission on the Space Shuttle Challenger Accident,* which buttresses its arguments with an effective series of drawings of the Solid Rocket Booster.

Exploded View. During stacking at the launch site, four segments were assembled to form the Solid Rocket Booster. The field

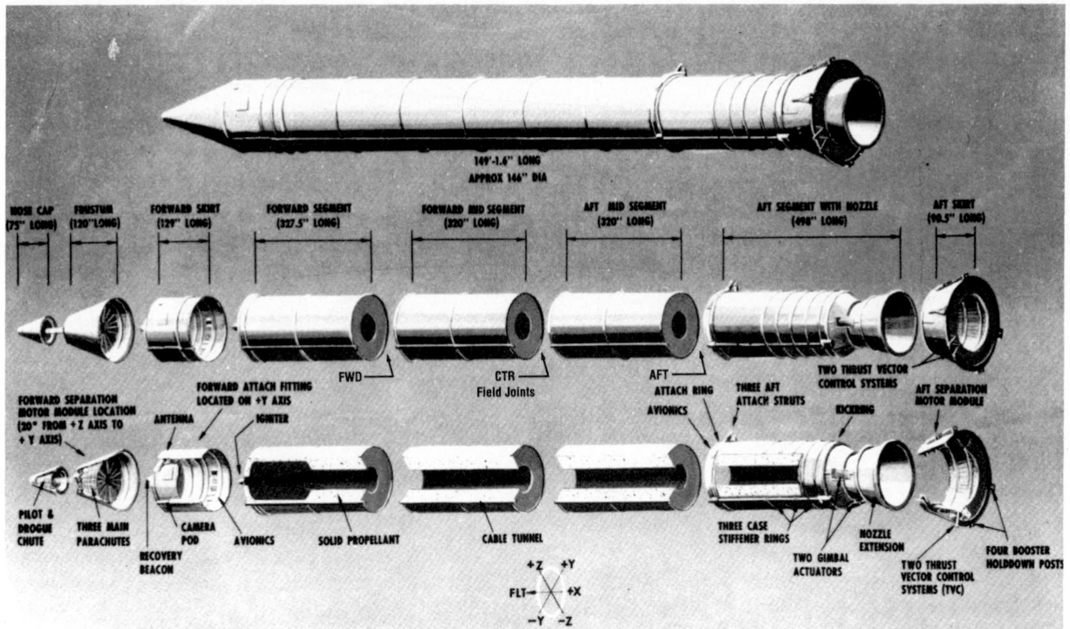

Figure 5.9 Exploded view showing motor segments of solid rocket booster. From *Report of the Presidential Commission on the Space Shuttle Challenger Accident.*

joints were sealed with two rubber O-rings. This step is described in the text, and illustrated with an exploded view (Figure 5.9).

Cross Section. The cross-sectional view (Figure 5.10) shows the position of the O rings. The zinc chromate putty was to act as a thermal barrier to prevent direct contact of combustion gas with the O rings. The O rings were to be actuated and sealed by combustion gas pressure displacing the putty in the space between the motor segments.

The report continues with a combination of text and figures, concluding: "The cause of the Challenger accident was the failure of the pressure seal in the aft field joint of the right Solid Rocket Motor." The interested reader had two avenues available to follow the argument—the two complementary paths made by the text and the supporting illustrations.

COMBINING PHOTOGRAPHS AND DRAWINGS

"Even in this age of photography," Sam Howard, art director at *Scientific American,* comments, "photographs have limits. There are things they can't show effectively. Medical illustration is a good

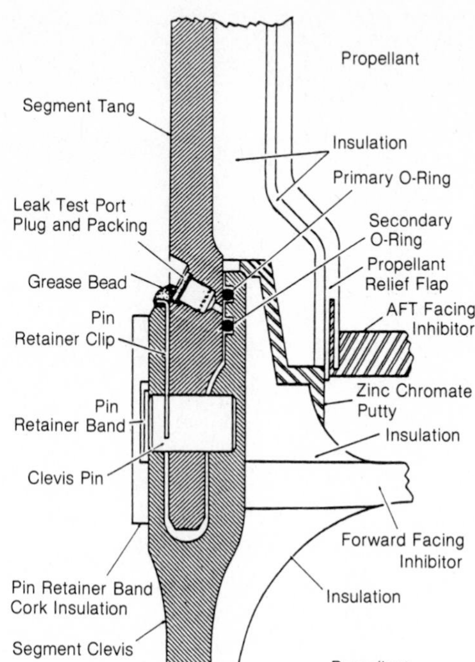

Figure 5.10 Cross-sectional view of solid rocket motor. From *Report of the Presidential Commission on the Space Shuttle Challenger Accident.*

example of the need for drawings, of one of the many places where photographs will only do a small fraction of the job."

There are times, though, when photographs can be combined effectively with drawings. Remember, however, that if you plan to make copies from the photographs, you should use originals that are high-quality, high-contrast glossies.

Example from a Manual for a Mechanical System

Woodward Governors produces many manuals in which it combines photographs used to introduce the equipment with drawings used to elucidate details. Its manuals have attracted complimentary notice from users who find the information easily studied and understood.

This series of photos shows a governor with reciprocating power cylinder, first with the cover on (Figure 5.11), and then with the cover removed (Figures 5.11a and 5.11b, left and right side views).

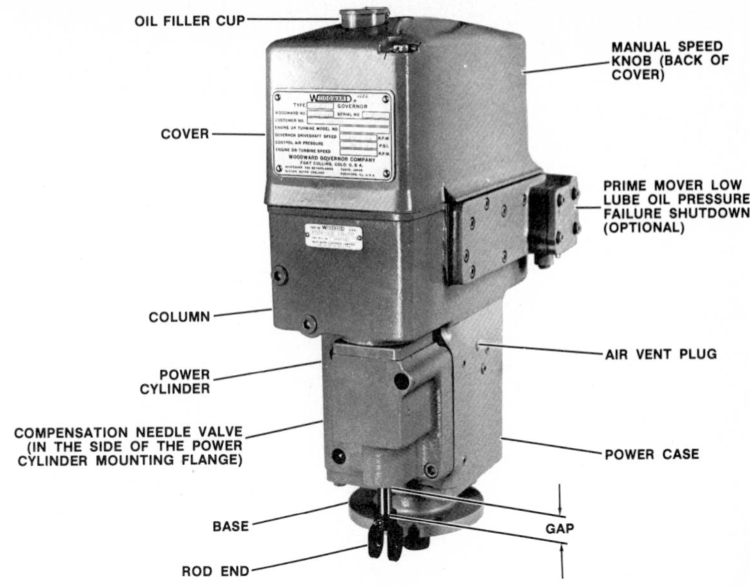

Figure 5.11 Example of photograph used for overview. Shown is a hydraulic governor. Courtesy Woodward Governor Company.

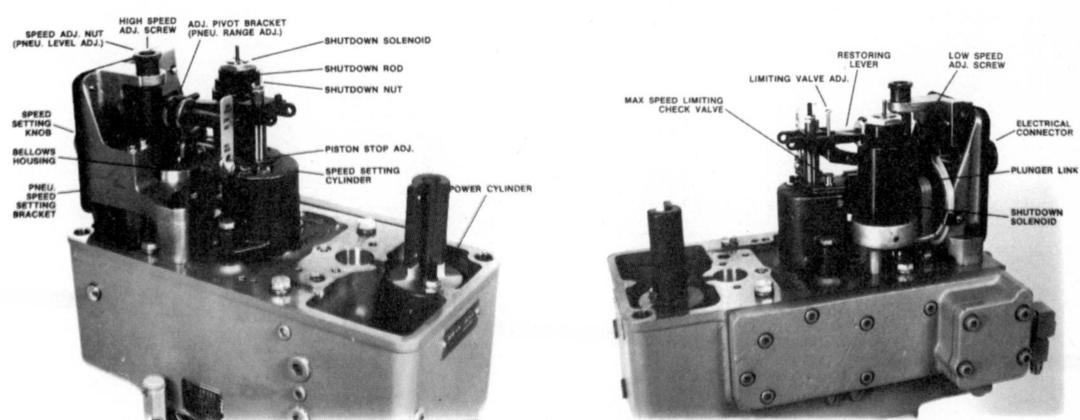

Figures 5.11a and b Photographs of left and right side views of opened governor (shown in part). Courtesy Woodward Governor Company.

Then details are developed with outline drawings (Figure 5.11c), and diagrams showing schematic and exploded views (Figures 5.11d and 5.11e).

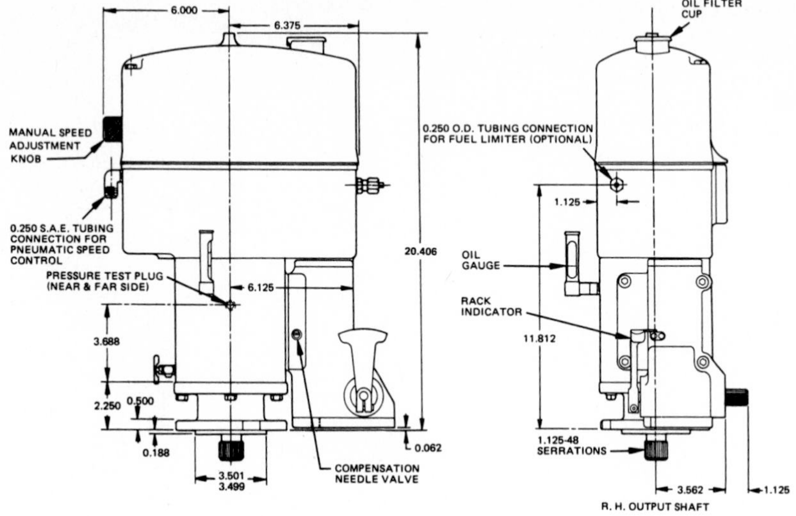

Figure 5.11c Example of outline drawing of governor (shown in part). Courtesy Woodward Governor Company.

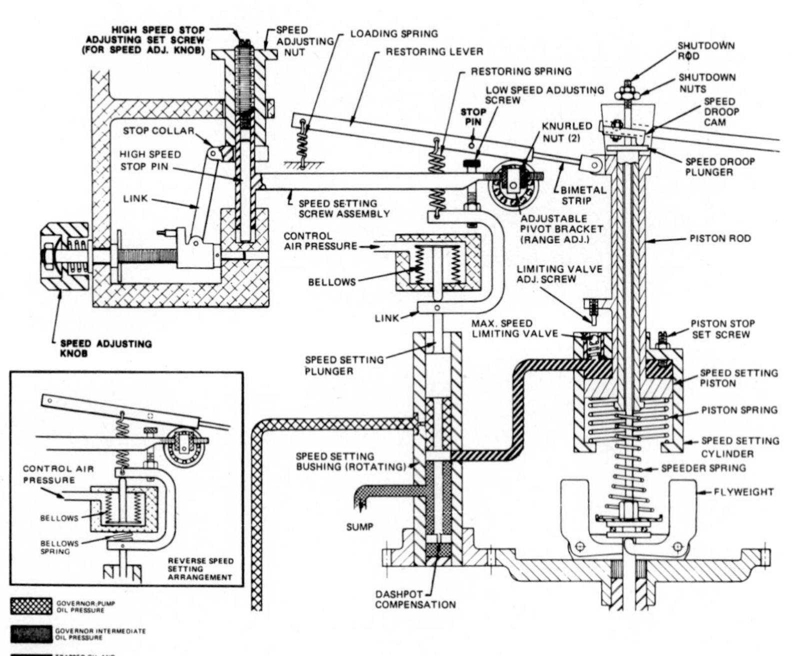

GOVERNOR-PUMP OIL PRESSURE
GOVERNOR INTERMEDIATE OIL PRESSURE
TRAPPED OIL AND GOVERNOR POWER CYLINDER OIL PRESSURE
SUMP OIL PRESSURE

Figure 5.11d Example of schematic diagram of governor (shown in part). Courtesy Woodward Governor Company.

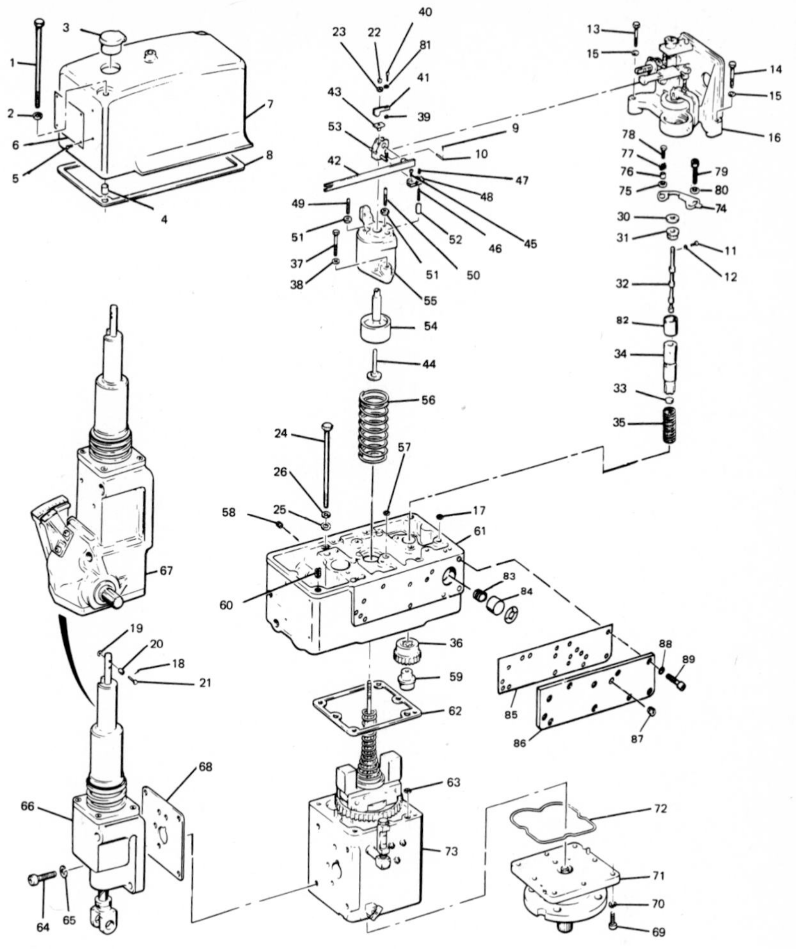

Figure 5.11e Exploded view of governor. Courtesy Woodward Governor Company.

Example from Sailing Directions

The Canadian Hydrographic Service is famous among those who sail northeastern waters for its *Sailing Directions Series.* These books have essential charts and accompanying narratives; they also use photographs very effectively as orienters, to be used as needed by the sailor.

Shown here are photographs and accompanying charts of Tusket Islands (Figure 5.12, photo; Figure 5.13, chart, reduced).

CARL A. RUDISILL LIBRARY
LENOIR-RHYNE COLLEGE

Figure 5.12 Photograph of Tusket Islands. Clear sailing directions combine an overview provided by the photograph with detailed information provided by the chart. (See Figure 5.13) Photo by J. R. Belanger, Staff Photographer, Bedford Institute of Oceanography. Courtesy Canadian Hydrographic Service (Atlantic Region).

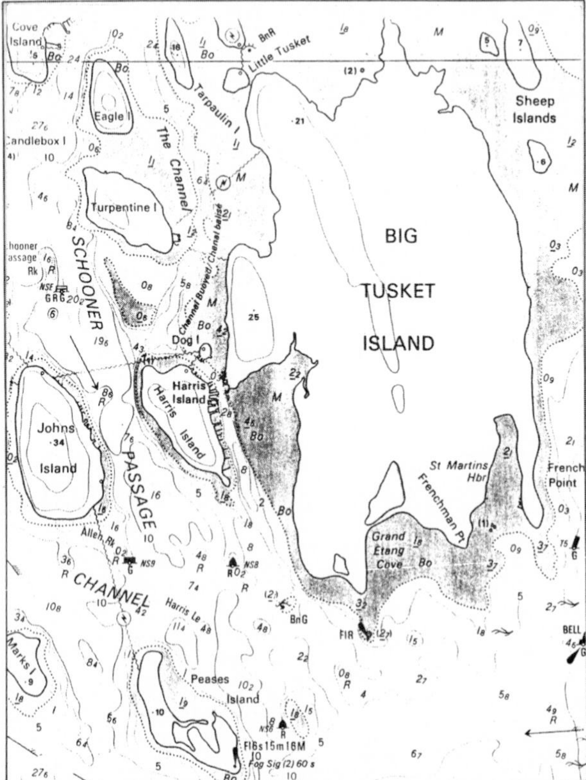

Figure 5.13 Chart of Tusket Islands. Courtesy of Canadian Hydrographic Service (Atlantic Region), Bedford Institute of Oceanography.

USING GRAPHS

Graphs help viewers visualize trends and make comparisons. They are usually either

- rectangular, using dots, lines, or bars
- circular
- pictorial, with quantities shown through size or number of pictured items

Line Graphs

These are superb for showing trends and comparisons. One of the closing visuals in the Challenger report (Figure 5.14), for instance, charts test results indicating that lower temperatures diminished the recovery of the O rings to their original shapes. (The O rings are flexible devices that were meant to seal in the booster rocket's gases during firing.)

Notice that the lines are clearly distinguished by the use of different symbols, to avoid the "vegetable soup" or "fruit salad" effect in which lines intermingle.

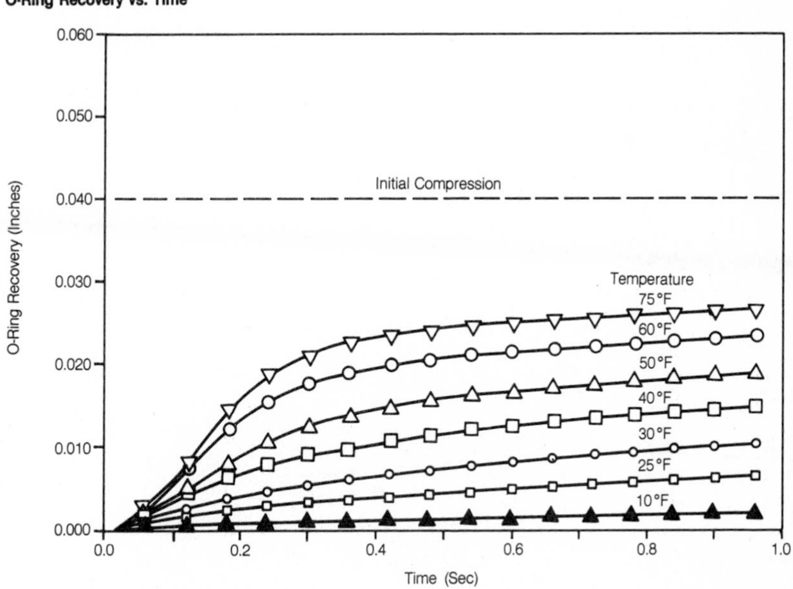

Figure 5.14 Lines in this graph from the report on the Space Shuttle Challenger are distinguished by different symbols.

Line graphs are effective, but they should not be used when the data are disconnected. For instance, if you are graphing rainfall vs. time and you have data taken every two years, the proper way to map the data is in bar graphs, as line graphs suggest there are data for the intervening years.

Logarithmic Graphs

Many physical quantities, such as radioactivity and fluorescence, decay exponentially with time. If such quantities are plotted directly against time, the curves grow very small. For instance, in Figure 5.15a, the closer we get to 100 nsec, the harder it grows to distinguish detail.

If you want to get a better idea of what is happening at longer times, you can make a semi-logarithmic plot in which the log of the quantity being measured is plotted against time. The curve will become a straight line, and any deviations will be easily discerned by curves in the line. Furthermore, any noise in the signal at longer times will be revealed (Figure 5.15b).

Bar Graphs

If absolute values, rather than contrast or trend, are what you want to show, then a bar graph has a more immediate visual effect than a line graph.

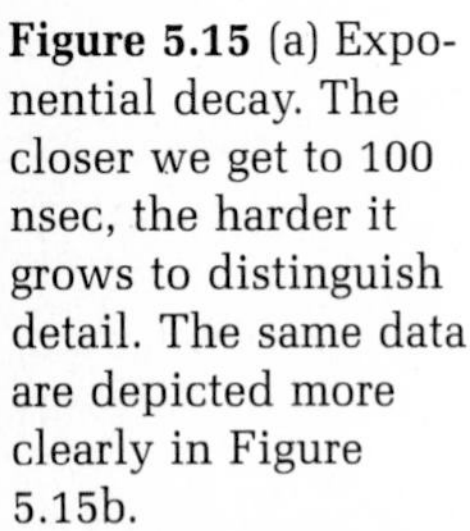

Figure 5.15 (a) Exponential decay. The closer we get to 100 nsec, the harder it grows to distinguish detail. The same data are depicted more clearly in Figure 5.15b.

(b) Semi-logarithmic graph. The curve of Figure 5.15a becomes a straight line, and any deviations are easily discernable as curves in the line. Note, too, that noise in the signal at longer times is revealed.

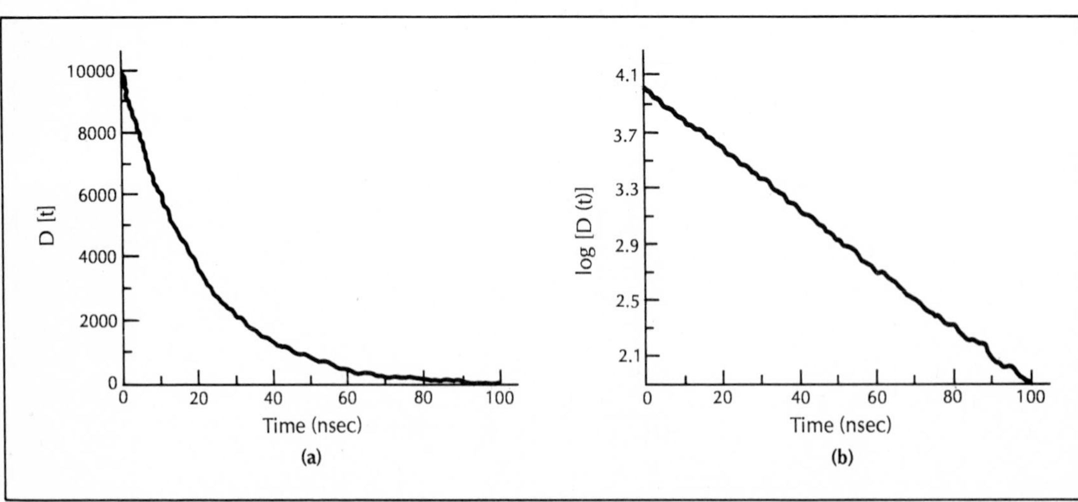

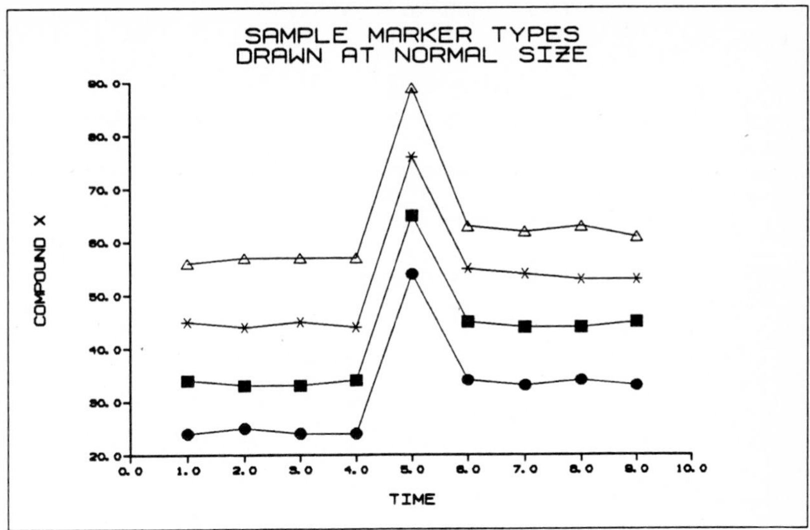

Figure 5.16 Lines in a graph should be clearly distinguished by the use of different symbols. The different data points in this software program are circles, squares, stars, x's, triangles, and diamonds.

In "The Crowded PBX Market" (Figure 5.17), the percentages are summarized at the end of each horizontal line. The viewer has both an image and numbers. If this data were in a graph, the reader would have to extrapolate—mentally extend a line to the horizontal scale and then imagine a value. The same data could be shown, but the visual impact would not be as striking.

Bar graphs can be vertical, stacked, or horizontal (Figure 5.18).

Composites

Line and bar graphs, as well as dot graphs, may show more than one set of data. To do this, some graphs employ both the left and the right edge (Figure 5.19). Others build vertically, stacking bar graphs on top of one another (Figure 5.18b). Some graphics combine lines and bars. Dot graphs may also be composites. Figure 5.20, a dot graph, shows not only closing value, but also high and low value for each day.

Circle Graphs

Circle graphs and pie charts are increasingly popular as graphics packages make them available to more users. They have what is called a low data density; that is, for the space they use, they

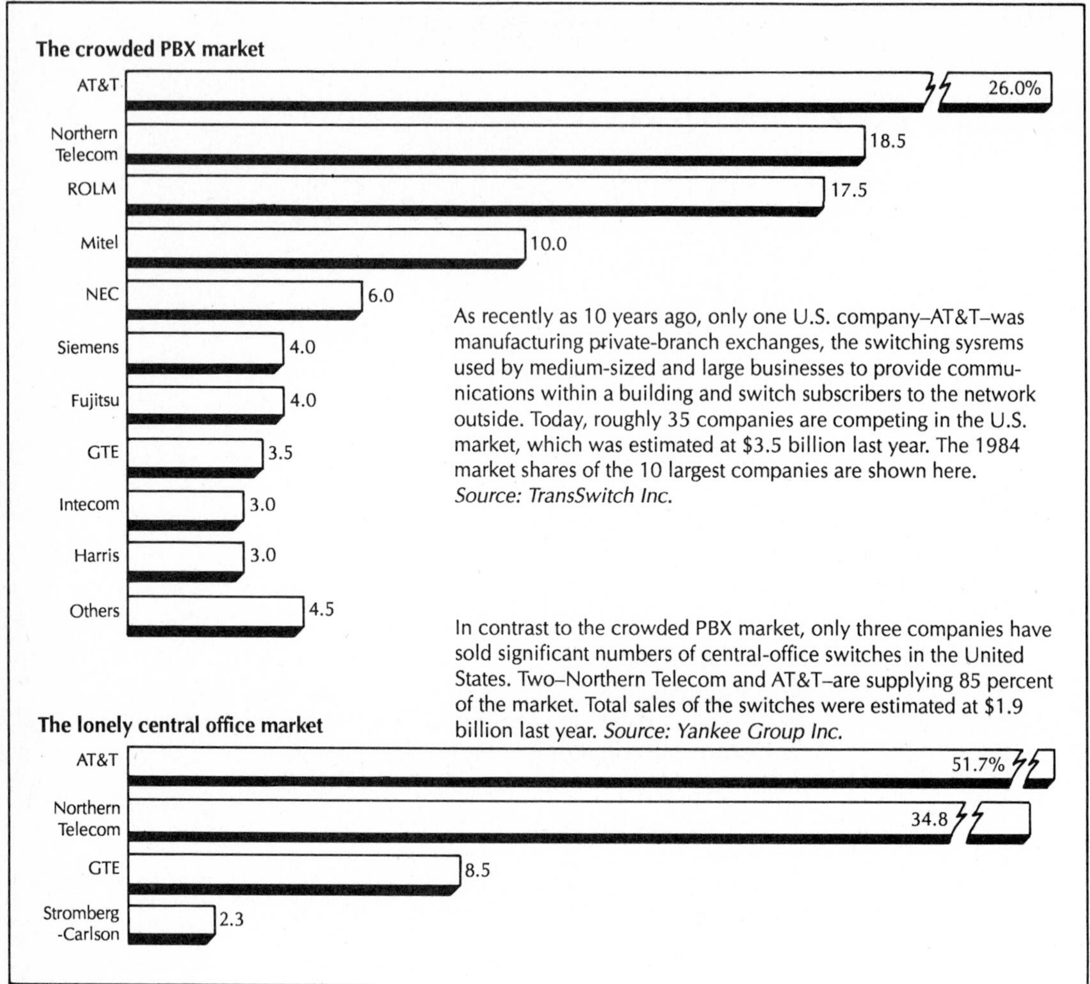

Figure 5.17 Percentages are summarized at the end of each bar. Note zig-zag to indicate graph has been cropped. From "The Crowded PBX Market," IEEE *Spectrum*, November 1985, p. 58. © 1985 IEEE.

illustrate little data. For instance, in Figure 5.21, the same data that occupies the first line of the table (Figure 5.21a) takes up an entire page using pie charts (Figure 5.21b).

"Given their low data density and failure to order numbers along a visual dimension, pie charts should never be used," comments Edward Tufte in *The Visual Display of Quantitative Information*. "A table is nearly always better." Why are pie charts so popular?

(Text continues on page 111.)

Figure 5.18 Bar graphs may be vertical (a), stacked (b), or horizontal (c).

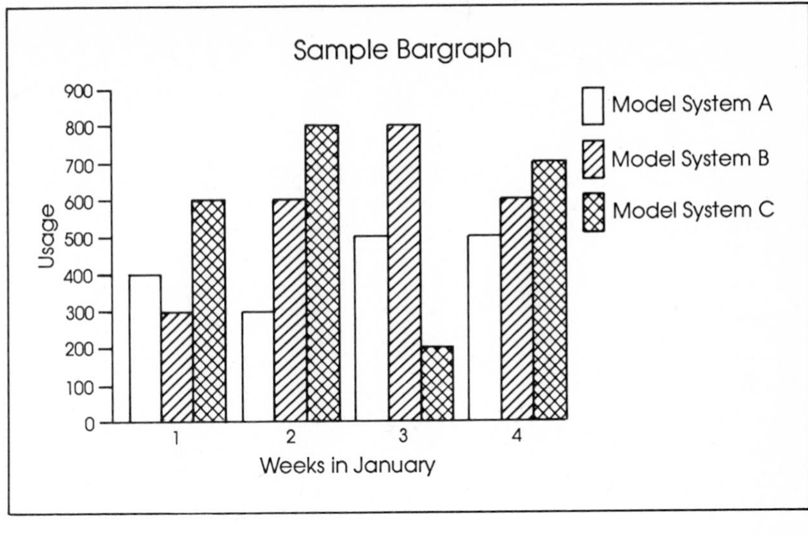

(a)

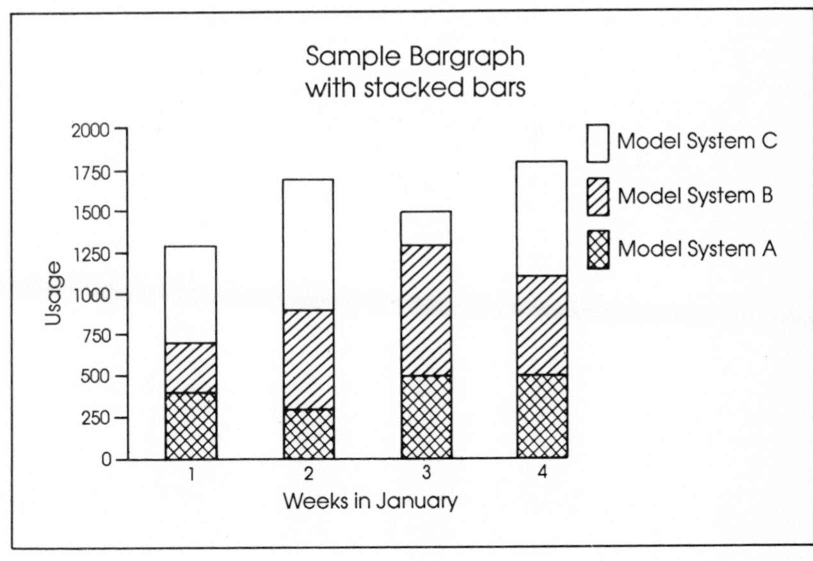

(b)

Figure 5.18 (*continues*)

Figure 5.18 *(continued)*

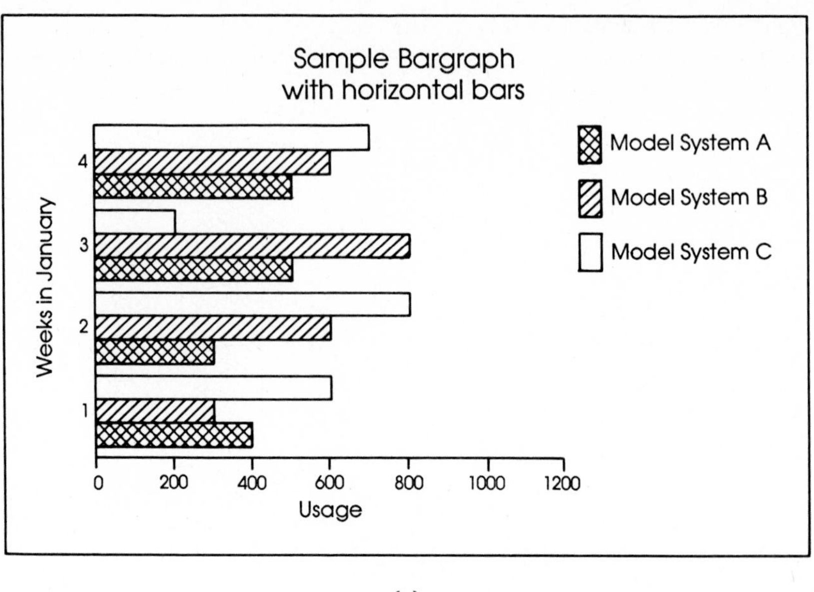

(c)

Figure 5.19 Composite line graph uses both left and right axes.

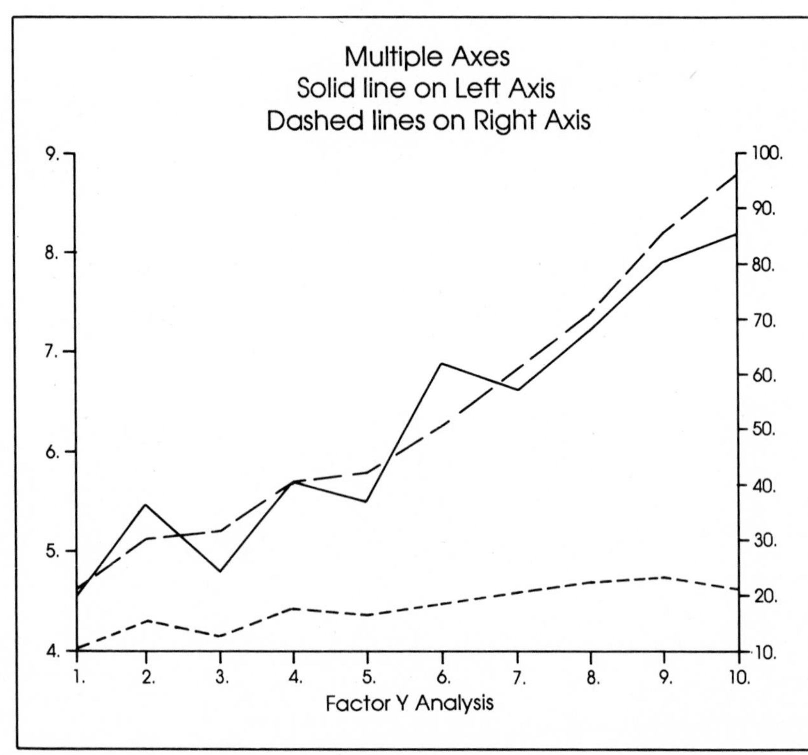

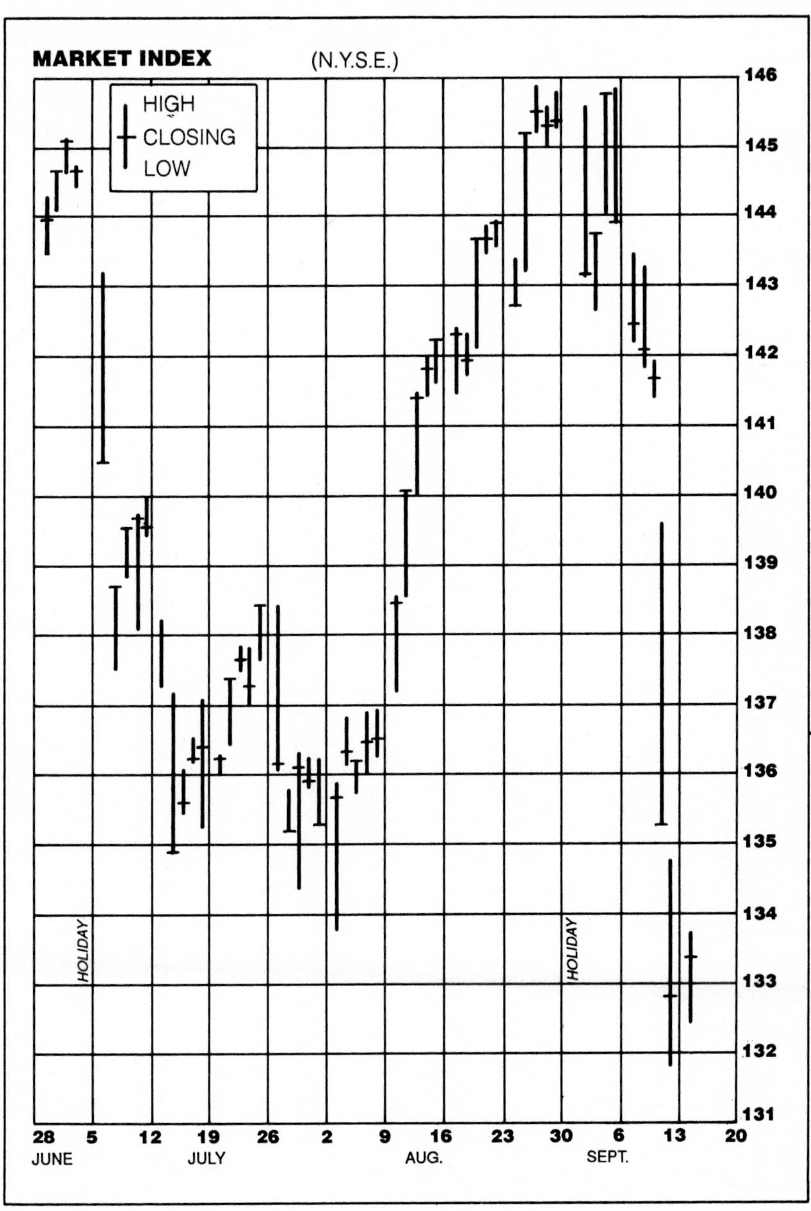

Figure 5.20 Composite graph.

Table 1a. Age and Marital Status of Women Business Owners, by Industry Division and Major Group: 1977
(For meaning of abbreviations and symbols, see Introduction. Item coverage rates appear in Appendix C. May not add to 100 percent because of rounding)

SIC code	Industry	Firms (number)	Owner's age as of December 31,1977 (percent)					Owner's Marital status (percent)			
			Less than 25 yrs	25 to 34 yrs	35 to 44 yrs	45 to 54 yrs	55 yrs or more	Never married	Married	Divorcd or sepd	Widowd
	All industries	701,957	5.3	17.0	17.3	18.8	41.7	18.8	27.0	26.6	27.6
	Construction	21,129	6.0	16.8	24.9	20.2	32.1	9.7	54.9	14.5	20.9
17	Special trade contractors	14,409	6.0	18.2	23.3	18.9	33.6	11.6	52.9	14.5	21.0
—	Other contract construction	6,720	5.9	13.6	28.7	23.2	28.6	5.1	59.5	14.6	20.8
	Manufacturing	18,914	3.0	19.1	15.5	20.1	42.2	16.2	44.5	16.0	23.3
23	Apparel and other textile products	2,725	4.6	19.3	17.4	23.5	35.1	14.7	44.5	24.1	16.7
27	Printing and publishing	4,914	2.5	18.2	13.5	20.9	44.9	25.0	38.6	16.8	19.6
32	Stone, clay, and glass products	2,166	2.3	34.6	17.6	12.9	32.6	21.6	29.1	27.9	21.4
39	Miscellaneous manufacturing industries	2,388	1.4	13.6	16.5	16.2	52.2	7.3	53.1	6.6	33.0
—	Other manufacturing	6,721	3.4	15.3	15.3	21.9	44.1	10.2	52.8	10.4	26.6
	Transportation and public utilities	11,874	3.2	14.9	17.3	24.2	39.7	11.1	45.5	17.7	25.8
42	Trucking and warehousing	4,804	3.8	12.5	19.1	24.0	40.6	40.1	52.1	13.3	30.4
47	Transportation services	3,233	2.4	15.6	18.6	24.8	38.6	18.4	43.5	21.5	16.6
—	Other transportation and public utilities	3,837	5.4	17.2	13.7	24.0	39.6	12.4	39.2	19.3	29.0
	Wholesale trade	16,133	4.3	13.3	17.2	21.2	44.0	12.8	52.3	12.2	22.6
50	Durable goods	7,446	3.8	13.4	15.9	19.1	47.8	8.6	55.9	12.0	23.5
51	Nondurable good	8,687	4.7	13.3	18.3	23.1	40.6	16.6	49.1	12.5	21.8
	Retail trade	211,723	4.2	13.7	16.9	21.1	44.1	12.0	33.4	22.8	21.9
52	Building materials and garden supplies	5,145	2.2	15.4	18.0	15.4	48.9	7.4	53.7	11.5	27.4
53	General merchandise stores	3,770	1.6	11.0	16.0	19.6	51.8	11.0	34.1	18.0	36.8
54	Food stores	21,309	1.9	8.1	16.6	17.7	55.8	9.3	32.6	17.7	41.0
55	Automotive dealers and service stations	8,186	2.0	8.1	17.6	28.9	43.4	5.7	51.4	10.3	32.6
56	Apparel and accesory stores	16,716	2.9	14.9	18.5	21.0	42.8	12.1	50.9	15.5	21.6
57	Furniture and home furnishing stores	8,949	2.7	12.5	13.0	29.9	41.9	7.3	39.4	25.5	27.8
58	Eating and drinking places	39,415	.7	6.0	20.5	34.0	38.9	6.7	37.2	21.4	34.5
59	Miscellaneous retail	108,233	6.5	17.6	15.7	16.6	43.6	15.2	26.7	26.9	31.2
	Finance, insurance, and real estate	66,257	2.3	a4.7	20.3	18.3	44.4	9.6	23.6	37.4	29.5
64	Insurance agents, brokers, and service	8,596	4.0	10.3	9.2	21.6	56.8	12.4	23.3	26.6	37.8
65	Real estate	55,093	2.1	15.5	22.2	17.5	42.8	9.0	22.8	39.8	28.4
—	Other finance, insurance and real estate	2,568	3.4	11.6	14.1	24.5	46.4	14.9	41.6	17.5	26.0
	Selected services	316,031	6.9	18.4	16.3	17.6	40.8	25.8	19.8	28.5	26.0
70	Hotels and other lodging places	12,590	1.0	2.5	10.7	23.0	62.7	6.6	30.7	24.4	38.3
72	Personal services	95,202	8.6	16.6	14.4	20.1	40.4	24.6	19.3	29.8	26.3
73	Business services	47,436	5.1	24.8	23.9	17.3	28.9	25.1	19.6	34.9	20.4
75	Automotive repair serices and garages	4,636	7.1	16.4	21.8	24.3	30.5	6.2	62.5	10.1	21.2
76	Miscellaneous repair services	4,670	3.4	17.5	18.6	19.5	41.0	9.9	52.1	17.2	20.8
78	Motion Pictures	1,439	6.2	29.7	21.4	16.6	26.0	28.6	33.4	21.4	16.5
79	Amusement and recreation services	16,763	20.6	23.8	20.9	15.6	19.0	37.9	24.6	24.8	12.5
80	Health Services	44,762	2.7	13.6	15.5	15.6	53.0	27.9	15.7	24.6	31.8
81	Legal Services	6,405	1.9	42.6	14.6	12.2	28.8	35.4	37.9	14.8	11.4
82	Educational services	23,148	2.9	12.8	15.0	17.2	52.1	32.4	8.3	25.7	33.6
83	Social services	2,780	3.6	23.1	30.6	25.5	17.2	8.1	47.1	31.7	13.0
89	Miscellaneous services	56,200	9.1	21.8	13.3	13.7	42.1	27.6	14.9	30.7	26.7
	Other industries	12,087	8.5	17.3	18.6	14.6	41.0	22.9	29.8	22.0	25.3
	Not classified by industry	27,809	2.2	21.9	20.0	14.3	31.6	25.6	19.4	35.4	19.6

Figure 5.21 (a)

CARL A. RUDISILL LIBRARY
LENOIR-RHYNE COLLEGE

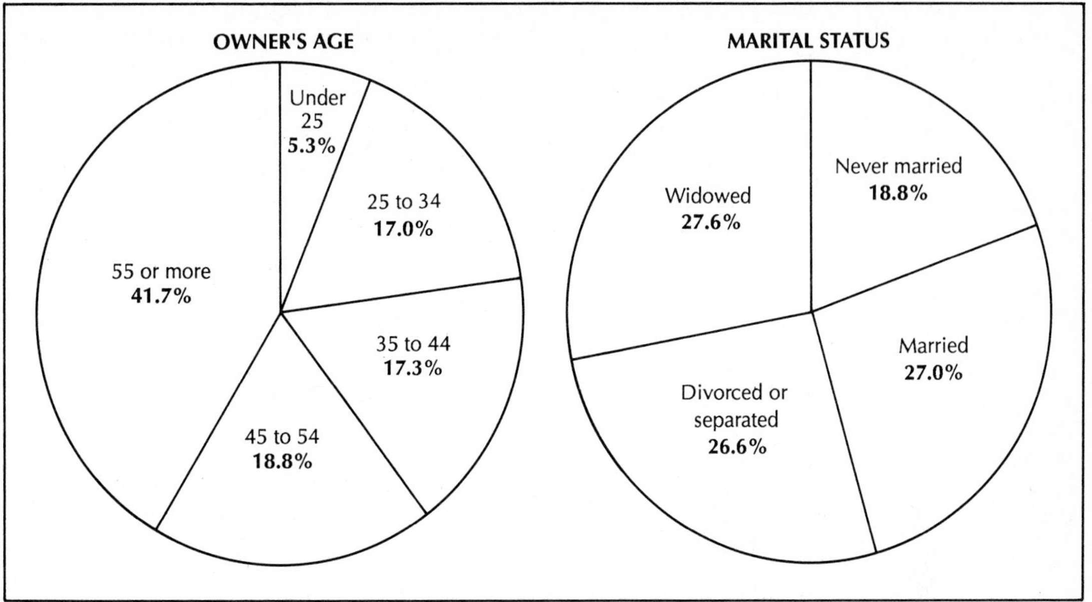

(b)

Figure 5.21 (*continued*) Same data shown in table (a) vs. pie chart (b). The same space taken by the first line of the table takes up nearly one half of a page using pie charts. (Figure 5.25b) From *Selected Characteristics of Women-Owned Businesses 1977*. U.S. Department of Commerce.

They are useful when you want to show the parts of the whole to people who are uncomfortable with tables, or who respond to simple graphics. They have an immediate visual impact, and are effective for lay audiences.

Tables

Tables are irreplaceable for supplying details that would otherwise clog the text. For instance, in a description of blood alcohol concentration, the point is made in the text: intoxication is caused by blood alcohol concentration. The table gives the specifics based on the number of drinks and body weight (Figure 5.23).

Note that while figures have captions *below*, tables have titles *above* the data. If you are using more than one table, number them consecutively.

Column headings are obligatory, except for the stub (far left) column, which sometimes needs no label. If you use a stub column and have divisions within it, indent to show subordination. Wording

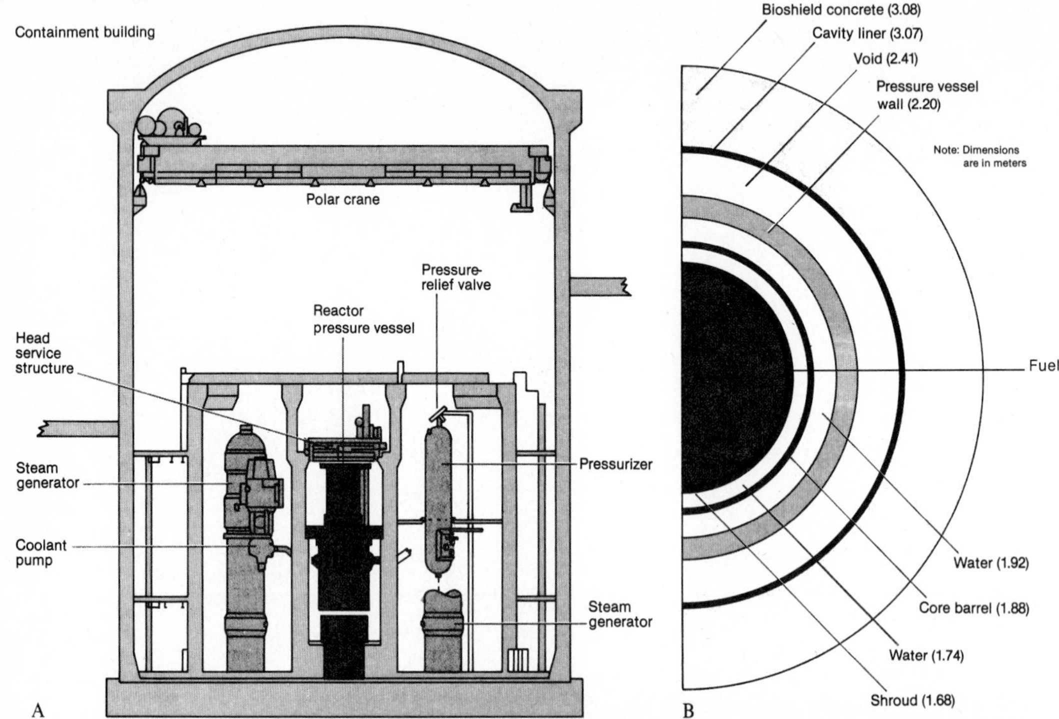

In decommissioning at any nuclear power plant, much of the work focuses on the reactor building, which typically contains the reactor vessel, steam generators, cooling equipment, and other hardware (A), all of which become radioactive. The highest concentration of radioactivity occurs in the structures that make up a typical reactor pressure vessel (B). The most radioactive and long-lived isotopes trapped in these structures are listed in Table II.

II. Radioactivity in major activated reactor components at reactor shutdown

Isotope	Half-life	Radioactivity, Ci/m³				
		Shroud	Core barrel	Vessel inner cladding	Vessel wall	Biological shield
^{95}Nb	35 day	2.0×10^{3}	7.6×10^{0}	5.6×10^{-3}	1.7×10^{-3}	5.4×10^{-6}
^{59}Fe	45 day	4.6×10^{4}	4.4×10^{3}	1.0×10^{2}	2.7×10^{1}	1.0×10^{0}
^{95}Zr	65 day	1.1×10^{-1}	6.2×10^{-3}	2.0×10^{-4}	7.2×10^{-4}	—
^{58}Co	72 day	1.5×10^{5}	1.0×10^{4}	3.3×10^{2}	6.6×10^{0}	2.2×10^{-2}
^{65}Zn	245 day	1.2×10^{2}	1.1×10^{0}	6.7×10^{-4}	3.5×10^{-5}	4.5×10^{-7}
^{54}Mn	300 day	6.8×10^{4}	3.7×10^{3}	1.2×10^{2}	4.7×10^{1}	1.7×10^{-1}
^{55}Fe	2.7 yr	1.3×10^{6}	1.5×10^{5}	3.5×10^{3}	7.2×10^{2}	3.0×10^{1}
^{60}Co	5.27 yr	9.6×10^{5}	9.3×10^{4}	2.5×10^{3}	7.5×10^{1}	6.7×10^{-1}
^{63}Ni	100 yr	1.2×10^{5}	1.5×10^{4}	3.6×10^{2}	3.8×10^{0}	1.4×10^{-1}
^{93}Mo	3500 yr	3.6×10^{-1}	5.2×10^{-2}	1.2×10^{-3}	1.3×10^{-3}	6.7×10^{-5}
^{14}C	5750 yr	1.5×10^{2}	1.8×10^{1}	4.0×10^{-1}	1.9×10^{-2}	6.9×10^{-4}
^{94}Nb	20 000 yr	5.4×10^{0}	2.6×10^{-1}	9.5×10^{-3}	—	—
^{59}Ni	80 000 yr	7.4×10^{2}	1.3×10^{2}	3.0×10^{0}	3.2×10^{-2}	1.2×10^{-3}
Total		2.97×10^{6}	3.07×10^{5}	7.73×10^{3}	9.04×10^{2}	3.21×10^{1}

Figure 5.22 In a report on reactors the cross section (a), schematic (b), and table (c) work together to illustrate technical detail. From "When Reactors Reach Old Age," IEEE *Spectrum*, February 1986. © 1986 IEEE.

Figure 5.23 Table of blood alcohol concentration.

APPROXIMATE BLOOD ALCOHOL PERCENTAGE

	Body Weight in Pounds							
Drinks	100	120	140	160	180	200	220	240
1	.04	.03	.03	.02	.02	.02	.02	.02
2	.08	.06	.05	.05	.04	.04	.03	.03
3	.11	.09	.08	.07	.06	.06	.05	.05
4	.15	.12	.11	.09	.08	.08	.07	.06
5	.19	.16	.13	.12	.11	.09	.09	.08
6	.23	.19	.16	.14	.13	.11	.10	.09
7	.26	.22	.19	.16	.15	.13	.12	.11
8	.30	.25	.21	.19	.17	.15	.14	.13
9	.34	.28	.24	.21	.19	.17	.15	.14
10	.38	.31	.27	.23	.21	.19	.17	.16

of headings should be considered carefully, as the number of characters affects column width.

Rules are usually used at the top, the bottom, and under column heads. Straddle rules are useful to separate levels of column heads (Figure 5.7).

Footnotes are usually set in a smaller typeface.

As a service to the reader, try to design your tables so they can be viewed upright, rather than sideways. If they don't fit vertically, reduce the type, but not below five points (about the size of print in the telephone book).

TYPOGRAPHY AND PAGE LAYOUT

Visual appeal is also measured by how text looks on the page—the net effect of many variables from the size of the print to the sorts of subheads that partition the text. Often these elements of page design are grouped together under the heading "layout" or "typography and page layout."

Typography and page layout lack the power to elucidate concepts—that job is occupied by graphs, charts, drawings, and photographs. But a well-designed page will make the text more attractive to the reader.

Figure 5.24 is a classic example of poor typography and page layout: The page is a "wall of print"; here are some of the elements of typography and layout that could improve it.

Type Size

Points. The type in Figure 5.24 is too small. Type size for reports should range from 9 points to 13 points. Ten-point type is the most common, one study finds.

Figure 5.24 Example of poor typography and page layout.

Type Size

Points. The type in Figure 5.24 is too small. Type size for reports should range from 9 points to 13 points. Ten-point type is the most common, one study finds (see Figure 5.25).

"Point" is a professional printer's unit of measurement. There are 12 points to a pica and 6 picas to the inch.

Pitch. Many printers classify type using the term "pitch" or width rather than points.

Pitches are measured in characters per inch (cpi), For most printers, pica has 10 characters per inch, elite 12 characters per inch, and compressed about 17 characters per inch. Figure 5.30 shows a summary of one desktop printer's pitches.

Proportionality. Proportional print uses space according to the type of letter; thus more space is used for a **w** or an **m** than for an **l** or **t** (see Figure 5.27).

Type Font

"Font" is a printer's term for the face of the type. Some basic variations in face and in darkness appear in Figure 5.28.

Leading

The lines of type in Figure 5.24 are too close together. The distance between lines of type is called "leading," and can be calculated in points the same way that type fonts can be measured (Fig 5.29).

White Space

Leading is one way to add white space or "air" to a page. One of the problems in the design of Figure 5.24 is its denseness. It can be lightened by wider margins, more space between lines, and the use of indentation. Such devices are not window dressing. White space gives the eye room to zoom in more efficiently and do its job.

Margins. These can be ragged right, or justified, depending on the effect you want (Figure 5.30)

Headings. They are among the most important visual aids in technical text. They divide the text, permitting readers to find the information they want. They permit quick access, a way to skim a report or memo for the necessary information. The reader, instead of being taken from beginning to end of a document, is invited to head for the most convenient spot. The use of headings and subheadings is yet another way that technical writing differs from academic writing. You would be unlikely to make much use of headings or subheadings in, for instance, an essay on Shakespeare, where you would be striving for a flow and unity that would be lessened by headings.

Headings are a convenience for both reader and writer. They help readers who want to speed through a document locate pertinent information. This convenievce helps make the document more accessible, and in business, accessibilty is important. Headings also help the writer. Deciding on the wording of a heading forces the writer to double check the information in the section and evaluate whether its placement and development are logical. Figure 5.31 shows a page that includes headings at four different levels. Many of the suggestions for wording of headings are the same as those for the wording of the table of contents:

•Make the headings at any one level parallel. This does not mean that all of your second level headings have to parallel throughout a report. It means that within a section, the headings should divide your text in a consistent manner.

•Make the language as informative as possible. Prefer "Use of Antiseptic on Compound Fractures" to a spare "Procedure". Remember, the heading is a preview of the contents for the reader. The more descriptive the language, the better the preview.

•Allow some text between headings—at least one sentence—or your material may begin to resemble a telephone book instead of a report.

Indentation Devices. Bullets, dots, or numbers may be used to lighten the page when you have a list or sequence you want to emphasize. Figure 5.32 uses squares, rather than bullets, to set off divisions.

Boldface or underscoring can be used to emphasize key items (Fig.5.33).

Figure 5.34 shows the same text as Figure 5.28, printed with more generous margins and leadings, and with type that is larger and darker.

TYPOGRAPHY OF CAPTIONS AND CALLOUTS

It's the rare picture that speaks a thousand words: most illustrations need some language to drive home their point.

Some popular labels include

•Callouts and legends. Words that describe parts of the drawing or photograph. These may be set either directly on the figure and connected with lead lines (Fig. 5.35) or beside in a legend (Fig. 5.36).

Figure 5.25 Example of 9/10 point type (a) and 12/13 type (b). Courtesy Photo-Lettering, Inc.

abcdefghijklmnopqrstuvwxyz
ABCDEFGHIJKLMNOPQRSTUVWXYZ

9/10 An adventure in typography. Texclusive faces from Photo-Lettering Inc. These selected styles with their beauty of composition will intrigue the virtuoso of typographic forms and will impart a touch of charm and un iqueness to the reader of your advertising message. Make this booklet a

(a)

abcdefghijklmnopqrstuvwxyz
ABCDEFGHIJKLMNOPQRSTUVWXYZ

12/13 An adventure in typography. Texclusive faces from Photo-Lettering Inc. These selected styles with their beauty of composition will intrigue the virtuoso of t ypographic forms and will impart a touch of charm

(b)

"Point" is a professional printer's unit of measurement. There are 12 points to a pica and 6 picas to the inch.

Pitch. Many printers classify type using the term "pitch" or width, rather than points (Figure 5.26).

Pitches are measured in characters per inch (cpi). For most printers, pica has 10 characters per inch, elite 12 characters per inch, and compressed about 17 characters per inch. Figure 5.26 shows a summary of one desktop printer's pitches.

Proportionality. Proportional print uses space according to the

Print sample | ← 1 inch → |	Density (in cpi)	Entry code
PICA PRINT	10	Default mode (Roman Pica)
ELITE PRINT	12	⟨ESC⟩"M"
COMPRESSED PRINT	17.16	CHR$(15) or ⟨ESC⟩CHR$(15)
EXPANDED PICA PRINT	5	⟨ESC⟩"W1" or CHR$(14) or ⟨ESC⟩CHR$(14)
EXPANDED ELITE PRINT	6	⟨ESC⟩"W1";⟨ESC⟩"M"
EXPANDED COMPRESSED PRINT	8.58	⟨ESC⟩"W1";CHR$(15)
| ← 1 inch → |		

Figure 5.26 Examples of print pitches.

Figure 5.27 Examples of non-proportional (a) and proportional (b) type.

The following paragraphs are an example of non-proportional type

The distance between lines of type is called "leading," and can be calculated in points the same way that type fonts can be measured (Fig 5.29).

Leading is one way to add white space or "air" to a page. One of the problems in the design of Figure 5.24 is its denseness. It can be lightened by wider margins, more space between lines, and the use of indentation. Such devices are not window dressing. White space gives the eye room to zoom in more efficiently and do its job.

(a)

The following paragraphs are an example of proportional type

The distance between lines of type is called "leading," and can be calculated in points the same way that type fonts can be measured (Fig 5.29).

Leading is one way to add white space or "air" to a page. One of the problems in the design of Figure 5.24 is its denseness. It can be lightened by wider margins, more space between lines, and the use of indentation. Such devices are not window dressing. White space gives the eye room to zoom in more efficiently and do its job.

(b)

type of letter; thus more space is used for a **w** or an **m** than for an **l** or **t** (Figure 5.27).

Type Font

"Font" is a printer's term for the face of the type. Some basic variations in face and in darkness appear in Figure 5.28.

Leading

The lines of type in Figure 5.24 are too close together. The distance between lines of type is called "leading," and can be calculated in points the same way that type fonts can be measured (Figure 5.29).

STANDARD	BLOCK SOLID FILLED	BLOCK NO FILL
Font 1	Font 1	Font 1
Font 3	Font 3	Font 3
Font 5	Font 5	Font 5

Figure 5.28 Fonts available in one graphics package.

White Space

Leading is one way to add white space or "air" to a page. One of the problems in the design of Figure 5.24 is its denseness. It can be lightened by wider margins, more space between the lines, and the use of indentation. Such devices are not window dressing. White space gives the eye room to zoom in more efficiently and do its job.

Margins. These can be ragged right, or justified, depending on the effect you want (Figure 5.30).

Headings. They are among the most important visual aids in technical text. They divide the text, permitting readers to find the information they want. They permit quick access, a way to skim a report or memo for the necessary information. The reader, instead of being taken from beginning to end of a document, is invited to head for the most convenient spot. The use of headings and subheadings is yet another way that technical writing differs from academic writing. You would be unlikely to make much use of headings or subheadings in, for instance, an essay on Shakespeare, where you would be striving for a flow and unity that would be lessened by headings.

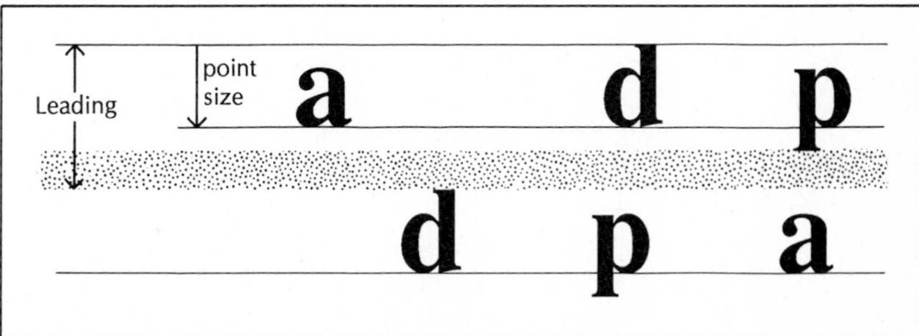

Figure 5.29 *Leading* is the amount of space between the top of one line and the top of the next line.

Figure 5.30 Examples of ragged right margin (a) and right-justified margin (b).

The distance between lines of type is called "leading," and can be calculated in points the same way that type fonts can be measured (Fig 5.29).

Leading is one way to add white space or "air" to a page. One of the problems in the design of Figure 5.24 is its denseness. It can be lightened by wider margins, more space between lines, and the use of indentation. Such devices are not window dressing. White space gives the eye room to zoom in more efficiently and do its job.

(a)

The distance between lines of type is called "leading," and can be calculated in points the same way that type fonts can be measured (Fig 5.29).

Leading is one way to add white space or "air" to a page. One of the problems in the design of Figure 5.24 is its denseness. It can be lightened by wider margins, more space between lines, and the use of indentation. Such devices are not window dressing. White space gives the eye room to zoom in more efficiently and do its job.

(b)

Headings are a convenience for both reader and writer. They help readers who want to speed through a document locate pertinent information. This convenience helps make the document more accessible, and in business, accessibility is important. Headings also help the writer. Deciding on the wording of a heading forces the writer to double check the information in the section and evaluate whether its placement and development are logical. Figure 5.31 shows a page that includes headings at four different levels. Many of the suggestions for wording of headings are the same as those for the wording of the table of contents:

- Make the headings at any one level parallel. This does not mean that all of your second level headings have to be parallel throughout a report. It means that within a section, the headings should divide your text in a consistent manner.
- Make the language as informative as possible. Prefer "Use of Antiseptic on Compound Fractures" to a spare "Procedure." Re-

Major Headings
- centered
- all capitals
- may be boldfaced

Minor Headings
- centered
- initial letters capitalized

Subheadings
- flush left
- underlined
- initial letters capitalized

Paragraph Headings
- indented
- underlined
- closed with a period
- initial letters capitalized

LETTERS AND CORRESPONDENCE

Letters and memos are both forms of correspondence. Letters are usually sent outside the company or agency; memos circulate within the organization. Occasionally these forms overlap; for instance, the president of a company may send a letter, rather than a memo, to convey holiday greetings to the staff.

Standard Format for Letters

Letters have a heading and inside address, a salutation, a body and closing, and supplementary lines. They may also have a subject line.

Heading

If you are using paper with a printed letterhead, the date is the only part of the heading you'll need to add. If your paper does not have a printed letterhead, type address and date, single-spaced, leaving a margin of 2" above and to the right.

Inside Address

At a minimum, double-space between heading and inside address. If the body of the letter is short, add double-spaces between the heading and inside address until the body of the letter is centered attractively between top and bottom margins.

Salutation

Format. Double-space between the last line of the inside address and the salutation. End the salutation with a colon; if you are quite friendly with the recipient, a comma is acceptable.

Wording of Honorifics. Whether you are writing the salutation of a letter, or filling in the recipient's name in a memo, you will often be faced with a choice in honorifics (Mr., Miss, Mrs., Dr., Ms.).

If you do not know the sex of the person you are addressing, it would be wise to spend a few minutes trying to find out, rather than settling for "Dear Sir."

When casting a letter or memo, don't assume that all engineers are male or, for instance, that all food technologists are female. If you are writing a form letter, include both sexes in the salutation. If it is a personal letter, inquire of the correct honorific in the salutation, just as you check the spelling of the name for correctness. It is one more example of a careful, professional approach to a problem.

Figure 5.31 Example of headings at four levels.

In September, 1969, two months after the initial lunar landing, a Space Task Group chaired by the Vice President offered a choice of three long-range plans:

- ■ A $8-$10 billion per year program involving a manned Mars expedition, a space station in lunar orbit and a 50-person Earth-orbiting station serviced by a reusable ferry, or Space Shuttle.
- ■ An intermediate program, costing less than $8 billion annually, that would include the Mars mission.
- ■ A relatively modest $4-$5.7 billion a year program that would embrace an Earth-orbiting space station and the Space Shuttle as its link to Earth.[1]

In March, 1970, President Nixon made it clear that, while he favored a continuing active space program, funding on the order of Apollo was not in the cards. He opted for the shuttle-tended space base as a long-range goal but deferred going ahead with the space station pending development of the shuttle vehicle. Thus the reusable

Figure 5.32 Example of squares used to emphasize items in a list.

member, the heading is a preview of the contents for the reader. The more descriptive the language, the better the preview.

• Allow some text between headings—at least one sentence—or your material may begin to resemble a telephone book instead of a report.

Indentation Devices. Bullets, dots, or numbers may be used to lighten the page when you have a list or sequence you want to emphasize. Figure 5.32 uses squares, rather than bullets, to set off divisions.

Boldface or underscoring can be used to emphasize key items (Figure 5.33).

Figure 5.34 shows the same text as Figure 5.24, printed with more generous margins and leadings, and with type that is larger and darker.

TYPOGRAPHY OF CAPTIONS AND CALLOUTS

It's the rare picture that speaks a thousand words; most illustrations need some language to drive home their point.

Figure 5.33 In this description, boldface gives visual emphasis to the main point and its divisions.

EMPLOYEE BENEFITS

The Tax Reform Act includes major changes in the rules for individual retirement arrangements (IRAs). Fewer people will be able deduct IRA contributions.

Full Deduction. You may make fully deductible IRA contributions of $2,000 annually ($2,250 for a spousal IRA) provided

•you do not belong to a company-funded retirement plan

•you do belong to a company-funded retirement plan, but your adjusted gross income (AGI) is between $25,000 and $35,000 if single, $40,000 to $50,000 if married. The AGI should be computed without subtraction of the IRA deduction.

Partial Deduction. The $2,000 IRA deduction is phased out over a $10,000 range for AGI above $25,000 if single, $40,000 if married. The deduction is reduced by $200 for each $1,000 of income above the lower limit. For example, if you are single and have an AGI of $26,000, your deduction is $1,800 rather than $2,000. If your AGI is $30,000, your deduction is $1,000.

No Deduction. You are not eligible if you are single with an AGI in excess of $35,000, or if you are married and filing joint returns on an AGI of $50,000 or more.

Some popular labels include

• **Callouts and legends.** Words that describe parts of the drawing or photograph. These may be set either directly on the figure and connected with lead lines (Figure 5.35) or beside in a legend (Figure 5.36).

• **Captions.** A caption is any text placed under a visual. Some publications repeat key information from the report or article beneath the figure; others add entirely new information. The figure and the caption together should tell a story. Don't assume that the reader can infer the story from the figure.

• **Arrows, circles, and other overlays.** Points in graphics are sometimes emphasized by outlines around key spots—for instance, a circle around a section of a photograph—or even arrows that point toward an important spot in a graph or photograph (Figure 5.37). Sam Howard says arrows are "cornball" devices; they certainly have

Figure 5.34 Effective layout and typography transform a "wall of print" into an attractive page. To compare this design with the original, see Figure 5.24.

Type Size

Points. The type in Figure 5.24 is too small. Type size for reports should range from 9 points to 13 points. Ten-point type is the most common, one study finds (see Figure 5.25).

"Point" is a professional printer's unit of measurement. There are 12 points to a pica and 6 picas to the inch.

Pitch. Many printers classify type using the term pitch or width rather than points.

Pitches are measured in characters per inch (cpi), For most printers, pica has 10 characters per inch, elite 12 characters per inch, and compressed about 17 characters per inch. Figure 5.30 shows a summary of one desktop printer's pitches.

Proportionality. Proportional print uses space according to the type of letter; thus more space is used for a **w** or an **m** than for an **l** or **t** (see Figure 5.27).

Type Font

"Font" is a printer's term for the face of the type. Some basic variations in face and in darkness appear in Figure 5.28.

Leading

The lines of type in Figure 5.24 are too close together. The distance between lines of type is called "leading," and can be calculated in points the same way that type fonts can be measured (Fig 5.29).

White Space

Leading is one way to add white space or "air" to a page. One of the problems in the design of Figure 5.24 is its denseness. It can be lightened by wider margins, more space between lines, and the use of indentation. Such devices are not window dressing. White space gives the eye room to zoom in more efficiently and do its job.

Headings

Headings are among the most important visual aids in technical text. They divide the text, permitting readers to find the information they want. They permit quick access, a way to skim a report or memo for the necessary information. The reader, instead of being taken from beginning to end of a document, is invited to head for the most con-

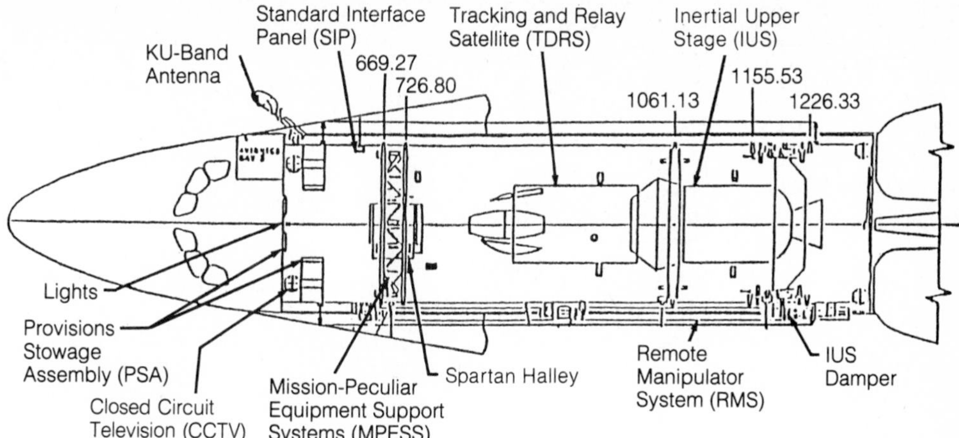

Overhead drawing of the Orbiter shows position of payload and other elements within the payload bay of the Challenger 51-L mission.

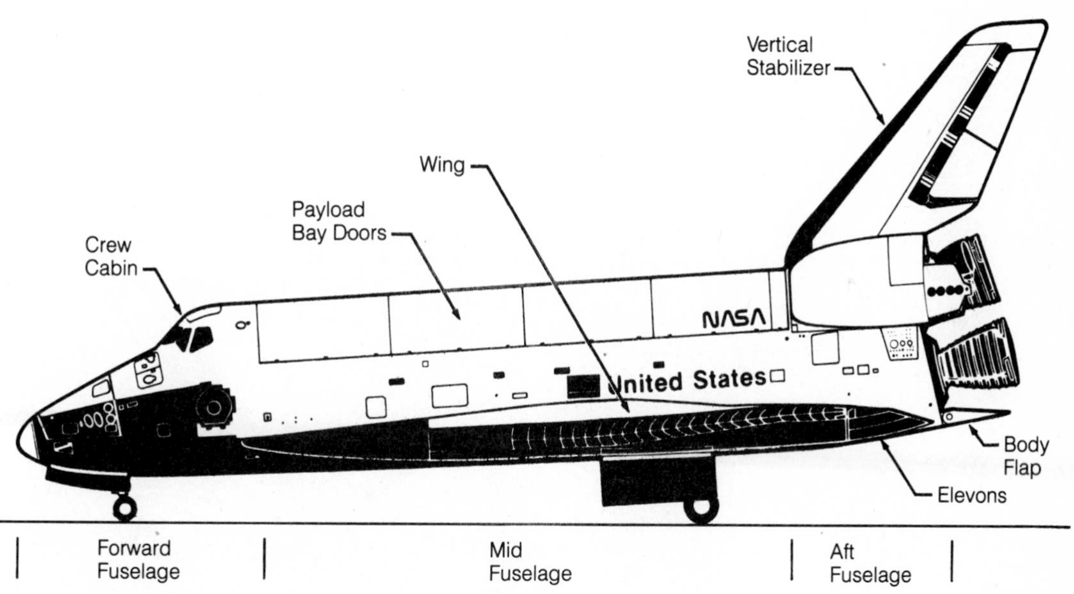

Sketch of Space Shuttle Orbiter in the landing configuration viewed from -Y position identifies aerodynamic flight surfaces.

Figure 5.35 Callouts may be set directly on the figure and connected with lead lines. From *Report of the Presidential Commission on the Space Shuttle Challenger Accident.*

Figure 5.36 Parts may be indexed and then identified in separate legends. From *Report of the Presidential Commission on the Space Shuttle Challenger Accident.*

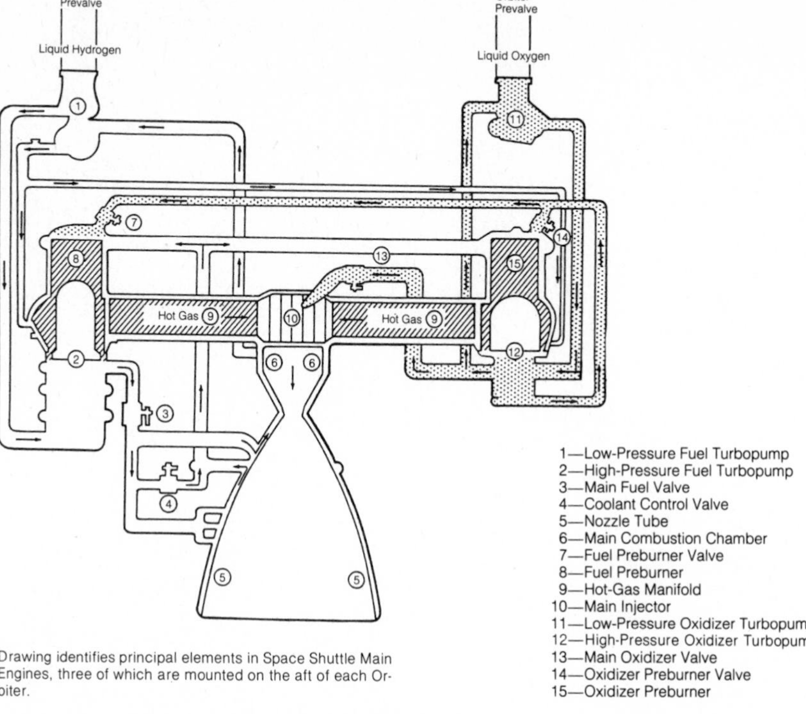

Drawing identifies principal elements in Space Shuttle Main Engines, three of which are mounted on the aft of each Orbiter.

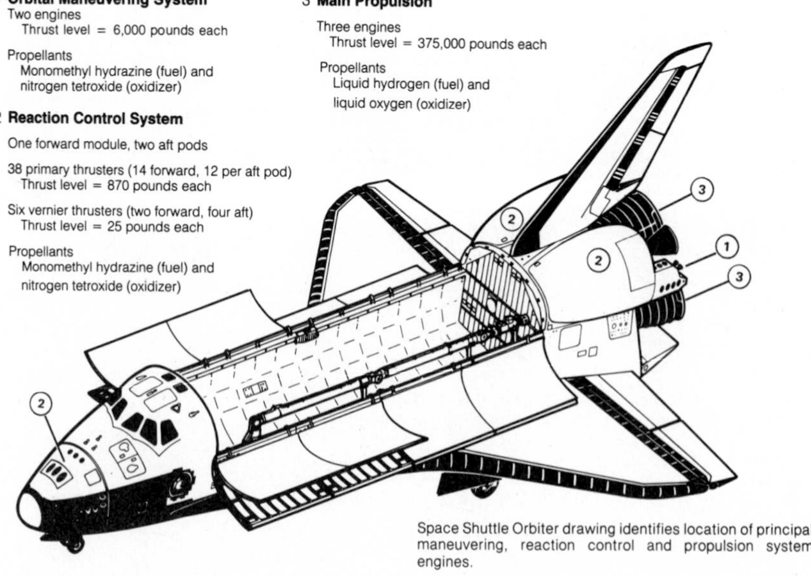

Space Shuttle Orbiter drawing identifies location of principal maneuvering, reaction control and propulsion system engines.

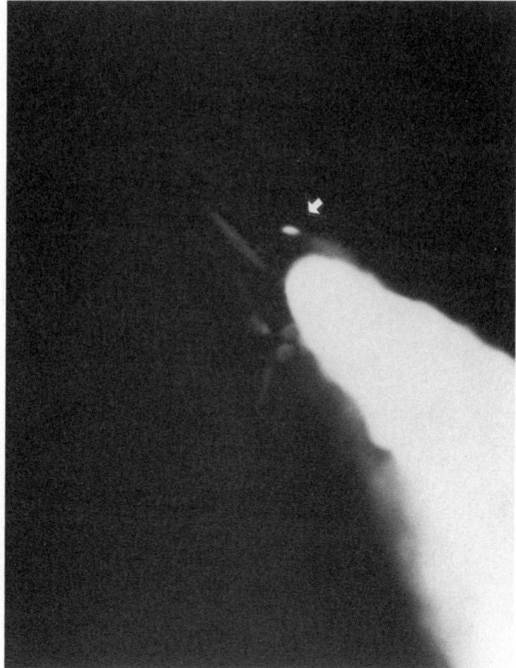

Figure 5.37 Arrows and circles focus attention. Here an arrow guides the eye to the first flame from the Space Shuttle Challenger.

a homey air. Others find them useful. Many graphics packages include arrows so that important spots on graphs can be marked (Figure 5.38).

Be careful of the type size used in legends, callouts, and labels. Figures are usually reduced for reproduction; if the type is small to begin with, it can shrink until it is incomprehensible.

VISUALS THAT DISTORT

Like language, visuals should tell the truth about the data. But, like language misused, visuals sometimes conceal or confuse this truth.

Suppressing the Zero

Figure 5.39a, Crude Oil Futures, has quite a bit of dispensable space in it. There is, after all, no change between 0 and 8 on the vertical scale. Why not crop the graph and have it start at eight? The new, cropped version is shown in Figure 5.39b.

But when we crop the graph, we introduce a distortion. A percent change on a full graph looks larger when the graph is trimmed,

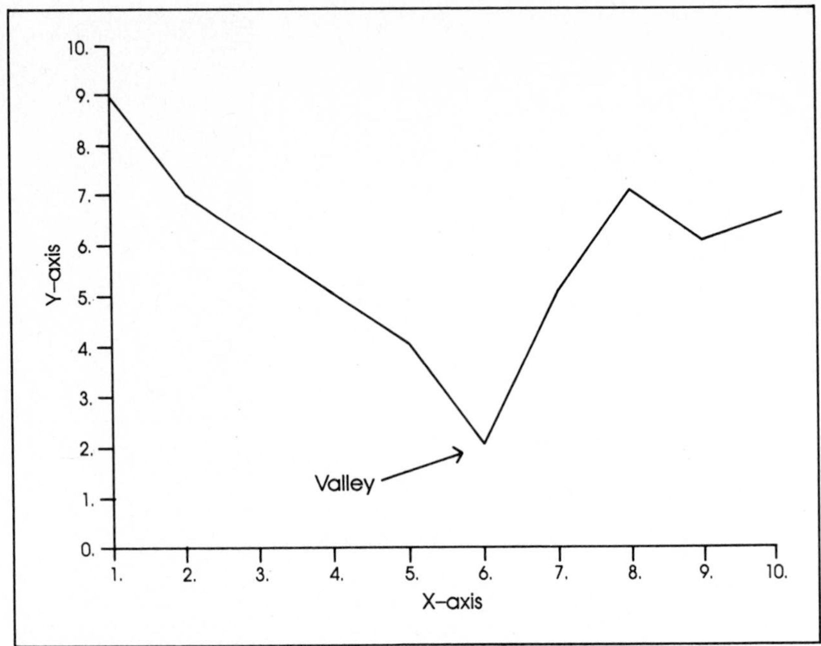

Figure 5.38 This graphics package includes an arrow to use for emphasis.

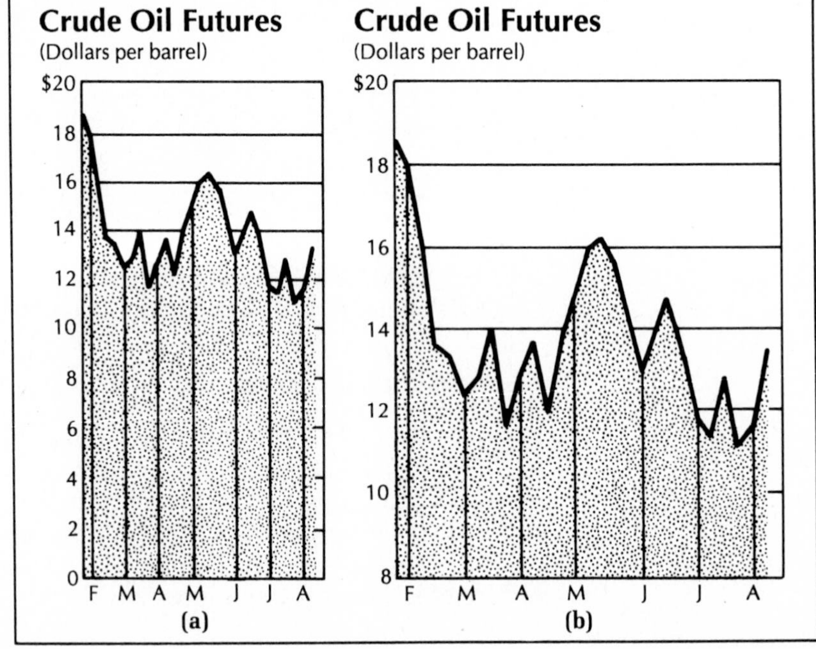

Figure 5.39 (a) Graph starts at zero on vertical axis. (b) Because the graph is cropped to start at 8, all changes are dramatized. For instance, the drop from February (F) to March (M) looks like a 40% change, as proportion is lost in the chopped graph.

because the proportion is lost. In the second version of Crude Oil Futures, for instance, the drop from February to March looks at a glance like a drop of 40 to 50 percent. If you look closer, though, and don't rely on your sense of proportion—the very sense the graph is made for—you see that the drop, from 19 to 13, is about 30 percent. When the graph is not cropped, the drop is far less dramatic.

If you do crop a graph, use a thunderbolt or zig-zag to show the deletion (see Figure 5.17).

Changing the Scale

Figure 5.40 has two horizontal scales. A half-inch on the 1986 weekly scale equals one month. A half-inch on the August daily

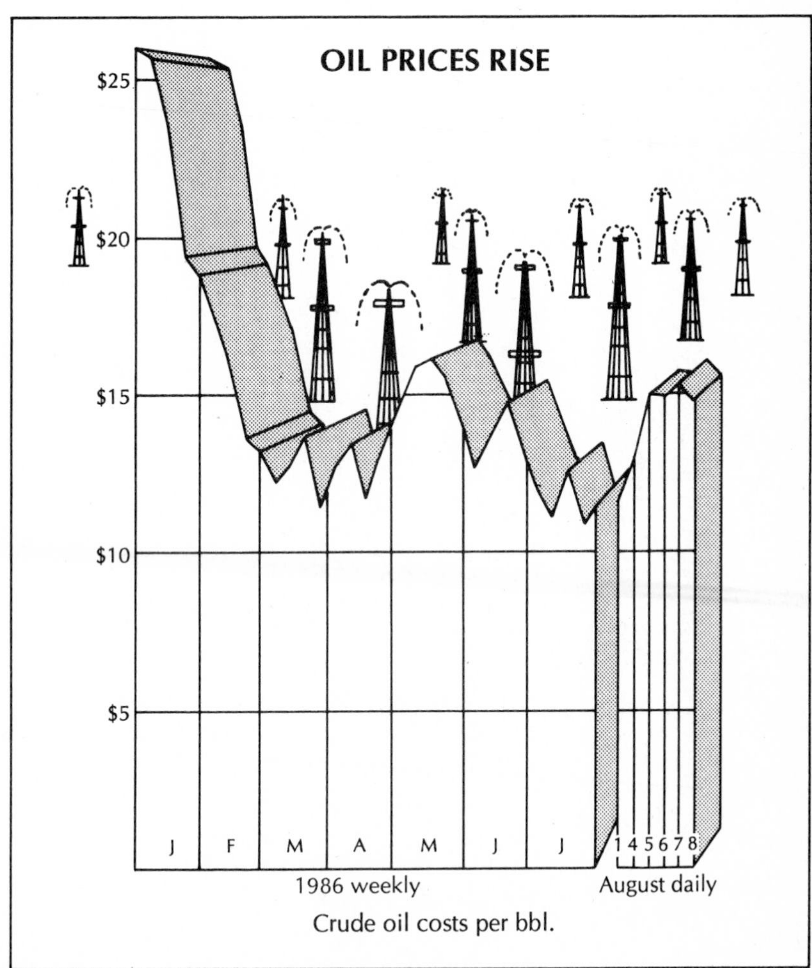

Figure 5.40 Because the figure has two horizontal scales, a half-inch on the 1986 weekly scale equals one month, but a half-inch on the August daily scale equals four days.

The oilwells are clutter, adding no information to the chart.

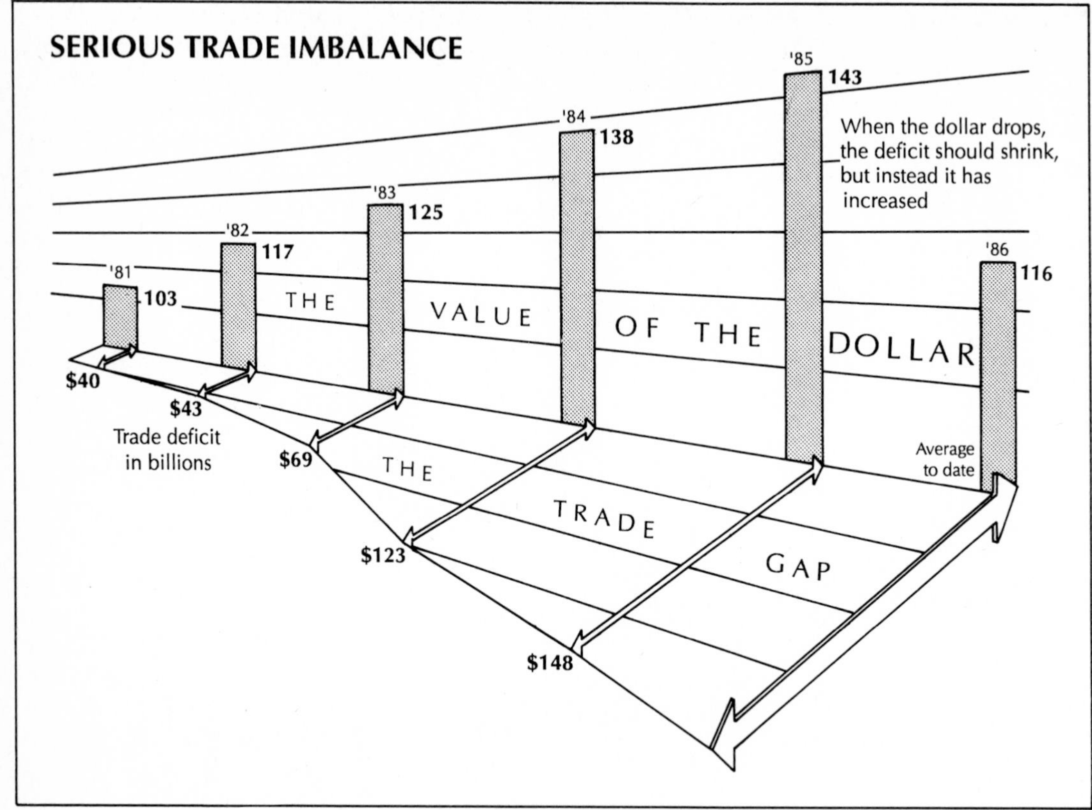

Figure 5.41 The figure has an unlabeled baseline. The careful viewer can find the baseline not at 0, but at $90. Note also the height of the vertical bar representing $116 in comparison to the bar representing $117.

scale equals four days. If you look closely, you can see the labels, but the eye sees proportion—that's the point of a graph. In this case, the eye sees a false proportion.

The Vanishing Baseline

Figure 5.41 has an unlabeled baseline. The careful viewer can finally find it—it must be $90. The fact that it is $90 instead of zero means that the viewer's sense of proportion is misled. (The viewer must also figure out that each unlabeled horizontal line represents $10.)

Distortion from Perspective

Figure 5.41 has another distortion. The introduction of perspective has made its own problems. The scale at the far left is completely different from the scale at the far right. Look, for instance, at the vertical bar for '82, representing $117. It is *half* the size of the vertical bar representing $116 in '86. The smaller amount looks twice as big.

Distortion When One Dimension Is Represented in Two Dimensions

Figure 5.42 uses the length of the credit card to indicate the debt. Thus the credit card is twice as long at 24 billion dollars as it is at 12 billion dollars. But the eye perceives the proportion not only of the length of the credit card, but its width—the eye perceives the *area* of the card—and when the length doubles, so does the width, giving us a four-fold increase in area. The debt has doubled, but our eye sees it quadrupled.

Clutter

An example of clutter is seen in the depiction of the oil wells in Figure 5.40. The wells are window dressing. They add no information to the graphic.

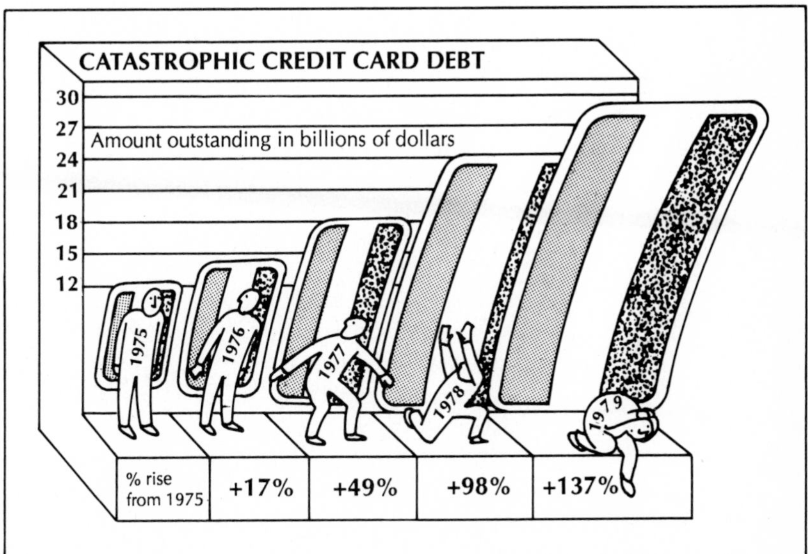

Figure 5.42 The credit card debt has doubled, but the depiction of the debt has quadrupled.

SUMMING UP

Good visuals help make technical text readable.

Graphic design includes not only illustrations, but page layout and typography.

Visuals are used to

- orient. The overview may be a photograph or drawing.
- reveal what lies beneath the surface. Examples include cutaways, exploded views and cross sections.
- select an aspect. Examples includes rear and side views, schematics, and outline drawings.
- emphasize a difference. Photographs, drawings, and graphs can all be used effectively for contrast.
- show a process. Line drawings, flow charts, and logic trees usually work best for procedural writing.
- give detail without congesting the text. Tables and charts are both excellent for the presentation for detail.

Illustrations are not window dressing. They should be used to lead people to an immediate visual understanding of what is crucial in the argument.

Cropping a graph introduces distortion. If you do crop, use a thunderbolt or zig-zag to show the deletion.

Changing the scale introduces distortion.

Using perspective in a graph leads to problems, particularly when a one-dimensional value is represented in two dimensions. The eye perceives not only length, but width.

Graphs may be

- rectangular, using dots, lines, or bars
- circular
- pictorial

Line graphs are superb for trends and comparisons. Lines should be clearly distinguished by the use of different symbols to avoid the fruit salad effect in which lines intermingle.

Bar graphs are good to show absolute values. They may be horizontal or vertical.

Composites are bar, line, or dot graphs that show more than one set of data. Line graphs do this by using both the left and right edge. Bar graphs do this by stacking.

Circle graphs have a low data density. Many authorities prefer tables.

Avoid laying out a page so that it looks like a "wall of print." Instead, use white space, type face and size, indentation, and other typographical aids to make the text more readable.

Professional printers specify type size in points. Ten-point type is the standard size for most reports.

Desktop printers often categorize type size by width, called pitch. Pitch is measured in characters per inch. Pica has about 10 characters per inch, elite about 12, compressed about 17. Pica and elite are both acceptable for reports.

Leading is the distance between lines of type, and can be calculated in points just as type fonts can.

Leading is an example of a way to add white space or air to a page. Other ways to lighten text include

- wider margins
- indentation devices such as bullets, dots, and hanging paragraphs
- first-, second-, and third-level headings
- boldface, underscoring and italics

Callouts for figures can be placed by the illustration and connected by lead lines; if the callout is complicated, the locations may be numbered and explained in a legend. Type size in callouts should be sufficiently large to remain readable if the figure is reduced.

Caption all figures. Don't assume the reader can infer the meaning from the figure.

EXERCISES

1. Choose an example of technical writing—a report, article, textbook excerpt, handbook selection, or set of instructions—in which the visuals are either effective or ineffective. What are the strengths or weaknesses of the visuals? Write a brief analysis. Include a photocopy of the visuals you are discussing.

2. As language is adjusted for audience, so too are the types of illustrations that give visual immediacy to language.

Examples 5.1 and 5.2 show selections from two sets of instructions and accompanying text. Are the language and the visuals more sophisticated in one set of instructions than in the other? What are the differences in the language and the visuals?

Example 5.1 Instructions for a slide test. Courtesy Roche Diagnostics.

Pregnosis SLIDE TEST—QUALITATIVE PROCEDURE

Controls: Known positive and negative *female* urine specimens should be used in quality control procedures periodically. They may be stored in small frozen aliquots and should be discarded after use.

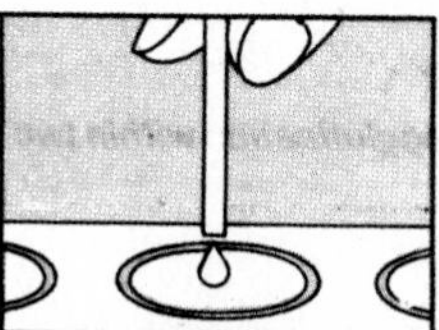

DROP:
1 drop of urine* in circle on glass slide with disposable pipette provided; hold pipette approximately one inch from the sealed end with the tip *perpendicular* to slide.

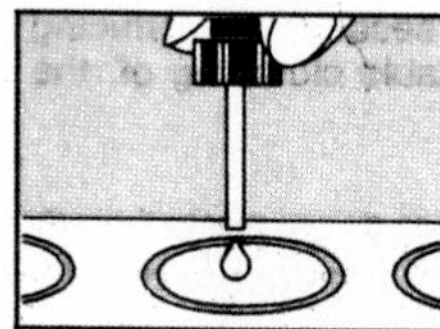

ADD:
1 drop of Antiserum Reagent (Black Cap) holding dropper tip *perpendicular* to slide.

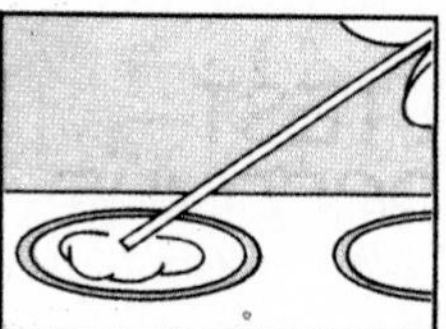

MIX:
with applicator stick and rotate slide gently for 30 seconds.

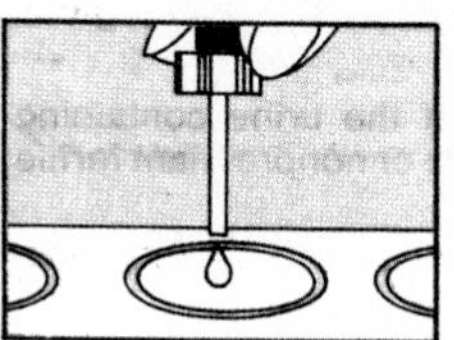

ADD:
1 drop of Antigen Reagent (White Cap) holding dropper tip *perpendicular* to slide; mix and spread urine/antiserum/antigen mixture over entire circle with applicator stick.

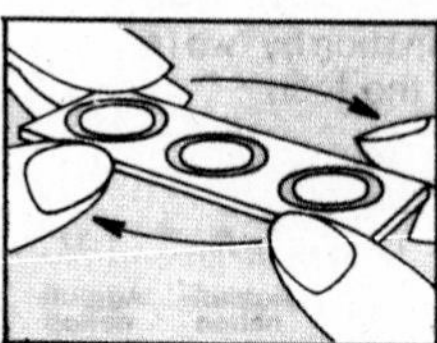

ROTATE:
slide for no longer than two minutes while observing for agglutination. A direct light source is recommended for this step.

*To withdraw urine, squeeze disposable pipette between thumb and forefinger (approximately 1 inch from sealed end), insert into urine and release pressure. Squeeze disposable pipette to deliver one drop.

NOTE: A special glass slide for performing 10 tests (in practically the same length of time it takes to do three) is available, without charge, by writing to Roche Diagnostics, Nutley, N.J. 07110.

Ribbon

Remove the ribbon cartridge from its packing materials. Holding the cartridge by the plastic fin on the top, guide the pair of tabs at each end of the cartridge into the corresponding slots in the printer frame (Figure 1). The cartridge should snap neatly into place.

With the paper bail resting on the metal platen, you can tuck the ribbon between the metal ribbon guide and the black print head.

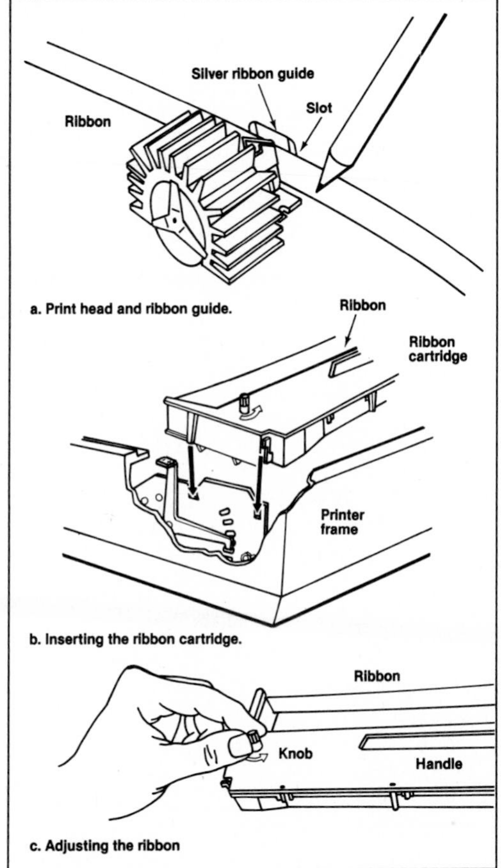

Figure 1.

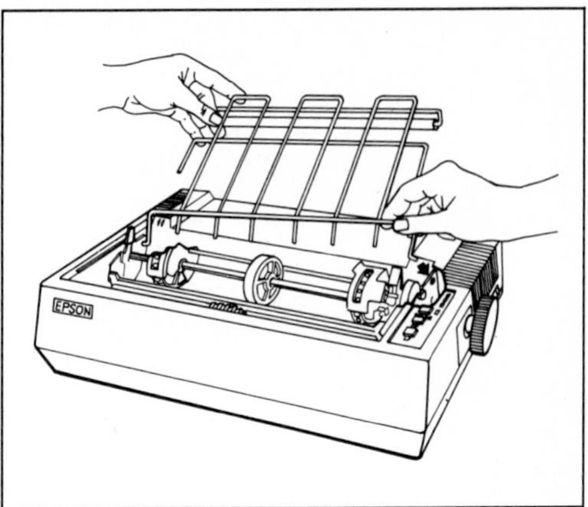

Figure 2.

As Figure 1 suggests, you can ease the ribbon into place with the deft application of a dull pencil (or a sharp finger). To remove any slack in the ribbon, turn the ribbon knob in the direction of the arrow.

Note: When you replace a ribbon, remember that the print head may be hot from usage; be careful.

Paper Guide

Next, install the wire-rack paper guide as shown in Figure 2.

On both versions of the RX-80, the tractor-feed mechanism can be adjusted for any width of paper, from fanfold 9-½-by-11-inch paper down to 4-inch mailing labels. Simply place your thumbs against the black tractor covers and pull the plastic locking levers forward (see Figure 3). Now you can move the tractors to approximate positions for your paper width.

Set a stack of paper on a flat surface under the desk or printer stand. If the paper is kinked or sitting off to one side, it may not feed through properly.

Example 5.2 Instructions for changing a ribbon. Courtesy Seiko-Epson Corporation.

Example 5.2 (*continues*)

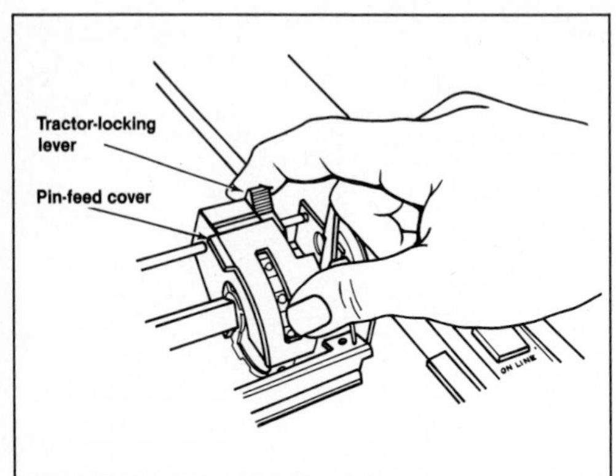

Be sure the RX is turned off so that you can move the print head to its middle position. Open both tractor covers and pull the paper bail toward the front of the printer. Guide the paper under the wire rack but above its roller, under the metal platen, behind the paper bail, and finally up to the tractors. Position the paper holes over the tractor pins. Now adjust the tractors to ease the paper; it should be held firmly but not tautly. Close the black tractor covers, pop the bail back against the paper, and roll the paper forward with the manual-feed knob.

Figure 3.

3. Data are poorly displayed when they are simply set out line by line in the narrative. A table, in contrast, will allow you to present the same information concisely and vividly. For example, consider the following excerpt from a study of two treatments for children with acute diarrhea:

> Rogers treated 224 children with acute diarrhea. The children were categorized by symptoms: mild, one to three moderately loose stools per day with no other symptoms; moderate, one to three watery stools per day; severe, four or more loose stools per day; very severe, four or more watery stools per day. One-hundred-twenty-seven were treated with calcium polycarbophil, 97 with kaolin-pectin.
>
> In the calcium polycarbophil group, those with very severe symptoms (67) reported the following results: 30 had an excellent response to treatment, 17 a good response, 14 a fair response, and 6 a poor response. Those with severe responses (29) reported as follows: 11, excellent; 8, good; 7, fair; 3, poor. Those with moderate intensity of symptoms (20) reported as follows: 7, excellent; 9, good; 4, fair; 0, poor. Those in the mild group (11) reported as follows: 6, excellent; 3, good; 1, fair; 1, poor.
>
> In the kaolin-pectin group, of those with very severe symptoms (56), eight reported excellent response to treatment; 14, good; 17, fair; 17, poor. Of those with severe symptoms (15), 1 reported excellent; good, 7; fair, 5; poor, 2. Of those with moderate symptoms

Table E.—*Imports of uranium ore and concentrates (Schedule A number 6270700) by country of origin: 1942 to1960*

[Quantities in thousands of pounds; values in thousands of dollars]

Year	Canada		Union of South Africa		Unidentified countries [1]		Total	
	Net quantity	Value	Net quantity	Value	Net quantity	Value	Net quantity	Value
1942	304	674	—	—	4,319	2,951	4,623	3,625
1943	80	154	—	—	13,629	2,124	13,709	2,278
1944	640	1,246	—	—	22,255	2,571	22,895	3,817
1945	300	582	—	—	17,574	2,486	17,874	3,068
1946	420	966	—	—	13,656	9,595	14,076	10,561
1947	540	3,067	—	—	6,027	8,972	6,567	12,039
1948	340	4,277	—	—	8,539	15,886	8,879	20,163
1949	500	7,558	—	—	5,140	8,248	5,640	15,806
1950	450	6,313	—	—	12,787	15,664	13,237	21,977
1951	450	6,885	—	—	15,038	23,881	15,488	30,766
1952	460	7,868	—	—	15,251	27,434	15,711	35,302
1953	837	11,895	522	5,006	9,844	24,265	11,203	41,166
1954	2,143	28,050	2,641	24,314	8,452	24,131	13,236	76,495
1955	2,362	26,168	6,125	58,296	5,485	21,798	13,972	106,262
1956	4,388	46,854	8,492	83,324	5,717	28,472	18,597	158,650
1957	14,788	134,942	10,773	105,602	5,765	31,807	31,326	272,351
1958	32,450	280,469	10,348	107,169	6,440	34,052	49,238	421,690
1959	34,271	310,484	8,493	89,830	4,023	19,673	46,787	419,987
1960	26,769	252,360	8,517	92,192	3,606	19,024	38,892	363,576

— Represents zero.

[1] Countries which could not be identified for publication purposes.

Example 5.3

SOURCE: Foreign Commerce & Navigation of the United States, 1946–1963. U.S. Dept. of Commerce, Bureau of the Census, Washington, DC 1965.

(16), 6 reported excellent results; good, 4; fair, 4; poor, 2. Of those with mild symptoms (10), 3 reported excellent response; good, 2; fair, 1; poor, 4.

Make a table that summarizes these data. Note that such a table frees the writer to discuss important findings in the narrative, without dragging the reader line-by-line through the data.

4. From the data in the table above (Example 5.3), make a line graph of quantity of uranium imported versus year, with separate symbols and lines for the three different sources. Then try representing the same data with a bar graph. Which is more effective?

GATHERING THE DATA

PART III

CHAPTER 6

Interviewing

Gathering information is at the heart of most technical writing jobs.

Technical materials are topical. In fast-changing areas—and that includes most of American technology—the library won't have all the answers.

True, when you begin a project a part of the job will be locating, reading, and mastering many documents. But some of the information will not be written down. It will exist only as oral history, and you will find yourself learning how to collect it through interviewing.

"Isn't interviewing just for journalists?" a student asked. Not today. One of the best ways—often the *only* way—to explore and develop technical information is through skilled interviewing.

EXAMPLES OF INTERVIEWING AT WORK

• George Ford inherits a new project at work—someone's incomplete documentation of a project. He collects the materials—specifications, flowcharts, and other papers—and studies them. The specifications and flowcharts are already dated, and many of the notations are cryptic. The solution? Hours of interviews with the people who have the answers—in this case, the programmers and engineers whose work Ford has to document.

• Florence Perigini is a part-time business student who works for a life insurance company. The company has hired a subcontractor to administer requests for one form of group insurance, and the subcontractor needs a set of printed procedures. Florence, who is assigned this job, studies other company manuals to see which sections she can use as "boilerplates"—standard informative paragraphs that can be moved intact from one document to the next. She finds no information on the newer ways developed in the past year for checking enrollment requests. To get the information, she interviews the managers who direct this job, elicits a procedure from them, and turns the interview notes into a document. She circulates the procedure, the managers respond with corrections and additions, and she revises until she has an up-to-date procedure for checking enrollment requests. All the primary information comes from interviews.

• Larry Bright is an engineer working for the central research laboratories of an international corporation. Like many other research facilities, this one often finds itself in a defensive position: In the eyes of many it produces no immediate product, and spends money instead of making it. A good way to counter this problem is for the lab to involve itself constantly in the essential business of the operating plants, to make itself useful to the corporation by solving real-life production problems. To this end, the lab dispatches Larry and his team to the company's largest manufacturing operation. Their assignment: to find ways to employ the lab's engineering and research expertise on behalf of the plant. The answers they seek are not in books. Instead, they schedule a series of interviews throughout the plant, looking for ways to involve themselves in plans for computer-integrated manufacturing. They sense that it is long overdue, and that it has been delayed because of little or no expertise in computer systems. They soon learn in interviews that this is, in fact, the case. The report they file is based entirely on data obtained in interviews.

ANATOMY OF A TECHNICAL INTERVIEW

Step one: Do your homework. The secret of good interviews is preparation. Read before you listen, and your ability to make sense of what you hear will improve dramatically.

For instance, if you are interviewing and the topic is technical, inevitably your subjects will use one new term after another. In such cases, you have to stop the speakers, or you will soon find yourself at a loss.

But you can minimize these interruptions by doing some background reading well before the actual interview. If your job is to document procedures for one kind of group insurance eligibility, for instance, read related procedures for background information. Read letters, proposals, and memos that describe the overall insurance program and its goals. If appropriate, take off for the library and track down related articles.

The more demanding the subject, the more reading you should do *before* the interviews.

Step two: Use a professional approach when you ask for someone's time. If you already know the people you are interviewing, getting in the door is no problem. Getting an interview with a stranger, though, is another matter. Ms. X will probably talk with you—but only if she's sure you aren't wasting her time.

As a rule, people are willing to talk about their professional experience and background, and to share their technical expertise with someone who has a legitimate need to know. But they have to be persuaded that your need is legitimate, that their time and attention are not being wasted.

You can do this in part by taking advantage of all that reading you did before asking for the interview. A good background gives you the language to frame your request so that you are specific instead of vague, focused instead of diffuse. If, for instance, you are asking for an interview by telephone, you'll be prepared by knowing the person's name, title, area of expertise, and specific places where he or she might be helpful. Instead of saying, "May I speak to someone about nuclear magnetic radiation devices?" you can say, "May I speak to Peter Mills?" When he comes on, you can say. "Mr. Mills, my name is . . . and I'm preparing an overview on NMR for a publication that goes to 4000 professionals in the imaging field. I understand you are familiar with spin-spin rotation. I'd appreciate some background information on how. . . ."

In other words, be ready to summarize clearly your reasons for the call. You must get down to business quickly, before your time is up.

If you are writing for an appointment, your query letter should be clear and to the point. Follow up with a telephone call.

Step three: Prepare a question list. You'll need a set of questions, some of which you will use to get your subject started, others to keep him or her on track. But when the time comes for the interview, be prepared to abandon many of your carefully thought-out gambits.

Dick Cavett, the talk show host, has a fund of stories about interviewing. One story he told during a broadcast concerned an inept interviewer. In this instance, the interviewer did his homework: He read widely and well before the appointed time, and made up a canny list of questions. Unfortunately, he had a tendency to cling to the security blanket of his question list, regardless of what his subject was saying. This was the result:

> Interviewee: Yes, it was a memorable experience. I found myself lifting the lid of the truck slowly. It rose gradually, the hinges groaning. And there, lying deep within the darkness, was . . . . . .
>
> Interviewer (obliviously): And what hobbies do you pursue when you're not at work?

The moral? Make up a list of questions, but be prepared to abandon them to follow conversational opportunities as they arise.

Step four: Collect and prepare equipment. Are you taping the interview? There's no substitute for the verbatim quotation that livens a report, or the security of being able to check for the exact wording. A tape is the route to word-perfect quotations for those among us, and our numbers are many, who've never mastered either shorthand or a homegrown form of speedwriting.

If you do decide to tape, there are some procedures to bear in mind.

- **Ask first.** Don't tape a phone interview and then afterwards announce, "Oh, by the way, I taped this. I hope you don't mind." They very well may. Ethics demand that you ask *at the beginning*. Some people will simply and flatly say *no*—particularly when you are doing the interview on the telephone and can't establish face-to-face rapport.
- **Do not rely on the tape.** Take notes, too, however raggedly. There are several reasons to write while you are interviewing. If you are doing an in-person interview, the sight of the tape recorder doing its job may make your subject nervous. Place it to the side (but not out of range) and start writing. Note-taking is usually a lot less

threatening to the subject, and at the beginning you will want to relax your subjects and get them talking.

You should also keep your pencil busy when you are doing a taped telephone interview. Taping is a process that is far from foolproof. Just as your subject is explaining a particularly mysterious diagram, the tape may run out, or break. By the time you put in a fresh cassette, you will have lost what the subject was saying. If you're taking notes, you can simply keep writing until the next conversational lull, and then replace the tape.

Notes are also a way of reviewing the interview quickly. Sometimes you may have the luxury of a transcription, but often you will find yourself writing from the tape itself. If the tape is a long one, you will want to be able to fast forward it to find a key quotation, an important idea, an elusive technical detail. Notes will tell you where on the tape to look.

If you are taping on the phone, an inexpensive coupler will do the job; the suction cup is attached to the telephone receiver, the lead to the tape recorder. Other kinds are available, as well.

If you are taping in person, bring lots of spare batteries and, just in case, a long extension cord.

Step five: A professional demeanor will help set your subject at ease. Remember, all the people you want to interview have something else they would just as soon be doing. These people need to sense that their time is not being wasted.

Whether by phone or in person, use the opening minutes to relax your subject. Show that you have done your homework, that you have a highly focused, coherent reason for being there.

The seriousness and care with which you prepare for the interview, and show this preparation from the very beginning, will help to reassure your subject.

If the interview is face to face, be sure to *look* professional. Could someone be so petty as to refuse to talk with you because of dandruff on your highly educated shoulders? You bet. Tidy up. Avoid any clothing or accessories that may backfire.

Once your professionalism has been established and civilities have been exchanged, you will be in business.

Step six: Pursue specific information. You will probably need the answers to difficult questions. But people are not always able to supply the answers, even when they know them.

Why? Some of the people you interview will be monosyllabic. They fit the old joke:

"Do you always answer questions with one word?"

"Yes."

"Are there any exceptions?"

"No."

You have to get these heirs to Calvin Coolidge talking. The secret is in the kinds of questions you ask. Avoid asking closed-ended questions of the taciturn. For instance, if you are interviewing users of a manual to find out what improvements need to be made, and one of your subjects is taciturn, the closed-ended question, "Have you had any bad experiences using the present manual?" will net you a "Yes," and no more. "Do you have any examples?"—another closed-ended question—may merit a simple, "No." Instead, try the open request, "Tell me about your worst experience with the manual," or, "Tell me about the time the manual was the most irritating."

There's a story of a woman who once made a bet at a Washington, D.C., dinner that she could get more than two words out of President Calvin Coolidge. She went up to Coolidge and explained the bet. "You lose," said the President.

Open-ended questions are good for expanding information, for getting people talking. Closed-ended questions are good for following up on specific details once the subject has talked.

If you want to avoid influencing suggestible subjects, keep away from leading questions. "How would you compare the quality of the two keyboards?" will get you a straighter answer than "Have you found that the Brand X keyboard is inferior to NOC's keyboard?"

Do not ask two questions at once, such as "What are your best and worst experiences with the keyboard?" The subject may answer the first question and forget the second one. Ask one question at a time.

Fish for anecdotes. Try for stories that will illuminate. To get people talking about specifics, try superlatives—the worst, the best, the funniest, the most serious—to spark their memories and get them talking. You can refine the information later.

There is also the other extreme in subjects, the person who talks and talks—but not about what you need to know. Once people like

Closed-Ended Question	Open-Ended Questions
Have you had any bad experiences with the manual?	Tell me about your worst experience with the manual. Tell me about the time the manual was most irritating.

Box 6.1

this are launched, they are capable of talking for twenty minutes about something that is fascinating but irrelevant—and then rising to terminate the interview with a brisk "Time's up. I really enjoyed this, though."

To direct a wandering respondent, try waiting for the first natural pause, then use a summary push: Wrap up whatever he's told you, no matter how irrelevant, and then turn the discussion in the necessary direction. Brief summaries are also useful when the respondent breaks off to answer a telephone call, and then turns back to you blankly with a "Where were we?"

Sometimes the respondent answers a question, but not the one you asked. Don't despair; rephrase the question and try again. Fortunately for the interviewer, most people fall in the middle of these extremes. After a few false starts, they open up as they talk about their professional ideas and problems. They are a goldmine of information.

Step seven: The last question should be, "Is there anything else you'd like to add?" Respondents who are relaxed, who feel the interview has gone well, will add useful information. Riding the easy current of an anecdote, they may provide examples, stories or insights that will help you tremendously when you sit down to write the report, memo, article, or manual.

Before you leave, make sure the lines of communication are open. Should the tape break or your notes prove incomprehensible, a follow-up call is indispensable. Even without a foul-up, though, you may want to talk to them again when the story is written, either to check facts, or to get additional details.

If, for instance, you have written a procedure for requesting additional evidence of insurability, you will probably want to send copies of the procedure to the people you interviewed. They, after all, are the experts whose technical knowledge you collected and used as the basis for the document. They will probably have many valuable comments, additions, and deletions, all of which will work to improve the accuracy and completeness of the document.

Most journalists object vehemently to having their draft copy read and edited by people they have interviewed. They consider it a form of prior censorship. Technical writing is a different matter, though.

You will probably find that input from the people you interview is quite valuable. They can become your fact-checkers and reviewers, and help you to produce a more accurate document. "I'd be delighted if you could check the draft copy for any additions or corrections," is usually a wise way to close an interview in which you are collecting technical information.

SUMMING UP

- Skilled interviewing is not just for journalists. It is one of a technical writer's three basic ways—along with the library and direct observation—to gather information.
- Prepare for the interview. The more demanding the subject, the more reading you need to do before you can listen well.
- Use a professional approach when you ask for someone's time. People always have something else they would just as soon be doing. You must persuade them yours is a legitimate use of their time. People are usually willing to share their professional knowledge and experience—but only if they are reassured their time and attention are not being wasted.
- Make a question list, but do not be rigid in following it. Abandon the questions and follow opportunities as they arise.
- If you are taping, ask first.
- Do not rely on the tape. Take notes, too.
- In the opening part of the interview, set the subject at ease by showing you are serious and prepared in your questions, and have a focused reason for being there.
- Use open-ended questions to get people talking, and closed-ended ones to follow up on specific details.
- End with, "Is there anything you'd like to add?" Riding the easy current of the interview, your subject may come up with invaluable anecdotes and insights that make all the difference.
- Leave the lines of communication open. You may want to talk to your subjects again, either to check facts or to get additional details.

EXERCISES

1. Spend about half an hour talking with the owner of a small business such as a photo shop, a graphics studio, a garage, or a pizza restaurant. Be sure that the person you interview is a stranger, not someone you already know, and that you are talking with the boss, not an employee.

Find out what the proprietor considers the worst business problem, and what he or she is doing to try to cope with it. Do not settle for platitudes and generalizations. Get details and examples. Make sure you understand the problems the owner has and how they are hurting the business.

Write an organized summary of the conversation.

- Do not try to write a polished article or story.
- Do not submit a verbatim transcript.

Include full-sentence direct quotations, partial quotations, and paraphrases.

2. Pick a recent technical innovation that interests you, perhaps in computing, compact discs, automobile fabrication, gene splicing, or high resolution graphics. If your interest lies in business or finance, consider recent changes in the tax code and how these changes have affected one area—for instance, energy credits or individual retirement accounts.

Do a roundup story on the latest in this area, gathering most of the information by telephone interview.

Include direct quotations as well as paraphrases. Attach to the story a list of the names, telephone numbers, and titles of the people you interviewed.

3. Cover a trade show or exhibition that occurs within the next few weeks in your city or town. You might investigate a home show or a boat show, an exhibition of vintage cars, or a farm equipment fair. Interview a range of exhibitors. Write a report summarizing strengths and weaknesses of the show, drawing on your interview material.

4. Do an extended interview with someone whose accomplishments in business, industry, science, or technology you think are valuable. Prepare by studying the background of the subject: If you've chosen a corporate figure, read about company as well as individual accomplishments; if you've chosen a professor whose research you admire, request reprints of papers, scour the library for related publications, and ask the librarian to help you do a computer search for publications by the faculty member. Allow at least an hour for the interview.

You'll find examples of extended interviews, known as profiles, in publications from *The Wall Street Journal* to *The New Yorker*. A short example is provided by the following profile of Phillip Morrison that appeared originally in *Physics Today* (Example 6.1).

Philip Morrison—a profile

Valued for his scientific contributions to the Manhattan Project, to theoretical physics and to astrophysics, he has also contributed to the public understanding of science and has been one of the most thoughtful advocates of arms control.

When Philip Morrison, Institute Professor at MIT, came to the Polytechnic Institute of New York recently to give the Sigma Xi lecture, a diverse group attended. The group included physicists, chemists, engineers; people who admired Morrison for his sustained fight against red-baiting in the 1950s (in 1953 a national newsletter called him "the man with one of the most incriminating pro-Communist records in the entire academic world"); and people in the humanities who had enjoyed his book reviews, films, articles and textbooks. The diversity of the audience reflected the diversity of Morrison's career.

Morrison is valued in the scientific community for his gift of language, for his wide-ranging intellect, and for his ability to pull together insights from different fields to shed light on a subject. Because he has spent considerable time writing about science—explaining and interpreting it for the public—he exists also in the imaginations of people outside science. He possesses what historian Alice Kimball Smith has called[1] a "rare sensitivity of spirit."

His career has included Los Alamos and Hiroshima in the 1940s, McCarthyism in the 1950s, the Peace Movement in the 1960s, and arms control from 1945 to the present. It began in Pittsburgh where he was reared and attended Carnegie Tech. After an initial interest in radio engineering, he majored in physics, and went on to do his doctoral work in theoretical nuclear physics with Oppenheimer at the University of California at Berkeley. They got along well; Morrison admired Oppenheimer and reminisces today about him: "There was only one difficulty most of us had with Robert. You had to be very careful with him, you couldn't give him too much of your problem, or he would solve it before you."

The Manhattan Project

Morrison had just gone to the University of Illinois at Urbana when the war broke out. Hired by the Manhattan Project, he went to Chicago to work with Fermi, and stayed there until 1944. Morrison became leader of the group that tested neutron multiplication in successive design studies for the Hanford reactors.

Then, in 1944, he was recruited for the Los Alamos effort by Robert Bacher. Morrison worked at Los Alamos in the group headed by Robert Frisch, who, with his aunt Lise Meitner, had discovered fission a few years earlier. His job at Los Alamos was to extend work done at Chicago at which he was expert. "We made small critical assemblies to test the neutron behavior of the new plutonium and uranium fission materials being produced at the main plants and shipped to Los Alamos, in preparation for use in the two bombs. Our job was to study chain reactions in that stuff."

It was here that Morrison and his group did

Example 6.1

the famous experiments later characterized by Feynman as "tickling the dragon's tail." "No one had ever made a chain reaction that had so many prompt neutrons in it," Morrison comments. "All the chain reactions of reactors are mediated in part by delayed neutrons; otherwise they aren't controllable at all. The bomb, on the other hand, is made by fast, prompt neutrons, which of course are uncontrollable."

Morrison was concerned with building up experience on the passage from the controlled state to the uncontrolled state. This meant keeping the reaction in a partially contained state under active control, instead of relying on the inherent stability of the system. "We moved the system so carefully, but so rapidly, that it had no chance to build up on us—we hoped. We came very close to making explosions, stopping just in time. Feynman said this was like tickling the tail of a dragon, and so it was."

In *Disturbing the Universe,* Freeman Dyson characterizes[2] the spirit of Los Alamos as the "shared ambition to do great things in science without any personal feeling of jealousy." Morrison says that for himself the motivation was not science, but victory over the Germans.

> In my group, two people died, We had the feeling of front-line soldiers with an important campaign at hand.
>
> To begin with, we felt we were well behind the Germans. Rightly or wrongly, we were seized by the notion of this terrible weapon in the hands of the Germans, whose scientists we respected, admired, and feared greatly because they had been the teachers of our teachers and colleagues.
>
> We felt ourselves a little like the English in 1940—a small band standing in the way. Could we possibly beat them? At first there was this terrible responsibility, and then in the end we became more and more flushed with the fact that we had overcome them. But it wasn't a question of science. It was one of victory. I remember very well.

Morrison conveyed this atmosphere to us with a story of John Wheeler in Chicago: "When noontime came and the 12:00 o'clock bell rang, most of us would go to lunch at the nearby cafeteria. We'd learned, though, not to bother Wheeler. He brought his lunch and when the bell rang he took it and his Princeton notebook out. Then he went ahead to do what he regarded as his 'real work'. He was so conscientious he would never do this during work hours, only during lunch. And that was the attitude at Los Alamos as well."

The absorption in the immediate task was complete. Only as work on the bomb drew to a climax did Morrison consider how it would be used against the Japanese. "We knew there would have to be a trial, but we thought suitable conditions could be made. For instance, I thought, as did many other people, that there was going to be a warning." But no explicit warning was given. The bomb was tested at Alamogordo 16 July and used on Hiroshima 6 August. Morrison says, "The military authorities rejected any demonstration as impractical. They felt Japan would not be deterred by the sight of a patch of scorched earth in the desert. The military had made up its mind. It would have taken a very powerful political presence—one that wasn't available—to sway them. The United States therefore gave no explicit warning. I think this was a moral failure."

Was Morrison surprised the scientists at Los Alamos were not more concerned with the implications of the bomb they were building? "Not at all. There was much discussion about this in the labs, quieter, of course, than those at the Met Labs in Chicago. But we were seized with a terrible responsibility, and our leaders were trying to make sure our attention was not diverted."

After the Trinity test, Morrison, who had

Example 6.1 (*continues*)

been responsible for the design and final deployment of the plutonium core, again prepared and packed the equipment, this time to go to the Mariana Islands. When the bombs were dropped on Japan he was on the island of Tinian, from which the planes for both atomic attacks set off.

He was among the first Americans to visit Hiroshima after the war. "I had earlier decided that the most useful thing one could do would be to try to go through the entire process as a historical witness." At the invitation of General Thomas Farrell, assistant to General Leslie Groves, Morrison joined the 12-man group that went to Hiroshima just 31 days after the explosion to determine the effects of the atomic bomb released by the *Enola Gay*. They arrived in Yokohama the day after MacArthur, and followed him to Tokyo. "For me," Morrison said in an interview[3] with Daniel Lang, "The first and main impact of Hiroshima's destruction had come . . . when we were flying down there from Tokyo. First we flew over Nagoya, Osaka, and Kobe, which had been bombed in the conventional manner, and they looked checkered—patches of red rust where fire bombs had hit intermingled with the gray roofs and green vegetation of undamaged sections. Then we circled Hiroshima, and there was just one enormous, flat, rust-red scar, and no green or gray, because there were no roofs or vegetation left."

Morrison walked through the city with Geiger counters and Lauritzen electroscopes and aided by an interpreter, a guide, and a policeman. "It had burst precisely at the spot we wanted it to, high over Hiroshima. There had been a minimum of radioactivity."

Arms Control

After the war, Morrison returned to the US to find himself at the heart of the movement for international arms control, whose advocates operated in diverse ways—in arenas ranging from guarded offices to hearing chambers and press conferences at the Senate Office Building; dispensing the message through coded teletypes and rushed press statements; disputing with colonels and reconnaissance experts; persuading congressmen and reporters. A large number of concerned scientists—many of them organized into groups such as the Manhattan Project Scientists, the Association of Los Alamos Scientists, the Association of Oak Ridge Scientists, Atomic Scientists of Chicago—met in Washington in the fall of 1945. Out of this meeting the Federation of American Scientists was eventually formed. The Federation began operating in January of 1946, with Morrison as a member of the administrative committee.

Morrison described their original goals to us:

> We—the people the press soon characterized as atomic scientists—wanted to turn over technical details of bomb production to a world authority under adequate controls. We sought to prevent a nuclear arms race by establishing this worldwide authority.

The Federation believed that a continuing monopoly of the atom bomb by the United States was impossible. Without staff or salary, Federation members worked in Washington preparing reports on how to establish a worldwide atomic authority.

> It seems to me that one finds in the story two distinct ways of meeting the sense of responsibility—indeed, of grave duty—that the Manhattan-project scientists as a whole felt then and feel still.
>
> One of these is the way of the "insider." Oppenheimer—lucid, persuasive, wonderfully analytical—worked in secret with generals and diplomats, trying in a thousand ways to demonstrate what

the facts implied. Szilard lived by the phone, buttonholing lobbyists and becoming himself the lobbyist *par excellence*. Both men acted inside the government, personally bringing their schemes before the individuals who had power, who wrote and passed laws.

And then there were the rest of us: younger, less famous and less able. Ours was the way of the dissenter. In the way we acted there was a sense less of knowledge than of commitment. William Higinbotham, Joseph Rush, Louis Ridenour, John Simpson and scores of others in Washington spoke and wrote publicly for 3000 scientists back home at the project laboratories or crowding back into the universities, and also for the physicists and chemists who had not been in the project at all but felt about as we did. From shabby rented offices overcrowded and littered with mimeographed statements and pamphlets the 'atomic scientists' floated in the eddying stream of American public opinion.[4]

Morrison comments that "mutual deterrence was not the vision of 1946. The scientists sought true stability then, not metastability, not the top-heavy balancing rock on which we all breathlessly sit."

Morrison played many roles during the period, roles that called both upon his fertile mind and upon his considerable ability as a speaker. He worked for the Bulletin of the Atomic Scientists, composed FAS policy drafts, and appeared as a principal witness at hearings on atomic bomb policy. He worked on a report of ways to detect atomic bomb laboratories, testing sites and assembly plants. But no matter how carefully he and others stressed how an international authority could operate under adequate controls—indeed, no matter how many times they explained what they meant by "under adequate controls"—they were accused of wanting to give away the bomb.

The arms race that Morrison had predicted grew as the scientists' movement for international controls waned after 1946. Morrison, who joined the faculty of Cornell University in 1946, remained in the fight for international arms control even as the public acclaim for scientists began to ebb.

McCarthyism

He was soon in need of defense himself. As an undergraduate at Carnegie Tech Morrison had joined the Communist Party, and he remained a member when he went to graduate school at the University of California at Berkeley, a school known at that time for its free-thinking, socialistic atmosphere. By 1941, Morrison was out of the party, but his political activities continued. At Cornell, he was deeply involved in the Peace Movement and in a variety of radical intellectual activities. It was not the involvement that was so noteworthy as much as the level of activity: a continuous string of speeches and appearances made Morrison one of the most politically active scientists throughout the fifties.

During this period there were many attempts to fire him. "What has Cornell University done about Morrison?" the right-wing newsletter *Counterattack* asked[5] in March 1953, answering "Nothing!" In part the attempts were foiled by his situation, because, as a private school, Cornell was not quite as vulnerable to pressure as public schools. Nonetheless, considerable forces were exerted on Cornell, where his promotion from associate to full professor was held up for so long that the Physics Department began to talk of refusing to submit any further proposals for promotions until Morrison's was acted on.

His promotion finally became an issue before the Cornell Board of Trustees, who had him summoned. Even in those times, with

Example 6.1 (*continues*)

Morrison the center of a series of attacks for such charges as "urging clemency for the Rosenbergs," the trustees were charmed by Morrison's intelligence and grace; they granted his promotion.

Morrison was also called before Senator William Jenner's Internal Security Subcommittee, where he talked frankly about himself and his early involvement with the Communist Party without naming other names; unsatisfied, the subcommittee continued to pry. For instance, they summoned another physicist for a special security clearance. This physicist was surprised but somewhat flattered to be called for special clearance. When he got there, he was taken aback to discover the committee had no interest in him; they were only using the occasion as an opportunity to pump him about Morrison.

Morrison spent 19 years on the Cornell faculty before going to MIT. At Cornell, Morrison was famous not only for his social activism but also for his teaching. "Phil's a born teacher," Dyson, who was a colleague of Morrison's at Cornell, comments. "Whenever one didn't know what to do with a student, one sent the student along to Phil. He had an infinite supply of patience." Dyson says that it often seemed as if half the graduate students in the Physics Department were taken care of by Morrison, who spent hours talking to them, finding out which research ideas they could tackle.

Astrophysics

It was while Morrison was at Cornell that his interest turned from theoretical physics to astrophysics. "I was always rather interested in astrophysics," he recalls. "As a graduate student I published several small papers in nuclear astrophysical problems with Oppenheimer. At Cornell, though, I was actually trying to be a nuclear physicist until I took a sabbatical leave in 1952."

While on leave, Morrison determined to work on some of Bruno Rossi's problems; he knew Rossi's work from their days together at Los Alamos. "Along with many other scientists in the cosmic-ray domain, the early 1950s found me pushed into astronomy. The cosmic-ray people had always used this natural phenomenon as a source for high-energy particles—mesons were first discovered in cosmic rays—but in the early fifties machines became powerful enough to rival cosmic rays. Then, as machines improved, the cosmic rays were simply outcompeted. So cosmic rays were no longer of central interest from the point of view of their intrinsic physics; the interest was more in where they came from, first considering possible sources within the solar system, and then beyond. That interest gradually drew me and other scientists farther and farther into astronomy." He is pleased with the work he did on the origin of cosmic rays. "I do consider it as rather a high point. I regarded myself as a specialist in cosmic rays during the 1950s. At that time I proposed no single origin for them, but instead suggested they were highly hierarchical." Morrison argued that different places make different cosmic rays and that the highest energy concentrations might come from quasar-like objects such as the nearby radio galaxy M87.

At Cornell Morrison worked with Hans Bethe, a long-term friend and supporter. In 1956 they wrote a textbook together, *Elementary Nuclear Theory*. "It was a useful and happy collaboration," Bethe says today. "He has ideas which are not obvious. His genius is to connect many different parts of physics." As an example, Bethe cites Morrison's discussion of the radiogenic origin of the helium isotopes in rocks. Morrison argued that the ratio of helium-3 to helium-4 is much greater in the atmosphere than it is in rocks, because in rocks helium-4 comes mainly from radioactivity, whereas in the atmosphere there is relatively more helium-3 produced by the cosmic ray-mediated disintegration of nitrogen. "It is a typical insight of Philip's to connect two opposite things—such as cosmic rays and terrestrial radioactivity—to determine the composition of samples taken from such places as hot springs."

Morrison is known not only for his ability to connect disparate elements, but for his willingness to challenge assumptions. His interpretation of M82, once touted as an example of an exploding galaxy, is one instance of this characteristic. Morrison suggested that what we were seeing is not an explosion, but rather an intergalactic dust cloud through which the galaxy is passing, the interaction giving rise to features that one might interpret as an explosion. "Although M82 looks superficially as though it were exploding in a mini-quasarlike way," Morrison comments, "in fact it seems pretty clear it isn't at all." Instead of there being one point-like center—a tiny engine that does everything for the device—the central object is the whole core of the galaxy, thousands of light years across, in which hundreds, even thousands or millions of new stars are suddenly formed. "The rapid bursts of star formations can create in some ways the same kind of activity as if there were a quasar-like object. In this case, however, the energy is primarily nuclear instead of primarily gravitational."

Paul Joss, a theoretical astrophysicist at MIT, comments on Morrison's work: "Both with M82 and with his supernova model, Morrison proposed testable models that gave us something to attack, challenging us and forcing us to rethink." Morrison's supernova model is an attempt to account for the visible light that comes from supernovae "without worrying too much about the causes of the explosion." The central idea of his theory is that the observed light from the supernova consists of two portions: those photons that reach the observer directly along a straight line and those that interact at least once, traveling along a dogleg path. Because the original outburst is so brief, even the small delays that arise from the somewhat greater length of the dogleg path are significant. Simple geometrical arguments show that the locus of the secondary emission points (places where light from the supernova is absorbed and then remitted as fluorescence) form a sequence of expanding ellipsoids whose focal points are the point of the supernova outburst and the position of the observer. Because fluorescence efficiencies are typically a percent or less, the total energy of the explosion is from 100 to 1000 times more than can be detected on earth in the visible region.

Joss says, "Phil's work on supernovae is a very good example of his impact on astrophysics. He has a way of looking at fundamental assumptions and asking, 'Why do we believe this?' In supernovae, for instance, there was a standard picture, one that was probably right in a primitive sense, that is, supernovae result from violent explosions in massive stars, causing in turn both a very large expulsion of matter into interstellar space and a very large amount of electromagnetic radiation. But Phil noted that if you take a star the size of the sun and blow it up, you are not going to get a tremendous amount of visible light. The energy that comes out is 10^{10} or 10^{11} times the luminosity of the sun, and if it radiates as a blackbody, then that energy is not going to come out as visible light, it is going to come out as x rays. The expansion of the exploding material, increasing the size of the radiating surface, won't help either, because by the time the material has expanded as much as it has to—through several orders of magnitude times its original size—it will have undergone such adiabatic cooling it will hardly radiate at all. So the reason that one can see this visible light has to be more complicated. What Phil did was come up with a very specific model. It's been controversial, but that's not the point. It was a testable model that made specific predictions and challenged astrophysicists to reconsider some of their basic assumptions about the supernova phenomenon."

Example 6.1 (*continues*)

Teaching

Morrison has been at MIT since 1964, first as Francis Friedman Visiting Professor, and then as a permanent faculty member since 1965. Morrison's interest in educational theory influenced his move to MIT. "Gerald Zacharias invited me to the school. He had an intense interest in science education, an interest he knew I shared." MIT was a center of educational innovation, and Morrison was associated with the Physical Science Study Committee at its inception and coauthor of its secondary-school text *Physics*. Morrison, together with Don Holcomb of Cornell, also wrote a physics text for college students, *My Father's Watch*. Although not widely used, it had a special appeal to teachers introducing adults to physics, perhaps because of Morrison's care to relate scientific arguments to history, art and philosophy.

Throughout Morrison's career he has interpreted science for the public in popular articles, in science films, and in monthly book reviews for Scientific American. These book reviews in all fields of science are particularly well-known. One hundred years ago, Charles Darwin wrote[6] of the scientist Robert Brown: "He was rather given to sneering at anyone who wrote about what he did not fully understand. I remember praising Whewell's *History of the Inductive Sciences* to him, and he answered, 'Yes, I suppose that he has read the prefaces of very many books.' " Morrison is vulnerable to the same sneer, yet few would comment so of his incisive performances each month in Scientific American. Instead, one senses a polymath interested in every nook and cranny, as Morrison somehow makes his way through the 500 books he receives each month, choosing and then reviewing thoroughly the handful he selects as interesting and instructive.

"I judge my job to see what's inside, and then to unpack it. The nice part of the book-review column in Scientific American, and what makes it different from others, is that I don't need to review all the *important* books. I am not obliged to say, this is a lousy book but we have to review it because it is the work of an important author." Instead, Morrison tries to take a variety of books representing either a good popular approach or an approach at an introductory level. The reviews—serious, generous, often more entertaining than the original volumes—are a reflection of the intellectual energy that consumes Morrison; they are also the result of the peculiar ability he has to read almost as rapidly as he can turn pages.

Throughout Morrison's book reviews, books and films, there is a stress on the evidence rather than on neatly packaged conclusions or indeed on the personality of the presenter. "The key thing in a science film is to show the evidence," Morrison says, "but the media believe more in testimony and atmosphere." Morrison tells an anecdote to illustrate this conflict. In his film "Whispers from Space," which Morrison considers his best, he spends half the program establishing and demonstrating experiments that are at least 100 years old. For instance, to illustrate one of the most important features of blackbody radiation, the viewers see a kiln loaded with dishes and piggybanks. These gradually heat up until all detail is lost: first the dishes disappear, then the piggybanks, until the viewer is left with a bland, smooth space. When the executive producer saw the clip, he exclaimed, "You're spending all this time and money on a thing you tell me was discovered 150 years ago. We can't do that old-hat stuff."

Morrison comments, "So long as science is seen largely as a personal view, so long as science films have a speaker who mainly ignores the evidence and presents the history of science as his own concoction of ideas and insights, it is possible to talk of Bermuda triangles and flying saucers. It's good enough if someone says it. If you invent myths and don't explain, people can't test the foundations of your beliefs, or be prepared to change when the foundation changes. Then another myth comes along and beats your myth. That's how the creationists can come along with their demands for equal time: as far as they are concerned, it is myth against myth."

Essentially Morrison was a radical as a youth and remains that way today. His deep involvement in arms control extends from 1945 to the present. Two years ago he, his wife, and four Boston-area colleagues published *The Price of Our Defense: A New Strategy for Military Spending*. The book aimed at limiting the upward-spiraling arms trade and thus lightening what Morrison calls "the thermonuclear sword hanging over all mankind, sharper and heavier each decade." The authors take a look at how much the US needs to spend to maintain its national security, and propose a program for decreasing land, sea and air forces to give a "prudent military structure prepared for eventualities short of all-out nuclear attack." Against an all-out nuclear attack, the authors argue, there can be no defense; one must rely on deterrence alone.

How well has the book done? "The Pentagon was interested," Morrison comments. "It sold quite well in bookstores in Washington. It's also been popular with people in the peace movement. But we are in a period when the Russians are perceived as standing 10 feet tall. There are no signs that the government is considering the nuclear-arms cuts we proposed. In fact, it's quite the opposite." Morrison continues to act as a gadfly to the defense establishment with an energy characteristic of all his political struggles. One of his targets is the Air Force, which he says is on the edge of obsolescence. "Of course, it can't accept that, and so it tries harder. As the largest industrial organization in the world, it is up to all the sorts of things you would expect from a huge organization that cannot face its own obsolescence. The MX system is a perfect example; its chief value lies in its ability to keep the Air Force in the strategic-missile business."

Morrison continues to have a deep concern about nuclear weapons. "It is one of the great failings of the American political process," he says, "that there is a huge hue and cry against nuclear reactors, and nothing much about bombs. I think to some extent this had to do with displacement. People can't deal with bombs, and they displace their concern onto reactors, which turn out to be vulnerable objects. It's a most important phenomenon, the absence of attention to one, and the irrational attention to the other. But since the summer of 1981 I see a decisive change."

One of Morrison's most striking characteristics is the immense energy he has spent writing about science for the public. Why do this? "In part," he replies, "I think it is simply that I have a flair for it. But I imagine it's more than that. I feel very keenly an obligation to maintain the social nexus in which I've learned and become a scientist. The one obligation society makes on you is that you must explain your craft, because that is the cultural treasure you can pass on. People in the future will need the information."

References

1. A. K. Smith, *A Peril and a Hope: The Scientists' Movement in America, 1945–1947*, U. of Chicago P., Chicago (1965).
2. F. Dyson, *Disturbing the Universe*, Harper & Row, New York (1979).
3. D. Lang, *From Hiroshima to the Moon: Chronicles of Life in the Atomic Age*, Simon & Schuster, New York (1959).
4. P. Morrison, Scientific American **213**, September 1965, page 257.
5. "Counterattack: Facts to Combat Communism," 6 March 1953, American Business Consultants, Inc., 55 West 42 Street, New York.
6. C. R. Darwin, *Autobiography of Charles Darwin, 1809–1882*, Norton, New York (1969).

Using the Library

CHAPTER 7

Exploring a technical subject usually means exploring in the library, whether you need historical background to a problem or results of the most recent research in your field.

Like the other steps in the writing process, a library search is rarely a linear, straightforward journey. Some people begin by browsing through the shelves and card catalog, other by doing a database search of the literature within a topic. Some narrow their search through the use of indexes, abstracts, and bibliographies before they head for primary sources. Others who know key works in a field begin with the references or citations in these key works and use them to fan out. People do the job in different ways, depending on what they already know, and the type of search they are doing.

This chapter suggests a variety of strategies, depending on your background knowledge of the technical area, the type of report you are preparing, and the possibility of help from a professional librarian.

FINDING BACKGROUND INFORMATION

If you are starting from scratch, you will want to acquire a framework for the subject. The best path to take is usually from the more general reference sources to the more specific ones; the more specific references lead in turn to the primary documents—the actual reports, books, and periodical literature you seek.

Encyclopedias

An encyclopedia article may provide the initial framework for your search; such articles vary from the general to the highly specialized. If, for instance, you are looking into an application of polymers, you might start with an article in a general encyclopedia to place the subject within a larger framework and then move on to a more specialized encyclopedia, such as the *Kirk-Othmer Encyclopedia of Chemical Technology*. There you'll find excellent articles, heavily referenced and cross-indexed.

Led by the invaluable citations in the specialized reference, your search may then focus on primary sources. Here is a listing of some of these introductory references, arranged from the general to the specific.

GENERAL

Encyclopedia Americana. New York: Americana.

Encyclopaedia Britannica. Chicago: Encyclopaedia Britannica.

The Columbia Encyclopedia. New York: Columbia University Press.

The World Book. Chicago: Field Enterprises.

General Science and Technology

Harper Encyclopedia of Science. New York: Harper & Row.

McGraw-Hill Encyclopedia of Science and Technology. New York: McGraw-Hill.

Van Nostrand's Scientific Encyclopedia. New York: Van Nostrand Reinhold.

SPECIALIZED

Specialized encyclopedias discuss topics in science and engineering. As these change rapidly, many of the encyclopedias listed below are supplemented regularly with newer volumes. The text of most entries is arranged in a logical order: definitions, historical relationship to related topics, discussion, and bibliography.

Astronomy

Larousse Encyclopedia of Astronomy. New York: Prometheus, 1959.

Biology

The Cambridge Natural History. New York: Macmillan, 1895–1909.

Encyclopedia of the Biological Sciences, 2d ed., Peter Gray (ed.). New York: Van Nostrand Reinhold, 1970.

Encyclopedia of the Life Sciences. New York: Doubleday, 1964–1965.

Hall, E. Raymond, *Mammals of North America*. New York: Ronald, 1959.

Business

Encyclopedia of Business Information Sources, 5th ed., Paul Wasserman, *et al.* (eds.). Detroit: Gal, 1983.

Encyclopedia of Management, 3rd ed., Carl Heyel. New York: Van Nostrand Reinhold, 1982.

Chemical Engineering and Chemistry

Chemical and Process Technology Encyclopedia, D. Considine (ed.): New York: McGraw-Hill, 1974.

Encyclopedia of the Chemical Elements, Clifford A. Hampel (ed.). New York: Van Nostrand Reinhold, 1968.

Encyclopedia of Chemical Process Equipment, William J. Mead (ed.). New York: Van Nostrand Reinhold, 1964.

Encyclopedia of Chemical Technology, 3d ed., R. E. Kirk and D. F. Othmer (eds.). New York: Wiley, 1980.

Encyclopedia of Chemistry, 3d ed., Clifford A. Hampel and G. G. Hawley (eds.). New York: Van Nostrand Reinhold, 1973.

Encyclopedia of Polymer Science and Technology: Plastics, Resins, Rubbers, Fibers, Herman F. Mark and Norman G. Gaylord (eds.). New York: Wiley, 1964–1972.

Kingzett's Chemical Encyclopedia: A Digest of Chemistry and Its Industrial Applications, 9th ed., D. H. Hey (ed.). New York: Van Nostrand Reinhold, 1967.

Modern Plastics Encyclopedia. New York: McGraw-Hill, 1941–.

Computer Science

Encyclopedia of Computer Science and Technology, Jack Beizer (ed.). New York: Dekker, 1980.

Economics

Encyclopedia of Economics. New York: McGraw-Hill, 1982.

Engineering

Encyclopedia of Engineering Materials and Processes, H. R. Clauser *et al.* (eds.). New York: Van Nostrand Reinhold, 1963.

McGraw-Hill Encyclopedia of Environmental Science. New York: McGraw-Hill, 1974.

Materials Handbook: An Encyclopedia for Purchasing Agents, Engineers, Executives and Foremen, 10th ed., George Stuart Brady (ed.). New York: McGraw-Hill, 1971.

Newnes' Concise Encyclopedia of Nuclear Energy, D. E. Barnes (ed.). London: Newnes, 1961.

Geology

Larousse Encyclopedia of the Earth. New York: Prometheus, 1961.

Mathematics

Universal Encyclopedia of Mathematics. London: Allen and Unwin, 1964.

Medicine

U.S. Pharmacopoeia and National Formulary. Rockville, Md.: U.S. Pharmacopial Convention, 1982.

Physics

Encyclopaedic Dictionary of Physics, J. Thewlis (ed.). Elmsford, New York: Pergamon, 1961–.

Encyclopedia of Physics, 2d ed., Robert M. Besancon (ed.). New York: Van Nostrand Reinhold, 1974.

GUIDES TO THE LITERATURE

Beyond encyclopedia articles lies an array of information:

- references such as handbooks and dictionaries
- secondary sources such as indexing and abstracting services
- primary sources such as books and journal articles

Before you step into these specific areas, you might want to look through an appropriate guide to the literature of the field that interests you. There are guides to the literature of mathematics, computer science, nursing, management, physics, chemistry, biology, astronomy, medicine, and geology, among others. There are guides to government and industrial publications. Most of the specialized guides are annotated, with descriptive comments included for most items. As with encyclopedias ranked from general to specialized, literature guides vary from the general to the specific. If you know something about the subject, but are unfamiliar with how knowledge within the subject is ordered, a guide to the literature will be helpful.

Specialized Guides to the Literature

Chen, Ching-chih. *Scientific and Technical Information Sources.* Cambridge, Massachusetts: MIT Press, 1977.

Economics Information Resources Directory. Detroit: Gale, 1984–.

Herner, Saul. *A Brief Guide to Sources of Scientific and Technical Information,* 2d ed. Arlington, Virginia: Information Resources Press, 1980.

Hyslop, Marjorie. *A Brief Guide to Sources of Metals Information.* Arlington, Virginia: Information Resources Press, 1973.

McInnis, R. G., ed. *Research Guide for Psychology.* Westport, Connecticut: Greenwood, 1982.

Malinowsky, H. Robert, Gray, Richard A., and Gray, Dorothy A. *Science and Engineering Literature,* 2d ed. Littleton, Colorado: Libraries Unlimited, 1976.

Mount, Eli. *Guide to Basic Information Sources in Engineering.* New York: Norton, 1976.

Schlessinger, Bernard, ed. *Basic Business Library Core Resources.* Phoenix, Arizona: Oryx, 1983.

Simonton, D. P. *Directory of Engineering Documentation Sources.* Santa Ana, California: Global Engineering Documents, 1974.

Strauch, K. P. and Brundage, K. *Guide to Library Resources for Nursing*. New York: Appleton, 1980.

Walford, A. J. *Guide to Reference Material*, 2d ed. Volume 1: Science and Technology. London: The Library Association, 1966.

There are also excellent general guides to the literature and bibliographies:

Ayers Directory of Publications.

Bibliographic Index: A Cumulative Bibliography of Bibliographies. New York: H. W. Wilson, 1938–.

Books in Print. New York: Bowker, 1948–.

Cumulative Book Index. New York: H. W. Wilson, 1900–.

Directory of Information Resources in the United States. U.S. Government Printing Office.

McCormick, Mona. *The New York Times Guide to Reference Books*. New York Times Books, 1985.

Paperbound Books in Print. New York: Bowker, 1955–.

Sheehy, Eugene P., ed. *Guide to Reference Books*, 9th ed. Chicago: ALA, 1976–.

Ulrich's International Periodicals Directory. New York: Bowker, 1973.

DICTIONARIES AND HANDBOOKS

On the shelves of the reference section in the library you will find dictionaries, handbooks, and other basic references for technical subjects.

Dictionaries

While encyclopedias treat topics, dictionaries define words and terms. Technical dictionaries, available for each subject, include *Hackh's Chemical Dictionary* (McGraw-Hill), *Chemist's Dictionary* (Van Nostrand), *Chamber's Dictionary of Science and Technology* (Macmillan), *Merck Index of Chemicals and Drugs* (Merck), Penguin

Dictionary of Science (Penguin), *Stedman's Medical Dictionary* (Williams & Wilkins), and many others.

Handbooks

These invaluable one-volume references to a field abound. Filled with tables, charts, graphs, and detail, handbooks are updated regularly by experts. Handbooks exist within every subject; recent guides to the literature list at least 200 handbooks within the categories of engineering and science. Among the most noted are *The Handbook of Chemistry and Physics* (Chemical Rubber Co.), *Handbook of Chemistry* (McGraw-Hill), *Engineering Manual* (McGraw-Hill), *Chemical Engineers' Handbook* (McGraw-Hill), and *Handbook of Physics* (McGraw-Hill), *International Critical Tables of Numerical Data, Physics, Chemistry, and Technology* (McGraw-Hill). *Metals Handbook* (American Metals Society), *ASME Handbook* (American Society of Mechanical Engineers), *Civil Engineering Handbook* (McGraw-Hill), *Integrated Circuits Applications Handbook* (Wiley), and *Composite Materials Handbook* (McGraw-Hill).

PERIODICAL LITERATURE

Once you've arrived at the category of periodical literature, you will probably need the help of secondary sources—indexes and abstracts—to get started. The original three scientific periodicals—*The Philosophical Transactions of the Royal Society of London,* the *Journal des Scavans,* and *Acta Eruditorum,* all in print by the 1680s—have been succeeded by a vast and spiraling serial literature: more than 60,000 scientific periodicals were published between 1900 and 1960 alone, and their numbers continue to multiply.

To keep up with this literature, periodical indexes and abstracts have emerged in all technical fields.

Indexes give bibliographic citations with title, author, subject, and related information (Figure 7.1).

Abstracts give descriptive or informative summaries. These invaluable secondary sources are available both on the shelf for a hands-on search, and on-line for a computerized search (Figure 7.2).

The method or search strategy you use to comb the indexes and abstracts for essential primary sources will depend on what information you already have, the time and money at your disposal, the type of document you are preparing, and the sorts of material you need.

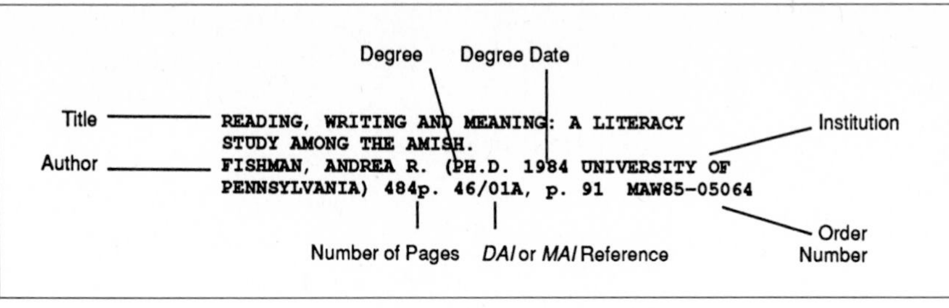

Figure 7.1 Indexes give bibliographic citations. This example is from University Microfilms International, which indexes masters' and doctoral theses. Courtesy University Microfilms.

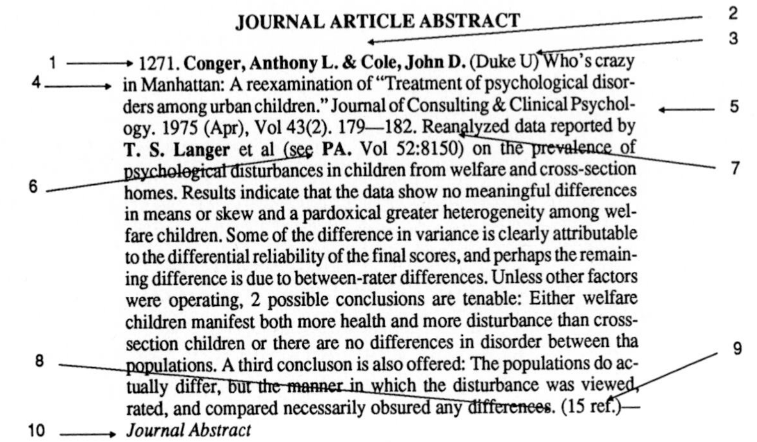

JOURNAL ARTICLE ABSTRACT

1271. **Conger, Anthony L. & Cole, John D.** (Duke U) Who's crazy in Manhattan: A reexamination of "Treatment of psychological disorders among urban children." Journal of Consulting & Clinical Psychology. 1975 (Apr), Vol 43(2). 179—182. Reanalyzed data reported by **T. S. Langer** et al (see **PA.** Vol 52:8150) on the prevalence of psychological disturbances in children from welfare and cross-section homes. Results indicate that the data show no meaningful differences in means or skew and a pardoxical greater heterogeneity among welfare children. Some of the difference in variance is clearly attributable to the differential reliability of the final scores, and perhaps the remaining difference is due to between-rater differences. Unless other factors were operating, 2 possible conclusions are tenable: Either welfare children manifest both more health and more disturbance than cross-section children or there are no differences in disorder between tha populations. A third concluson is also offered: The populations do actually differ, but the manner in which the disturbance was viewed, rated, and compared necessarily obsured any differences. (15 ref.)—
Journal Abstract

1 Record number.
2 Author(s) or editor(s). As many as four are listed; if there are more than four, the first is listed followed by "et al." Succession marks (i.e., Jr., II, III, etc) are not given.
3 Affiliation of first-named author/editor only.
4 Article title, including subtitles: If the original article was written in a foreign language, the original article title is given followed by a slash and the translated title as provided by the source document, PsycINFO's translation is given in brackets. The language of the original atricle is indicated in parentheses.
5 Primary publication title and bibliographic data.
6 Text of abstract.
7 Reference to a previous entry in **Psychological Abstracts.** If an abstract number is available at time of publication, the issue number is given. Consult the Author Index of the **PA** issue noted in order to determine the record number for the article.
8 Summaries included in the primary publication are listed when in language(s) other than that of the article.
9 Number of references is included.
10 Abstract source.

Figure 7.2 Journal article abstract from *Psychological Abstracts.*

This citation is reprinted with permission of the American Psychological Association, publisher of *Psychological Abstracts* and the PsycINFO Database (Copyright © 1967–1988 by the American Psychological Association), and may not be reproduced without its prior permission.

In general, you will want to develop a strategy based on whether

- you are searching a topic, and don't yet have an inclusive list of authors and their works
- you are searching for specific articles, and do know titles and authors

This section lists five strategies, each useful according to the particular circumstances of the search:

1. Computerized Search.
2. On-Shelf Abstracting and Indexing Services.
3. Manual Indexing and Abstracting Services
4. Science Citation Index.
5. Table of Contents Search.

Computerized Search

If you have a complete citation, it's possible to go directly to the shelves and retrieve the articles you want. If, on the other hand, you have a topic, but lack specific references, or have only a few references, the computerized search may be your next step. The advantages are tremendous speed and thoroughness; the disadvantages are expense and the need to word requests so that the resulting citations are not too broad.

Computerized retrieval systems offer databases that can quickly search the literature of business, chemistry, physics, engineering, the life sciences, geology, astronomy, and many other fields. These searches may cover not only periodicals, but new books, government reports, proceedings, patents, and technical reports produced both by industrial and government researchers.

Usually the librarian does the search, using information supplied by the customer, who is charged for the on-line time. Here are two examples of on-line searches, the first done with the help of a librarian, the second done solely by the student.

1. For a Business Law assignment, the student, Sandra Chen, was doing research on laser infrared tomography, a technique in which infrared light is used to probe the interior of human tissue or other material. She wanted to know what articles and patents existed on the subject. She made an appointment with the librarian, and, before she went, checked her budget: She decided to spend no more than $20. At the computer station in the library, the librarian entered the name of an appropriate database for the subject (Chemical Abstracts) and asked Sandra for key words. Sandra rejected a

combination of "infrared," "laser" and "tomography," as the combination was too restrictive: Many articles might exist that would use two of the words in the title, but not three. She decided on the key words "IR or infrared" and "Tomography." Figure 7.3a shows the printout. Line S1 shows 20,408 citations under the word "IR"; Line S2 shows 16,274 citations under the word "infrared," Line S3 shows that 281 entries appeared under the word "tomography." Two hundred and eighty-one were still far too many. Such a list would be both expensive and inaccurate: Its broadness would guarantee that most of the citations would be outside her area of interest. Therefore Sandra combined the key words "IR or infrared" and "tomography." This tactic was successful: The printout listed two citations.

She had narrowed the search from over 20,000 items to 2. The librarian printed the two citations (Figure 7.3b), Sandra was charged $20 for the on-line time, and the search was done.

2. George Lorca did his own search on a computer equipped with a modem and connected to a search service, Dialog Information Retrieval Services. One of the many databases Dialog searches is Chemical Abstracts. George entered the system, using a password and series of commands, and asked for information on nonlinear optical properties of diacetylenes. The screen showed him there were more than 900 citations. He narrowed the search to a keyword combination of nonlinear and diacetylenes and found the actual citations he needed. There were five. He asked for a printout of citation plus abstract. The cost was about $20. Part of the printout appears in Figure 7.4.

Dialog is one of many commercial search services available. The scientific and technical data bases served by Dialog are listed below. Their descriptions will give you an idea of the breadth of data available through computerized indexing and abstracting services.

BUSINESS

MOODY'S CORPORATE NEWS Business news and financial information on U.S. Corporations.

STANDARD & POOR'S CORPORATE DESCRIPTIONS Corporate descriptions of 7800 publicly held U.S. companies.

CHEMISTRY

CA SEARCH Source information and Chemical Abstracts Registry Numbers for all documents covered by the Chemical Abstracts Service, including patents, reviews, journal articles, reports, books, dissertations, and proceedings.

Figure 7.3a Printout of literature search.

```
File 311:CA SEARCH
(Copr. 1985 by the Amer. Chem. Soc.)

Set   Items   Description

S1    20408   IR (INFRARED)
S2    16274   INFRARED
S3      281   TOMOGRAPHY
 4        2   (IR OR INFRARED) AND TOMOGRAPHY
```

Figure 7.3b Printout of citations.

```
4/2/1
103029287    CA: 103(4)29287e    JOURNAL
Light scatttering tomography
AUTHOR(S): Moriya, Kazuo
LOCATION: Res. Lab. Electron. Mater., Mituse Min.
and Smelting Col, Ltd., Ageo, Japan, 362
JOURNAL: Kogaku  DATE: 1985  VOLUME: 14  NUMBER: 2
PAGES: 97-107  CODEN: KOGAD5  ISSN: 0389-6625
LANGUAGE: Japanese
SECTION:
CA173000 Optical, Electron, and Mass Spectroscopy,
and Other Related Properties
CA175XXX Crystallography and Liquid Crystals
IDENTIFIERS: light scattering tomog review,
crystal defect light scatttering review, IR
scattering tomog review, luminescence tomog review

4/2/2
99046204    CA: 99(6)46204s    JOURNAL
Observation of lattice defects in gallium arsenide
crystals by infrared light scattering tomography
AUTHOR(S): Moriya, Kazuc; Ogawa, Tomoya
LOCATION: Dep. Phys., Gakushuin Univ. Mejiro,
Tokyo, Japan, 171
JOURNAL: Jpn. J. Appl. Physl, Part 2  DATE: 1983
VOLUME: 22  NUMBER: 4  pages 207-9  CODEN: JAPLD8
LANGUAGE: English
SECTION:
CA175003 Crystallography and Liquid Crystals
CA176XXX Electric Phenomena
IDENTIFIERS: lattice defect semiconductor IR
tomog, microdefect silicon IR scattering tomog,
dislocation gallium arsenide IR tomog
```

```
TI Nonlinear optical materials and processes employing dialectylenes
AU Garito, Anthony Frank
CS University Patents, Inc.
LO USA
PI Eur. Pat. Appl. EP 21695, 7 Jan 1982, 75 pp.
AI US Appl. 52007, 25 June 1979
CL G03C1/68, C07C011/22, C08F38/00, G02F1/00
SC 73-2 (Spectra by Absorption, Emission, Reflection, or Magnetic
Resonance, and Other Optical Properties)
SX 76
DT P
CO EPXXDW
PY 1981
LA Eng
AN CA95 (6):52501v
AB Novel nonlinear optical, piezoelec., pyroelec., waveguide, and
   other materials are disclosed, together with processes for their
   employment and articles formed from them. Such materials, proces-
   ses, and articles comprise diacetylenes and polymers formed from
   diacetylenic species, the polymers being amenable to close
   geometric, steric, structural, and electronic control. According
   to a preferred embodiment of the invention, diacetylenes which are
   crystallize into crystals having a non-centrosym. unit cell may
   form crystals or may be elaborated into a thin film upon a sub-
   strate by the Langmuir-Blodgett technique. Such films may, option-
   ally, be polymd. either thermally or by irradn. for use in
   nonlinear optical and other systems. According to other preferred
   embodiments, diacetylenes are covalently bonded to substrates
   through the employment of silane species and subsequently polymd.
   to yield nonlinear optic and other efficiencies and effects.
AN CA90(6):39368j
TI Third-order nonlinear mixing in polydiacetylene solutions
AU Shand, M.L.; Chance, R.R.
CS Corp. Res. Cent., Allied Chem. Corp.
LO Morristown, N.J., USA
SO J.Chem. Phys., 69(10), 4482-6
SC 35-5 (Synthetic High Polymers)
DT J
CO JCPSA6
IS 0021-9606
PY 1978
LA Eng
AN CA90(6):39368j
AB Third-order mixing was measured in solns. of polydiacetylenes
   [e.g., I, R = (CH2) 4O2CNHCH2CO2Bu [68810-61-7]]. The Raman mol.
   vibration contribution to the third-order susceptibility was
   characterized completely. A surprizingly strong polymer two-photon
   absorption contribution to third-order susceptibility was found.
   The cross section for the two-photon absorption at 2.omega.1 =
   32,000 cm-1 is 6.times. 10-47 cm4 s photon-1 (polymer repeat unit)-
   1, where .omega.1 is the frequency of the incident laser beam. The
   polymer nonresonant susceptibility could not be detd. because of
   the strong two-photon contribution to third-order susceptibility.
   For diagram(s), see printed CA Issue.
```

Figure 7.4

CHEMICAL BUSINESS NEWSBASE (CBNB) Abstracts of international trade and business journal articles, market research and stock broker reports, government and other documents on the chemical industry-track products, production quantities, plant capacity, and industry regulations.

HEILBRON Complete text of *Dictionary of Organic Compounds* and *Dictionary of Organometallic Compounds* with chemical substance, properties, and identification information.

MEDICINE, BIOSCIENCES

BIOSIS PREVIEWS Major source for research in the life sciences.

MEDLINE (MEDLARS) Comprehensive biomedical literature and research summaries provided by the U.S. National Library of Medicine.

SCIENCE, TECHNOLOGY

AEROSPACE DATABASE Summaries of scientific and technical documents on aerospace research and development.

COMPENDEX Synopses of worldwide engineering publications and articles.

FEDERAL RESEARCH IN PROGRESS Descriptions of current, multidisciplinary research under the sponsorship of U.S. government agencies.

GEOARCHIVE Indexing of publications covering geophysics, geochemistry, geology, paleontology, and mathematical geology.

INSPEC One of the largest English-language databases in the fields of physics, electrical engineering, electronics, computers, and control engineering.

ISMEC Research in mechanical engineering.

MATHSCI Reviews and summaries of publications in pure mathematics, applied mathematics, and related fields.

MENU—THE INTERNATIONAL SOFTWARE DATABASE Directory of over 60,000 commercially available software packages for micro, mini, and mainframe computers.

NTIS Catalog of government-sponsored research, development, and engineering.

SCISEARCH An index of citations given at the conclusions of articles.

On-Shelf Abstracting and Indexing Services

The library will have a range of technical indexes on the shelf to help you locate periodical literature. Looking under either the author's name or the subject, you will find citations and, in some cases, abstracts. The Engineering Index (EI), for instance, lists magazine and journal articles under main subject headings and subheadings (see Figure 7.5). The Engineering Index includes periodicals, serials, symposia, books, and reports. Copies of the text of most articles can be bought for a fee from the Engineering Societies Library, 345 East 47 Street, New York, NY 10017.

Figure 7.5 Sample entry in *The Engineering Index Annual*. Copyright *The Engineering Index® Annual* (with permission).

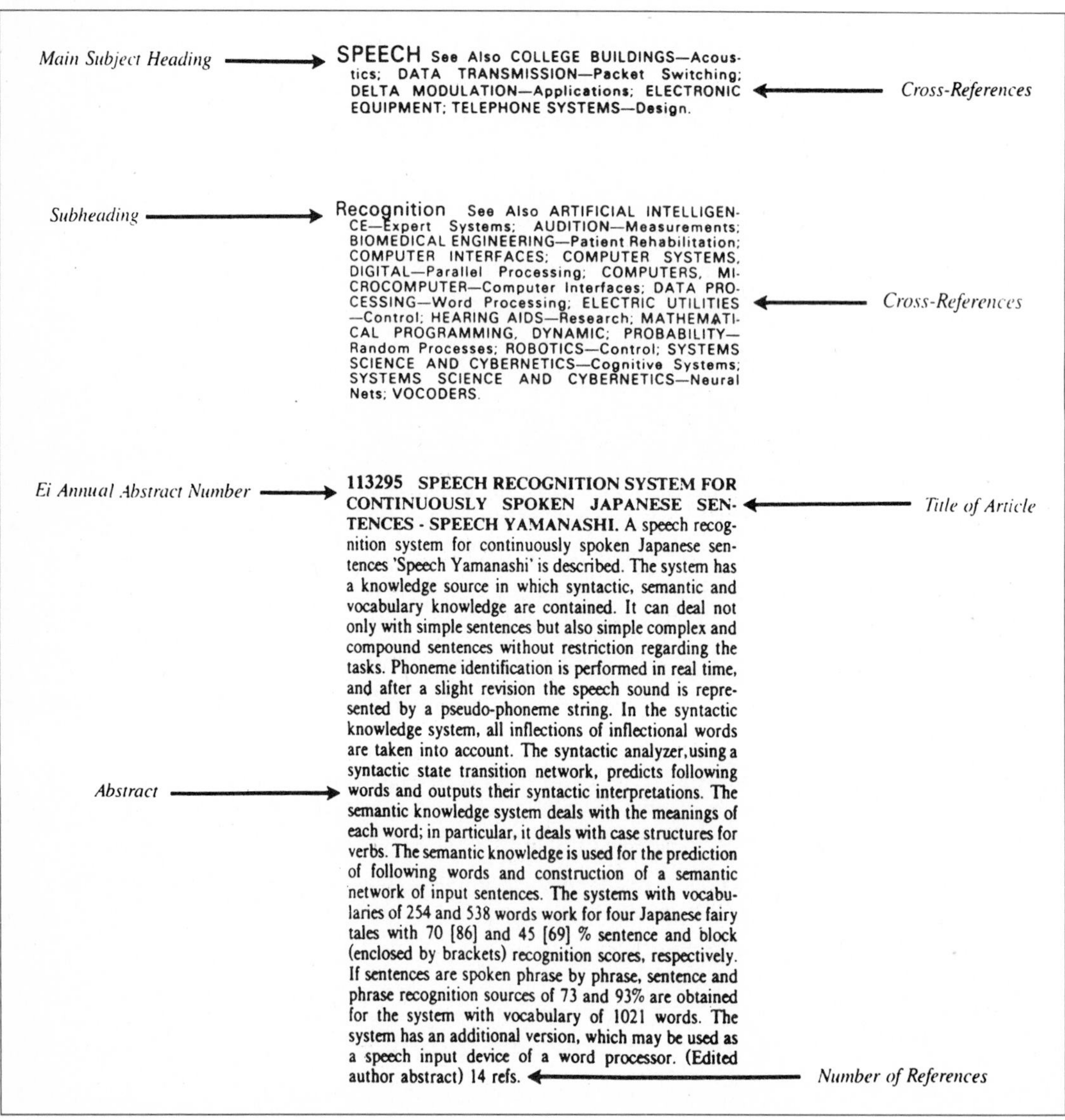

The Engineers Index, like many other abstracting and indexing services, is now available on-line (as COMPENDIX). However, using it and other indexes on the shelf rather than on-line has some advantages: historical depth, economy, and leisure. Databases for most publications only begin in the mid-1960s. For information before that, a manual search is the only strategy. Also, on-the-shelf indexes are accessible to browsers in a way that many computerized searches are not.

Be sure to study the key to the index, the instructions, the table of abbreviations, and the list of special terms that preface the index. For instance, if you were using the *Bibliographic Guide to Technology*, you would need to familiarize yourself not only with the instructions, but with the abbreviations LC, ISBN, DDC, and NYPL (Figure 7.6). If you were using an Engineering Index publication, you would need to read the prefatory pages on subject headings to understand the controlled vocabulary.

Here is a sampling of manual indexing and abstracting services. For exhaustive lists, see The Library of Congress's *A Guide to the World's Abstracting and Indexing Services in Science and Technology*, and *Abstracts and Indexes in Science and Technology* (Owen and Hanchey).

Manual Indexing and Abstracting Services

Accountants' Index. American Institute of Certified Public Accountants, 1921–.

Air Pollution Abstracts. U.S. Environmental Protection Agency, 1970–present.

Applied Science and Technology Index. New York: Wilson, 1913–present. Only English literature is covered. No author's index or abstracts.

ASCE Publications Abstracts. New York: American Society of Civil Engineers, 1966–present. Abstracts of all papers appearing in all the ASCE journals and in *Civil Engineering*.

Biological Abstracts. BioSciences Information Service of Biological Abstracts, Philadelphia, 1927–present. Abstracts of periodicals in zoology, botany, biochemistry, physiology.

Business Periodicals Index. New York: H. W. Wilson, 1958–. Includes *Business Week*, *Fortune*, and other business publications.

Biological and Agricultural Index. New York: H. W. Wilson, 1947–.

Figure 7.6 Key to a bibliographic guide. Courtesy G. K. Hall, *Bibliographic Guide to Technology.*

G.K. Hall *Bibliographic Guides* are comprehensive annual subject bibliographies. They bring together publications cataloged by The Research Libraries of The New York Public Library and the Library of Congress for thorough subject coverage. Included are works in all languages and all forms—non-book materials as well as books and serials.

Bibliographic Guides provide complete LC cataloging information for each title, as well as ISBN and identification of NYPL holdings. Access is by main entry (personal author, corporate body, names of conference, etc.), added entries (co–authors, editors, compilers, etc.), titles, series titles, and subject headings. All entries are integrated into one alphbetical sequence. Full bibliographic information, including tracings, is given in the main entry, with abbreviated or condensed citations for secondary entries. Subject headings appear in capital letters in bold-face type. Cataloging follows the *Anglo-American Cataloging Rules*. Following is a sample entry with full bibliographic information:

(a)**Rappaport, Anatol, 1911** - (b) N - person game theory; (C) concepts and applications. (D) Ann Arbor, (e) University of Michigan Press (f) [1970] (g) 331 p. (h) illus. 22cm. (i) (Ann Arbor science library) (j) Bibliography: p. 317–320. (k) ISBN 0 - 472–00117 - 5 (l) LC Card 79 - 83451 (m) DDC 512/.8 (n) 1. Games of stragey (Mathematics). (o) I. Title (p) QA270 .R34 1970 **(q) NYPL (r) [JSD 72-370]**

(a) Author's name.
(b) Short, or main title
(c) Sub-title and/or other title page information.
(d) Place of publication.
(e) Publisher.
(f) Date of publication.
(g) Pagination.
(h) Illustration statement.
(i) Series.
(j) Note(s).
(k) ISBN.
(l) LC Card number.
(m) DDC Card member.
(n) Subject heading.
(o) Added entry.
(p) LC Call number.
(q) NYPL indicator.
(r) NYPL Classmark.

Chemical Abstracts. Columbus, Ohio: American Chemical Society, 1907–present. Considered the foremost technical and scientific abstracting service, it covers new books, journals, patents, conference proceedings, and technical reports.

Computing Reviews. New York: Association for Computing Machinery, 1960–present.

Cumulative Index to Nursing and Allied Health Literature. Glendale, CINAHL, 1956–present.

Dissertation Abstracts International. Ann Arbor: University Microfilms, 1938–present.

Engineering Index. New York: Engineering Index, Inc., 1884–present. Includes periodicals, serials, symposia, books, and reports; regarded as the outstanding general index to engineering literature. Available on-line as COMPENDEX.

Electrical and Electronics Abstracts. London: Institution of Electrical Engineers, 1898–present. (Part B of *Science Abstracts;* covers all areas of electrical engineering.)

Energy Index. New York: Environment Information Center, 1973–present.

Government Reports Announcement and Index. Springfield, VA.: U.S. National Technical Information Service, 1946–present. For technical reports, indexed by author, agency, report number, contract number, and subject. Reports done on behalf of most government agencies available.

INIS Atomindex. Vienna: International Atomic Energy Agency, 1970–present. Nuclear energy: books, patents, technical reports.

International Aerospace Abstracts. New York: American Institute of Aeronautics and Astronautics, 1961–present. Aeronautics; space science. A companion volume to Scientific and Technical Aerospace Reports (STAR).

Mathematical Reviews. Providence, R.I.: American Mathematical Society, 1940–present.

Metals, Abstracts. London: Institute of Metals, 1968–present. Comprehensive abstracting service.

Monthly Catalog of United States Government Publications. Superintendent of Documents, 1895–present. For technical reports by government agencies.

New York Times Index. New York: New York Times, 1913–present.

Physics Abstracts. London: Institution of Electrical Engineers, 1898–present. Part A of *Science Abstracts*. Outstanding coverage of physics and astronomy.

Science Citation Index. Institute for Scientific Information, Philadelphia, Pa., 1961–present. An index of citations given at conclusion of articles.

Wall Street Journal Index. New York: Dow Jones, 1913–present.

Science Citation Index

A citation index is an interesting and useful variation on the classic indexes. If users have one important early citation in the history of the subject—say, for instance, Frances Crick's 1971 article on chromosomes of higher organisms—they can use this one citation to find all the important literature that follows, on the reasonable premise that any further work in the area would of necessity cite this study. *Science Citation Index* is available both on the shelf and on-line within the database SCISEARCH.

Table of Contents Search

Many libraries use a table of contents service to help readers keep up with periodicals. In this service, users scan a list of periodicals the library receives and circle the periodicals that interest them.

The library staff photocopies the table of contents for each journal the user chooses, and forwards the photocopies to the user. The user then scans the tables of contents for articles that may be of interest.

There are also professional table of contents periodicals—periodicals made up entirely of tables of contents for journals within fields such as engineering, technology, life science, physical sciences, and chemistry. *Current Contents* is the leading commercial table-of-contents publication.

DOCUMENT DELIVERY

If the books or articles you seek are sitting on the shelf of your library, available to you when you need them, so much the better. For instance, when the library has open stacks—shelves that users are free to walk among—you might go to the section of the collection that includes your topic, and, through browsing on the shelves, find just the books you need. For instance, you might be interested in Charles Babbage's early work on computers. You look him up in the card catalogue, and find a listing for what looks like a promising biography, *Irascible Genius*. You copy down the Library of Congress call number, and head for this section of the stacks. On the shelf you find not only *Irascible Genius,* but several other books about Babbage, and you browse until you find the ones that are right for you.

Often, though, the library will not have the information you need on the shelf. Many libraries simply cannot afford to keep up with the rising tide of books, journals, serials, and monographs available, particularly when the demand for the more specialized items is small. Consider, for example, serials related to computers. By the beginning of 1986, this one category held 414 magazines and newspapers, 197 newsletters, 150 journals, and 59 annuals and directories. Few libraries, if any, could stock all of these titles.

To deal with the problem of proliferating literature, library networks began to develop in the 1980s. One of the leaders, OCLC (OnLine Computer Library Center), for instance, is the largest cooperative network of automated libraries in existence. It has more than 3,000 members. Both the large networks like OCLC and RLIN (Research Libraries Information Network) and smaller ones like Metro provide many useful services if the material you want is not available on the shelves of your local library. Today most libraries are able to deliver the document, article, or book you seek, even when it is not part of their permanent collection.

Book Searches and Interlibrary Loans

One OCLC subsystem will search for any book within the holdings of the member libraries. The member library enters the request, locates the book electronically, and then requests an interlibrary loan. Charges for the search time vary—there is usually a fee for the search whether the book turns out to be available or not.

Hard Copy of Journal Articles

The user fills out a complete citation giving author, title, volume, and page numbers. The librarian enters the citation electronically to check the shared, computerized list of periodical holdings of libraries in the network, holdings that vary from business collections to medical archives. When the item is located, the library is notified of nearby locations. Thus the library unequipped with the *Harvard Journal on Legislation* or *Highway and Heavy Construction* can still help you with your research. If you are not able to travel to the library that possesses the journal, the librarian can request a photocopy. Charges vary, from "no charge for requests within reason" to $10 plus 15 cents/page. Turnaround time averages 6 days.

Libraries are not the only institutions that now provide what is called "document delivery"—the book, article, or photocopy you seek. Such delivery has become a burgeoning business. Many commercial services provide photocopies of the bulk of U.S. periodical literature, whether it is a government document or a journal article on paper and pulp technology. Many will add rush service for a surcharge. These services vary from National Technical Information Service (NTIS) in Springfield, VA, and the Michigan Information Transfer Source (MITS), Ann Arbor, MI, to most commercial database services.

PUBLICATIONS OF PROFESSIONAL SOCIETIES

Professional societies have many roles in research; among many other activities, they produce conferences, computerized data bases, abstracting and indexing services, monographs, symposia, and periodicals.

The professional society may therefore be one of your best sources for information. These are the major professional organization in science and technology.

Selective List of Professional Organizations

American Association for the Advancement of Science (AAAS)
1515 Massachusetts Avenue, NW
Washington, DC 20005

American Chemical Society (ACS)
115 16th Street, NW
Washington, DC 20036

American Geological Institute (AGI)
1201 M Street, NW
Washington, DC 20037

American Geophysical Union (AGU)
2100 Pennsylvania Avenue, NW
Suite 435
Washington, DC 20037

American Institute of Aeronautics and Astronautics (AIAA)
1290 Avenue of the Americas
New York, NY 10019

American Institute of Biological Sciences (AIBS)
104 Wilson Boulevard
Arlington, VA 22209

American Institute of Chemical Engineers (AICHE)
345 East 47th Street
New York, NY 10017

American Institute of Physics (AIP)
335 East 45 Street
New York, NY 10017
AIP Member Societies
American Physical Society
Optical Society of America
Acoustical Society of America
Society of Rheology
American Association of Physics Teachers
American Crystallographic Association
American Astronomical Society
American Association of Physicists in Medicine
American Vacuum Society

American Mathematical Society (AMS)
PO Box 6248
Providence, RI 02940

American Meteorological Society (AMS)
45 Beacon Street
Boston, MA 02108

American Nuclear Society (ANS)
244 East Ogden Avenue
Hinsdale, IL 60521

American Society for Engineering Education (ASEE)
1 Dupont Circle
Suite 400
Washington, DC 20036

American Society for Metals (ASM)
Metals Park, OH 44073

American Society for Testing and Materials (ASTM)
1916 Race Street
Philadelphia, PA 19103

American Society of Civil Engineers (ASCE)
345 East 47th Street
New York, NY 10017

American Society of Mechanical Engineers (ASME)
345 East 47th Street
New York, NY 10017

American Statistical Association (ASA)
806 15th Street, NW
Washington, DC 20005

Association for Computing Machinery (ACM)
1133 Avenue of the Americas
New York, NY 10036

Chemical Society (CS)
Burlington House
London W1V OBN
England

Engineers Joint Council (EJC)
345 East 47th Street
New York, NY 10017

Federation of American Societies for Experimental Biology (FASEB)
9650 Rockville Pike
Bethesda, MD 20014

Geological Society of America (GSA)
3300 Penrose Place
Boulder, CO 80301

Institute of Electrical and Electronics Engineers (IEEE)
345 East 47th Street
New York, NY 10017

Institute of Environmental Sciences (IES)
940 East Northwest Highway
Mt. Prospect, IL 60056

Institution of Chemical Engineers (ICE)
165-171 Railway Terrace
Rugby CV21 3HQ
England

Institution of Civil Engineers (ICE)
26-34 Old Street
London EC1V 9AD
England

Institution of Electrical Engineers (IEE)
PO Box 8
Southgate House
Stevenage Herts SG1 1HQ
England

Institution of Mechanical Engineers (IME)
1 Birdcage Walk
London SW1
England

Internationl Astronomical Union (IAU)
c/o Astronomy Department
Panepestimiopolis, Athens 621
Greece

International Union of Biological Sciences (IABS)
51 Boulevard de Montmorency
F–75016 Paris
France

National Society of Professional Engineers (NSPE)
2029 K Street, NW
Washington, DC 20006

Royal Society (RS)
6 Carlton House Terrace
London SW1Y 5AG
England

Society for Industrial and Applied Mathematics (SIAM)
33 South 17th Street
Philadelphia, PA 19103

Society of Automotive Engineers (SAE)
2 Pennsylvania Plaza
New York, NY 10001

Society of Chemical Industry (SCI)
50 East 41st Street
Suite 92
New York, NY 10017

SUMMING UP

- Many complex writing jobs begin in the library.
- If you are starting from scratch, encyclopedia articles may provide the initial framework for your search.
- Specialized encyclopedias are a good starting place for background and for citations that lead to primary documents.
- Guides to the literature will introduce you to the many categories of information within your field, including specialized handbooks and dictionaries.

- Indexing and abstracting services will help you locate periodical literature. Indexes give bibliographic citations; abstracts give summaries.
- If you are researching a topic, but don't yet have specific references, a computerized search may help. The advantages are speed and thoroughness; the disadvantages are expense and the need to word requests so that citations are on target.
- On-shelf indexing and abstracting services offer convenience, economy, the opportunity to browse, and historical depth—most computerized data bases only begin in the late 1960s.
- Many libraries offer interlibrary loans through membership in a network such as OCLC (On-Line Computer Library Center) or RLIN (Research Libraries Information Network). Such networks can be used to search not only for books, but for periodical holdings.
- Professional societies are an invaluable source for information within a specialized field.

EXERCISES

1. Assume that students at your school would benefit from a clearly written introductory guide to the library's services in their particular field, be it electrical engineering or business law.

   Pick a subject area with which you are familiar—perhaps the area in which you are majoring. Alone, or in a small group, prepare a guide to your library's services in this field. Your audience is students who will use the guide as they prepare papers and research assignments in their technical specialty.

   For the field you choose, be sure to include these features:

   - database searches
   - table of contents services
   - interlibrary networks
   - document delivery networks
   - abstracting and indexing services
   - special options offered by professional librarians at your school

2. Prepare an annotated list of outstanding reference books within a technical field for first-year students majoring in this specialty. For instance, you might select five reference classics from Transportation, Hydraulic Engineering, Transportation, or Chemical Technology.

CHAPTER 8

Direct Observation: Using a Notebook

Part of the information you use in your writing may come from the library, part of it from interviews.

It's also likely that a significant part will come from direct observation. The place to record these observations is in a notebook.

The notebook occupies a cornerstone in scientific method. Why? Human memory is entirely too fallible to be trusted. Chemists and civil engineers, physicists and doctors—all are enjoined in their professional training to use a notebook to preserve and protect data.

The notebook has another, widely recognized function. The insistence that students learn to observe—and sharpen these observational skills through writing—is a basic part of technical training. Keeping a notebook hones the eye, develops the writer's descriptive powers, and adds to the writer's skill at ordering evidence and drawing inferences. (For a related discussion, see Chapter 10.)

This chapter takes a look at some outstanding features of notebooks from Hippocrates to Charles Darwin, from ancient case studies to modern litigations for patent rights.

SHARP IN OBSERVATIONAL DETAIL, SCRUPULOUS IN RECORDING

One of the earliest examples in Western literature of a technical notebook comes from Hippocrates (460 to c. 377 B.C.; see Example 8.1). Hippocrates, unlike the priestly doctors who were his fellows,

A Case with Cheyne-Stokes Breathing

from the case records of Hippocrates

Philiscus lived by the wall. He took to his bed with acute fever on the first day and sweating; night uncomfortable.

Second day General exacerbation, later a small clyster moved the bowels well. A restful night.

Third day Early and until mid-day he appeared to have lost the fever, but towards evening acute fever with sweating, thirst, dry tongue, black urine. An uncomfortable night, without sleep, completely out of his mind.

Fourth day All symptoms exacerbated; black urine; a more comfortable night, and urine of a better color.

Fifth day About midday slight epistaxis of unmixed blood. Urine varied with scattered, round particles suspended in it, resembling semen; they did not settle. On the application of a suppository the patient passed, with flatulence, scanty excreta. A distressing night, snatches of sleep, irrational talk; extremities everywhere cold, and would not get warm again; black urine; snatches of sleep towards dawn; speechless; cold sweat; extremities livid.

About mid-day of the sixth day the patient died. The breathing throughout, as though he were recollecting to do it, was rare and large.

Example 8.1

believed that illness came from natural causes, not divine ones. He kept case records that are sharp in observational detail.

In the account of the patient Philiscus, who breathed at widely spaced intervals "as though he were recollecting to do it," we read of Cheyne-Stokes breathing, a respiratory irregularity that, except for another mention by Hippocrates, was not noticed again until the eighteenth century.

Early examples of notebooks already embody the style of technical description that developed in the West—detailed, accurate, economic presentation coupled with extreme sobriety of statement. It is easy to see, too, the gradual development of specialized vocabulary within disciplines. (For a related discussion of technical vocabulary, see Chapter 4.) In the following excerpt from Meriwether Lewis's writing (Example 8.2), for instance, technical terms are already taking their place ("croop," "chap," "convex," "covert," "imbricated"). Lewis learned the vocabulary on his own, to help him with what a fellow observer called "an approach to scientific exactness."

Example 8.2 From *Original Journals of the Lewis and Clark Expedition*. New York: Dodd, Mead & Co., 1904–05.

It is reather larger than the quail, or partridge [*viz.*, bobwhite, *Colinus virginianus*] as they are called in Virginia. It's form is precisely that of our partridge tho' it's plumage differs in every part. the upper part of the head, sides and back of the neck, including the croop [*i.e.*, crop] and about 1/3 of the under part of the body is of a bright dove coloured blue, underneath the under beak, as high as the lower edge of the eyes, and back as far as hinder part of the eyes and thence coming down to a point in front of the neck about two thirds of it's length downwards, is of a fine dark brick red. between this brick red and the dove colour there runs a narrow stripe of pure white. the ears are covered with some coarse stiff dark brown feathers. just at the base of the under chap [lower beak] there is [a] narrow transverse stripe of white. from the crown of the head two long round feathers extend backwards nearly in the direction of the beak and are of a black colour. the longest of these feathers is two inches and a half, it overlays and conceals the other which is somewhat shorter and seems to be raped in the plumage of that in front which folding backwards colapses behind and has a round appearance. the tail is composed of twelve dark brown feathers of nearly equal length. the large feathers of the wings are of a dark brown and are reather short in proportion to the body of the bird in that rispect very similar to our common partridge. the covert of the wings and back are of a dove colour with a slight admixture of redish brown. a wide strip which extends from side to side of the body and occupyes the lower region of the breast is beautifully variagated with the brick red white and black which p[r]edominate in the order they are mentioned and the colours mark the feathers transversely. the legs are covered with feathers as low as the knee; these feathers are of a dark brown tiped with dark brick red as are also those between and about the jointing of the legs with the body. they have four toes on each foot of which three are in front and that in the center the

Example 8.2 (*continues*)

longest, those one [on] each side nearly of a length; that behing[d] is also of good length and are all armed with long and strong nails. the legs and feet are white and imbrecated with proportionably large broad scales. the upper beak is short, wide at it's base, black convex, curved downwards and reather obtusely pointed. it exceeds the under chap considerably which is of a white colour, also convex underneath and obtusely pointed. the nostrils are remarkably small, placed far back and low down on the sides of the beak. they are covered by a thin protuberant elastic, black leatherlike substance. the eyes are of a uniform piercing black colour. this is a most beautiful bird. I preserved the skin of this bird retaining the wings feet and head which I hope will give a just idea of the bird. it's loud note is single and consists of a loud squall, intirely different from the whistling of our quales or partridge. it has a cherping note when allarmed something like ours. today there was a second of these birds killed which precisely resembled that just described. I believe these to be the male bird the female, if so, I have not yet seen.

Lewis is credited with exceptional powers of observation, with the ability to describe clearly and exactly the new empire his eyes beheld. Example 8.2 is a part of his description of a mountain quail. At the time he saw it, the bird was unknown to his readership, yet subsequently many who read this description had no difficulty whatever with identification.

To study the writing of scientists and engineers is to read their notebooks, past and present. These notebooks are among the richest sources of technical writing we have.

Charles Darwin's *Journal and Remarks, 1832–1836, Vol III of the Voyage of the Beagle,* for instance, was based on *eighteen* notebooks he kept. On his return to England, he started writing *On the Origin of Species,* deciding that "by following the example of Lyell in Geology, & by collecting all facts which bore in any way on the variation of animals & plants under domestication & nature, some light might perhaps be thrown on the whole subject." These facts were then recorded in four notebooks.

In Darwin's notebooks you will find innumerable descriptions of places, of states, and of organisms in which precise language and evocative detail work together. Here is a typical description by Darwin of a place (Chatham Island, in the Galapagos), taken from his notebooks of the Beagle (September 17, 1835):

> Nothing could be less inviting than the first appearance. A broken field of black basaltic lava is everywhere covered by a stunted brushwood, which shows little signs of life. The dry and parched

> surface, having been heated by the noonday sun, gave the air a close and sultry feeling, like that from a stove: we fancied even the bushes smelt unpleasantly.

Here is his description of a new subdivision of the genus Planaria (a type of worm), written in his notebooks after collecting outside Rio de Janeiro.

> This like the last was caught in the forest crawling on soft decayed wood. It is quite a different species. Back snow white, edged each side by very fine lines of reddish-brown.—also within are two other approximate ones of same colour.—sides on foot white nearer to the exterior red lines thickly clouded by 'pale blacking purple'; animals beautifully coloured.—foot beneath with white specks—but a few black dots on edge and more on head.—length of body one inch, not so narrow in proportion as other species; anterior extremity not nearly so much lengthened—the body in consequence of more uniform breadth. . . . Having found two species is fortunate as it more firmly establishes this new subdivision of the genus Planaria.

Thomas Edison left behind 3.5 million pages of work notes. Within them there are 3,600 notebooks. The notebooks were where Edison sketched, recorded direct observations, and occasionally practiced calligraphy. While developing the phonograph, for instance, Edison sketched a disc phonograph, made a list of parts to be designed, and wrote about applications of the phonograph. (He also doodled.) (See Figures 8.1 and 8.2.)

FOR PATENT CONTESTS

Properly kept notebooks become prime evidence if there is litigation or a contest for patent rights. Notebooks are corroboration should the inventor or research have to prove origin or substantiate statements and conclusions.

The case of Daniel Drawbaugh v. Alexander Graham Bell is a famous example where a notebook was needed to substantiate statements.

Bell filed his patent application for the telephone in 1875. Drawbaugh promptly sued, claiming the invention was his own. Drawbaugh was able to produce witnesses at court who testified Drawbaugh had discussed a crude telephone with them. This personal testimony, though, was not enough for the Supreme Court, which rejected Drawbaugh's claims largely on the basis of his ina-

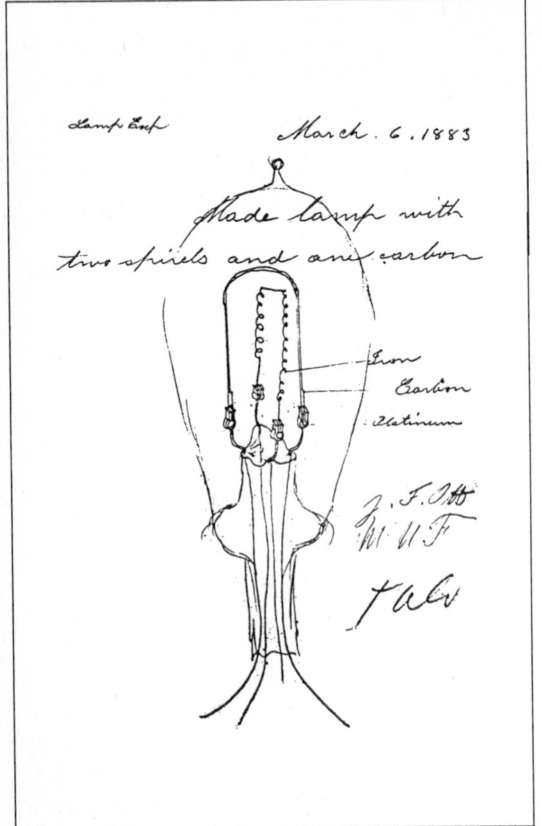

Figure 8.1 A page from Edison's notebook showing lightbulb, 1883.

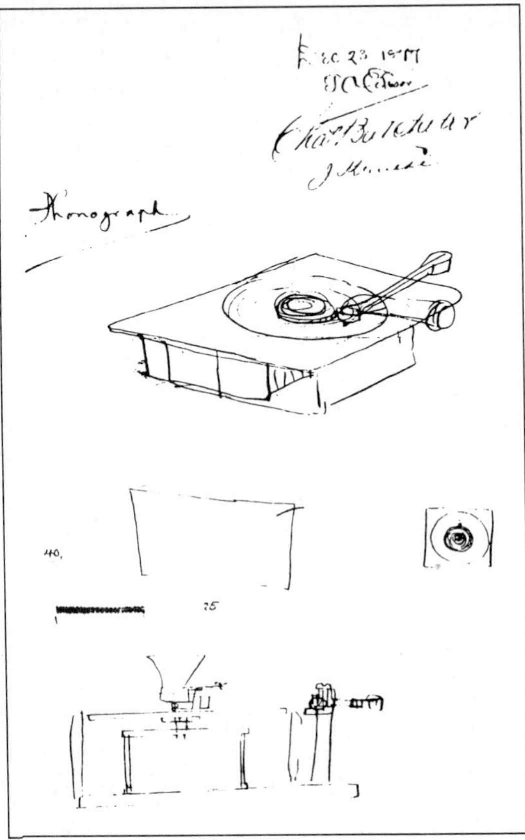

Figure 8.2 A page from Edison's notebook showing phonograph, 1877.

bility to produce a single properly dated piece of paper describing the invention.

Equally celebrated is the case of Gordon Gould, who went to court armed with his laboratory notebooks after failing to get a basic laser patent. Gould based his successful challenge on his dated, witnessed research notebook that showed a sketch, a statement of the main idea, and a derivative of the acronym LASER.

RULES FOR NOTEBOOK FORMAT

The popular formulation of rules for keeping a notebook is "Record it. Date it. Sign it. Have it witnessed." These rules are expressed more formally in these general guidelines:

1. Use a bound notebook.

2. Record all entries promptly in solvent-resistent black ink.

3. Include all original work in the notebook.

4. If you make a mistake, *do not erase*. Never use correction fluid or pasteovers of any kind. If you decide to correct an entry, place a single line through the mistake, sign and date the correction, and give a reason for the error. Make sure the underlying type can still be read. (Even the practice of drawing a line through errors is discouraged in many companies. Instead, people are asked to make a new entry correcting the error.)

5. Do not leave blank spaces on any page. Draw lines through any portion of the page you don't use.

6. Do not remove any pages, or any parts of pages. Pages missing from a notebook will weaken a case in the Patent Office or in cases that go to court for litigation.

7. If you must use a looseleaf notebook, number the pages in advance and keep a record if pages are given to laboratory workers. This is done to rebut any inference that a worker may have inserted a page at a later date.

8. Define the problem or objective concisely. Make entries consistently as the work is performed.

9. Date and sign what you have written on the day of entry. Have each page read, signed, and dated by a qualified witness—someone who is not directly involved in the work performed, but who understands the purpose of the experiment and the results obtained.

10. If it is unavoidable that you insert graphs and charts, have them read, signed, and dated by witnesses in the same way as other entries.

11. Identify all apparatus; use schematic sketches when appropriate.

12. Head each entry with a title. If you continue on the next page, say so at the bottom of the page before you continue.

WHO CAN BE A WITNESS?

The role of witnesses may be crucial in cases of Interference Proceedings. These proceedings occur when two applications are filed in the United States Patent and Trademark Office disclosing similar inventions, and the Patent and Trademark Office acts to determine which of the inventors is entitled to the patent. There have been many instances in Interference Proceedings where the inventor's own testimony, supported by thorough sets of notebooks, unquestionably prepared and dated, did not alone serve to establish date of conception. In these cases, the witnesses' backgrounds became crucial.

It was necessary that the witnesses understood all the entries. Merely witnessing them on the date did not suffice in a contest where the other claimant also had properly substantiated testimony. The point is that a witness is called upon to substantiate the facts and nature of the work performed at the date that the person signed as a witness, and not merely the fact that an entry was made by the inventor on that date.

It is common to believe anyone who is nearby—the technician down the hall, the worker in the next office, the office helper at hand—can be a witness. This is not so. Witnesses must be adults, preferably over 21, with the technical competence to understand the details of the subject.

DUE DILIGENCE AND REDUCTION TO PRACTICE

Patent law places great emphasis on date of conception and diligence in pursuing the idea.

Notebooks are particularly important in establishing due diligence—proof that the inventor pursued the idea, rather than setting it aside. New York patent attorney Philip Furgang related cases where an inventor who was first to have the idea and first to reduce it to practice still lost the patent for lack of properly signed, dated, and witnessed records demonstrating due diligence. Thus inventor Smith had the idea first, entered it properly in her notebook, and then set about reducing it to practice for two years. However, she was busy during these two years and kept the notebook poorly; not a single entry after the one establishing the date of conception was properly witnessed. Then a second person, inventor Jones, had the same idea. He reduced it to practice and filed his patent just before Smith. Between contesting inventors, the burden of providing who is entitled to the patent falls on the inventor last to file. Thus in this instance the burden was on Smith to prove her right to the patent. She went to court and there was no question of witnesses who could testify to origin; clearly she was first. Unfortunately, she could not prove due diligence because she did not have witnessed, signed notebook entries for the period in question, and thus Jones won the interference.

Proof of date of conception and due diligence can be guarded with the following steps in the notebook:

1. Get the idea into the notebook as quickly as possible. If it is

written down after some delay, relate the date and place of conception of the idea and the circumstances that stimulated the idea.

2. In the initial record, stress why the idea is novel. If you made notes on scrap paper, the original of the notes should be preserved, but the contents transcribed into the permanent notebook as soon as possible.

3. Continue to record every instance when you return to the idea so there is ample evidence of due diligence.

SUMMING UP

- Out of nothing comes nothing. To write, you will gather—by interview, by research, and by direct observation. Observational writing is woven into the history of science and technology, from the earliest accounts of Hippocrates to the present.
- Keeping a notebook hones the eye, develops the writer's descriptive powers, and adds to the writer's skill at ordering evidence and drawing inferences. Besides, human memory is entirely too fallible to be trusted. The notebook is the place to preserve and protect data.
- Notebooks embody the style of technical description that developed in the West—detailed, accurate, economic presentation.
- Properly kept notebooks become prime evidence if there is litigation or a contest for patent rights.
- To keep the notebook properly, make sure it is bound. Do not erase entries; do not remove pages or any parts of pages. Date and sign entries. Then have each page read, signed, and dated by a qualified witness.
- It is common to believe that anyone who is nearby can be a witness. This is not so. Witnesses must be adults, preferably over 21, with the technical competence to understand the details of the subject. The witness may be called upon to substantiate the facts and nature of the work—not just the fact that an entry was made by the inventor on that date.
- Notebooks are particularly important in establishing due diligence—proof that the inventor pursued the idea, rather than setting it aside. Proof of due diligence can be guarded if you get the idea into the notebook quickly. Continue to record every instance when you return to the idea.

EXERCISES

1. Keep a notebook for a short period—five or six weeks. Use it to record observations of behavior, procedures, objects, interactions—be these observations in the cafeteria line or the laboratory, a classroom, or a department store. Does knowing you are going to write in a notebook change the way you observe? Are there circumstances under which writing seems to crystallize your understanding?

2. Two readings follow this exercise:

• In "Sic Transit Transistor," by Robert Friedel (Example 8.3), Professor Friedel gives a lively and informative account of a technical development, using a page from a laboratory notebook as an illustration.

• "Microscopical observations about animals in the scurf of the teeth" (Example 8.4), is a letter from Antony van Leeuwenhoeck, written in 1684. Van Leeuwenhoeck, who built microscopes as a hobby, records some early observations of bacteria.

Using these pieces as examples, locate other discussions of specific developments within the history of technology. Your library will have general histories of science and technology, and histories that specialize within divisions such as transportation, energy, medicine, communications, information, food, and warfare.

Prepare a short (5- to 6-page) report on one development in the history of technology. Whether you choose the nautical sextant or marine chronometers, industrial robots or the development of Braille, steam engines or phonographs, try to include in the report an account of the notebooks or letters of the developers.

Example 8.3 © American Heritage, a subsidiary of Forbes Inc. Reprinted with permission from American Heritage of Invention and Technology, Summer 1986.

Sic Transit Transistor

It began as a crude, poorly understood device. Within ten years it was changing the world. And today it's history.

by ROBERT FRIEDEL

On the greeting-card racks this past Christmas could be seen a minor technological miracle—a Christmas card that upon opening showed a small yellow light that glowed while the card played a tinny but recognizable version of "Jingle Bells." The yellow light was the latest addition to a novelty that made its first appearance a couple of years ago—the electronic greeting card. The thing that is most notable about this minor miracle is just how ordinary it seems, how easily it is shrugged off by a public that accepts it as simply another application of microelectronics, perhaps to join the ranks of Pong (the early video game), talking scales, and wristwatches that incorporate calculators and thermometers.

Most of those picking up the singing Christmas card have at least a vague idea of how it works. There is, of course, a battery-powered "chip"—an integrated circuit, combining hundreds or thousands of electronic functions on a fingernail-sized slice of crystal. This device, at the heart of modern electronics technology, from guided missiles to computers to pocket calculators, is the result of the harnessing of a class of substances called semiconductors.

Our mastery of these materials began in earnest almost forty years ago, with the invention of the transistor. This most people realize; less well understood is the real role of the transistor in the microelectronics revolution. As some saw at its introduction, in 1948, the transistor brought to electronics new capabilities, presenting to engineers challenges and opportunities hitherto unknown and pointing ahead to a "solid-state" technology that would have a very different look and feel from vacuum-tube electronics. Yet the transistor as it first appeared and as it was applied over the first decade of its existence, was also the final stage in an older technical tradition, an electronics conceived and built around circuits of individual components. In hindsight, it is now possible to appreciate the single discrete transistor as a truly transitional technology. It was distinctively different from older ways of doing things, but it was not fully a part of the revolution to come.

In the years after World War II, few scientists were looking for a revolution in electronics. The war had seen the flourishing of radio and related technologies beyond anything that could have been imagined before the conflict began. While there were indeed wartime technical and scientific achievements that were more spectacular—the atomic bomb, for instance—to many observers it was electronics that gave the Allies the real technical margin of victory. Radar, sonar, field-communications systems, and the proximity fuze were recognized by soldiers and civilians alike as technical accomplishments of the first order, the creations of engineers and physicists intent on demonstrating how advanced technology could make the critical difference in the war effort.

Many people were also aware of the creation of another new device, the electronic digital computer, which saw limited wartime use but promised to open up whole new realms of technical possibilities. The years after the war were expected to see the consolidation of the great technical strides that had been taken under military pressure and the application of these new capabilities to areas that would improve civilian life. The world eagerly awaited the war-delayed spread of televison and FM radio, and industry was hungry to see what the new electronics could do in the massive conversion to a consumer economy. No fundamental technical change was sought or anticipated; everyone would be busy making the best use of the still bright, new technologies of the day.

Only a few years after the war's end, however, there began the most profound technical revolution since the beginning of electronics itself at the turn of the century. At the end of 1947, three physicists working at the Bell Telephone Laboratories in Murray Hill, New Jersey, learned how to make a piece of semiconductor crystal behave like a vacuum-tube amplifier. The story of John Bardeen, Walter Brattain, William Shockley, and the invention of the transistor is a well-known tale of the successful application of science in an industrial setting, a research program that earned the investigators the Nobel Prize for physics and the

Example 8.3 (*continues*)

BELL LABORATORIES/COURTESY OF AMERICAN INSTITUTE OF PHYSICS, NIELS BOHR LIBRARY, NEW YORK CITY

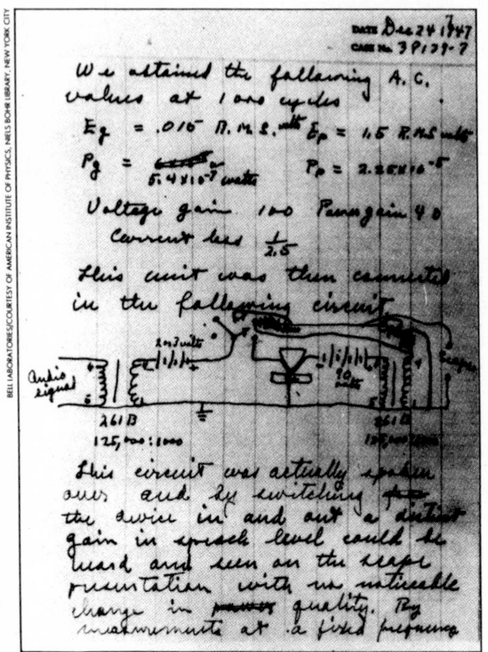

DATE Dec 24 1947
CASE No. 38139-7

We obtained the following A.C. values at 1000 cycles

Eg = .015 R.M.S. volts Ep = 1.5 R.M.S. volts

Pg = 5.4x10⁻⁷ watts Pp = 2.25x10⁻⁵

Voltage gain 100 Power gain 40

Current loss 1/2.5

This unit was then connected in the following circuit

This circuit was actually spoken over and by switching the device in and out a distinct gain in speech level could be heard and seen on the scope presentation with no noticeable change in quality. By measurements at a fixed frequency

A laboratory notebook entry by Walter Brattain records the events of December 23, 1947, when the transistor effect was first observed.

Bell System its most valuable patent since the one Alexander Graham Bell took out for the telephone in 1876.

Not so well known is just how the transistor came to have the revolutionary impact it did. It was one thing to bring together advanced scientific knowledge and sophisticated technical insight to create a new component that worked on novel principles and in distinctly different ways from older ones. It was quite another to make that component a useful—not to say profoundly important—part of the technological world. Doing so presented a challenge to the engineers and designers who were the shapers of that world, a challenge that initially went unrecognized and unappreciated by people who were comfortable and confident with the established technical order of things. It took close to a decade from the time of the transistor's invention at Murray Hill to the point where the new technology began to have a real impact on modern life—and began to lead to other, more far-reaching inventions.

The source of the electronics engineers' confidence at war's end was mastery of the vacuum tube. The history of the electronic vacuum tube began in 1883 with Thomas Edison's observation of the passage of an electric current from the hot, glowing filament of one of his incandescent lamps, across the high vacuum of the bulb, to an extra electrode introduced into the bulb's glass envelope. Edison thought this effect might be somehow useful but couldn't think of a good application and so did nothing more with it. An Englishman, John Ambrose Fleming, figured out two decades later that this was just the kind of device needed by the infant technology of radio.

Since the Edison-effect current will flow only in one direction, from the filament (or cathode) to the other electrode (or anode) but not the reverse, Fleming called his invention a valve. It was useful for radio because the response of the valve to a radio-wave signal, which is a kind of alternating current, was a direct current (hence the valve is also called a rectifier). Fleming's valve gave the radio pioneers a reliable and sensitive detector of signals and thus began displacing the "cat's-whisker" detector—a semiconductor crystal, ironically—that had been depended on up to then.

The real possibilities of vacuum-tube technology, however, were revealed by the American Lee de Forest in 1906. De Forest discovered that if he placed a third electrode between the cathode and the anode and allowed the radio signal to determine the voltage in this element (called the grid), the vacuum tube could be used as a powerful amplifier of the signal. De Forest called his invention an audion, but it became more generally known as a triode (just as the valve was referred to as a diode). The subsequent history of radio, and indeed of all electronics technology, for the next four decades involved the design of still more complex tubes—tetrodes, pentodes, hexodes, and even higher "odes"—and more complex circuits to make the best use of them. The "odes" were shortly joined by the "trons," such as the pliotron (a very-high-vacuum triode), the ignitron (for very-high-power applications), and the magnetron (for the generation of very-high-frequency signals).

When, in 1933, the electronics industry sought to show off its wonders at the Chicago World's Fair, one of the features of the display was a giant vacuum tube designed to explain the basic phenomenon to fairgoers. Other exhibits showed off some of the literally hundreds of different tubes that were used for a great variety of purposes, from ever-improving radio receivers to machine controls, phonograph amplifiers, short-wave radio sets, and even "electric eyes." Despite the Great Depression, the 1930s saw the vigorous expansion of electronics from the already impressive displays of Chicago in 1933 to the truly spectacular show put on at New York's World's Fair in 1939, where fully electronic television began regular broadcasting to the American public for the first time.

The coming of war in the 1940s, therefore, found the American electronics engineers well equipped to make important contributions to the nation's defense. Physicists and engineers at the MIT Radiation Laboratory devised tubes that made radar a reliable and widely applicable tool. Researchers at the Applied Physics Laboratory outside Baltimore built another, less famous but also strategically vital device, the proximity fuze. This consisted of a radar unit reduced in size to fit into an antiaircraft or artillery shell and set to detonate the shell at the desired distance from its target. It required the design of vacuum tubes that were both rugged enough to withstand being fired from a gun and small enough to fit into a circuit placed in the head of a five-inch shell. Engineers thus began speaking of "ruggedization" and "miniaturization" as new and important design challenges.

These challenges redirected the attention of some to the possible uses of semiconductors. Ever since the vacuum tube had replaced the cat's whisker, there had been only limited interest in these materials. Semiconductors, as the name implies, do not conduct electric currents as well as common metallic conductors (like copper, silver, or aluminum) but they perform better than true insulators (like glass, rubber, or most plastics). The contact between a semiconductor and a conductor can be designed so that current will pass easily in only one direction, acting as a rectifier. Before the war, numerous applications were found for rectifiers made of the semiconductor materials copper oxide and selenium. Ruggedization and miniaturization compelled many engineers to pay closer attention to these materials, but there was little theory to guide the applications of any of the semiconductors, and further uses would require more knowledge about how they really behaved.

Even before the war, and again during it, some electronics engineers asked if a semiconductor crystal couldn't be made to act like a triode amplifier as well as a diode rectifier. Isolated efforts showed that this was not, in fact, easily done. Besides, the ever-increasing power and versatility of vacuum tubes made the question of a semiconductor amplifier seem relatively unimportant. But tubes had their limitations, and as more applications were explored, these limitations became increasingly apparent. In high-frequency applications, tube designers

Example 8.3 (*continues*)

often ran into severe difficulties in combining power, signal response, and reliability. When long-term reliability was necessary, as in telephone networks or undersea cables, tubes fell short of what was needed. And when thousands of tubes were used together, as in the new giant digital computers, the problems of keeping such large numbers of heat-producing devices going caused numerous headaches. Considerations such as these were at work at Bell Labs when plans were made for the research that led to the transistor.

Even in the 1930s, some engineers had perceived the limitations of the mechanical switching technology that was at the heart of the telephone network. People like Mervin Kelly, Bell's vice-president for research, saw that electronic switching, with its instantaneous response to signals, held out a possible answer. The roadblock was in the nightmare of depending on huge banks of hot vacuum tubes, each liable to break down after a few months of constant use. As Kelly saw it, a solid-state switching device, requiring no power at all most of the time and capable of responding immediately to an incoming signal, was not only desirable, it was becoming necessary if the telephone system was not to drown in its own complexity and size.

So, with the war winding to a close in the summer of 1945, a research group in solid-state physics was organized at Bell Labs. From the summer of 1945 until the end of 1947, the solid-state physics team under William Shockley performed experiments on silicon and germanium crystals, aided by the superior materials-handling capabilities of Bell chemists and metallurgists. The primary goal of their experiments was a semiconductor amplifier, a replacement for the vacuum triode. The transistor effect was first observed in December of 1947, and the people at Bell knew instantly that they had an important invention on their hands. It was so important, in fact, that they didn't allow word to get out for six months; patent documents were drawn up and further experiments were pursued to make sure that the investigators knew just how their device worked.

On June 30, 1948, an elaborate public announcement and demonstration was staged in New York City, complete with a giant model of the point-contact transistor and a radio and a television with the new devices substituted for tubes. The public response was muted—*The New York Times,* for example, ran the story on page 46, at the end of its "News of Radio" column. Perhaps it was the sheer simplicity of the device that deceived onlookers. The *Times* described it as consisting "solely of two fine wires that run down to a pinhead of solid semi-conductive material soldered to a metal base." The first transistor, housed in a metal cylinder a little less than an inch long and a quarter-inch across, was not much smaller than some of the miniature tubes that had been made for the military.

Nonetheless, some people were impressed, even if they could only dimly perceive the implications. The popular magazine *Radio-Craft,* for example, featured the Bell announcement in an article headlined ECLIPSE OF THE RADIO TUBE: "The implications of this development, once it emerges from the laboratory and is placed in commercial channels, are staggering." This comment was evoked less by the transistor's size than by its most conspicuous other feature, its lower power requirement. "No longer will it be necessary to supply power—whether it be by batteries or filament heating transformers—to heat an electron-emitting cathode to incandescence." Since all vacuum tubes begin with the Edison effect, they must be "lit" like light bulbs and thus require considerable power. The transistor required far less power and would not burn out, so it promised a reliability and longevity that tube designers could only dream about.

But the new point-contact transistor was in

fact a very frustrating device. For a component that was supposed to be based on sound understanding of physical theory, it instead behaved more like the whimsical devices jury-rigged by amateurs in the first decades of radio. The first transistors were "noisy," easy to overload, restricted in their frequency responses, and easily damaged or contaminated. Transistor makers had great difficulty producing what they wanted. Some spoke of the "wishing in" effect, where the tester of a finished device was reduced to simply hoping it would work right. And there was the "friendly" effect—the tendency of a transistor to acknowledge on a testing oscilloscope a simple wave of the hand in its direction. At first only 20 percent of the transistors made were workable. One engineer was quoted later as saying, "The transistor in 1949 didn't seem like anything very revolutionary to me. It just seemed like another one of those crummy jobs that required one hell of a lot of overtime and a lot of guff from my wife."

To complicate things further, the transistor might do the same sorts of things a tube could do, but it didn't do them in the same way. This meant that circuit designers—the engineers responsible for turning components into workable tools and instruments—would have to reconfigure their often complex circuits to accommodate the transistor's special characteristics. For many, it hardly seemed worth it. But others made enormous efforts to get the transistor working and out into the world of applications.

Besides the Bell System itself, the most important impetus to transistor development was provided by the U.S. military. The value of miniature electronics had been clear to some military men at least since the 1930s, when the Army Signal Corps made the first walkie-talkies. By the beginning of World War II, the corps had produced the Handie-Talkie, a six-pound one-piece radio that was essential to field communications. Experience with this device, however, highlighted the limitations of vacuum-tube technology. Under rugged battlefield conditions, especially in extreme heat and cold, the instruments were prone to rapid breakdown. Small as they might be, the virtues of even greater reductions in size were apparent to anyone who had used them in the field, and the problems of heavy, quickly exhausted batteries were a constant annoyance.

The Signal Corps thus expressed interest in transistors right away and even started manufacturing them on a very small scale as early as 1949. Other branches of the military quickly followed, and a number of the first applications of the transistor were in experimental military devices, particularly computerlike instruments. The military services pressed Bell to speed up development and to begin quantity production, and by 1953 they were providing half the funding for Bell's transistor research and underwriting the costs of manufacturing facilities.

With this kind of encouragement, the researchers at Bell Labs were able to make considerable improvements in transistor technology. The most important of these was the making of the first junction transistor in 1951. Relying on the theoretical work of William Shockley and very sophisticated crystal-growing techniques developed by Gordon Teal and others, the junction transistor utilized the internal properties of semiconductors rather than the surface effects that the point-contact device depended on. In the long run this was a much more reliable technology, and it provided the fundamental basis for all subsequent transistor development. While the making of junction transistors was initially extremely difficult, the result was a true marvel, even smaller than the original transistors, using even less power, and behaving as well as the best vacuum tubes in terms of noise or interference. Soon after the announcement of the junction transistor, Bell began making arrangements to license companies that wished to ex-

Example 8.3 (*continues*)

ploit the new technology, holding seminars and training sessions for outside engineers, and compiling what was known about transistors into fat volumes of technical papers.

By 1953 Bell and the military had finally begun to make use of the transistor, but there was still no commercial, publicly available application. The transistor still meant nothing to the public at large. This changed suddenly that year, not in a way that affected a large number of people, but in a way that a relatively small number were very grateful for. In December 1952 the Sonotone Corporation announced that it was offering a hearing aid for sale in which a transistor replaced one of the three vacuum tubes. The device wasn't any smaller than earlier models, but the transistor saved on battery costs. This opened the floodgates to exploitation of the new technology by dozens of hearing-aid manufacturers, and within months there were all-transistor hearing aids that promised to rapidly transform a technology that had been growing steadily but slowly since the 1930s.

As manufacturers redesigned circuits and found other miniature components to fit into them, hearing aids shrank radically. Just as important, however, was the savings in batteries. Despite the fact that transistors cost about five times what comparable tubes did, it was estimated that a hearing-aid user saved the difference in reduced battery costs in only one year. It is no wonder that the transistor took the hearing-aid industry by storm. In that first year, 1953, transistors were found in three-quarters of all hearing aids; by 1954, fully 97 percent of all instruments used only transistors. At the same time, hearing-aid sales increased 50 percent in one year (from 225,000 to 335,000). The transistor was showing a capacity for transforming older technologies that only the most prescient could have foreseen.

When the Regency radio made its appearance in time for Christmas of 1954, this first all-transistor radio was an instant success at $49.95. Other radios quickly followed, and the "transistorization" of consumer electronics was under way. Still, the early transistor radio was largely a novelty, only slowly displacing older, more familiar instruments. Much more important was the application of the transistor to digital computers. The first truly electronic programmable digital computer, the ENIAC, had been completed in February 1946. It contained 17,468 tubes, took up 3,000 cubic feet of space, consumed 174 kilowatts, and weighed 30 tons. Many engineers saw from the beginning the great promise that the transistor held out for computer technology, but it was not until 1955 that IBM began marketing its first transistorized machine (the 7090). Size was reduced, the air conditioning that the tubes required was gone, and power consumption was reduced 95 percent. It was obvious to all that the marriage of the transistor and the digital computer would be an important and fruitful one.

Still, in 1955 the transistor remained a minor element in the larger scheme of things. Total transistor production up to that time totaled less than four million units (over about seven years). That many vacuum tubes were being made in the United States every two working days. And as exciting as the computer might be, it represented a tiny market—only 150 computer systems were sold in the United States that year. Over the next few years, however, transistor production and use expanded rapidly. As more companies entered the business, fabrication techniques were improved, and engineers continued to design circuits that used semiconductor devices.

Nevertheless, the microelectronics revolution might have remained a quiet and limited affair but for two great events, one within the industry, the other a dramatic intrusion from the outside world. On October 4, 1957, the

Soviet Union put Sputnik I into orbit around the Earth, ushering in an era of tense and vigorous competition for the mastery of space technology. The already considerable interest of the U.S. government in microelectronics now reached a fever pitch.

That same year, a small group of physicists and engineers, former employees of the transistor pioneer William Shockley, set up their own shop and laboratory in a warehouse in Mountain View, California, at the northern end of the Santa Clara valley. At the head of this new enterprise, called Fairchild Semiconductor, was the twenty-nine-year-old Robert Noyce. Within a couple of years Noyce and his colleagues had put together the elements of a new semiconductor technology that would shape the future in ways that the transistor itself never could have done.

In 1958 Fairchild's Jean Hoerni invented the planar technique for making transistors. Building on processes developed at Bell and General Electric, the planar technique allowed for simplification and refinement of transistor manufacture beyond anything yet possible. It consisted of making transistors by creating layers of silicon and silicon dioxide on a thinly sliced "wafer" of silicon crystal. This layering allowed careful control of the electronic properties of the fabricated material and also produced a quantity of very small, flat transistors in one sequence of operations. Much more important, however, the process led Robert Noyce to another, much more fundamental invention, the integrated circuit.

Noyce was neither the first to conceive of putting an entire electronic circuit on a single piece of semiconductor material nor was he the first to accomplish the feat. The idea had been around for some years, and in the summer of 1958 Jack Kilby, an engineer at Texas Instruments in Dallas, built the first true integrated circuit, a simple circuit on a piece of germanium. Kilby's device was a piece of very creative engineering, but it depended on putting together the circuit components by hand. Noyce's great contribution was to show how the planar process allowed construction of circuits on single pieces of silicon by the same mechanical and chemical procedures formerly used to make individual transistors. Noyce's process, which he described in early 1959 and demonstrated a few months later, allowed for the design and manufacture of circuits so small and so complex that the elements could not even be seen by the unaided eye. Indeed, in Noyce's circuits, the transistors were reduced to little dots under a microscope. Finally the microelectronics revolution had revealed its true form.

No one in 1948 could have predicted what that form would be. Indeed, even in 1962, when integrated circuits first came to market, the extraordinary nature of the changes to which they would lead was only dimly perceived by a few. The integrated circuit overcame what some engineers referred to as the "tyranny of numbers." By the late 1950s, large computers had as many as two hundred thousand separate components, and even with transistors the construction and testing of such large machines was becoming an engineering nightmare. Still larger computers, with millions of components, were envisioned, especially by military and space planners. The integrated circuit, with its ability to combine hundreds and later even thousands of elements into a single unit, made such complex machines possible. Just as importantly, however, it made them cheap. By the end of the 1960s, computer memory chips that could handle a thousand bits of information were available.

In 1971 the first microprocessor was introduced on the market by Intel Corporation. Sometimes called a "computer on a chip," the microprocessor demonstrated that sufficiently complicated integrated circuits could be mass-

Example 8.3 (*continues*)

produced and could bring large, sophisticated computers down to such a tiny size and small cost that they could be inserted into every area of life. The home computer, the modern automobile's instrumentation, the digital watch, the bank's electronic teller machine, and even the electronic greeting card all testify to the power and pervasiveness of the integrated circuit's triumph over size and complexity.

The appearance of the integrated circuit little more than ten years after the announcement of the transistor's invention marked the onset of the next stage in the evolution of electronics, a stage that a quarter-century later has still not run its course. The transistor as a discrete device, as a replacement for the vaccum tube, is still important, but in a more profound sense, the transistor was a transitional technology. It showed engineers and others the way beyond the tube, even though the truly revolutionary paths of electronics in the late twentieth century were to be blazed by other far more complex inventions.

Example 8.4

Microscopical Observations about Animals in the Scurf of the Teeth

ANTONY VAN LEEUWENHOEK

. . . Tho my teeth are kept usually very clean, nevertheless when I view them in a Magnifying Glass, I find growing between them a little white matter as thick as wetted flower: in this substance tho I do not perceive any motion, I judged there might probably be living Creatures.

I therefore took some of this flower and mixt it either with pure rain water wherein were no Animals; or else with some of my Spittle (having no Air bubbles to cause a motion in it) and then to my great surprize perceived that the aforesaid matter contained very many small living Animals, which moved themselves very extravagantly. the biggest sort had the shape of A. their motion was strong & nimble, and they darted themselves thro the water or spittle, as a Jack or Pike does thro the water. These were generally not many in number. The 2d. sort had the shape of B. these spun about like a Top, and took a course sometimes on one side, as is shown at G. and D. they were more in number than the first. In the 3d. sort I could not well distinguish the Figure, for sometimes it seemed to be an Oval, and other times a Circle. These were so small that they seem'd no bigger than E. and therewithal so swift, that I can compare them to nothing better than a

SOURCE: An abstract of a Letter from Mr. Anthony Leeuwenhoeck at Delft, dated Sep. 17, 1683. Containing some Microscopical Observations, about Animals in the scrurf of the Teeth. . . . *Philosophical Transactions of the Royal Society of London*, Vol. 14, May 20, 1684, no. 159, pages 568–574, 1 pl.

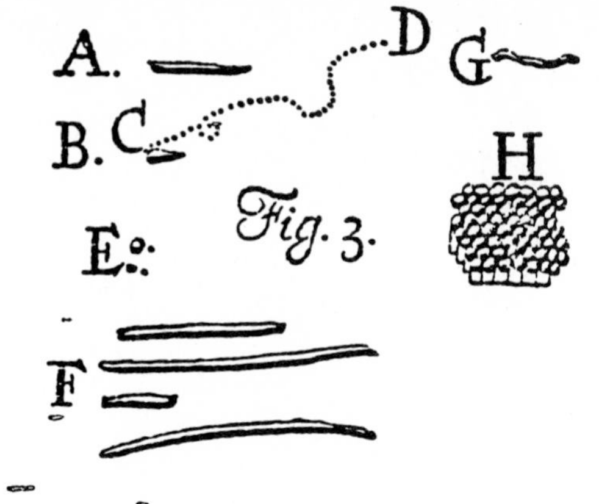

swarm of Flies or Gnats, flying and turning among one another in a small space [Brownian movement?]. Of this sort I believe there might be many thousands in a quantity of water no bigger than a sand, tho the flower were but the 9th part of the water or spittle containing it. Besides these Animals there were a great quantity of streaks or threds of different lengths, but like thickness, lying confusedly together, some bent, and others streight, as at F. These had no motion or life in them, for I well observed them, having formerly seen live Animals in water of the same figure.

I observed the Spittle of two several women of whose Teeth were kept clean, and there were no Animals in the spittle, but the meal between the teeth, being mixt with water (as before) I found the Animals above described, as also the long particles.

The Spittle of a Child of 8 years old had no living Creatures in it, but the meal between the Teeth, had a great many of the Animals above described, together with the streaks.

The Spittle of an old Man that had lived soberly, had no Animals in it; But the substance upon & between his Teeth, had a great many living Creatures, swimming nimbler then I had hitherto seen. . . . The Spittle of another old man and a good fellow was like the former, but the Animals in the scurf of the teeth, were not all killed by the parties continual drinking Brandy, Wine, and Tobacco, for I found a few living Animals of the 3d. sort, and in the scurf between the Teeth I found many more small Animals of the 2 smallest sorts.

I took in my mouth some very strong wine-Vinegar, and closing my Teeth, I gargled and rinsed them very well with the Vinegar, afterwards I washt them very well with fair water, but there were an innumerable quant'ty of Animals yet remaining in the scurf upon the Teeth, yet most in that between the Teeth, and very few Animals of the first sort A.

I took a very little wine-Vinegar and mixt it with the water in which the scurf was dissolved, whereupon the Animals dyed presently. From hence I conclude, that the Vinegar with which I washt my Teeth, kill'd only those Animals which were on the outside of the scurf, but did not pass thro the whole substance of it. . . .

The number of these Animals in the scurf of a mans Teeth, are so many that I believe they exceed the number of Men in a kingdom. For upon the examination of a small parcel of it, no thicker than a Horse-hair, I found too many living Animals therein, that I guess there might have been 1000 in a quantity of matter no bigger then the 1/100 part of a sand.

APPLICATIONS

PART IV

CHAPTER 9

Definitions

When a large technical school in a midwestern city installed two dozen personal computers in its administrative offices, the school found itself fielding one complaint after another. One of the most persistent complaints was that the manual that came with the computers failed to define terms properly. Sometimes the user could overcome the problem by reading between the lines, using context clues—hints from the surrounding words—to figure out what the author had in mind. But with two or three undefined terms in a paragraph, even the most adroit reader could go no further. And definitions in the glossary were so skimpy, so shorn of example and elaboration, that they were only marginally better than the alternative—no definitions at all.

The computers went to many people, from those who knew several programming languages to those who had never heard the term "cursor." Few of the recipients were happy with the manual;

when the time came to make a large-scale order, the school decided to use another brand. The reason lay not with the computer—that was a success—but with the problems most people had using the manual.

The incident provides both a dramatic example of poor writing and a moral: Don't assume your readers know your terminology. When you address a diversified group, define your terms.

The brevity or depth of definitions, indeed their very presence, depends upon whom you are writing for, what they know before you begin, and what you want them to know at the end. Definitions depend upon audience; they don't work by themselves, but only, as with so many other skills in writing, in relation to the reader.

In this chapter, there are two parts: first, examples of definitions written at various levels according to the audience; second, techniques for writing such definitions.

FOR A HOMOGENEOUS AUDIENCE

If your audience is sophisticated—that is, trained and conversant in the area you are discussing—you will probably sidestep definitions of terms you know to be understood.

Look for instance at Example 9.1, which is a summary of a research paper submitted to a national health institute for funding. Here the author has attempted no definition of "fibronectin," "soluble protein," "plasma" or "adhesion," for this summary is written solely for readers within a particular research field. Such definitions of basic terms would be irrelevant, even distracting, for a group of people with similar training.

FOR A SLIGHTLY DIVERSIFIED AUDIENCE

Example 9.2 is an excerpt from the introduction to a report in the *New England Journal of Medicine,* which is shown in its entirety in Chapter 2 (Example 2.2). As you read it, notice which terms are defined, and which are not.

Notice that in the title the terms "azoospermatic" and "nulliparous" are *not* defined, although both terms are important in grasping the method of study. The authors assume that their readers know these basic terms; however, they do define "fertility" and "fecundity" to show that they are using these two words in a restricted or special sense. Notice that the two terms are defined with a brevity typical of scientific or technical writing aimed at those within a specialized group.

Example 9.1 Summary for homogeneous audience.

Summary: Electron Microscope Studies of Fibronectin Interactions

Fibronectin circulates as a soluble protein in the plasma and is known to mediate adhesion by virtue of its binding affinities for a variety of biological materials, including those involved in coagulation and woundhealing. We propose that fibronectin is a flexible molecule with a shape that is dependent upon solution and surface conditions. In this investigation, we will determine the relationship between different molecular shapes of fibronectin and their biological roles. Transmission electron microscopy (TEM) and High Resolution Scanning Transmission Electron Microscopy (STEM) will be used to investigate systematically the role of solution and surface effects on fibronectin shape. This study should enable us to establish the relationship between fibronectin shape and the expression of biological functions.

Example 9.2 Introduction for diverse technical audience.

Female Fecundity as a Function of Age:

Results of Artificial Insemination in 2193 Nulliparous Women with Azoospermic Husbands

The decrease in the fecundity of women who have passed a certain age is generally acknowledged, but supporting data on natural reproduction are scarce. (We use the term "fecundity" in the sense of "capacity for procreation"; "fertility" denotes actual procreation.) In an analysis of data from three large studies, Leridon has suggested that fecundity is decreased in women over 30 years of age. However, this group includes women with a wide range of ages. Furthermore, it was not possible to determine from the data whether the decrease in fecundity was biologic or simply the consequence of diminished sexual activity.

Artificial insemination with donor semen (AID) seems to present an opportunity to con-

Example 9.2 (*continues*)

trol certain variables in the study of female fecundity over time, but the few studies published to date have been carried out in small populations. Moreover, the husbands of most women who have received AID cannot be considered to have been totally sterile. The probability that these men will procreate increases as the fecundity of their wives increases. Accordingly, among women treated by AID, there are very probably some with reduced fecundity; the degree of reduction increases as the time without conception ("exposure time") increases. Therefore an observed decrease in fecundity with age in these women could simply be due to a sampling bias.

We studied 2193 women who were receiving AID and whose husbands were totally sterile. The curve of the cumulative success rate for the women 25 years of age or younger was similar to that for women 26 to 30 years old. However, this curve showed a significant decrease in the cumulative success rate for women 30 to 35 years of age. This decrease was even greater for those over 35. Similar decreases with age were observed for the mean conception rate per cycle. Our data therefore provide evidence of reduced fecundity with age, which begins at some point after the age of 30 years.

FOR A BROAD-BASED AUDIENCE

For any heterogeneous group, define technical terms promptly, both on first mention and in the glossary. If you think the audience needs more than a compact definition, expand with an example, contrast, comparison, or analogy.

Here are two excerpts from manuals written to accompany personal computers (Examples 9.3a and 9.3b). Both sections give instructions on how to "scroll," or read quickly through text displayed on a computer screen; in the second example, however, the definition of "scrolling" is expanded to suit the audience.

In Example 9.3a, the definition of scrolling is brief—"scrolling is like turning the pages of a book." In the second definition (Example 9.3b), the author has gone much further than simply saying "scrolling is like turning the pages of a book." He expands the definition with a long comparison in which the screen is compared to a window and the text a long sheet of paper that we view through the window.

This expanded definition would be inappropriate for a sophisticated audience who might well be irritated by its lengthy explanation of the obvious. However, this set of instructions is written for the beginning user, and the author, very sensibly, is making as few assumptions as possible about the reader's background.

Example 9.3a

Moving Through a File

Let's say you want to look at the contents of a file using your screen. To do this, use the commands that "scroll" the file. Scrolling is like turning the pages of a book.

- Moving Forward or Backward

  Two PF keys are set to commands that scroll the file. PF8 scrolls forward one screen. PF7 scrolls backward one screen.

  If your file is many screens long, you press either of these keys repeatedly to go forward or backward through the file.

- Moving to the Bottom or Top

  Suppose your file is many screens long, and the current screen is somewhere in the middle of the file. To go forward to the end of the file, you could press PF8 repeatedly. Or, you can use the BOTTOM command. The BOTTOM command makes the last line of the file the new current line.

Example 9.3b

Scrolling

Scrolling may be a new term for you, so we will define it before going on. Think of material you have entered into the computer as being on one long sheet of paper—a scroll. And think of the screen as an opening or window through which you can see only a small section of that long sheet of paper. When the window moves up that long sheet of paper and you see the material you have first written, you are scrolling up, and when the window moves down the sheet toward the end of the text, you are scrolling down.

The scrolling commands are very useful when you are reading a document on a screen. There are four of these commands:

Z—scroll down line
R—scroll down screen
W—scroll up line
C—scroll up screen

To many people the references to *up* and *down* seem backward, but if you think of the screen as a window moving up or down over the material you type in, then it makes sense.

The use and extensiveness of definitions depend on the homogeneity of your audience. In general, the more diversified the audience, the more you'll need to weave definitions into your introductory text, expanding these definitions with examples, comparisons, and analogy. By the same token, people within a field expect short definitions, often given only in cases where a term is being used in a special or restricted sense.

TECHNIQUES FOR WRITING DEFINITIONS

Synonyms and Phrases

The briefest way to insert a definition when you address a diversified audience is to set off a synonym or rephrasing of the term with dashes, commas, or parentheses, or the expressions "that is," or "called a."

> Patients who experience shortness of breath or chest pain (angina) at least a month before their first heart attack also had an elevated risk of dying from a second heart attack.
>
> One project goal is to provide a network—a group of interconnected computers—with almost unlimited memory and computational power.
>
> As you type, the cursor automatically moves to the next space available. There are times, however, when you'll want to move it to a different place on the screen. Different keys move the cursor one space forward or backward, or one line up or down. These keys are typamatic, that is, they keep repeating as long as they are held down. When your system is on and available for use, a small square called a cursor appears on the screen. The cursor shows you where the next character you type will be displayed on the screen.

Often the term within parentheses or dashes gives etymological information—that is, information related to the origin of the word.

> The modem (for modulator-demodulator) is a device that can be attached to convert the computer's digital signals into signals for transmission over telephone lines.
>
> Bits (short for binary digits) are units describing the information content of any message.

Logical Definitions

If you are defining in a sentence or two, a standard technique is to assign the item to a class, and then to differentiate it from all other items in the class. This manner of distinguishing items was developed by Aristotle, and is called the "logical definition."

Term	*Class (Genus)*	*Differentiation*
hydraulics	The branch of physics	that concerns itself with the mechanical properties of water and other liquids in motion and with the application of these properties in engineering.

The definition may combine a brief physical description with functional or operational information.

Term	*Class (Genus)*	*Differentiation*
Diaphragm (in Anatomy)	A dome-shaped partition	which is composed of muscle and connective tissue; separates the abdominal and thoracic cavities in mammals.

The definition may be developed by cause and effect.

Term	*Class (Genus)*	*Differentiation*
Scintillation	rapid changes of brightness of stars or other distant celestial objects	caused by variations in density of the air through which the light passes.

A logical definition is often expanded by the use of an example. The following definition of diaphragm defines by form ("thin, flexible sheet") and function ("can be moved by sound or can produce sound waves when moved"). The examples (microphone, loudspeaker) help the reader connect the general to the concrete.

Term	*Class (Genus)*	*Differentiation*
diaphragm (in engineering acoustics)	a thin, flexible sheet	that can be moved by sound waves, as in a microphone, or can produce sound waves when moved, as in a loudspeaker.

In technical writing, where an introduction often entails defining a handful of terms, dashes are useful to set off the short terms you are defining within the larger ones. Example 9.4 is an excerpt from a report. Within the operational definition of "computer" the author has embedded a definition of "binary numbers."

Expanded Definitions

Sometimes a brief definition, no matter how logical and to the point, is not enough for readers, particularly if you are introducing them to an entirely new subject. In cases like this, when you realize that far from belaboring a point, you will aid the reader's comprehension, you may expand your definitions by a variety of rhetorical techniques.

A computer is essentially a machine that receives, stores, manipulates and communicates information. It does so by breaking a task down into logical operations that can be carried out on binary numbers—strings of zeroes and ones—and doing hundreds of thousands or millions of such operations per second.

Term	*Class (Genus)*	*Differentiation*
Computer	a machine that receives, stores, manipulates, and communicates information	does so by breaking a task down into logical operations that can be carried out on binary numbers and doing millions of such operations per second.
Binary Numbers	strings of zeroes and ones.	

Example 9.4

We define a personal computer as a stand-alone system that has all the following characteristics:

1. The system either includes or can be linked to secondary memory in the form of floppy or hard disks.

2. The microprocessor can support a primary-memory capability of 640 kilobytes or more. (A kilobyte is equal to 2^{10} or 1,025 bytes. A byte is a string of eight bits, or binary digits. One byte can represent one alphabetic character or one or two decimal digits. A 640-kilobyte memory can store 655,360 characters, or some 100,000 words of English text.)

3. The computer can handle at least one high-level language such as Basic, Fortran, or Cobol. In a language of this kind instructions can be formulated in a fairly high level of abstractions and without taking into account the detailed operation of the hardware.

4. The operating system facilitates an interactive dialogue: The computer responds immediately (or at least quickly) to the user's actions and requests.

5. Distribution is largely through mass-marketing channels, with emphasis on sales to people who have not worked with a computer before.

6. The system is flexible enough to accept a wide range of programs serving varied applications; it is not designed for a single purpose or a single category of purchasers.

Example 9.5

A. A definition may be expanded by a list of characteristics or properties, or by categorizing its parts, as in Example 9.5, an excerpt from a report.

B. You may develop a definition by comparing or contrasting the new term to one that is already familiar to the reader.

In Example 9.6a, for instance, the author starts with a definition of a term "spectral lines," and then decides to expand it by comparing the new term to something the readers already know—Newton's experiments with light (Example 9.6b).

As much as the author admired brevity, he felt the definition was too compressed to make its point accessible to the audience. He therefore added a contrast.

In Example 9.7, the author introduces definitions of two new terms, read-only memory (ROM) and random-access memory (RAM) using comparison and contrast. Note the embedding of a definition of the term "system programs" in the larger definition.

Example 9.6a

Author's First Definition of "Spectral Lines"

Atoms of pure elements in the gas phase can emit light when heated. The emitted light consists of narrow bands or lines of color (spectral lines) with dark spaces between them. Each element has a characteristic pattern of spectral lines, called an atomic spectrum, that serves as a "fingerprint" to identify the element. Spectral lines are emitted not only in the visible region of the electromagnetic spectrum, but also in the ultraviolet region, at wavelengths shorter than those of visible light, and in the infrared region, at wavelengths longer than those of visible light.

Example 9.6b

Author's Revised Definition of "Spectral Lines"

Atoms of a pure element in the gas phase can emit light when heated. When studies of this phenomenon began in the mid-nineteenth century, it was expected that the results would be similar to those obtained in an experiment done two centuries earlier by Isaac Newton.

Newton passed ordinary sunlight through a prism. He found that a prism breaks the sunlight into a continuous spectrum that contains the colors of visible light. But the same sort of continuous spectrum is not obtained when light is emitted by gaseous atoms of a pure element. Instead, the emitted light consists of a relatively small number of narrow bands (or lines) of color, with large dark spaces between them.

These lines are called **spectral lines.** Each element has a characteristic pattern of spectral lines, called an **atomic spectrum,** that serves as a "fingerprint" to identify the element. Spectral lines are emitted not only in the visible region of the electromagnetic spectum, but also in the ultraviolet region, at wavelengths shorter than those of visible light, and in the infrared region, at wavelengths longer than those of visible light.

Example 9.7

> There are two kinds of primary memory: read-only memory (ROM) and random-access memory (RAM). Read-only memory is for information that is "written in" at the factory and is to be stored permanently. It cannot be altered. For a single-application computer such as a word processor the information in ROM might include the application program. In the case of a versatile personal computer it would include at least the most fundamental of the "system programs," those that get a computer going when it is turned on or interpret a keystroke on the keyboard or cause a file stored in the computer to be printed. ROMs are used extensively to store applications programs in lap-top computers, usually in the form of interchangeable plug-in modules.
>
> Random-access memory is also called read/write memory: New information can be written in and read out as often as it is needed. RAM chips store information that is changed from time to time, including both programs and data. For example, a program for a particular application is read into RAM from a secondary-storage disk; once the program is in RAM its instructions are available to the microprocessor. A RAM chip holds information in a repetitive array of microelectric "cells," each cell storing one bit.

C. You may expand a definition with figures. In Example 9.8, for instance, the writer elaborates on the definition of "floppy disk system" with an illustration which together with the callout (the list of parts on the figure) help to introduce the new term effectively to a person unfamiliar with computer systems.

SUMMING UP

- The use and extensiveness of definitions depend on the homogeneity of your audience.
- The more diversified the audience, the more you'll need to weave definitions into your introductory text. By the same token, people within a field expect short definitions, given only in cases where a term is used in a special or restricted sense.
- If your audience is homogeneous—trained and conversant in the very area you are discussing—you will probably sidestep definitions of common terminology. Definitions of basic terms are irrelevant, even distracting, for a group of people with similar training.

Example 9.8

FLOPPY DISK SYSTEM records large quantities of information on a flexible plastic disk coated with a ferromagnetic material. The disk rotates at 300 revolutions per minute in a lubricated plastic jacket. An electromagnetic head is moved radially across the surface of the disk by a stepper motor to a position over one of the concentric tracks where data are stored as a series of reversals in the direction of magnetization. The head can read or write: sense the reversals to retrieve information or impose magnetization to store information. An index mark, whose passage is sensed by a photoelectric device, synchronizes the recording or reading with the rotation of the disk. This is a schematic drawing of a double-sided disk drive made by Qume. There are two gimballed heads, which read and write information on both sides of the 5¼-inch disk. On each side of the disk some 160 kilobytes of information can be stored in 40 concentric tracks.

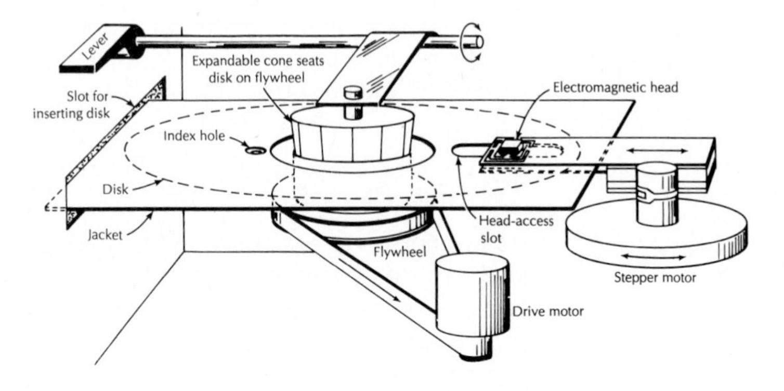

• For any heterogeneous group, define technical terms promptly, both on first mention and in the glossary. If you think the audience needs more than a compact definition, expand with an example, contrast, comparison, or analogy.

• The briefest way to insert a definition is to choose a synonym and set it off with dashes, commas, parentheses, or the expression "that is."

• If you are defining in a sentence or two, try assigning the item to a class and then differentiating it from all other items in the class. This is called a "logical definition."

• The definition may combine a brief physical description with functional or operational information.

• A logical definition is often expanded with an example to help the reader connect the abstract to the concrete.

EXERCISES

1. Find two or three definitions that you think are done effectively. For a source, you might want to look through an introductory textbook in economics, engineering, science, business, political history, philosophy, foreign languages, sociology, history, or psychology.

Prepare a brief oral report on the techniques that the authors used in writing the definitions. Are there contrasts? Examples? Rephrasing? Are the elaborations helpful in understanding the new terms?

Make photocopies of the definitions you've chosen and distribute them to the class before you discuss the authors' techniques. Allow time for the class to read the definitions before you present your analysis.

2. Write definitions of two terms within your major field of study. Avoid definitions that are too general ("A spoon is used for eating"), too narrow ("A spoon is a device made of plastic or metal that is used for eating soup") or circular ("A spoon is used for spooning)." Include examples or contrasts within the definitions. To get you started, here are a few possibilities:

dubbing (related to recordings)

"reader friendly"

word wrap

muffler (related to automobiles)

amniocentesis

gross national product

counterpoint (related to musical composition)

price-earning ratio

formatted disk

vaporware

3. Example 9.9 discusses a technical subject in a broad way appropriate to a highly diversified audience of people with varying backgrounds. Which terms are defined? What are their definitions? How has the author tailored the definitions to the audience?

HOW MUCH IS TOO MUCH?

People and situations differ, so there is no easy answer to this question. Although recent studies seem to indicate that up to two drinks a day is a "safe" level of consumption for most healthy people, one thing is certain . . . **getting drunk is never safe.**

GETTING DRUNK

Everyone knows what getting drunk means, right? Maybe, maybe not. Anyone can get drunk. All you have to do is drink too much alcohol. You may not "look" drunk, or "sound" drunk or even "act" drunk. But if you drink too much for you or your situation, you will get drunk. At that point, you put yourself and others around you at risk, possibly without even realizing it.

There are two factors about drinking that do not vary. Alcohol is a depressant drug that deadens

the central nervous system. Beer, wine, liquor, cordials and cocktails have a common denominator — alcohol. A 12-ounce beer, five ounces of wine or a 9-ounce wine cooler, 1½ ounces of liquor or two cordials **all** contain approximately the same amount of alcohol (0.6 ounces).

When you drink, alcohol goes to your stomach and small intestine. There the alcohol directly enters the bloodstream without being digested. This alcohol in your bloodstream circulates to your brain and affects your body systems.

Your body eliminates alcohol at the rate of about one standard drink an hour. The liver breaks down about 90 percent of the alcohol you drink. The rest is eliminated through the lungs and kidneys.

Intoxication is caused by the amount of alcohol in your bloodstream — your blood alcohol concentration (BAC). When you drink alcohol at a rate faster than your body can get rid of it, your BAC increases. The chart below provides an **estimate** of your BAC based on the number of drinks you consume and your body weight.

APPROXIMATE BLOOD ALCOHOL PERCENTAGE
Body Weight in Pounds

Drinks	100	120	140	160	180	200	220	240
1	.04	.03	.03	.02	.02	.02	.02	.02
2	.08	.06	.05	.05	.04	.04	.03	.03
3	.11	.09	.08	.07	.06	.06	.05	.05
4	.15	.12	.11	.09	.08	.08	.07	.06
5	.19	.16	.13	.12	.11	.09	.09	.08
6	.23	.19	.16	.14	.13	.11	.10	.09
7	.26	.22	.19	.16	.15	.13	.12	.11
8	.30	.25	.21	.19	.17	.15	.14	.13
9	.34	.28	.24	.21	.19	.17	.15	.14
10	.38	.31	.27	.23	.21	.19	.17	.16

At a BAC of as low as .05 percent, your judgment, coordination and possibly your vision are impaired. If you get behind the wheel of a car, your risk of being involved in a crash doubles.

As you drink more and your BAC increases, your chances of being involved in a fall, a fire or a fight increase significantly. Alcohol is involved in 85 percent of accidental deaths, 68 percent of drownings and over 50 percent of fatal auto crashes.

Example 9.9 Alcohol: How Much Is Too Much?

Remember that coffee, fresh air, cold showers and a variety of home remedies for "sobering up" do absolutely nothing to reduce your BAC. Only time can accomplish that.

DRINKING SAFELY

While it's true that anyone can get drunk, it is equally true that most people don't — most of the time.

- 32 percent of American adults don't drink at all.
- 61 percent drink less than two drinks a day; most of them less than once a week.

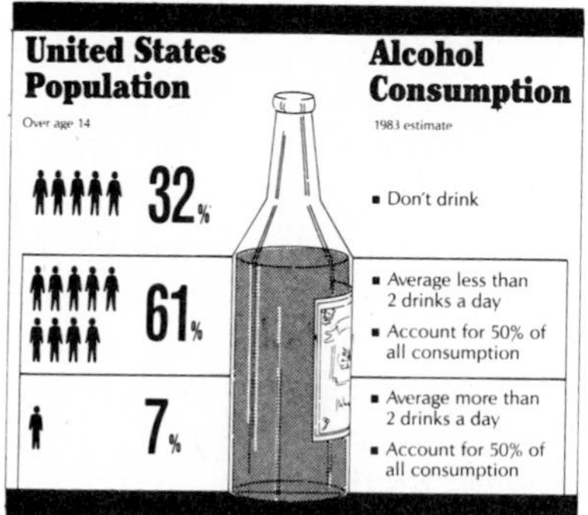

Among Americans, just 7 percent drink 50 percent of all the alcohol sold, and it's likely that these individuals do, or will, have a problem. But what of the 61 percent who drink "moderately"? They must be alert to the fact that anyone can drink too much, and that there are certain high-risk groups and times when drinking is not safe.

As a general rule, a **maximum** daily average of two drinks a day seems safe for most people. That's an average of zero to two drinks a day, not 14 drinks "saved up" for a one-night binge.

It's also true that individual reactions to alcohol can vary dramatically because of an array of factors that include your weight, eating habits, sex, age, health, family background and consumption rate. In other words, a risk-free drinking pattern for someone else could be an absolute disaster for you. Some people are at high risk for alcohol-related problems and would be advised not to drink at all.

HIGH-RISK GROUPS

Certain individuals, because of special medical conditions or unusual sensitivity to alcohol, may be unable to drink alcohol safely. Diabetics, heart patients and persons with diseases of the digestive and nervous systems should consult their physicians about drinking. People with a family history of alcoholism are four times more likely to become alcoholic themselves.

Others fall into the high-risk category only temporarily, such as young people and pregnant women. Children and adolescents differ from adults in terms of body size and the liver's ability to handle alcohol. It takes less alcohol for a young person to become intoxicated. As new drinkers, their tolerance will be low.

As people age, their bodies change and it takes longer to process alcohol. Elderly people find they are more intensely affected by the same amount of alcohol they consumed in earlier years. They may also experience complications when alcohol is mixed with other medications.

Throughout pregnancy the use of alcohol poses a serious risk to the developing fetus. Low birth weight, spontaneous abortion, mental retardation, hearing defects and physical abnormalities are among the variety of fetal alcohol effects. Because it is impossible to precisely estimate how much alcohol will damage a developing fetus, the safest course is total abstinence during pregnancy. **Other high-risk people should consider a similar approach and avoid alcohol.**

4. What techniques are used in the following definitions? Are these definitions effective?

a. Different compounds that have the same molecular formulation are called isomers (Greek *isos*, equal; *meros*, part). They contain the same numbers of the same kinds of atoms, but the atoms are attached to one another in different ways. Isomers are different compounds because they have different molecular structures.

This difference in molecular structure gives rise to a difference in properties; it is the difference in properties which tells us that we are dealing with different compounds. In some cases, the difference in structure—and hence the difference in properties—is so marked that the isomers are assigned to different chemical families, as, for example, ethyl *alcohol* and methyl *ether*. In other cases the difference in structure is so subtle that it can be described only in terms of three-dimensional models. Other kinds of isomerism fall between these two extremes.

b. In the case of the gypsy moth, a virus is believed to cause periodic "crashes" of the moth population. When gypsy moth caterpillars feed on oak leaves, the neighboring leaves increase their production of tannins, which inhibit the growth of the caterpillars. However, tannins also increase the caterpillars' resistance to viruses, which allows the gypsy moth population to slowly build. Eventually, though, the concentration of viruses in the caterpillars increases to a level that overrides the protection from tannins, and the gypsy moth population crashes.

c. Passive income is rental income you receive and money you get from activities in which you do not materially participate. The law defines material participation as involvement in the operations of a business activity on a regular, continuous, and substantial basis. Generally, conventional limited partnerships that have equity stakes in properties generate passive income.

d. Aerobic exercise is "isotonic" or dynamic, which means that it needs movement. Unlike isometric activity (such as weightlifting), which involves static muscle contraction, it does not raise blood pressure without the release of movement, so it does not boost pressure and strain the heart.

e. Nastic movements are one of two kinds of growth movements plants make in response to outside stimuli. A nastic movement is a response that is independent of the direction from which the external stimulus strikes the organism. For instance, the opening of flowers after sunrise is a nastic movement, for illumination from any direction will trigger the response.

Figure 9.1

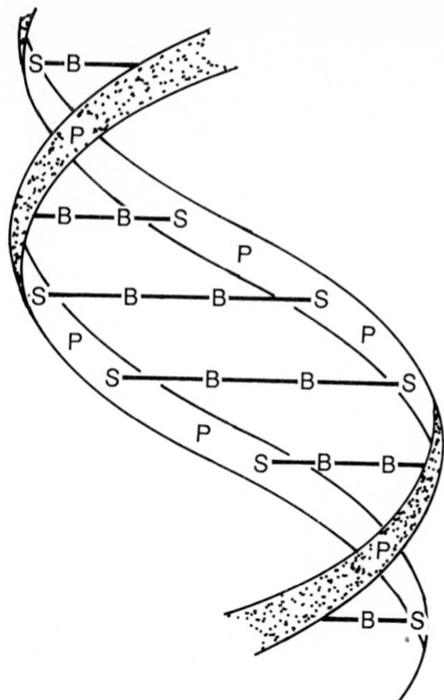

The response is not oriented with respect to the direction of the stimulus.

f. Chromosomes are giant, double-chain molecules of DNA on which genetic information is encoded in a sequence of paired chemical units known as nucleotide bases.

g. The nucleotides of a DNA strand are bonded together so that the sugar and phosphate portions form a long chain, or backbone. The organic bases stick out from this backbone and are bonded weakly to those of the second strand. The resulting structure is something like a ladder in which the uprights represent the sugar and phosphate backbones of the two strands, and the crossbars represent the organic bases. In addition, the molecular ladder is twisted to form a double helix (Figure 9.1).

h. DNA molecules are long, twisted ladders of chemicals called nucleotide bases: adenine, thymine, guanine, and cytosine. Adenine pairs with thymine, cytosine with guanine. These chemical connections are called base pairs; a single gene, a selection of DNA, is typically made up of 10,000 to 20,000 base pairs. Human beings have between 100,000 and 200,000 genes, or up to 4 billion base pairs, organized on 46 chromosomes.

CHAPTER 10

Descriptive Writing

Science begins with the observation of selected parts of nature. . . . Observation leads to description.

E. Bright Wilson, *An Introduction to Scientific Research*

In the essay, "Look at Your Fish," Samuel Scudder recounts his first lesson with the naturalist Louis Agassiz. Scudder appeared as an eager student; Agassiz presented him with a fish called a hamelon, said, "observe it," and left Scudder to his own devices. After Scudder had spent 10 minutes looking at the fish, inhaling the pungent solution in which it was kept, he felt he'd seen all that could possibly be seen of it. Agassiz disagreed, and insisted Scudder spend three days "looking at the fish."

During those three days Scudder finally learned how to look in the sense Agassiz intended. Scudder discovered he could use his eyes and pencil to observe new aspects of the hamelon, to order these observations, and to draw inferences from them. Agassiz was training him, and as part of the training he was teaching Scudder to observe, and *to crystallize and order the observations through writing and drawing.*

Description is at the heart of observation.

There are times when the line, the plane, the circle, the cylinder, the cone, the sphere will be your natural descriptive allies. If you are describing a mechanism, for instance, words are a prelude to the drawings whose form and dimension constitute their own universal language.

But often language stands alone, with no need of the supporting drawing. Geometry is by no means the only way to three-dimensional reality.

We are indebted to the descriptive powers of Ada Lovelace, for instance, for the only complete accounts of Babbage's Analytical Engine, precursor of the modern computer. Babbage (1792–1871) was virtually incapable of giving a coherent account of his Analytical Engine during the last forty years of his life. We know about it through the fine descriptive writing of Ada Lovelace, the mathematician and friend of Babbage.

This chapter takes a look at ways to describe technical information, using examples ranging from a book on how to build scientific apparatus to patents. The examples show the interrelated importance of logic, organizational patterns, language, voice, and visuals in effective description.

RIVETS AND THREADED FASTENERS: Descriptions of Objects in a Manual for a Specialized Audience

Building Scientific Apparatus is a book for a highly specific audience: people in laboratory research in the physical and engineering sciences who want to assemble their own equipment. Their job is made a lot easier by this manual, which skillfully employs techniques of language, layout, and visual display to make the information in the text readable.

Here are two excerpts from the manual, each followed by comments on the writing and illustration (Examples 10.1 and 10.2).

In the brief selection "Rivets" (Example 10.1), the reader quickly gets the answers to the questions, "What are they?" "What do they look like?" "How do they work?"

The organization of this piece is to define, then to describe in general, and then to give details and exceptions. The authors use the passive voice, placing the focus on the rivets, not the user.

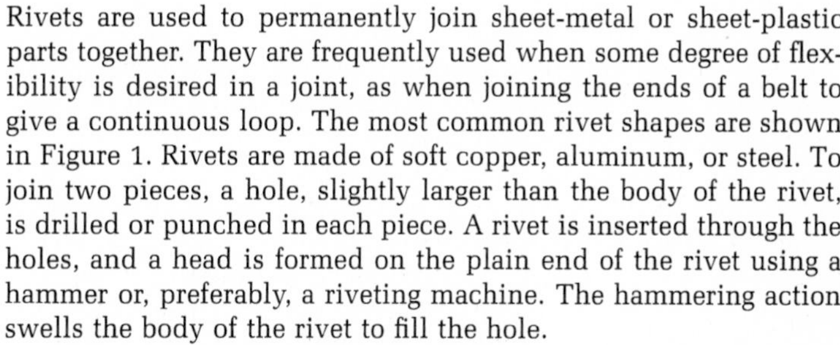

Rivets

Rivets are used to permanently join sheet-metal or sheet-plastic parts together. They are frequently used when some degree of flexibility is desired in a joint, as when joining the ends of a belt to give a continuous loop. The most common rivet shapes are shown in Figure 1. Rivets are made of soft copper, aluminum, or steel. To join two pieces, a hole, slightly larger than the body of the rivet, is drilled or punched in each piece. A rivet is inserted through the holes, and a head is formed on the plain end of the rivet using a hammer or, preferably, a riveting machine. The hammering action swells the body of the rivet to fill the hole.

"Pop" rivets, illustrated in Figure 1, are useful in the lab. These can be installed without access to the back side of the joint. The mandrel is grasped by a special rivet gun, the rivet is inserted into the hole, and the mandrel is pulled back until it breaks. The head of the mandrel rolls the stem of the rivet over to form a head, and the remaining broken portion of the mandrel seals the center hole of the rivet.

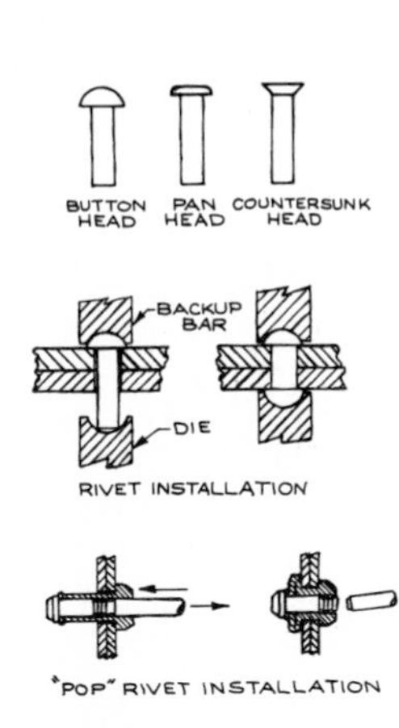

Figure 1

Example 10.1 Description of rivets. From John H. Moore, Christopher C. Davis, Michael A. Coplan, *Building Scientific Apparatus*, Addison-Wesley Publishing Company, 1983.

A second selection from *Building Scientific Apparatus* (Example 10.2) is more complicated in its subject matter: It describes threaded fasteners.

Here is a short analysis of this passage on threaded fasteners organized by key categories in technical writing.

Definitions. If you can't postpone definitions—and often this is neither possible nor appropriate—state them promptly. In this case, the writers define their terms both in the text and with a supporting illustration. They've italicized key terms in the text to match the figure callout.

Organizational Patterns and Logic. Detail is introduced *after* the category is established. Strong lead sentences provide structure. There are many examples to reduce the abstraction of the text. For instance, see the specification of threaded parts.

Language. Direct voice is unnecessary here, where the writers want the light to shine on the fasteners, not the user.

Visual Display and Design. Tables are used for detail, freeing the text. The tables have white space to help the eye move in on what it needs. Figures are used effectively for contrast and comparison, and are accompanied by captions and callouts. Italics are used to reinforce figure callouts.

Threaded Fasteners

Threaded fasteners are used to join parts that must be frequently disassembled. A thread on the outside of a cylinder, such as the thread on a bolt, is referred to as an external or male thread. The thread in a nut or a tapped hole is referred to as an internal or female thread. The terminology used to specify a screw thread is illustrated in Figure 2. The *pitch* is the distance between successive crests of the thread. The pitch in inches is the reciprocal of the number of threads per inch (tpi). The *major diameter* is the largest diameter of either an external or an internal thread. The *minor diameter* is the smallest diameter. In the United States, Britain, and Canada the form of a thread is specified by the Unified Standard. Both the *crest* and *root* of these threads are flat, as shown in Figure 2, or else slightly rounded. The *thread angle* is always 60°.

There are two thread series commonly used for instrument work. The *coarse-thread series*, designated UNC for Unified National Coarse, is for general use. Coarse threads provide maximum strength. The *fine-thread series*, designated UNF, is for use on parts subject to shock or vibration, since a tightened fine-thread nut and bolt are less likely to shake loose than a coarse-thread nut and bolt. The UNF thread is also used where fine adjustment is necessary.

The fit of threaded fasteners is specified by tolerances designated as 1A, 2A, 3A, and 5A for external threads and 1B, 2B, 3B, and 5B for internal threads. The fit of 2A and 2B threads is adequate for most applications, and such threads are usually provided if a tolerance is not specified. Many types of machine screws are available only with 2A threads. The 2A and 2B fits allow sufficient clearance for plating. 1A and 1B fits leave sufficient clearance that dirty and scratched parts can be easily assembled. 3A and 3B fits are for very precise work. 5A and 5B are interference fits such as are used on studs that are to be installed semipermanently.

The specification of threaded parts is illustrated by the following examples:

1. An externally threaded part with a nominal major diameter of $\frac{3}{8}$ in., a coarse thread of 16 threads per inch, and a 2A tolerance is

   $\frac{3}{8}$-16 UNC-2A.

2. An internally threaded part with a nominal major diameter of $\frac{5}{8}$ in., a fine thread of 18 tpi, and a 2B tolerance is designated

   $\frac{5}{8}$-18 UNF-2B.

3. Major diameters less than $\frac{1}{4}$ in. are specified by a gauge number; thus a thread with a nominal major diameter of 0.164 in. and a coarse thread of 32 tpi is designated

   8-32 UNC.

The specifications of the UNC and UNF thread forms are listed in Table 1.

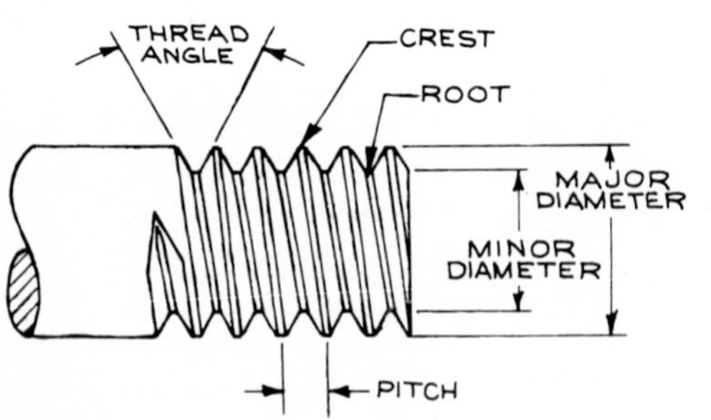

Figure 2

Example 10.2 Description of threaded fasteners. From John H. Moore, Christopher C. Davis, Michael A. Coplan, *Building Scientific Apparatus*, Addison-Wesley Publishing Company, 1983.

Table 1 AMERICAN STANDARD UNIFIED AND AMERICAN NATIONAL THREADS

Size (nominal diameter)	Coarse (NC, UNC)		Fine (NF, UNF)	
	Threads per Inch	Tap Drill[a]	Threads per Inch	Tap Drill[a]
0 (0.060)			80	3/64
1 (0.073)	64	No. 53	72	No. 53
2 (0.086)	56	No. 50	64	No. 50
3 (0.099)	48	No. 47	56	No. 45
4 (0.112)	40	No. 43	48	No. 42
5 (0.125)	40	No. 38	44	No. 37
6 (0.138)	32	No. 36	40	No. 33
8 (0.164)	32	No. 29	36	No. 29
10 (0.190)	24	No. 25	32	No. 21
12 (0.216)	24	No. 16	28	No. 14
1/4	20	No. 7	28	No. 3
5/16	18	Let. F	24	Let. I
3/8	16	5/16	24	Let. Q
7/16	14	Let. U	20	25/64
1/2	13	27/64	20	29/64
9/16	12	31/64	18	33/64
5/8	11	17/32	18	37/64
3/4	10	21/32	16	11/16
7/8	9	49/64	14	13/16
1	8	7/8	12	59/64

Note: ASA B1.1-1960.
[a] For approximately 75% thread depth.

Pipes and pipe fittings are threaded together. Pipe threads are tapered so that a seal is formed when an externally threaded pipe is screwed into an internally threaded fitting. The American Standard Pipe Thread, designated NPT for National Pipe Thread, has a taper of 1 in 16. The diameter of a pipe thread is specified by stating the nominal internal diameter of a pipe that will accept that thread on its outside. For example, a pipe with a nominal internal diameter of $\frac{1}{4}$ in. and a standard thread of 18 tpi is designated

$$\tfrac{1}{4}\text{-18 NPT}$$

or simply

$$\tfrac{1}{4}\text{-NPT.}$$

The American Standard Pipe Thread specifications are listed in Table 2.

Table 2 AMERICAN STANDARD TAPER PIPE THREADS

Nominal Pipe Size	Actual O.D. of Pipe	Threads per Inch	Normal Length of Engagement by Hand	Length of Effective Thread
1/8	0.405	27	0.180	0.260
1/4	0.540	18	0.200	0.401
3/8	0.675	18	0.240	0.408
1/2	0.840	14	0.320	0.534
3/4	1.050	14	0.340	0.546
1	1.315	11½	0.400	0.682
1¼	1.660	11½	0.420	0.707
1½	1.900	11½	0.420	0.724
2	2.375	11½	0.436	0.756
2½	2.875	8	0.682	1.136
3	3.500	8	0.766	1.200

Note: ASA B2.1-1960.

The common forms of machine screws are illustrated in Figure 3. *Hex-head* cap screws are ordinarily available in sizes larger than $\frac{1}{4}$ in. They have a large bearing surface and thus cause less damage to the surface under the head than other types of screws. A large torque can be applied to a hex-head screw, since it is tightened with a wrench.

Slotted-head screws are available with *round, flat,* and *fillister* heads. The flat head is countersunk so that the top of the head is flush. Fillister screws are preferred to round-head screws, since the square shoulders of the head provide better support for the blade of a screwdriver.

Socket-head cap screws with a hexagonal recess are preferred for instrument work.

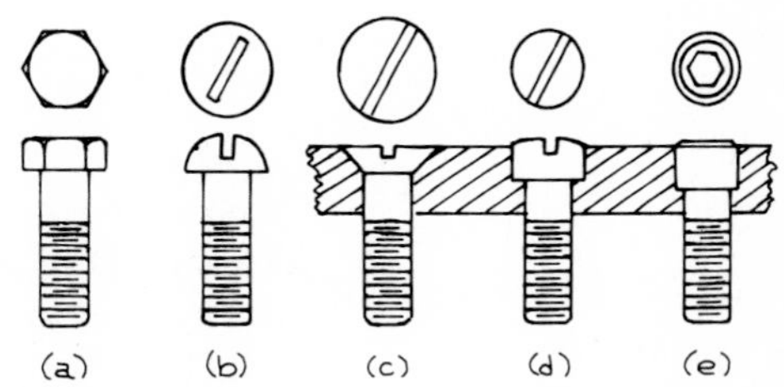

Figure 3

These are also known as *Allen* screws. L-shaped, straight, and ball-pointed hex drivers are available that permit Allen screws to be installed in locations inaccessible to a wrench or screwdriver. An Allen screw can only be driven by a wrench of the correct size, so the socket does not wear so fast as the slot in a slotted-head screw. These screws have a relatively small bearing surface under the head and thus should be used with a washer.

Setscrews are used to fix one part in relation to another. They are often used to secure a hub to a shaft. In this application it is wise to put a flat on the shaft where the setscrew is to bear; otherwise the screw may mar the shaft, making it impossible for it to be withdrawn from the hub. Setscrews should not be used to lock a hub to a hollow shaft, since the force exerted by the screw will deform the shaft. In general, setscrews are suitable only for the transmission of small torque, and their use should be avoided if possible.

Machine screws shorter than 2 in. are threaded their entire length. Longer screws are only threaded for part of their length. Screws are usually only available with class 2A coarse or fine threads.

DISKETTES AND A PC: Describing Objects and Systems in a Manual for a Broad-Based Audience

This description of a diskette is from a manual written for a broad-based audience assumed to be absolute beginners (Example 10.3). The description is organized in a way similar to the earlier excerpts on rivets and threaded fasteners (definition, division of whole into parts, figure reference, description of parts); there are, however, striking differences in the language and the design.

Logic and Organizational Pattern. The excerpt starts with an operational definition, and then divides the whole into parts. Detail is introduced well after the category is established. There are many examples to make the abstract more concrete. The new is compared to the old. For instance, diskettes are compared to records and recording tape.

Language. The writers strenuously attempt to limit their use of specialized vocabulary. For instance, instead of naming the diskette's material, they call it "a flexible material with a surface that can be magnetized." The language is simplified throughout the text.

The reader is addressed directly ("Diskettes are used in your system"; "Your computer can read information stored on a diskette").

(Text continues on page 230)

All About Diskettes

What Are Diskettes

Diskettes are used in your system to store information.

Each diskette is permanently encased in a square protective jacket which measures 5¼ in. (133 mm).

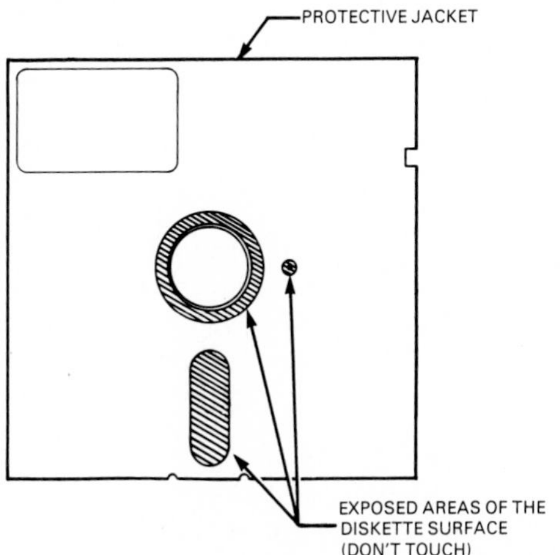

The diskette is *similar* to a phonograph record in shape and function. Like a record, the diskette *spins* when being used. Both have "information" stored on them. The "information" on a record can be music. The music or information stored on a record is permanent and cannot be changed.

The diskette can store computer programs or files of business information. Your computer can "read" information stored on a diskette, and you can change that information. Your computer can also "write" new information onto a diskette. The computer's ability to *read and write* makes the diskette an essential part of your system.

Let's take a closer look at the different parts of a diskette.

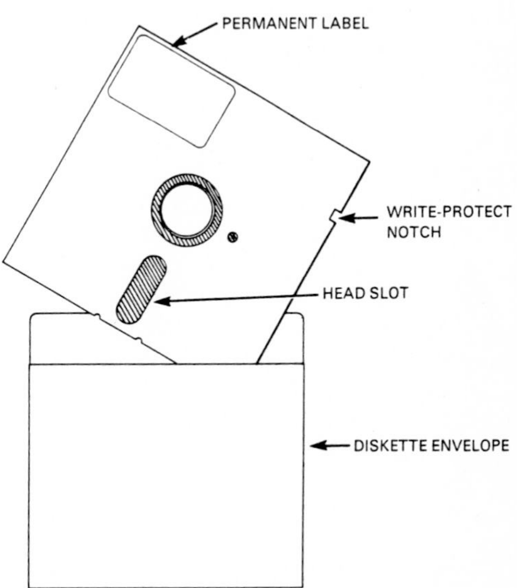

The diskette is composed of a flexible material which has a surface that can be magnetized. In this way, diskettes are similar to magnetic recording tape. Single-sided diskettes have information stored on only one side while double-sided diskettes can have information on both sides. A black, protective jacket permanently encloses the diskette. The surface of a diskette is very sensitive to contamination by things like finger smudges, dust, cigarette ashes, food particles, and scratches. The jacket protects *most* of the diskette surface from the environment.

Several areas of the diskette surface are exposed. One of these is the *head slot*. When the

Example 10.3 Description of diskettes. Courtesy Texas Instrument.

Example 10.3 (*continues*)

diskette is in the diskette drive, the computer reads and writes information by using a *read/write head*, which comes in contact with the diskette surface through the head slot.

Most diskettes have a notch cut out of the protective jacket along one edge. This is called the *write-protect notch*. When this notch is covered with foil tape (that comes with new diskettes), the computer can read the information from the diskette but can't *write* anything on it. Covering this notch is a convenient way to protect information that you don't need to change very often. Then you can't *accidentally* write on this diskette and destroy this information. Another way to protect yourself from this happening is to make a copy, or *backup* of a diskette. The procedure for backing up is discussed in detail in Chapter 3.

Diskettes come with a manufacturer's label permanently attached to the upper left-hand corner. This label may or may not give you an idea of what's on the diskette. Most often, you attach a temporary label next to the permanent label. We recommend that you write, on the temporary label, any information that helps you identify what's on the diskette.

The diskette envelope is used to hold the diskette when it's not being used. It helps protect the exposed surfaces of the diskette.

Some Tips on Handling Diskettes

Very often diskettes contain months of your work. Therefore, handling them carefully is a good practice. Recreating information from a damaged diskette is typically a time-consuming task that could have been prevented.

We suggest you follow these guidlines when handling your diskettes:

1. Don't touch the exposed areas of the diskette. Keep these areas away from dust, smoke particles, and other contaminants.

   When you are not using your diskettes, keep them stored upright in their envelopes to prevent damage.
2. Store your diskettes away from magnetic devices (such as magnets, televisions, tape recorders, motors, etc.) which can cause distortion of data stored on the diskette.
3. Don't place your diskettes in very hot or cold environments. Diskettes left in a parked car on a warm day can be permanently damaged. In general, if you store your diskettes in a place that's comfortable enough for you, your diskettes will be safe.
4. Whenever you want to write on a label already on a diskette, use *only* a soft, felt-tipped pen. Otherwise, you might damage the diskette surface.
5. Handle your diskettes *gently*, especially when inserting them or removing them from the diskette drive. It's also a good practice to remove diskettes from the drives before turning off power to your system.

   Don't remove or insert diskettes from diskette drives when the red in-use lights are on. This may cause permanent damage to your diskettes or diskette drives.
6. Diskettes *eventually wear out*. Make copies (back-ups) of all important diskettes on a regular basis.

Taking care of your diskettes will save you both time and headaches. *We strongly recommend treating your diskettes like your most valuable personal or business records.* With reasonable care, your diskettes will deliver trouble-free operation.

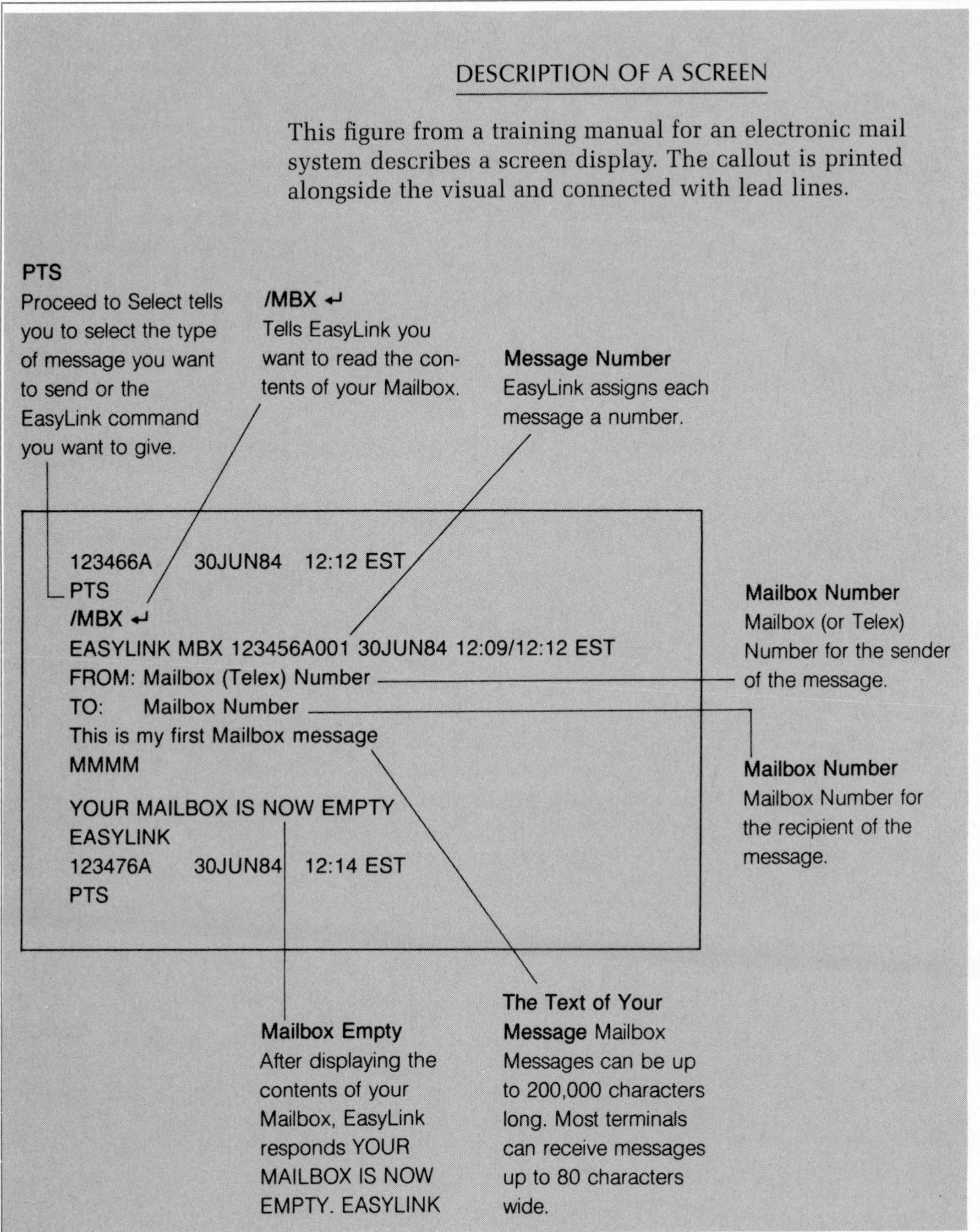

Box 10.1 Description of a screen. Courtesy Western Union *Easy Link Training Manual*.

Direct questions are used for subheadings ("What are diskettes?"). In the main, the verbs are passive in the first section to keep the spotlight on the diskettes ("diskettes are used"; "each diskette is encased"; "the diskette is composed"). When the time comes for instruction, the writers switch entirely to active voice.

Visual Display and Design. The paragraphs are short, permitting much white space. All key words are italicized. Italics are even used to emphasize "accidentally," "write-protect notch," and other words and phrases that would never be emphasized if this test were for a more sophisticated audience. Separate figures with labels are used to introduce the parts of the whole, and to give a warning. If the audience were more sophisticated, these two figures might be combined.

LOGIC AND ORGANIZATIONAL PATTERNS IN DESCRIPTIONS

Obviously the structure of a description depends in part on the subject. For example, a description that distinguishes between species of the same genus of trees usually entails a listing of characteristics for leaves, flowers, fruits, bark, size, silhouette, and habitat. This list becomes the organizational skeleton for the description. If, on the other hand, you are describing the housing in an island village, you might use a spatial pattern: first the roofs of the huts, then the walls, and so on.

It's the writer's job to emphasize the logic of the material with a clear organizational pattern that

- makes a structure within which details can be introduced
- creates a path the reader can follow

The description of a steel trap (Figure 10.1) has a spatial pattern: The writer chooses three views of the object, then describes each. "In the top view (Figure 10.1a) a circular base underlies the long basal bar to which two springs are riveted. The ends of the toothed jaws are supported in jaw posts, each with a single pin. When the springs are depressed, the jaws may be opened. A pliable, circular

Figure 10.1 Views of a trap: (a) Top view of a set trap; (b) Side view of a closed trap; (c) End view of jaws. From L. Mascall, *A Booke of Fishing with Hook and Line,* John Wolfe, 1590.

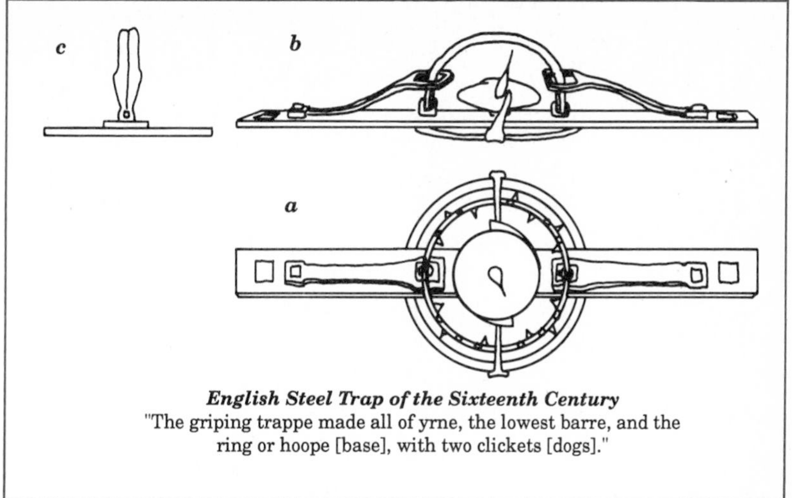

English Steel Trap of the Sixteenth Century
"The griping trappe made all of yrne, the lowest barre, and the ring or hoope [base], with two clickets [dogs]."

Box 10.2

HONING OBSERVATIONAL SKILLS: WRITING A DESCRIPTION DEVELOPS "CLARITY OF THOUGHT"

Mitchell R. Malachowski, who teaches Organic Chemistry at the University of San Diego, develops students' observational skills by having them write descriptions of unfamiliar equipment. For instance, before the class does a fractional distillation, they are shown the unfamiliar apparatus and asked to spend 5–10 minutes describing it and discussing what function it might serve. The next week he shows the equipment again, this time explaining its use. After completing the experiment, students again spend 5–10 minutes writing about the equipment. "Invariably the students show a fundamental understanding of the equipment. . . . [The process] engages students in writing assignments in the sciences and therefore shows them the clarity of thought that this task develops," the professor comments in the *Journal of Chemical Education.*

pan has a central mounting and receives at its edge the two 'clickets,' or 'dogs,' which restrain the open jaws. Meat or other bait may be impaled on a sharp iron pin which projects upward from the center of the pan. Since the pan rotates on its central mounting, the dogs may be released either by the tread of an animal's foot or by the disturbance caused when an animal tugs at the bait."

In the side view (Figure 10.1b) there are at the ends of the long barre two square holes, "which holes are made to pinne the long barre fast to the ground, when yee set or tyie him in any place at your pleasure."

Figure 10.1c is an end view of the jaws, with springs removed.

Example 10.4 shows another approach to the use of an organizational pattern for a description. This time the author uses an overview/listing pattern. The parts of the whole are set off typographically with boldface headings and hanging paragraphs.

Example 10.4 Description of a commercial laser. From Ronald J. Sanderson, *Lasers in Metalworking*, Tech Tran Corp., 1983.

Commercial industrial lasers consist of a basic laser plus a number of supporting elements. Depending on the nature of the supporting elements used, a commercial laser can range in configuration from a simple package consisting of little more than the laser cavity itself to a complete turnkey material processing system. When a laser is purchased for use in a specific manufacturing operation, it is generally installed as a complete system. Most leading laser equipment suppliers offer lasers in the form of machining or processing systems. A basic laser system generally consists of the following six components:

Laser The laser includes the laser cavity, gas or rod, optics, and power supply (that is, the elements necessary to generate a laser beam).

Support Structure A sturdy base is required to support the laser, the work station, and the beam delivery system, so as to prevent vibration.

Beam Delivery System Since laser beams can be reflected, focused, or modified using the same principles as conventional light, a laser processing system generally employs some form of optical delivery system to define the shape of the laser beam and direct it to the proper location to work on the material.

System Controls Operator controls for the laser are generally mounted onto a control panel located at a remove from the laser so the operator does not need to work next to the laser.

Exhaust System Both harmful fumes and debris from processing must be removed from the laser system through use of an exhaust system.

Laser Enclosure Another standard safety feature is a system enclosure over the laser and workstation to protect the operator from moving equipment, reflected laser beam energy, and the effects of harmful fumes from processed materials.

OBSERVATIONAL CHECKLIST FOR AN ENGINEERING INSPECTION

People in the market for a new home often call in a civil engineer to inspect the building they are thinking of buying. The engineer reports on the physical condition, using an observational checklist. The checklist is a highly focused form of descriptive writing, with the categories dictated by the nature of the inspection.

Example 10.7 is an excerpt from the completed checklist. It is prefaced by two important documents: a letter of transmittal (Example 10.5), and a glossary (Example 10.6).

- *The letter of transmittal* (Example 10.5) functions as an abstract, giving the reader the important points in a nutshell. It provides the overview so important in technical writing.
- *The definitions of terms* (Example 10.6) help the reader cope with the specialized grading system of the report.

OBSERVATIONAL CHECKLIST FOR A PHYSICAL EXAMINATION

You may use your observational skills to study a fish, a three-story building, a crying infant, a sugar pine. The categories you will use depend on the subject area, but the emphasis on clear language and accurate detail will remain constant.

The following list and informal diagrams are from a set of guidelines to medical students for writing the description of a physical examination. The author advises students to be "quite detailed, since doing so is the only way to build your descriptive skills, vocabulary and speed—admittedly a painful, tedious process." Students are also advised to record *all* data, both positive and negative, and not to trust to memory, no matter how vivid the detail at the moment of observation. Data not recorded are data lost. (For a related discussion in laboratory notebooks, see Chapter 8.)

These data are recorded on printed forms with categories (Figure 10.2 on page 238). The observer may supplement the data with diagrams (Figure 10.3 on page 238). "Diagrams add greatly to the speed and ease with which a record communicates its message," the author advises.

Example 10.5 Letter of transmittal introduces the checklist.

HOUSING INSPECTION INC.

INSPECTORS OF
HOMES, APARTMENTS, &
COMMERCIAL BUILDINGS

117 E. 47 Street
Fairfield, New Jersey

1 August 1987

Dr. George Sinclair
147 E. 18 Street
Fairfield, New Jersey

Dear Dr. Sinclair:

Attached is the inspection report for the house at 87 Beech Street, Fairfield, New Jersey, which was inspected by us on Friday, 26 July 1987.

The house is being used as a three-family unit. It is a four-story, semi-attached, mortar-faced structure that rests on a cellar. There is no garage.

The house is serviced by sewers and municipal water. It is supplied with piped gas. Electric service comes in underground.

Considering the age of the house, its original construction, and its present condition, it is our opinion that the exterior is in moderately poor condition and the interior is in moderately good condition.

Repairs and local upgrading as indicated in the report can significantly improve the condition of the house.

The more prominent items of concern are starred in the report.

Sincerely,

(Name, Title)

Example 10.6 Definitions of terms preface the checklist.

DEFINITION OF TERMS

TERM	DEFINITION
GOOD (G)	MATERIAL AND WORKMANSHIP ARE QUITE SATISFACTORY. POSSIBLE SUPERFICIAL DEFECTS.
MODERATELY GOOD (MG)	MINOR DEFECTS OR MINOR WEAR. WORKMANSHIP GENERALLY SATISFACTORY.
FAIR (F)	MODERATE DEFECTS OR AGING WHICH MAY REQUIRE REPAIR AND/OR UPGRADING.
MODERATELY POOR (MP)	MAJOR DEFECTS AND/OR WEAR WHICH PROBABLY WILL REQUIRE REPAIR OR REPLACEMENT.
POOR (P)	DEFECTS AND/OR WEAR ARE OF SUCH A NATURE THAT REPAIRS AND/OR REPLACEMENT ARE NOW INDICATED.

Example 10.7 Checklist for engineering inspection.

BUILDING EXTERIOR:

FRONT WALL: PAINTED BRICK WITH STONE TRIM
    CONDITION: GENERALLY FAIR

RIGHT SIDE WALL: ROUGH MORTAR-FACED
    CONDITION: FAIR TO MODERATELY POOR

LEFT SIDE WALL: ATTACHED OR ABUTS THE ADJACENT BUILDING

REAR WALL: MORTAR FACED
    CONDITION: GENERALLY FAIR

REAR ONE-STORY EXTENSION: MORTAR-FACED
    CONDITION: FAIR

    LEFT SIDE OF EXTENSION WAS NOT OBSERVABLE.

    MANY CRACKS OBSERVED ON SIDE WALL

Example 10.7 *(continues)*

ALSO, MUCH SPALLING AND SOME LOOSE BRICK SEEN ON THIS WALL.

SOME CRACKS OBSERVED ON REAR WALL. CRACKS ARE USUALLY DUE TO SETTLEMENT.

CRACKS AND SPALLING MAY PERMIT PENETRATION OF WATER OR MOISTURE. PROPER REPAIR AND CRACK SEALING IS RECOMMENDED.

REPAIR OF LOOSE BRICKWORK ALSO RECOMMENDED.

SOME CRACKING AND DETERIORATION ALSO OBSERVED ON WALLS' STONE TRIM.

STONE CRACKING AND DETERIORATION ARE USUALLY DUE TO AGE AND WEATHERING AND REPAIRS ARE RECOMMENDED.

MANY MORTAR JOINTS ON EXTERIOR FRONT WALL APPEAR SOFT AND PART OF THEM ARE LOOSE AND ERODED.

THIS CONDITION MAY BE DUE TO AGING, WEATHERING OR QUALITY OF MORTAR INITIALLY USED. SOFT OR ERODED JOINTS WILL OFTEN BE CONDUCIVE TO THE PENETRATION OF MOISTURE THROUGH THE WALLS.

SINCE THE MORTAR JOINTS IN THE BRICKWORK ARE SOFT AND/OR LOOSE, IT IS RECOMMENDED THAT THE JOINTS BE TUCK-POINTED TO RESTORE THEIR STRUCTURAL INTEGRITY AND ALSO TO WATERPROOF THE WALLS. VARIOUS CLEAR COATINGS MAY THEN BE APPLIED FOR ADDED RAIN RESISTANCE.THIS LIQUID WILL THEN SEAL POROUS MORTAR OR BRICKWORK IN SOUND MATERIAL.

THE EXTERIOR WALLS HAVE BEEN COATED OR FACED AS NOTED ABOVE. THIS IS OFTEN DONE TO WATERPROOF THE WALLS WHEN THE MORTAR JOINTS BECOME SOFT AND/OR POROUS INSTEAD OF DOING THE MORE EXPENSIVE JOB OF TUCK POINTING OF THE JOINTS.

VARIOUS DEGREES OF WARPING WERE OBSERVED ON THE SIDE WALL.
THE WARPING MAY BE INDICATIVE OF HIDDEN DAMAGE BELOW THE MORTAR FACING.

THE OWNER INDICATES THAT THERE USED TO BE AN ATTACHING HOUSE ON THE RIGHT SIDE THAT HAS SINCE BEEN REMOVED. THE CLIENT IS ADVISED THAT ATTACHED HOUSES NORMALLY SUPPLY LATERAL SUPPORT TO EACH OTHER, AND

WHEN ONE IS REMOVED THE AMOUNT OF LATERAL STABILITY IN THE REMAINING HOUSE MAY DIMINISH UNLESS ADEQUATE NEW BRACING WAS PROVIDED. ALTHOUGH A FEW WALL BRACES WERE OBSERVED, IT CANNOT BE DETERMINED IF ADEQUATE LATERAL BRACING WAS PROVIDED. FURTHER INVESTIGATION IS RECOMMENDED

SLIGHT WARPING AND BULGING OBSERVED ON THE FRONT AND REAR WALLS.

THE CORNICE ON THE FRONT WALL, TOP, APPEARS TO HAVE SUFFERED DAMAGE AND REPAIRS ARE RECOMMENDED.

BUILDING TRIM: PAINTING CONDITION: FAIR

WINDOW AND DOOR FRAMES:

THE EXTERIOR WOOD TRIM AROUND THE WINDOWS AND DOORS WHERE OBSERVABLE IN FAIR TO MODERATELY POOR CONDITION.

MUCH SPLITTING AND SOME DRY ROT OBSERVED.

SOME OF THE WINDOW FRAMES APPEAR TO BE SOMEWHAT OUT OF PLUMB. THIS IS PROBABLY INDICATIVE OF THE MANNER IN WHICH THE HOUSE HAS SETTLED.

EXTERIOR DOORS: CONDITION: FAIR

CAULKING: PARTIAL; MUCH RECAULKING IS INDICATED.

STORM WINDOWS: MOSTLY INSTALLED ALUMINUM

CONDITION: FAIR

SOME PUTTY REPAIR IS REQUIRED.

PATENT DESCRIPTIONS

Article I, Section 8, of the Constitution of the United States says that "The Congress shall have Power . . . to promote the Progress of Science and useful Arts by securing for limited Times to Authors and Inventors the exclusive Right to their respective Writings and Discoveries." Samuel Hopkins received the first patent, for an improvement "in the making of Pot ash and Pearl ash by a new Apparatus and Process," in 1790 (Figure 10.4).

Figure 10.2 Descriptive data are recorded on printed medical forms.

Date problem entered	No.	Active problems	Inactive problems
	1		

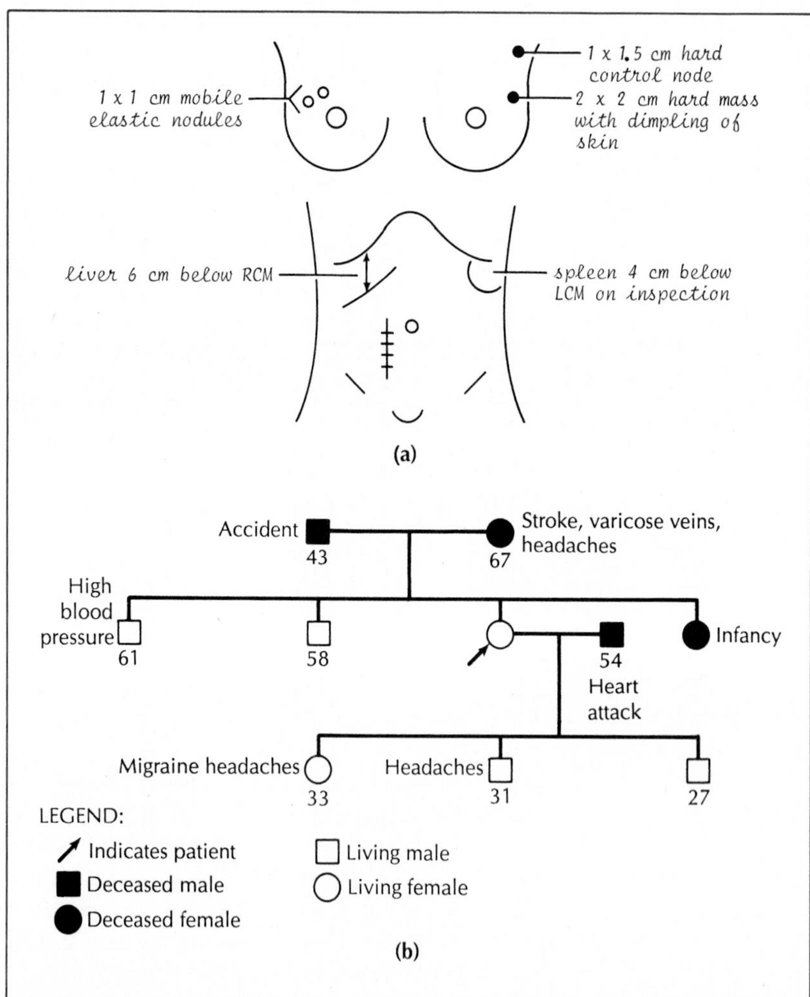

Figure 10.3 Observers may supplement data with informal drawings (a) and diagrams (b).

The United States.

To all to whom these Presents shall come Greeting.

Whereas Samuel Hopkins of the City of Philadelphia and State of Pensylvania hath discovered an Improvement, not known or used before such Discovery, in the making of Pot ash and Pearl ash by a new Apparatus and Process; that is to say, in the making of Pearl ash 1st. by burning the raw Ashes in a Furnace, 2d. by dissolving and boiling them when so burnt in Water, 3rd. by drawing off and settling the Ley, and 4th. by boiling the Ley into Salts which then are the true Pearl ash; and also in the making of Pot ash by fluxing the Pearl ash so made as aforesaid; which Operation of burning the raw Ashes in a Furnace, preparatory to their Dissolution and boiling in Water, is new, leaves little Residuum; and produces a much greater Quantity of Salt: These are therefore in pursuance of the Act, entituled "An Act to promote the Progress of useful Arts", to grant to the said Samuel Hopkins, his Heirs, Administrators and Assigns, for the Term of fourteen Years, the sole and exclusive Right and Liberty of using, and vending to others the said Discovery, of burning the raw Ashes previous to their being dissolved and boiled in Water, according to the true Intent and Meaning of the Act aforesaid. In Testimony whereof I have caused these Letters to be made patent, and the Seal of the United States to be hereunto affixed. Given under my Hand at the City of New York this thirty first Day of July in the Year of our Lord one thousand seven hundred & Ninety.

G. Washington

City of New York July 31st. 1790.–

I do hereby Certify that the foregoing Letters patent were delivered to me in pursuance of the Act, entituled "An Act to promote the Progress of useful Arts"; that I have examined the same, and find them conformable to the said Act.

Edm: Randolph Attorney General for the United States.–

Figure 10.4 The first U.S. patent issued under the patent bill was for the making of potash, a chemical used in the manufacture of soap. The patent was signed by President Washington and Attorney General Edmund Randolph.

Since then, business has been brisk. In the first years of patents, the examiners required proof of newness or novelty. But the Board, composed of the Secretary of State, the Secretary of War, and the Attorney General, was soon swamped with applications. In 1793, a new act placed patent issues under the Secretary of State in the Department of State. The requirements did not include proof of novelty. Inventors needed only provide a description, a drawing, a model, and the fee.

Without any examination for novelty, there were cases of one conflicting patent after another; finally an act was passed in 1836, introducing the examination system calling for proof of novelty.

Anyone interested in learning about descriptive technical writing will be well rewarded by studying patents; they are a neglected literary form of technical writing. To look through sample patents is to see how writers struggled in their descriptions and drawings for exactness. The patents make very good reading. Figure 10.5, for instance, is Hunt's patent of 1849 for a "new and useful improvement in the Make or Form of Dress-Pins." Hunt's physical description of the device is a prose model of accuracy and clarity.

With Leo Szilard, Albert Einstein obtained 19 patents, most related to improvement in heat pumps. The papers Einstein published in the patent literature were important and extensive.

Some other fascinating patents include these:

Samuel F. B. Morse. Telegraph signs. "Improving in the Mode of Communicating Information by Signals by the Application of Electro-Magnetism." #1,647

Abraham Lincoln. A device for "Bouying Vessels over Shoals." #6,469

Alexander Graham Bell. Telephone. "Improvement in Telegraphy." #174,465

Thomas A. Edison. Phonograph. "Improvement in Phonograph or Speaking Machines." #200,521

Thomas A. Edison. Incandescent lamp. "Electric Lamp." #223,898

Nikola Tesla. Induction type of electric motor. "Electrical Transmission of Power." #382,280

Herman Hollerith. Punch card accounting. "Art of Compiling Statistics." #395,782

Charles M. Hall. Process for the "Manufacture of Aluminum." #400,665

Thomas A. Edison. "Speaking-Telegraph." #474,230

Thomas A. Edison. Motion picture projector. "Apparatus for Exhibiting Photographs of Moving Objects." #493,426

Guglielmo Marconi. Wireless telegraphy. "Transmitting Electrical Signals." #586,193

Thomas A. Edison. "Kinetographic Camera." #589,168

Rudolf Diesel. Diesel engine. "Internal-Combustion Engine." #608,845

Orville Wright; Wilbur Wright. Airplane with motor. "Flying-Machine." #821,393

George Washington Carver. Cosmetics. "Cosmetic and Process of Producing the Same." #1,522,176

Enrico Fermi; Leo Szilard. "Neutronic Reactor." #2,708,656

UNITED STATES PATENT OFFICE.

WALTER HUNT, OF NEW YORK, N. Y., ASSIGNOR TO WM. RICHARDSON AND JNO. RICHARDSON.

DRESS-PIN.

Specification of Letters Patent No. 6,281, dated April 10, 1849.

To all whom it may concern:

Be it known that I, WALTER HUNT, of the city, county, and State of New York, have invented a new and useful Improvement in the Make or Form of Dress-Pins, of which the following is a faithful and accurate description.

The distinguishing features of this invention consist in the construction of a pin made of one piece of wire or metal combining a spring, and clasp or catch, in which catch, the point of said pin is forced and by its own spring securely retained. They may be made of common pin wire, or of the precious metals.

See Figure 1 in the annexed drawings (which are drawn upon a full scale, and in which the same letters refer to similar parts,) which figure presents a side view of said pin, and in which is shown the three distinct mechanical features, viz: the pin A, the coiled spring B, and the catch D, which is made at the extreme end of the wire bar C, extended from B. Fig. 2 is a similar view of a pin with an elliptical coiled spring, the pin being detached from the catch D and thrown open by the spring B. Fig. 3 gives a top view of the same. Fig. 4 is a top view of the spring made in a flat spiral coil. Fig. 5 is a side view of the same.

Any ornamental design may be attached to the bar C, (see Figs. 6, 7 and 8,) which combined with the advantages of the spring and catch, renders it equally ornamental, and at the same time more secure and durable than any other plan of a clasp pin, heretofore in use, there being no joint to break or pivot to wear or get loose as in other plans. Another great advantages unknown in other plans is found in the perfect convenience of inserting these into the dress, without danger of bending the pin, or wounding the fingers, which renders them equally adapted to either ornamental, common dress, or nursery uses. The same principle is applicable to hair-pins.

My claims in the above described invention, for which I desire to secure Letters Patent are confined to the construction of dress-pins, hair-pins, &c., made from one entire piece of wire or metal, (without a joint or hinge, or any additional metal except for ornament,) forming said pin and combining with it in one and the same piece of wire, a coiled or curved spring, and a clasp or catch, constructed substantially as above set forth and described.

WALTER HUNT.

Witnesses:
JOHN M. KNOX,
JNO. R. CHAPIN.

Figure 10.5 Hunt's patent for the Dress Pin. Patent writers seek exactness in their descriptions.

Arthur L. Schawlow; Charles H. Townes. Laser (optical maser). "Masers and Maser Communications System." #2,929,922

Lorenzo Dow Adkins. Water wheel. "Spiral—Bucket Water—Wheel." #1,154

John Rand. Collapsible tube (as for toothpaste). "Improvement in the Construction of Vessels or Apparatus for Preserving Paint, etc." #2,252

Linus Yale. Cylinder lock. "Door—Lock." #3,630

Elias Howe, Jr. Sewing machine. "Improvement in Sewing Machines." #4,750

Richard M. Hoe. Rotary printing press. "Improvement in Printing-Presses." #5,188

John Gorrie. Ice-making machine. "Improved Process for the Artificial Production of Ice." #8,080

Horace Smith; Daniel B. Wesson. "Improvement in Fire-Arms." #10,535

Isaac M. Singer. "Improvement in Sewing-Machines." #13,661

R. J. Gatling. Machine gun. "Improvement in Revolving Battery-Guns." #36,836

Joseph F. Glidden. Barbed wire. "Improvement in Wire Fences." #157,124

Thaddeus Hyatt. Reinforced concrete. "Improvement in Composition Floors, Roofs, Pavements, etc." #206,112

Simon Lake. Even keel submarine. "Submarine Vessel." #581,213

Felix Hoffmann. Aspirin. "Acetyl Salicylic Acid." #644,077

King C. Gillette. Safety razor. "Razor" (title for both patents). #775,134 and #775,135

Robert H. Goddard. Rocket engine. "Rocket Apparatus." #1,103,503

Mary Phelps Jacob (Caresse Crosby). "Brassiere." #1,115,674

Harry Houdini. "Diver's Suit" (permitting escape). #1,370,376

Clarence Birdseye. Packaged frozen foods. "Method of Preparing Food Products." #1,773,980 and #1,773,981

Edwin H. Land; Joseph S. Friedman. "Polarizing Refracting Bodies." #1,918,848

Charles B. Darrow. Monopoly. "Board Game Apparatus." #2,026,082

Charles P. Ginsburg; Shelby F. Henderson, Jr.; Ray M. Dolby; Charles E. Anderson. Videotape recorder. "Broad Band Magnetic Tape System and Method." #2,956,114

SUMMING UP

• You may use your observational skills to study and describe a fish, a three-story building, a crying infant, a sugar pine. The categories you'll use to develop the description will depend on the subject area, but the emphasis on clear language, accurate detail, and the drawing of conclusions remains a constant imperative.

• Observation leads to description. In science and technology, the two actions are closely linked. Good technical descriptions grow out of the interplay of logic, organizational pattern, language, and visuals.

• The pattern of technical description is often definition, figure reference, division of parts, details.

• If you include a figure callout, coordinate the wording with the wording in the text. If you are describing an object, try italicizing or underlining words in the text that echo the callout.

• Use tables for detail; leave white space so the eye can move in on what it needs.

• Try figures for contrast and comparison. Accompany figures with captions and callouts.

• For a broad-based audience, use examples to make descriptions more concrete. Limit specialized vocabulary when possible. Address the reader directly.

• It's the writer's job to emphasize the logic of the material with a clear organizational pattern that

- makes a structure within which details can be introduced
- creates a path the reader can follow

The pattern depends on the material. It may be spatial or chronological. It may be overview/listing. It may be definition, summary statement, division into parts.

EXERCISES

1. Write a description of a place of business. It may be a hamburger restaurant near your school or a local hardware store. Whichever place you choose, spend some time there observing. Pay attention to details. Collect bits of dialogue. Don't interview people for this assignment—do not go around collecting facts about when the place was founded, who owns it, and what its gross annual revenues are. Instead, use your powers of observation and description to convey a sense of what the place is like, what it might feel like to be there as an employee or a customer. Use the active voice; try to do more showing than telling. Limit the description to one or two doublespaced pages.

2. Prepare a brief technical description of an object, system, or behavior with which you are familiar. Some suggestions:

a spreadsheet

a boomerang

a keyboard for a word processor

a fugue

the lighting board for a stage production

a prospectus

a doublecrostic puzzle

speed skates

a sewing machine

3. Orientation talk. Prepare a brief oral description of an object or system. Assume the talk is part of an orientation for new students or employees. Some possible topics:

An introduction to

- special services in the school library
- exercise equipment available to students
- soundstage and lighting equipment for theatrical productions
- equipment in a laboratory
- institutional cooking equipment
- mailroom services
- photoduplicating services

4. Locate and photocopy two technical descriptions that you think are either well or poorly done. In a small group, discuss the strengths or weaknesses of the pieces you choose. Make copies so that other people in the group can read the descriptions.

To get you started, here are several excellent descriptions. The first is R. J. Petri's description of his dish (Example 10.8). The second is a selection of passages from *Arctic Dreams*, by Barry Lopez (Example 10.9).

5. These definitions, taken from Tracy Kidder's *House* (Example 10.10), are descriptive. In a small group, discuss the descriptive techniques Kidder uses in his glossary. The writing provides many examples of the overlap between definition and description.

Example 10.8 Descriptive writing: a minor modification of the plating technique of Koch.

In order to perform the gelatin plate technique of Koch, it is necessary to have a special horizontal pouring apparatus. The poured plates are then placed over one another in layers on small glass shelves in a large bell jar. In many cases it would be desirable to carry out the procedure with less equipment, especially without the pouring apparatus. Since the first of the year I have been using flat double dishes of 10–11 cm. in diameter and 1–1.5 cm. high. The upper dish serves as a lid as usual and has a somewhat larger diameter. These dishes are sterilized by dry heat as usual and after cooling the nutrient gelatin containing the inoculum is poured in. The upper lid is lifted only slightly and used as a shield while the tube containing the gelatin, its edge previously flamed and cooled in the usual manner, is emptied into the bottom of the dish. Under these conditions contamination from airborne germs rarely occurs. The poured layer of gelatin soon hardens into a layer several millimeters thick which can be kept and observed for a long time because of the protecting upper lid. In studies of soil samples, sand, earth, and similar substances, it is advantageous to place the material in the dish and then pour the liquid gelatin over it. The material is well mixed with the gelatin by rotating the dish with short, intermittent movements. With the dimensions given, every spot on the gelatin surface is accessible with the low power microscope. Only when high power lenses are used is the area at the edge of the dish no longer accessible. The gelatin dries in these dishes quite slowly. They can be kept moist longer if 5–6 dishes are placed on top of one another on a disc of moist filter paper in a flat dish over which a bell jar is inverted. These dishes can be especially recommended for agar-agar plates, since agar-agar sticks poorly to simple glass plates unless special means are used. In addition, it is quite simple to count the colonies that have grown on the plates. The upper lid is replaced by a glass plate that has etched on it squares of known area. The colonies are then counted against a black background using a magnifier. The total area of the plate can be calculated from the diameter.

Freshwater ice usually begins to crystallize at 39.2°F, the temperature at which fresh water is densest. Sea ice does not achieve its maximum density, or start freezing, until it is cooled to 28.6°F. In its initial stages, the crystalline structure of sea ice incorporates brine and is not solid. It will therefore bend under a load before it fractures, while newly formed freshwater ice, brittle and also more transparent, will fracture suddenly, like a windowpane. (Because of its elasticity, even sea ice four inches thick is unsafe to walk on, while freshwater ice only half as thick will support a human being.)

In the absence of any wind or strong current, sea ice first appears on the surface as an oily film of crystals. This frazil ice thickens to a kind of gray slush called grease ice, which then thickens vertically to form an elastic layer of ice crystals an inch or so thick called nilas. Young nilas bends like watered silk over a light ocean swell and is nearly transparent (i.e., dark like the water). When it is about four inches thick, nilas begins to turn gray and is called young ice, or gray ice. When gray ice finally becomes opaque it is called first-year ice. And in these later stages it thickens more slowly.

By spring, first-year ice might be four to six feet thick. If it doesn't melt completely during the summer, it becomes second-year ice in the fall, tinted blue and much harder. (Brine in the upper layers has drained out during the summer and fresh ice crystals have filled the interstices.) Second-year ice continues to thicken, until it stabilizes after a few years at about 10 to 12 feet. If it remains unmelted through a second summer, it is simply called multiyear ice, or polar pack ice, to distinguish it from first- and second-year pack ice.* A more formidable version of multiyear ice, paleocrystic ice, forms in the open polar sea and may be 50 feet thick.

Pack ice may be consolidated in great expanses of rubble called field ice, or broken up by lanes of open water (leads) to a lesser or greater degree, creating various types of close and open pack ice—for example "close pack" (seven-tenths to nine-tenths coverage of the sea).†

Winds and currents almost always affect the formation of sea ice. If a swell comes up in a sludge of grease ice, for example, the crystals congeal in large, round plates that develop upturned edges from bumping against each other—a stage called pancake ice. If nilas is broken up by the wind, the separate sheets often ride up over each other in a characteristic interlocking pattern called finger-rafting.

* The term "pack ice" is used in a wide sense to include any accumulation of sea ice other than fast ice (ice attached to the shore), no matter what form it takes or how disposed.

† Because sea ice presents such a grave danger to ships, precisely defined terms are critical for accurate reporting. Terms that refer to the type and extent of ice coverage are standardized in the Scott Polar Institute's *Illustrated Glossary of Snow and Ice* and the Canadian Hydrological Services' *Pilot of Arctic Canada.*

Example 10.9 Descriptive writing: from Barry Lopez, *Arctic Dreams*, Charles Scribner's Sons, 1986.

A Note on Terminology

Many of the terms in the housebuilder's technical vocabulary are serviceable heirlooms handed down through the centuries, and some are so well worn as to be universally understood. Other terms are relatively obscure, though. For the uninitiated, talk among professionals—the sort of talk you overhear at lumberyards—can be painfully mysterious. With apologies to those who wield a hammer, I offer a short glossary:

A *gable* is a triangular section of wall formed by the two sloping sides of a pitched roof. In the Souweine house, the *temple end* is the *gabled end* that contains the front door and that has *pilasters* (square, ornamental columns set into a wall) at the corners. A *pediment* is a wide, low-pitched gable. An *ell*, as every New Englander knows, is a wing of a building attached at right angles to the main structure. A *lintel* is the horizontal upper part of a door or window frame; it supports part of the structure above the door or window. Bill and Jim used *lintel* for the slab of granite laid horizontally across the top of the back of the hearth.

Lumber dealers usually sell moldings by the *lineal* or *running foot*—in that system of measurement only the length of the piece of wood is described. But in most cases, lumberyards deal in *board feet:* one board foot consists of a piece of wood that is nominally one inch thick, one foot wide, and one foot long. Baffling arrays of terms describe the quality of lumber. *Clear* or *select* lumber should be knotless, and lumber described as *number two* will contain some live knots. *Waney* lumber, like a waning moon, is missing pieces of its edges.

Housebuilders turn wood into these things, among many others: *sills* (the timbers that rest on the foundation walls, the first course of wood); *floor joists* (timbers laid horizontally to make the frames that support floors); *studs* (timbers standing upright to make the frames of walls); *collar ties* (timbers laid horizontally below the peak of a roof and fastened to rafters on either side of a roof, each collar tie tying one pair of rafters together); *gussets* (small pieces of wood or steel, usually triangular, applied to the corners of construction for strength). *Furring strips* are narrow boards applied to a surface—a wall or ceiling, say—to provide a base for some covering material such as plasterboard. *Clapboards* are beveled pieces of wood applied in horizontal, partially overlapping rows to the sides of buildings, and *casings* are the mostly decorative frames that surround doors and windows. *Bedmolding* is molding that is fitted in the joint between a vertical and overhanging horizontal surface. Staircases like the one Jim built have several sturdy *newel posts* and in between many thinner *balusters,* all of which hold up the *banisters* that children like to slide down. On such a stair, the *treads* that people walk upon have rounded *nosings* on their outward-facing edges, and for decoration, underneath each nosing, a strip of the delicately sculpted *scotia* pattern of molding. For the bottommost section of the banister, Jim used a curved section of handrail called an *easement.*

Example 10.10 Descriptive writing: from Tracy Kidder, *House,* Houghton Mifflin Company.

Explaining a Process

CHAPTER

11

What do the following four examples have in common?

Example 11.1

In electronic imaging, a scene is recorded, converted to an electronic signal, stored on a disk or magnetic tape, and then displayed.

Example 11.2

Right now, sunlight is heating the atmosphere and setting it in motion. Hot moist air is being pulled up through the cold dry air, and a storm at some point is almost inevitable. As the hot air reaches higher and cooler altitudes, its moisture will condense into drops of rain and particles of ice. The heat lost in condensation will provide energy for the storm to grow. Soon, the storm will display a definite structure: a column of rising moist air (the "core"), led by a curtain of rain (the "precipitation shaft").

Example 11.3

The human heart has four chambers, the left and right atria and the left and right ventricles. The atria collect blood as it returns from the body (in the case of the right atrium) or from the lungs (in the case of the left atrium). During diastole, one-way valves open between the atria and the ventricles and blood pours into the ventricles, filling them. Then, during systole, the ventricles contract, closing the valves to the atria and sending blood through other one-way valves to the lungs (in the case of the right ventricle) or to the body (in the case of the left ventricle).

Example 11.4 From A. Knop and L. A. Pilato, *Phenolic Resins*, Springer Verlag, 1985.

Novolak Resins

Oxalic acid, MP 101 °C, as dihydrate is the predominate catalyst for the manufacture of novolaks. Oxalic acid sublimes in vacuum at about 100 °C and at normal pressure at 157 °C with decomposition. At a higher temperature (180 °C) it decomposes to carbon monoxide, carbon dioxide and water so that a removal process is unnecessary. Due to its reducing behavior very light resins are obtained. Other less frequently used acid catalysts are HCl, H_2SO_4, toluene sulfonic acid and H_3PO_4. Historically hydrochloric acid was used because of its low cost but is being used sparingly because of its high corrosiveness. It should be noted that the use of hydrochloric acid in phenolic resin manufacture can result in the generation of a hazardous intermediate. When formaldehyde and hydrochloric acid are present in the gas phase in concentrations of more than 100 ppm, 1,1-dichlorodimethyl ether is obtained, a highly hazardous and carcinogenic compound. Maleic acid has been recommended for the production of high melting novolaks.

The molar ratio of phenol to formaldehyde is normally within the range of 1:(0.75–0.85). While novolaks with a MP of 70–75 °C are used for the production of foundry resins for shell cores, those with MP's between 80–100 °C are used for all other application areas. These novolaks are flaked to an appropriate size, mixed with HMTA, and ground to powder in special mills and processed in this form.

In the batch process phenol, which is stored in tanks of alloyed steel at approximately 60 °C, is transferred to the reaction vessel via a scale and heated to 95 °C. After catalyst addition formaldehyde solution is introduced with stirring at a rate so that

Figure 1. Batch processing of novolak resins.

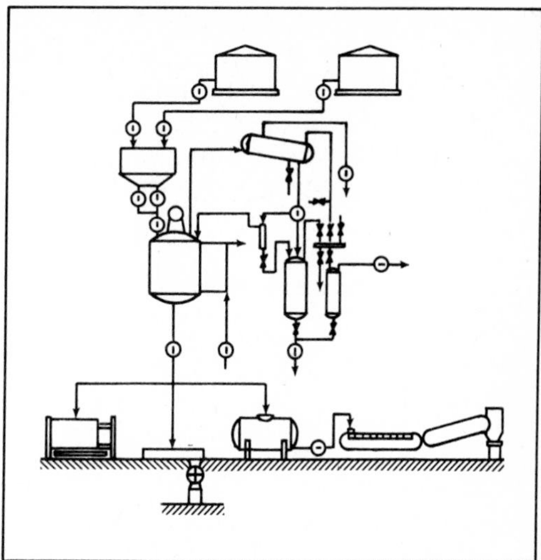

the mixture is boiling gently. When all the formaldehyde is added, the temperature is maintained until the formaldehyde is consumed. Then, the water is removed at normal pressure and, by further heating to 160 °C with vacuum, the unreacted phenol. Removal of volatiles can be facilitated by introducing steam. As soon as the desired melting point is reached, the resin is transferred to a heated vessel and then flaked on a continuous cooling belt.

The answer? All four selections are descriptions or explanations of a process.

Process description is probably the most common pattern in technical exposition. Were you to analyze the text of most technical communication—from a lecture by the physicist Richard Feynman to a procedures manual for group insurance plans; from textbooks in chemistry, physics, economics, and engineering to new product releases, specification sheets, and design proposals—you would find that centerstage goes to this special kind of narrative, the description or explanation of a process.

Describing processes—telling what happens and why—is a job you are likely to do quite regularly in your work, for process is the key to any technical task. It is never easy to force language to describe what you have observed. Process descriptions are particularly devilish, and for good reason. The description of a process may be a sequential account of stages, but it may also include the chain of causes that produces these stages. It takes a deft writer to delineate not only the steps, but the underlying, intertwined reasons for the steps.

You'll find that writing a clear process description is a challenge, but that the final product can have the kind of elegance that all good writing shares, the elegance of a clear explanation.

AUDIENCE ANALYSIS

Audience analysis, logic and organization, direct language, and visual display are all intertwined in effective presentation. This chapter examines how these four factors come into play in process descriptions.

Audience background is the first consideration.

The examples of process description at the beginning of the chapter are aimed at different audiences, starting with the broad readership of a newspaper and ending with the highly specialized readership of an academic monograph.

Audience analysis affects choice of language, organizational patterns, and visual display.

• The more sophisticated the audience, the fewer the definitions.

• The more knowledgeable the audience, the denser the detail. This includes detail in illustrations. Illustrations for the knowledgeable audience make assumptions about the viewer's ability to integrate new, highly specialized detail into an already established image. Illustrations for the broader audience must start with fewer assumptions.

• The closer the audience is to a peer group, the more compact the prose can be. Examples and elaborations are pruned.

Short Process Description from a Newspaper

> In electronic imaging, a scene is recorded, converted to an electronic signal, stored on a disk or magnetic tape, and then displayed.

This sentence is from a roundup article on advances in electronic imaging that appeared in the business section of a newspaper. It is aimed at an extremely diversified readership. The process is described in broad strokes, with no attempt at technical detail. Few technical terms are used.

Note, however, that the *description is accurate. Simplification should not cause distortion.* Anything that is simplified is reduced—in this case, the level of detail is reduced. But the process description is still correct. The writer for the broad audience always has this worry: how to steer between the two extremes of oversimplification and incomprehensibility. This trip is one of the essential journeys in technical writing.

Note also that the writer has used continuous, narrative prose rather than tabular layout. The chronology is handled by simply listing the steps in the sequence, with one transitional word ("then"). A tabular layout would look like this:

Electronic imaging:

1. Scene recorded.
2. Record converted to electronic signal.
3. Signal stored on disk or magnetic tape.
4. Signal displayed.

Tabular layout, flow diagrams, outlines, or simple listings are very effective for instructions, where the reader is called upon to enact the procedure step by step (see Chapter 12). Sometimes procedures and instructions are two sides of the same coin, with the instructions the active, detailed version of the description. In the example of electronic imaging, however, the user is simply given a procedural description, not a series of steps for enactment. Tabular presentation is not necessary here.

Process Description from a General Interest Magazine

> Right now, sunlight is heating the atmosphere and setting it in motion. Hot moist air is being pulled up through the cold dry air, and a storm at some point is almost inevitable. As the hot air reaches higher and cooler altitudes, its moisture will condense into drops of rain and particles of ice. The heat lost in condensation will provide energy for the storm to grow. Soon, the storm will display a definite structure: a column of rising moist air (the "core"), led by a curtain of rain (the "precipitation shaft").

This example of a process description is from an article on tornados by William Hauptman, "On the Dryline," that appeared originally in *The Atlantic Monthly*. The audience is assumed to be a bit more homogeneous, at least in terms of education, than newspaper readers. They can be expected to tolerate more detail and some technical terminology.

Hauptman takes a complicated causal chain and lays it out in language this is direct and clear. Technical terms ("core," "precipitation shaft") are delayed until the general point is established. The reader is helped to see the organizational pattern by the use of transitional words—"now," "as the hot air reaches higher," and "soon." The use of a narrative rather than tabular format helps to link the chain of cause and effect smoothly. The colon in the last sentence also helps the fluid, logical flow of the paragraph.

The technical details in this example are slightly more demanding than those in Example 11.1. They are handled by postponed definitions, introduction of detail *after* the main point is established, direct language, and transitional words. The content remains accurate; correctness has not been sacrificed at the altar of readability.

Description from *Scientific American*

> The human heart has four chambers, the left and right atria and the left and right ventricles. The atria collect blood as it returns from the body (in the case of the right atrium) or from the lungs (in the case of the left atrium). During diastole, one-way valves open between the atria and the ventricles and blood pours into the ventricles, filling them. Then, during systole, the ventricles contract, closing the valves to the atria and sending blood through other one-way valves to the lungs (in the case of the right ventricle) or to the body (in the case of the left ventricle).

This example is from *Scientific American*. The audience is assumed to be more specialized—that is, readers are assumed to have an interest in physics and chemistry. But although the magazine is specialized, it is not for specialists. People of many different technical backgrounds read it.

The style of this paragraph is more sophisticated than that of the two earlier examples:

- The density of detail is greater.
- Some terms aren't defined (such as "diastole" and "systole"), but can be deduced from their context.
- Sentences are longer and have parenthetical inserts. This makes them a bit harder for the reader to process.

The format is narrative, with transitional words used to make the sequence clear: "during diastole," "then," and "during systole."

What about the physicians and biologists who read the publication? They will probably skim past this paragraph—they already know the information. But the physicist who needs a 30-second refresher may give it a glance before getting into the details of the article. The background information will be helpful.

Process Description for Peer Group

Example 11.4 is from a highly specialized publication, *Phenolic Resins.*

- A grasp of shared terminology is assumed.
- Details are very densely packed.
- The illustration is detailed.

These characteristics, however, do not mean the excerpt is unreadable. On the contrary, it is well organized and clearly presented for its audience level. There are transitions for each paragraph, and for each sentence. The illustration amplifies and explains the text.

REVISION: Working Definitions into a Process Description

One of the big problems in technical writing is fitting definitions into text. "There are six or seven terms I want to define," a student said, "but I can't define all of them at the beginning. I feel as though I'm juggling seven tennis balls at once." Fitting new terms into a process description is a kind of juggling act, in which the writer introduces the first ball, then after a few repetitions the second ball, then the third. Since all of the terms are necessary to explain the process, the writer must develop the skill to dispense the terms one by one.

In Example 11.5a the author is explaining a phenomenon, the photoelectric effect. In the first draft, he confronts the familiar problem of defining both the general terms and the terms within the process. Like most of us, he has problems doing this. The first draft is on the following page.

The explanation of the process is compressed but clear for someone who understands the definitions and distinctions between "frequency," "energy," and "intensity." But the author decided his audience might not be that sophisticated, so he revised to include definitions of these key terms. Example 11.5b is the second version.

Example 11.5a First version of process description.

FIRST VERSION

Einstein applied the quantum idea to a phenomenon called the photoelectric effect, which had been unexplained until then. The photoelectric effect is the emission of electrons from the surface of a metal that is struck by light or other electromagnetic radiation. The wave theory could not explain the photoelectric effect, in which once the threshold frequency is reached, the number of electrons given off by the metal does not change if the frequency is increased still higher. What does change is the energy of the electrons that are emitted. As the frequency of the incident radiation increases, only the energy of the emitted electrons increases. As the intensity of the incident radiation above the threshold frequency is increased, only the number of electrons emitted increases.

Example 11.5b Second version of process description.

SECOND VERSION

It was Einstein's paper on the nature of light that was cited when he was awarded the Nobel Prize.

In this paper Einstein applied the quantum idea to a phenomenon called the photoelectric effect, which had been unexplained until then. The photoelectric effect is the emission of electrons from the surface of a metal that is struck by light or other electromagnetic radiation.

It is found that the emission of electrons simply depends on the frequency—that is, the energy—of the radiation. If the electromagnetic radiation that strikes the surface of the metal is of low frequency (which means that each of its quanta has only a relatively small amount of energy), no electrons are emitted by the metal. If energy of progressively higher frequency is used (which means that the energy of each quantum is increased) one suddenly arrives at a "threshold"—a frequency at which electrons are first emitted by the metal. No electrons are emitted at frequencies lower than the threshold frequency, no matter how much radiation of that frequency strikes the metal. That is, below the threshold frequency, increasing the intensity of the radiation has no effect.

Once the threshold frequency is reached, the number of electrons given off by the metal does not change if the frequency is increased still higher. What does change is the energy of the electrons that are emitted. As the *frequency* of the incident radiation increases, only the energy of the emitted electrons increases. As the *intensity* of the incident radiation above the threshold frequency is increased, only the *number* of electrons emitted increases.

These results were a puzzle in 1905, because they could not be reconciled with the wave theory of light. Einstein solved the puzzle by proposing that radiant energy exists in quanta called **photons.** In doing so, Einstein made the first practical application of Planck's quantum theory.

The author continued in this way, adding definitions and then revising so there was some space between each new definition—so that definitions were not all clumped together in one indigestible lump at the beginning of the article. Doing this took some juggling, but was worth the effort: The author preserved the specialized vocabulary he needed for the discussion.

It's worth noting that the final version (Example 11.5c) did not spring out of his head fully grown. Instead, it developed through two revisions as he adjusted the content to suit his audience.

Example 11.5c Third version of process description.

THIRD VERSION

It was Einstein's paper on the nature of light that was cited when he was awarded the Nobel Prize.

In this paper Einstein applied the quantum idea to a phenomenon called the photoelectric effect, which had been unexplained until then. The photoelectric effect is the emission of electrons from the surface of a metal that is struck by light or other electromagnetic radiation.

It is found that the emission of electrons simply depends on the frequency—that is, the energy—of the radiation. If the electromagnetic radiation that strikes the surface of the metal is of low frequency (which means that each of its quanta has only a relatively small amount of energy), no electrons are

Example 11.5c (*continues*)

emitted by the metal. If energy of progressively higher frequency is used (which means that the energy of each quantum is increased) one suddenly arrives at a "threshold"—a frequency at which electrons are first emitted by the metal. No electrons are emitted at frequencies lower than the threshold frequency, no matter how much radiation of that frequency strikes the metal. That is, below the threshold frequency, increasing the intensity of the radiation has no effect.

Once the threshold frequency is reached, the number of electrons given off by the metal does not change if the frequency is increased still higher. What does change is the energy of the electrons that are emitted. As the *frequency* of the incident radiation increases, only the energy of the emitted electrons increases. As the *intensity* of the incident radiation above the threshold frequency is increased, only the *number* of electrons emitted increases.

These results were a puzzle in 1905, because they could not be reconciled with the wave theory of light. Einstein solved the puzzle by proposing that radiant energy exists in quanta called **photons.** In doing so, Einstein made the first practical application of Planck's quantum theory. Einstein not only used the same term, quantum, to describe a "bundle" of light, but he also used Planck's constant, h, in the formula for the energy of a quantum of light. The energy of a quantum of light, Einstein said, is $E = hv$; that is, the energy is the frequency times Planck's constant.

The idea that light is quantized was revolutionary. The quantum description of light conveys the idea that light consists of particles or corpuscles, an idea that had long been discarded in favor of the wave theory. Nevertheless, the wave theory could not explain the photoelectric effect, but the quantum theory could.

To say that light consists of small bundles—quanta—of energy is the same as saying that light consists of particles, the photons. One can visualize electrons being knocked loose from a metal by a beam of photons. We can picture an electron as being held by the nucleus of an atom with a certain amount of energy, called the *binding energy,* E. Only a photon with enough energy to overcome the binding energy can knock the electron loose. In this visualization, if the photon that hits the electron does not have enough energy, it will bounce away and the electron will remain in the atom.

As in many other examples, this one has grown longer. This does not mean that longer is better. But the extremely compressed style of the *Phenolic Resins* example won't do when you are explaining something complex to those who do not yet know the basics. If you are explaining a phenomenon and you want to retain the detail, then you must insert definitions and examples, adjusting your prose to the readership in the same way a fish adjusts its strokes to the condition of the water.

ILLUSTRATION IN PROCESS DESCRIPTION

Process descriptions are explanations of sequences, and that means the information has to proceed in a logical order. If you are concerned not only about the order, but about the underlying chain of causes, you may find the best way to make the point is to combine text with figures or other illustration. This is true whether the audience is very general or very specialized; causal chains lend themselves to graphics. Can such information be explained without graphics? Of course, but text and illustration together will greatly help the audience.

The three-paragraph example on the following page (Example 11.6), which combines process description with an illustration of a causal chain, is from an introductory chemistry text for college students.

Could you use the illustration without the accompanying text? Yes, but the combination makes the information accessible to more readers—those who grasp the point from the illustration, and those who use both illustration and text.

Contrasting illustrations are especially useful in explaining a technical point. Contrast in illustration is a mainstay of advertising and marketing applications of technical information, as well as presentation of technical information for the public.

For instance, in the public reports on the crash of the space shuttle Challenger, contrasting figures played an important explanatory role. The investigation concentrated on the rocket joints and their O rings. O rings are precisely engineered versions of washers. Unlike washers, though, which tend to leak under pressure, O rings form a tighter seal as pressure increases by deforming to close off any avenues of escape. They are used to isolate materials; in the Challenger they were used to prevent gases from escaping from the solid-fuel booster rockets.

The resiliency of the O rings was crucial, for the joint design

Carbon Dioxide and the Carbon Cycle

Life on earth is based on carbon. The major source of carbon for the compounds that make life on earth possible is the carbon dioxide in the atmosphere. The group of reactions by which CO_2 is removed from and returned to the atmosphere is called the carbon cycle.

Photosynthesis in plants removes carbon dioxide from the atmosphere in a process that converts solar energy to chemical energy. In photosynthesis, sunlight is used to produce glucose, a sugar, and other organic compounds from carbon dioxide and water. The plants that carry on photosynthesis and the animals that eat those plants use the chemical energy produced by photosynthesis, converting the organic substances back to carbon dioxide in the process. When organisms die, their carbon compounds are returned to the atmosphere in the form of CO_2 as part of the process called decay. Some dead plants are converted to fossil fuels. When we burn these fuels, CO_2 is added to the atmosphere.

The atmosphere contains a relatively small amount of carbon dioxide. Only 0.0325% of the atmospehre—325 parts per million (ppm) by volume—is CO_2. But in recent decades, the concentration of CO_2 in the atmosphere has been going up by about 1 ppm per year. The increase appears to be caused primarily by the burning of fossil fuels. On a global scale, the rising CO_2 content of the atmosphere could have serious long-term consequences. If the century-long pattern of rising consumption of fossil fuels continues, the amount of CO_2 in the atmosphere could double by early in the twenty-first century. Climatologists calculate that such an increase could raise the global temperature by an average of between 0.5 K and 1 K. For the first time, human activity will be on a scale great enough to change the climate, with unknown consequences.

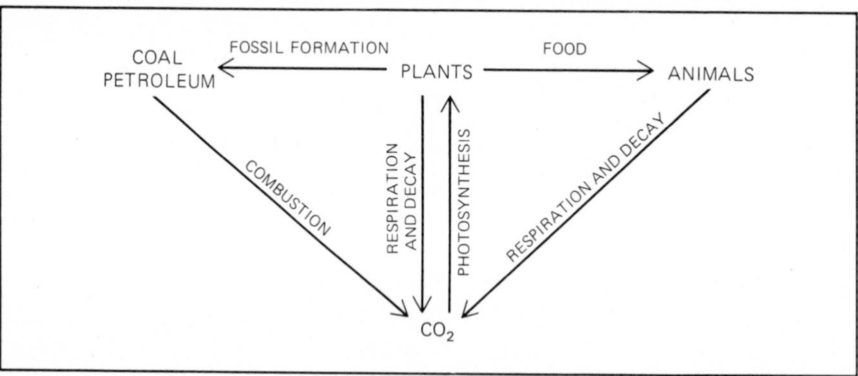

The carbon cycle. Plants obtain CO_2 from the atmosphere for photosynthesis. The carbon is returned to the atmosphere by respiration, decay, or combustion of organic material.

Example 11.6 Process description and illustration. From Robert S. Boikess and Edward Edelson, *Chemical Principles*, Harper & Row, 1981.

called for the ring to jam into a narrow slot within milliseconds of ignition, to stop hot gases from escaping.

Physicist Richard Feynman conducted an impromptu experiment at the Challenger hearings to show that the O rings lost resiliency when dipped into ice water. The results of Feynman's impromptu experiment were borne out in temperature tests, in which data showed that a ring at the temperature of the joint at launching on January 28, 1986 (25° Fahrenheit), had less than a fifth of the resiliency of the same ring at 75° F. The joint sealed only at 55° F or above. Below that temperature the seal was too hard to seat properly in the channel. Serious leakage occurred at 50 to 40° F.

A series of contrasting figures depicted the O ring process throughout the investigation. Figure 11.1 shows a cross section of the joint. Figure 11.2 shows a drawing of how the tang/clevis joint deflects during pressurization to open a gap at the location of O ring slots.

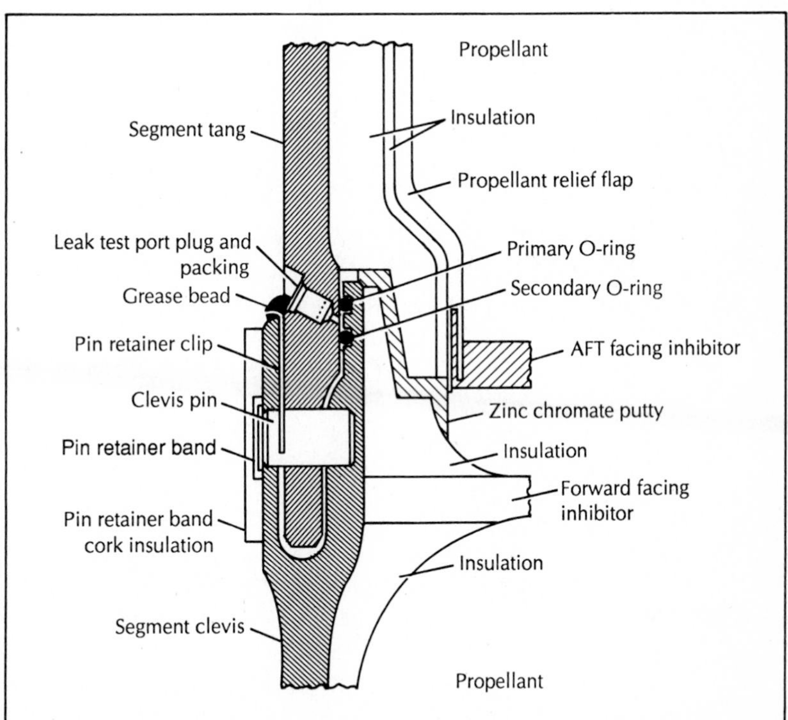

Figure 11.1 Cross section of joint. *From Report of the Presidential Commission on the Space Shuttle Challenger Accident.*

Figure 11.2 From *Report of the Presidential Commission on the Space Shuttle Challenger Accident.*

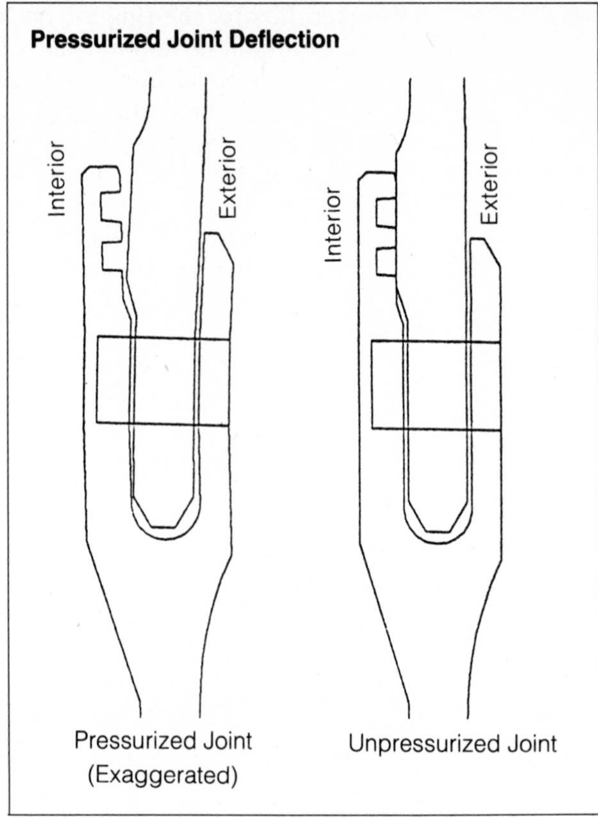

After the investigation, NASA picked a new rocket design for the space shuttle that addressed, among other things, the deficiencies in the field joint: the gap that opened wider as gas pressure rose, putty that interfered with operation of the joint, and the O ring seals that lost resiliency at low temperatures and failed to plug the gap. Figure 11.3 shows NASA's contrasting illustrations of the design.

THE MIGHTY CAPTION

In Example 11.7, the author uses two contrasting illustrations to make a point about closest-packed structures. Example 11.7a is the excerpt with the illustrations alone.

Later the author decided to add captions to the illustrations. Is the sequence more effective when text and illustrations are reinforced with captions? (See Example 11.7b.)

Figure 11.3 Courtesy NASA.

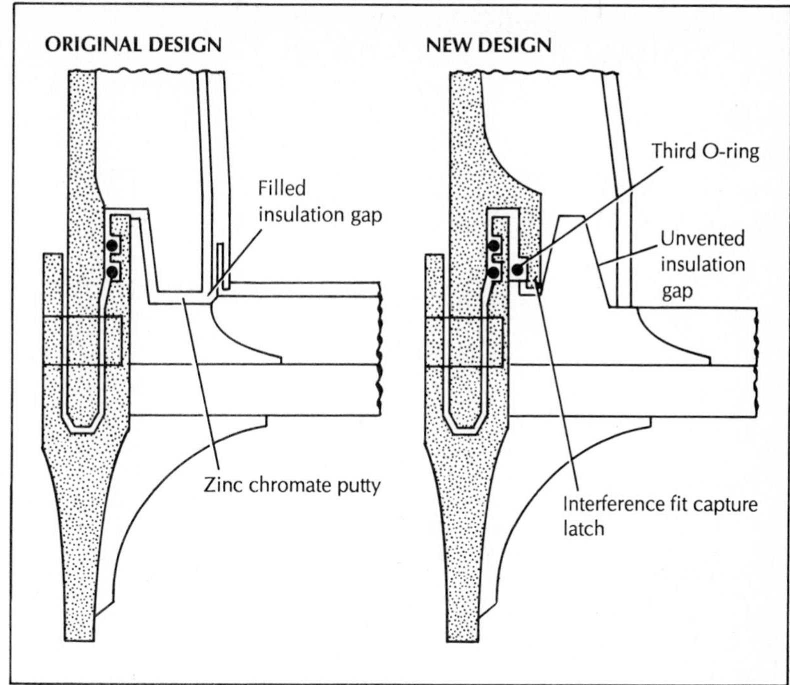

Captions provide an explicit way to state the essential contrast, the contrast that will help the viewer understand the concept.

Don't hesitate to make the captions deep—that is, one or even two sentences long. The illustrations in Figure 11.4 use contrast to compare the Chernobyl reactor to the U.S. reactors. Note how the deep captions describing the two processes interact with and amplify the illustrations.

NARRATIVE DESCRIPTIONS

Many examples of process descriptions follow the order of

- definitions
- figure references
- division into parts
- details or process
- supporting illustrations and tables

Closest-Packed Structures

The structure of any crystal, no matter how complex, can be described by lattice points and unit cells. The method works for a crystal of a complicated organic molecule, such as a protein or a nucleic acid, as well as it does for a crystal of a simple monatomic substance, such as a metal.

There is a simpler, alternative method of picturing the crystal structure of many substances. We can describe the crystal as an assembly of closely packed spheres. This alternative method works very well when the basic structural unit is an atom. Some molecular substances can also be described in this way, but only if the molecules are simple enough to be roughly spherical. Many minerals also lend themselves to a description of this sort.

The basic approach is to see how spheres of identical size can be packed to fill a volume as completely as possible. This problem is encountered in everyday life, as when we try to pack oranges or other shperical objects in a crate.

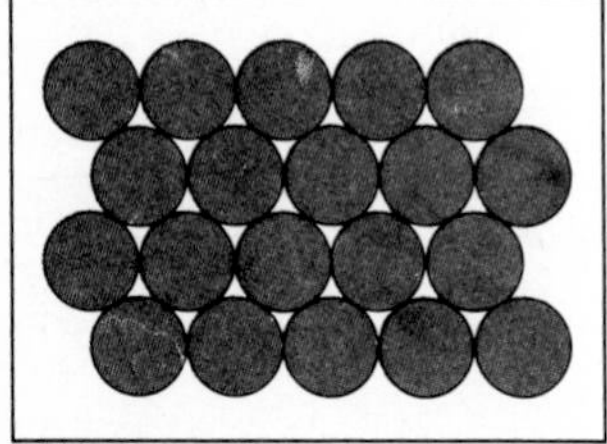

Figure 1

There are two ways to pack spheres most efficiently while preserving long-range order. ("Efficiently" means that the smallest possible fraction of the volume is empty space.)

Figure 1 shows one layer of spheres packed as closely as possible. There are small empty spaces or *holes*, between spheres. These holes, sometimes called *interstices*, are almost as important as the spheres for understanding close-packed structures.

In Figure 1, each sphere is surrounded symmetrically by six other spheres, its nearest neighbors. In three dimensions, close-packed arrangements are made by layers of spheres stacked on one another. As Figure 2 shows, this stacking is done most efficiently when the spheres of the top layer are fitted into the holes of the bottom layer. Figure 2 also shows that only half of the holes in the bottom layer are covered by spheres in the top layer.

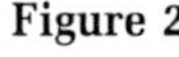

Figure 2

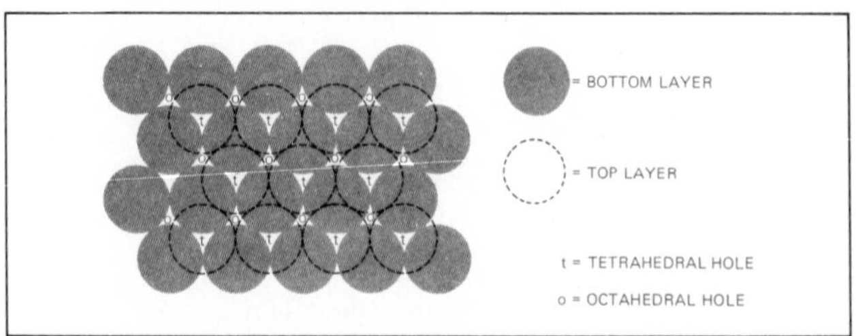

Example 11.7a From Robert S. Boikess and Edward Edelson, *Chemical Principles*, Harper & Row, 1981.

Closest-Packed Structures

The structure of any crystal, no matter how complex, can be described by lattice points and unit cells. The method works for a crystal of a complicated organic molecule, such as a protein or a nucleic acid, as well as it does for a crystal of a simple monatomic substance, such as a metal.

There is a simpler, alternative method of picturing the crystal structure of many substances. We can describe the crystal as an assembly of closely packed spheres. This alternative method works very well when the basic structural unit is an atom. Some molecular substances can also be described in this way, but only if the molecules are simple enough to be roughly spherical. Many minerals also lend themselves to a description of this sort.

The basic approach is to see how spheres of identical size can be packed to fill a volume as completely as possible. This problem is encountered in everyday life, as when we try to pack oranges or other shperical objects in a crate.

There are two ways to pack spheres most efficiently while preserving long-range order. ("Efficiently" means that the smallest possible fraction of the volume is empty space.)

Figure 1 shows one layer of spheres packed as closely as possible. There are small empty spaces or *holes*, between spheres. These holes, sometimes called *interstices*, are almost as important as the spheres for understanding close-packed structures.

In Figure 1, each sphere is surrounded symmetrically by six other spheres, its nearest neighbors. In three dimensions, close-packed arrangements are made by layers of spheres stacked on one another. As Figure 2 shows, this stacking is done most efficiently when the spheres of the top layer are fitted into the holes of the bottom layer. Figure 2 also shows that only half of the holes in the bottom layer are covered by spheres in the top layer.

Figure 1

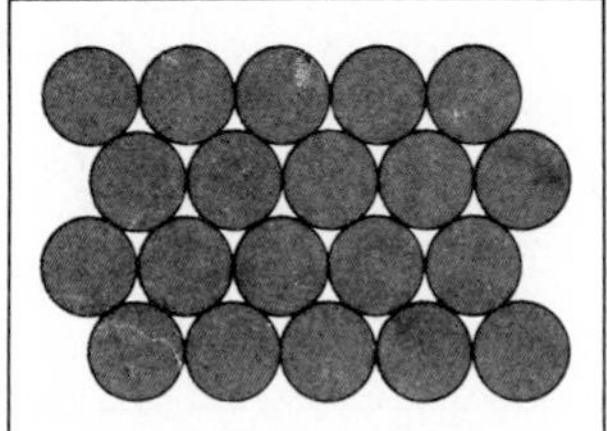

One layer of spheres packed as closely as possible. Most of the volume is occupied by the spheres, but an irreducible volume is represented by the holes, or interstices, between the spheres.

Figure 2

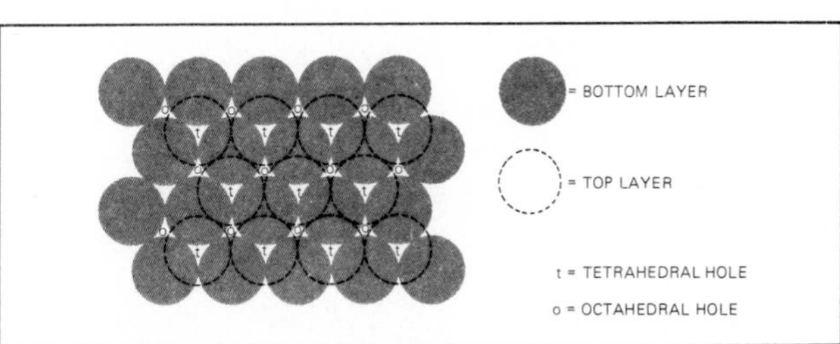

When one layer of close-packed spheres is placed atop another, two types of holes are found in the bottom layer: tetrahedral (*t*) holes, which are covered by spheres in the top layer, and octahedral (*o*) holes, which are not coverd by spheres in the top layer.

Example 11.7b From Robert S. Boikess and Edward Edelson, *Chemical Principles*, Harper & Row, 1981.

Comparing the Chernobyl Reactor to the Standard U.S. Reactor

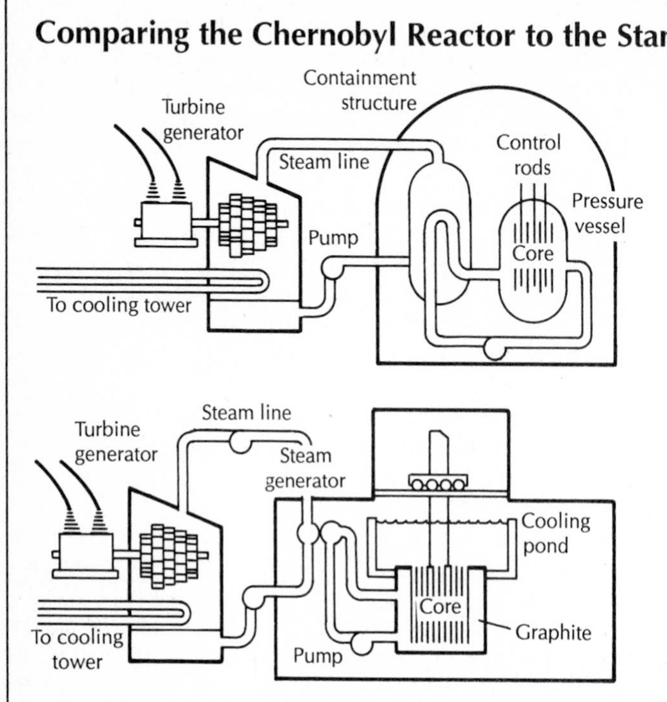

Pressurized Water Reactor
Commercial nuclear reactors in the U.S. are usually the pressurized-water type. The water that moderates, or slows, the nuclear reaction within the core heats a steam generator, which drives turbines to create electricity. In case of accident, water in contact with the radioactive fuel is contained within a pressurized dome to prevent leak of radioactive isotopes into the atmosphere.

Graphite-Moderated Reactor
In a graphite-moderated reactor, such as the one at Chernobyl, nuclear fuel is encased in rods, and lowered into block of graphite, which slows the nuclear fission to a controllable speed. Water pipes carry off the heat generated by the reaction. In case of a loss of pressure in the cooling water system, the core could overheat, resulting in a meltdown. This reactor has no containment structure to keep radiation from escaping in case of an accident.

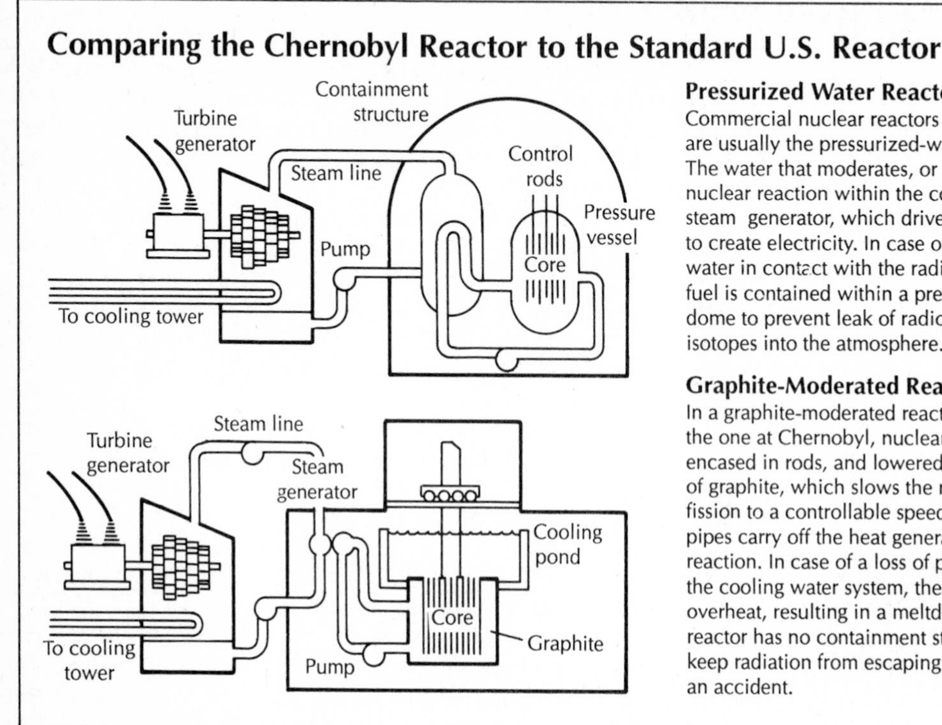

Figure 11.4 Use of contrast in a figure caption.

Details in the text are divided by underlined subheadings, boldfaced type, and italics.

This is indeed a common format, but it is not the only one. Some writers, for instance, do not care for divided text—text that clearly distinguishes its parts through headings, subheadings, and tabular display; they consider such patent divisions a predigested, lower form of writing in which the content is packaged in bite-sized pieces. "I never use those dreadful bullets or dots for lists," one instructor commented. "I think they cheapen the text. People should be able to understand a list without having each item displayed separately."

Obviously, this speaker is not going to use a boldfaced, divided-text method. Such writers prefer continuous narrative text without the ordering and emphasis provided by headings, subheadings, and tabular display of lists. Writers who aspire to the ideal of "straight narrative form," however, should be aware that such a form places serious demands on both writer and reader. Not every reader is willing to go through page after page of unblemished narrative; and not every writer is capable of the expository style necessary to sustain such a narrative. Still, there are people who are superb at the task. One of them is Jeremy Bernstein.

Here is an example of his writing from *Science Observed* where the text alone provides the information with no help from page design or visuals. It is a description of a reactor designed for the production of short-lived, medically useful isotopes. Its market is universities and large medical centers. The reactor goes by the acronym TRIGA (Example 11.8).

If you do favor tabular layout for procedures, remember that this layout in itself is no magic guarantor of readability. For instance, the following procedure from the Tax Reform Act of 1986 is in tabular rather than narrative layout. It is still, however, quite difficult

Example 11.8 Example of narrative process description. From Jeremy Bernstein, *Science Observed*, Basic Books, Inc., 1982.

Example of Narrative Process Description

. . . I should like to describe briefly in what way the TRIGA is a safe reactor. To explain this in a schematic way, let me first review how a reactor works. Reactors are powered by nuclear fission. In particular, the so-called fuel elements of the common reactors consist of long, thin rods containing pellets of uranium that have been "enriched" so as to contain more of the fissionable isotope U 235 than is normally found in nature. In nature more than 99 percent of the uranium is found in the U 238 isotope, which is not readily fissionable. In a TRIGA, for example, the fuel elements contain an enrichment of 20 percent of the U 235 isotope. When a slow neutron encounters a U 235 nucleus, it has a good chance of splitting it and releasing energy and additional neutrons which can then carry on the chain reaction that produces energy.

If the neutrons are to be made effective, they must be slowed down. This is accomplished by putting the fuel rods in a so-called moderator. In the TRIGA, as in most of the common reactors, the moderator is ordinary water. The fuel elements are in a small swimming pool which glows a pleasant incandescent blue when the reactor is operating. The power generated by the TRIGA is so low that one can stand safely above the pool while the reactor is working. Reactors used by the power industry are at least ten thousand times more powerful, and one cannot approach them when they are in operation. The water in the pool serves the double function of slowing down the neutrons and cooling the fuel elements so they won't melt—the "melt-down." The fissions generate heat, and in a power reactor this heat produces steam, which drives electric turbines. In the power reactors the essential thing, from a safety point of view, is to make sure that under no circumstances does one lose cooling water and thus expose the heated fuel elements.

to follow because of the compactness of the prose, the absence of examples, the wordiness of the sentences, and the smallness of the type (see Figure 11.5).

SUMMING UP

• Often a process description includes not only a sequence of events, but the underlying, intertwined reasons for the sequence. This causal chain is often difficult to capture in words. Audience analysis, organizational pattern, direct language, and visual display all come into play to do the job effectively.

• The closer the audience is to a peer group, the denser the prose, the more sophisticated the illustrations, the fewer the definitions, examples, and elaborations. The more diversified the audience, the more it needs definitions, examples, elaborations, and illustrations to portray the process.

• Simplification should not cause distortion. Detail may be limited and technical terms postponed, but points should remain accurate. The writer must steer between the Scylla of oversimplification and the Charybdis of incomprehensibility.

• Don't clump all new terms and definitions together at the top in one indigestible lump. Try metering them, if you want to preserve the specialized vocabulary you need for discussion of detail.

• Polished process descriptions don't usually spring, full blown, out of the author's head, like Athena out of Zeus. Instead, they come about in a series of revisions the author makes while considering the audience's ability to follow certain paths.

• Illustrations and captions are very useful in deepening the viewer's understanding of a process. Causal chains and contrasts are particularly effective techniques when you want immediate visual links. Don't hesitate to use informative captions. They are an explicit way to amplify illustrations, to state the contrast which will help the viewer understand the main idea.

EXERCISES

1. Find and photocopy one example of a process description that you either like or dislike. Write a brief analysis of it, discussing audience focus, logical organization, use of examples and elabora-

Figure 11.5 Tabular layout is no guarantor of readability. Here compact prose, absence of example, and small type make the process of description difficult to follow. From *Tax Reform Act of 1986*.

(d) TAX TREATMENT OF DISTRIBUTIONS.—

(1) In General.—Except as otherwise provided in this subsection, any amount paid or distributed out of an individual retirement plan shall be included in gross income by the payee or distributee, as the case may be, in the manner provided under section 72.

(2) Special Rules for Applying Section 72.—For purposes of applying section 72 to any activity described in paragraph (1)—

(A) all individual retirement plans shall be treated as 1 contract,

(B) all distributions during any taxable year shall be treated as 1 distribution, and

(C) the value of the contract, income on the contract, and investment in the contract shall be computed as of the close of the calendar year with or within which the taxable year ended.

For purposes of subparagraph (C), the value of the contract shall be increased by the amount distributions during the calendar year.

(3) Rollover Contribution.— An amount is described in this paragraph as a rollover contribution if it meets the requirements of subparagraphs (A) and (B).

(A) In General.—Paragraph (1) does not apply to any amount paid or distributed out of an individual retirement account or individual retirement annuity to the individual for whose benefit the account or annuity is maintained if—

(i) the entire amount received (including money and any other property) is paid into an individual retirement account or individual retirement annuity (other than an endowment contract) for the benefit of such individual not later than the 60th day after the day on which he receives the payment or distribution;

(ii) the entire amount received (including money and any other property) represents the entire amount in the account or the entire value of the annuity and no amount in the account and no part of the value of the annuity is attributable to any source other than a rollover contribution of a qualified total distribution (as defined in section 402(a) (5)) from an employee's trust described in section 401(a) which is exempt from tax under section 510(a), or an annuity plan described in section 403(a) and any earnings on such sums and the entire thereof is paid into another trust (for the benefit of such individual) or annuity plan not later than the 60th day on which he receives the payment or distribution; or

(iii)(I) the entire amount received (including money and other property) represents the entire interest in the account or the entire amount of the annuity,

(II) no amount in the account and no part of the value of the annuity is attributable to any source other than a rollover contribution from an annuity contract described in section 403(b) and any earnings on such rollover, and

(III) the entire amount thereof is paid into another annuity contract descibed in section 403(b)(for the benefit of such individual) not later than the 60th day after he receives the payment or distribution.

Clause (ii) shall not apply during the 5 year period beginning on the date of the qualified distribution referred to in such clause if the individual was treated as a 5-percent owner in respect to such distribution under section 402(a)(5)(F)(ii).

from Tax Reform Act of 1986

tion, and use of illustrations, captions, and page layout. You will find process descriptions in, among other places, technical textbooks, handbooks, and reference works.

2. Describe a process with which you are familiar. Assume that your audience includes people unfamiliar with your field. Some possibilities:

how blood flows through the body

how compact discs are made

how enzymes work

how a spread sheet program works

3. Write a short (2- to 3-page) procedural paper in which you explain a process at your school or place of employment. Assume your audience is a group of aides who will be hired to do administrative work, and need written procedures to help them learn the guidelines of their new job. You might want to write:

Screening rules for an organization that offers a service at your school or job, such as a tutorial center: Who is eligible? On what basis? Write guidelines that would help aides to determine eligibility.

Processing rules for an office, such as the registrar's or bursar's office: What is the process for changing grades? For completing registration after the official registration period has ended? Write a procedure for one set of circumstances.

CHAPTER 12

Instructions: Writing for the User

At school, students learn how to solve technical problems; at work, they often discover that once the problem is solved—the plans drawn, the strategy determined, the routine developed—the inevitable moment comes when the solution must be put into language for a user—the person who will actually analyze the sample, fill in the regulatory form, run the program.

Who writes all the instructions for users that a technically oriented society demands? Sometimes the job is done by a professional technical writer, but most of the time the same people who solve the technical problems end up doing the writing, too.

That means that writing a set of instructions may very well be one of your first assignments at work. When might you be called upon to write instructions? Here are a few examples.

• **Assignment: Safety Instructions.** You are assigned to a paper mill where safety procedures are mainly "oral history"—that is, the rules for safety are passed along from one crew to the next in conversation. Your job is to document the instructions so that workers on the production shift will have written directions that can be quickly and easily understood.

• **Assignment: Instructions for Laboratory Notebook Entries.** You've decided on an orderly way to make entries in the laboratory notebook. Your manager notices and suggests you write up your method so the entire research team can use it to standardize entries. A set of instructions will also be handy for new workers who come on the team. They'll have a checklist to follow as they make their first entries and get used to the company way.

• **Assignment: New Billing Procedure.** Your company has developed a new, computerized billing procedure for its product. You write the instructions the office staff will follow to enter each transaction.

AUDIENCE ANALYSIS: Good Instructions Are Imaginative

Instructions are typically written by those who *know* for those who *don't.* This can create problems, for the person who is skilled at a task often forgets what it was like to be a beginner, and ends up writing over the heads of the audience. This is called by many "assumptive" writing; that is, it assumes too much about the reader's background and experience.

It's difficult to pitch information at the right level. The writer has to figure out what the user needs—a difficult assignment for someone who already knows how to do the task. The expert is unlikely to be stopped by the sorts of confusions—"hmmm . . . is it left or right? up or down? I wonder what they meant by . . . ?"—that confront the beginning user.

The instruction writer must therefore be not only methodical, but *imaginative.* "Imaginative" is a word usually applied to literary writing; it's another of those words, like "persuasive" and "voice," which you might not at first associate with technical writing.

But technical writing is indeed imaginative. It is prose where the writer imagines the readers' needs—even though those needs do not exist for the writer—and then supplies them.

With instructions, you must imagine a person you no longer are—a beginner. What does this person need to know? What is the best order for introducing this information? What about places where the user may be stopped cold, and need a selection of possible paths? Are key terms defined? "Use Merge R in this procedure" will not be helpful to the novice unless a definition of "Merge R" is included.

GUIDELINES FOR INSTRUCTIONAL WRITING

The writer of instructions must

1. understand the task thoroughly. If you are already an expert, fine. If not, this is the time for research.

2. orient the reader to the task. The user usually needs a framework. What is the job? Why is it necessary? What equipment must be on hand? What object, process, or outcome should the user expect?

3. present the information in the correct order. Usually this means a chronological pattern, with one instruction per step.

4. be sure the information is accurate.

5. use direct address and specific language.

6. use an effective page layout.

7. coordinate illustrations and text.

8. give the instructions repeated trial runs. Once you have feedback from your users, you can revise for completeness, accuracy, and clarity in the light of their comments.

These are general guidelines for writing instructions. To give them more immediacy, let's look at these suggestions in relation to a specific set of instructions.

The object pictured in the accompanying illustration (Figure 12.1) is a portable screen. Ten students who knew how to use it were asked to write instructions. Then ten students who had never set up a portable screen were asked to follow the instructions the first group had written.

Here is a set of instructions written by one member of the group.

INSTRUCTIONS FOR USE OF PORTABLE SCREEN

Tripod should be opened and pulled out. Extend the support bar to proper height. Rotate the screen shaft from it's(sic) verticle(sic) position to a horizontal one. Hook the screen to rod. To close screen, reverse directions.

The users who tried to follow these instructions had a difficult time. Those who had a vague idea of how to assemble the screen—of what it should look like when it was mounted—managed to get through the job without dropping the screen, bending its legs, or

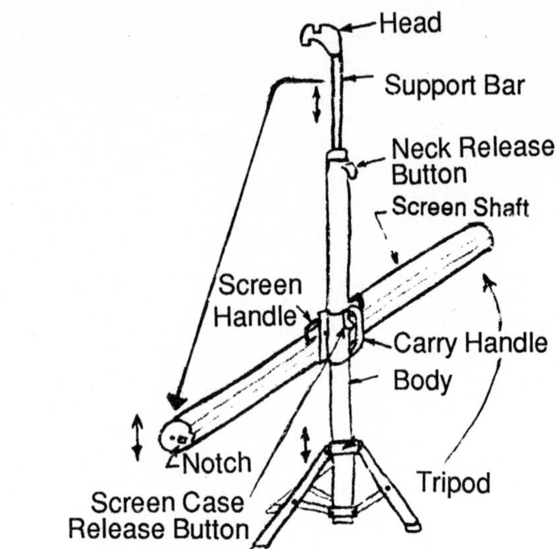

Figure 12.1 Portable screen (open).

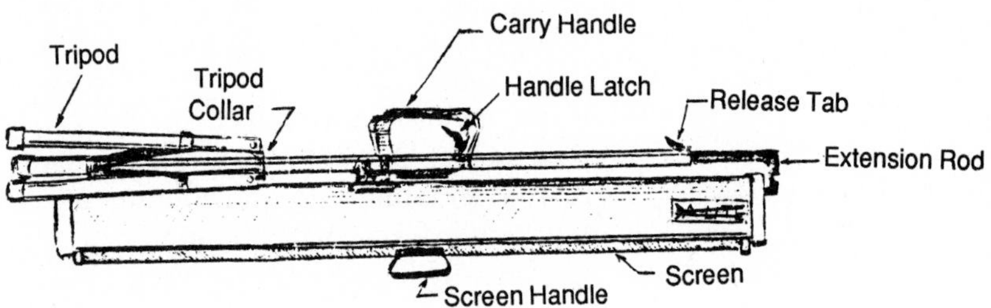

Figure 12.2 Portable screen (closed).

pinching their fingers. But those who were forced to rely on the written words had all these problems. What is the matter with these instructions?

Let's look at them in the light of the general guidelines for writing instructions.

Understanding the Task

Most of the time, the writer of instructions is an expert at the job and finds this sophisticated background a hindrance to imagining the audience's level of skill. In the portable screen instructions, for instance, the writers knew how to use the screen; this knowledge led them initially to underestimate many difficulties beginners would have.

When you write instructions at work, the chore you are called on to describe may be a familiar one. On the other hand, it may be a task that is new to you.

If this is the case, you'll find that a fresh eye can be an asset rather than a handicap. The person unfamiliar with the task brings to it much of the perspective of the new user. In fact, in some software businesses the staff prefer that the documentation be written by the newest person, one who is initially unfamiliar with the job, and therefore more likely to bring the eye of a novice to the task.

But if you don't have an intrinsic knowledge of the procedure, you have to be willing to learn it thoroughly before you write. Directions written by someone who hasn't bothered to master the task are an outrage—at least to the unfortunate user.

Orienting the User

New users need to know which is the top, which the bottom, before getting started. They need an *image* of where they are heading, right from the start. For instance, when the sample instructions for the screen were given to the new users, several read the instructions ("Tripod should be opened and pulled out") and promptly tried to open the screen on the floor; this didn't work, and eventually they picked up the screen with one hand and held it vertically before trying to open the tripod.

"How could they be so stupid?" one of the writers asked in irritation. "Don't they *know* that they have to hold the screen vertically to begin?" The answer is, "No, they didn't." That's why they were reading the instructions.

"Tripod should be opened and pulled out" told them neither *where* nor *how* to hold the screen. The users would also have benefited from a *figure with labeled parts* to help them locate each section of the screen before beginning. Such a figure would have oriented them to the task.

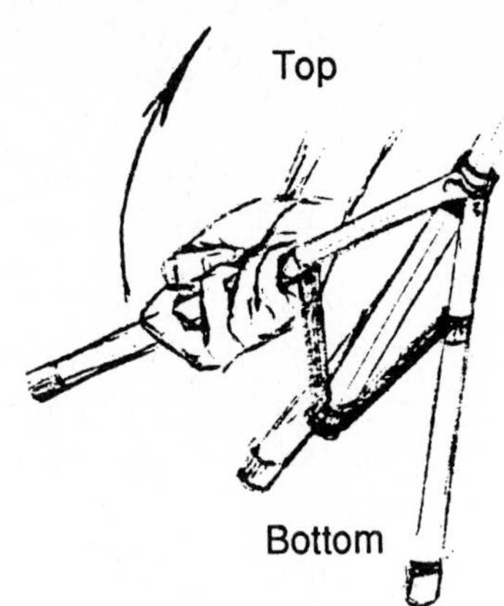

Figure 12.3 Pull tripod leg outward.

POOR

Tripod should be opened and pulled out. (Where is the screen in relation to the floor? How is it supported?)

REVISED TO INCLUDE ORIENTATION

Hold screen vertically with legs down. (See Figure *x.x.*)
or
Hold screen vertically with tripod down. (See Figure *x.x.*)
or
Hold screen with one hand, legs down. (See Figure *x.x.*)

Organizing the Sequence

Usually the information is presented chronologically, with one instruction per step.

In the portable screen instructions, the writer collapsed several steps into one command. Instead, one should break the information into a series of chronological steps that the beginner can follow.

POOR. INSTRUCTIONS TOO COMPRESSED

Tripod should be open and pulled out.

REVISED

1. Hold screen with one hand, legs down. (See Figure *x.x.*)
2. With other hand, loosen tripod collar by pushing it downward one inch.

3. Pull one leg outward. All legs will open fully at once.
4. Set on floor.

Checking for Accuracy

The second sentence from the original portable screen instructions ("Extend the support bar to proper height") has a serious error: A step is omitted. In order to raise this support bar, the user must first press a release tab. *The step is completely left out.* People trying out the instructions failed at first to lift the screen at all. After experimenting with all the latches, they finally found the right one, and the screen opened. But they certainly didn't get the information from the instructions; opening the screen became more of an IQ test—a challenge to their ingenuity and intelligence—than a test of reading skill.

Why did the writer leave out a step? "It's obvious that the latch releases the extension rod," was the reply. "Anyone who can't figure *that* out shouldn't be doing this job." But it wasn't obvious to the users, only to the writer.

Sometimes a step is omitted because the writer gives up, defeated by language. "Oh well," the writer decides, "they can figure it out anyway." One writer commented, "Explaining how to release the extension rod was surprisingly complicated, for such an easy task. Doing it was a snap, but *saying* how to do it was another story. I found myself growing more and more wordy. Finally I crossed out that part of the directions entirely, and simply summarized."

Many of the writers struggled with the opening step. Some wrote three or four sentences, decided that the language was excessive, attempted to prune the sentences, and gave up, settling for "open tripod." They knew "open tripod" was inadequate, but wrote it anyway, abandoning what E. B. White calls the "unequal fight against the odds of English syntax."

Omitting or misstating steps can have serious consequences for the unlucky people who must follow poor instructions. Recently, two schoolchildren doing a chemistry experiment from a textbook were hurt by an explosion. The court found that the instructions in the textbook had led to the accident. The publisher, Rand McNally & Co., was found liable.

In the case of the portable screen assignment, inaccurate instructions did result in a small injury to one user—a cut finger. One of the peculiarities of the screen is that you must hold the extension bar firmly before you depress the latch to lower it. Otherwise, when the tension is released by the screen latch, the bar will hurtle back into place, pinching any fingers in its way.

The solution? A clearly worded warning, placed in advance of the step itself and set off typographically from the running text.

- Use the term "caution" for possible damage to equipment, the term "warning" for possible damage to people.
- Set off each part of the warning or caution typographically, by such devices as boldface, capitals, underscoring, and boxes.
- Place the caution or warning *before* the step that warrants it.

Using Direct Voice

One problem with the opening sentence in the sample instructions for the portable screen ("Tripod should be opened and pulled out") is that the sentence is passive rather than active. It is indirect rather than direct.

PASSIVE VOICE	**ACTIVE VOICE**
Tripod should be opened	Open the tripod . . .
	The user opens the tripod . . .

Why use a direct rather than indirect approach? It is one way to give instructions more immediacy, to strengthen the voice of the instructor. Another way is to use commands, called "imperatives." Imperatives are not the most polite verb forms—few writers want to sound like drill sergeants barking out orders most of the time—but they are perfect for instructions where the user needs to know at each stage what to do. Imperatives ("Do this," rather than, "One does this," or "You do this by . . .") give necessary accent to the all-important verbs, which are the engines of instructions.

THIRD-PERSON	**IMPERATIVE**
One should open the screen by . . .	Open the screen by . . .
A person opens the screen by . . .	

Note that you should maintain the voice and person you choose, rather than switching within a sequence. Such switches are hard on the reader who must, in turn, switch point of view.

SWITCH IN VOICE	CONSISTENT VOICE
Tripod should be opened and pulled out. Extend bar to proper height.	Open tripod and pull out legs. Extend bar to proper height.

Using Effective Page Layout

Page layout is particularly important for instructions, where the user often swivels from text to task, and then back to text. A numbered list usually makes it easier to find the place. It also helps the user divide and distinguish each step.

Here's a contrast between listing and narrative formats using the opening steps of the portable screen instructions.

NARRATIVE FORMAT

Hold screen with one hand, tripod legs down. With other hand, loosen tripod collar by pushing it downward one inch. Then pull one leg outward. All legs will open fully at once. Set on floor.

LISTING FORMAT

To set up the screen, follow these steps:

1. Hold screen with one hand, tripod legs down.
2. With other hand, loosen tripod collar by pushing it downward one inch.
3. Pull one leg outward. All legs will open fully at once.
4. Set on floor.

In writing instructions, take advantage of

- headings
- subheadings
- bold and underlined keystrokes
- generous indentations

The combination of tabular layout, white space, headings, subheadings, and typographical aids will help the reader focus on each step.

Examples 12.1 and 12.2 show the difference between instructions written in a congested, wall-of-print style, and those presented with a page design that aides their readability:

Are all instructions written as lists? Absolutely not. It is common to find instructions written in paragraph format when the audience

COMPARISON OF THREE FORMATS

Instructions Written with Passive Verbs

These instructions should be followed whenever the microscope is used:

1. The condenser should be raised to its highest point.
2. The diaphragm should be closed somewhat to prevent an overly bright field.
3. The low power objective should be locked into position on the nosepiece.
4. The adjustment knobs should be turned to focus the image. Then the slide may be moved around on the stage by turning the appropriate mechanical stage knobs.

Instructions Written with Verbs in the Imperative

Follow these instructions when using the microscope:

1. Raise the condenser to its highest point.
2. Close the diaphragm enough to reduce brightness on the field.
3. Lock the low power objective into position on the nosepiece.
4. Turn the adjustment knobs to focus the image.
5. Turn the mechanical stage knobs to view different areas of the slide.

Instructions Written in Paragraph Format

Using the Microscope

Raise the condenser to its highest point. Partially close the diaphragm to reduce brightness on the field. Lock the low power objective into position on the nosepiece. Turn the adjustment knobs to focus the image. Turn the mechanical stage to view different areas of the slide.

Box 12.1 Comparison of three formats.

Example 12.1 Version A: Poor design.

INSTRUCTIONS FOR ASSEMBLY OF SCREEN

To set up screen, hold the case with one hand, legs down. With the other hand, push collar down one inch. Pull one leg outward. All legs should open fully at once. Set the screen on floor. Press the release tab at the top of the tube, and raise extension rod. Swing case to horizontal position, and place screen hanger on hook. Release handle lock and raise screen to desired position. To close the screen, grasp extension rod firmly, press release tab, and gently drop the rod to its lowest position. Unhook screen and reroll. Swing screen to vertical position. press release tab and push tab back into slot. Raise screen from floor. Close tripod legs by first pressing each in until engaged. Then push tripod collar down to lock them.

is assumed to be more sophisticated. For instance, these instructions to use to the Oxi/Ferm Tube are in paragraph format (Figure 12.4). Note that imperatives begin each paragraph.

The U.S. Navy guide to the writing of instructions has a useful example of the difference in visual impact between instructions that are written in narrative vs. listing format (Figure 12.5).

Coordinating the Illustrations with the Text

As in other forms of illustration (see Chapter 5 for a more extensive discussion), the figures in directions

- orient the reader
- show the parts of the whole
- provide closeups and details
- show contrast and process

Illustrations are not essential for all instructions. Sometimes words alone will do the job; at other times, figures are helpful but can be dropped if budget or space is tight.

Illustrations done for the portable screen helped many new users, but only when the writers doublechecked to make sure the *words used in the callout were identical to the words in the text.*

(Text continues on page 286)

Example 12.2 Version B: Improved design.

OPERATING INSTRUCTONS

TO SET UP THE SCREEN

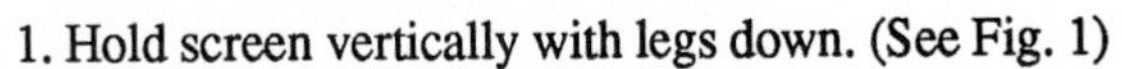

1. Hold screen vertically with legs down. (See Fig. 1)
2. With other hand, loosen tripod collar by pushing it downward one inch.
3. Pull one leg outward (See Fig. 2). All legs will open fully at once.
4. Set screen on floor.
5. Press the release tab at top of tube and raise extension rod (See Fig. 3)
6. Swing case to horizontal position and place screen hanger on hook. (See Fig. 4)
7. Release handle latch and raise case until it locks into place.

TO CLOSE SCREEN

WARNING

The extension rod is under tension, and will snap down abruptly when released. To avoid injury, keep fingers away from path of the extension rod when you collapse it.

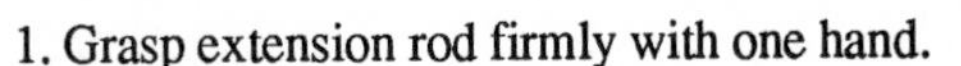

1. Grasp extension rod firmly with one hand.
2. With the other hand, press release tab, and gently drop the rod to its lowest position.
3. Unhook screen.
4. Swing case to vertical position and snap hook back into slot.
5. Holding screen with one hand, raise it slightly and close tripod legs.
6. Lock tripod by raising tripod collar.

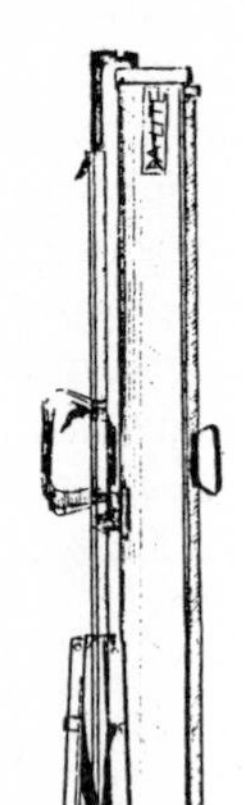

(1)

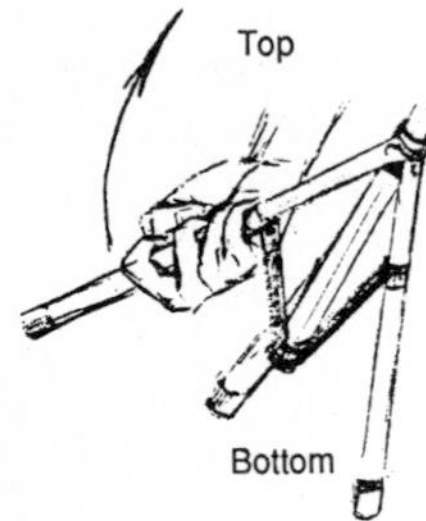

(2)

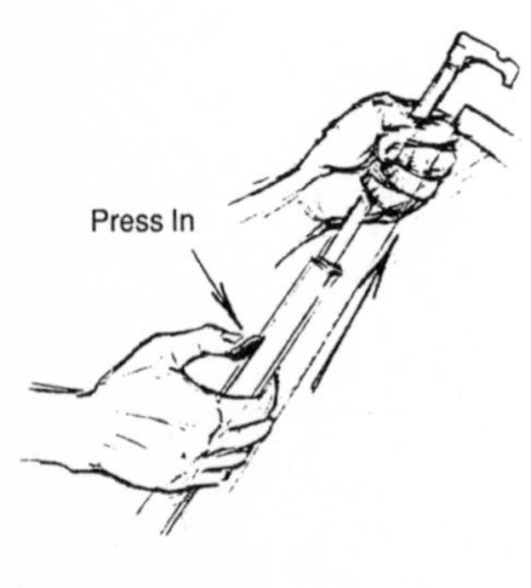

(3)

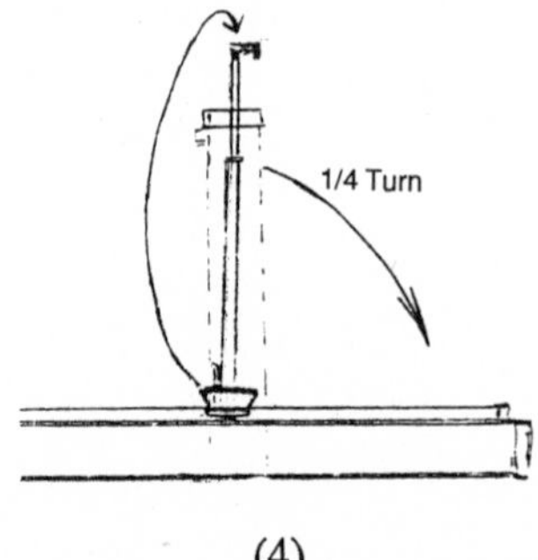

(4)

PROCEDURE
Steps must be followed precisely as shown.

To Use The OXI/FERM® TUBE

—for identification of oxidative-fermentative (OF) **gram-negative** rods:

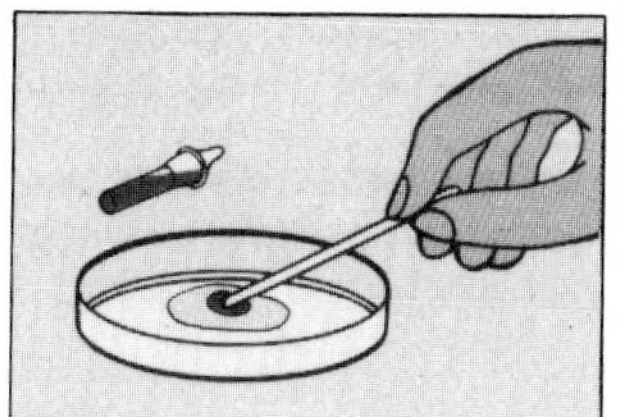

Perform oxidase test.
On a piece of Whatman No. 2 filter paper in a petri dish place several drops of oxidase reagent (see Materials Not Provided Section for preparation). Remove a loopful of the growth from the agar plate and smear on a small area of the reagent-impregnated paper. A positive reaction is denoted by the development of a dark purple color in 10 to 15 seconds.

Do not use oxidase strips to perform this test. Make certain that the loop being used does not react with the reagent to give a false-positive test—a wooden applicator stick may be used. Organisms grown on media which cause dye to be incorporated into the colony will often produce equivocal oxidase test results.

If the isolate is oxidase positive, or oxidase negative but not capable of fermenting anaerobic dextrose, inoculate the OXI/FERM TUBE (see flow diagram in Specimen Collection and Preparation Section).

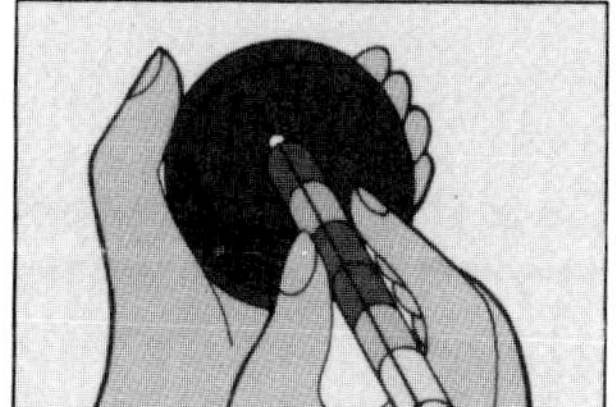

Remove both caps from the OXI/FERM TUBE. DO NOT FLAME NEEDLE.

Pick a large inoculum directly from the plate with the tip of the inoculating needle.
The inoculum must be readily visible on the needle. For some organisms where growth is sparse, several colonies may be required. Be careful to select pure, well-isolated colonies.

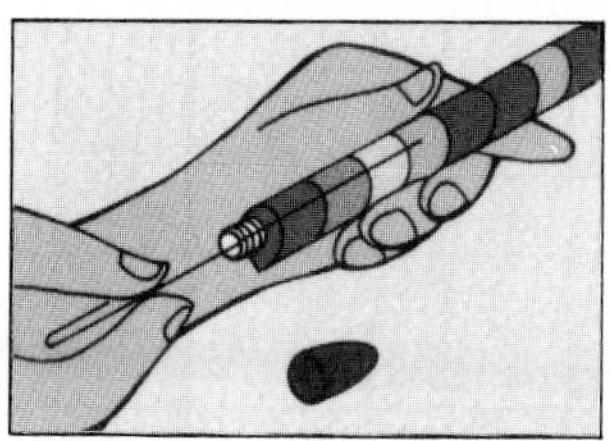

Inoculate the OXI/FERM TUBE by first twisting the needle, then pulling it through all eight compartments.
Remove the needle completely but **DO NOT FLAME.**

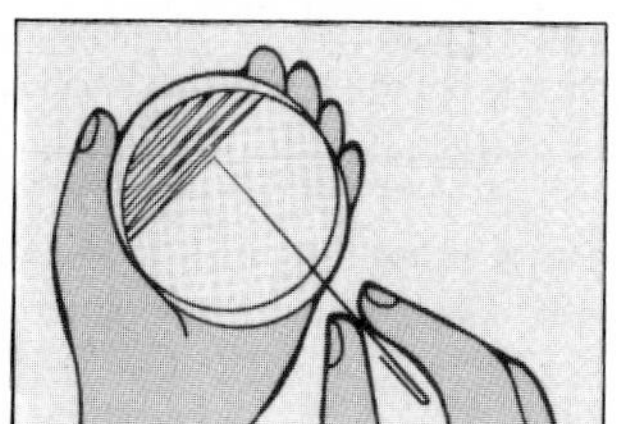

Streak a TSA plate.
Use the needle just removed from the now inoculated OXI/FERM TUBE. **DO NOT FLAME.** Incubate the TSA plate at 35° to 37°C for 18 to 24 hours, followed by 24-hours incubation at room temperature. The TSA plate will be used as a check for culture purity, pigmentation, and lack of special growth requirements.

Note: Other media may be used as a substitute for the TSA plate if the following limitations are recognized: Mueller-Hinton Agar (MHA) will provide a check for culture purity and lack of special growth requirements; however, some formulations of MHA may alter pigmentation. Blood Agar will provide a check for culture purity and pigmented colonies; however, Blood Agar will support the growth of fastidious organisms which cannot be identified using the OXI/FERM TUBE.

Figure 12.4 Instructions in paragraph format. Courtesy Roche Diagnostics.

Figure 12.4 (*continues*)

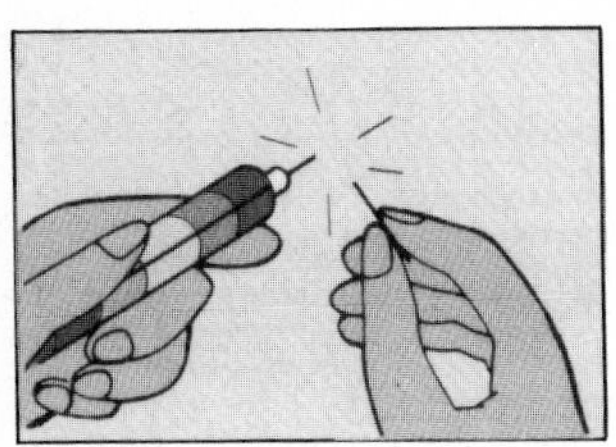

Partially reinsert the needle into the inoculated OXI/FERM TUBE using a turning motion.
Insert the needle through the anaerobic dextrose, arginine and nitrogen gas compartments. The tip of the needle should be seen in the H_2S/indole compartment. The portion of the needle remaining in the tube maintains anaerobic conditions necessary for true fermentation of dextrose, detection of arginine dihydrolase and production of nitrogen gas.

Snap needle at notch.
Discard handle and replace caps on tube.

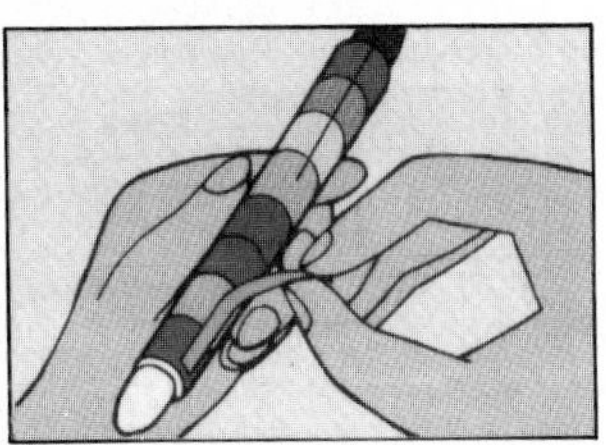

Strip off the blue tape.
This provides aerobic conditions in the OF xylose, OF aerobic dextrose, urea and citrate compartments.

Incubate the OXI/FERM TUBE at 35° to 37°C for 48 hours with the tube lying on its flat surface.
Check for a possible rapid urea reaction after 18 to 24 hours incubation.

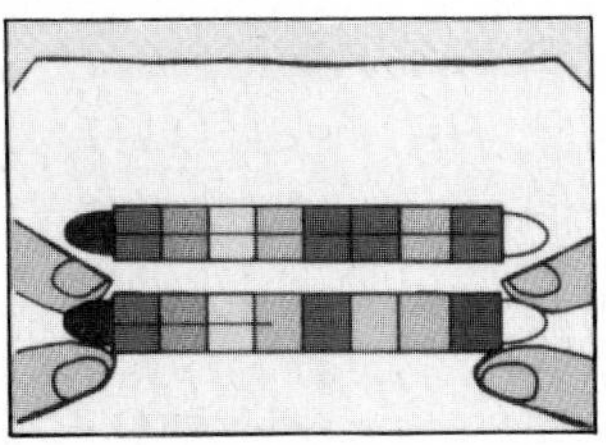

Interpret and record all reactions with the exception of indole.
Interpret the reacted tube comparing the colors with those of an uninoculated control tube. Record all reactions using the coding pad provided (see Results Section). All compartments must be interpreted before Kovacs' reagent is added to perform the indole test.

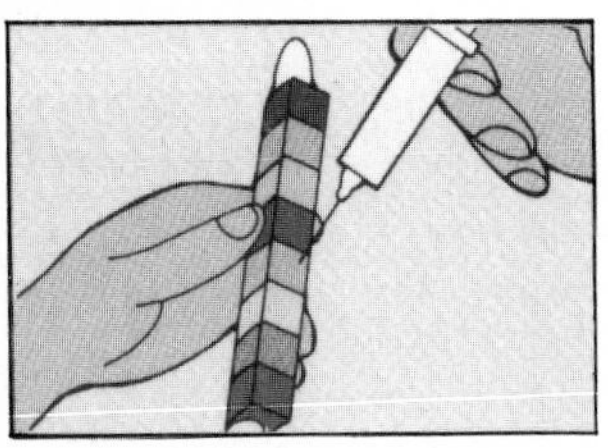

Perform the indole test.
Add 2 or 3 drops of Kovacs' indole reagent (see Materials Not Provided Section) through the plastic film of the H_2S/indole compartment using a needle and syringe. Release reagent directly onto the inside surface of the plastic film and allow it to drip down onto the agar. A positive test is indicated by development of a red color on the film or in the added reagent within one minute (see Reagents Section).

Record indole reaction on the coding pad.
Dispose of OXI/FERM TUBE by autoclaving.

5-7. CLEANING.

Materials List

Solvent, Trichloroethane MIL-T-81533
Lubricating Oil, Aircraft Instrument, Low Volatility MIL-L-6085

CAUTION

In the following procedure, do not use cleaning solvent of any kind to clean electrical parts.

5-8. Clean all electrical parts and nonmetallic parts using a stiff, nonwire brush. Blow the loose foreign particles from the part, using dry, filtered, low-pressure compressed air. Wipe the parts with a clean, lint-free cloth.

WARNING

When trichloroethane is used with a vapor degreaser, the cleaning should be performed only with adequate ventilation. Avoid prolonged breathing of vapor, avoid spilling, and avoid contact with skin.

5-9. Clean all other parts using an ultrasonic cleaner (UCW-200, or equivalent) or vapor liquid degreaser. When ultrasonic cleaner is used, use cleaning solvent approved by manufacturer. When vapor liquid degreaser is used, use trichloroethane (MIL-T-81533). Dry parts with dry, filtered, low-pressure compressed air. Wipe dry with a clean, lint-free cloth. Apply a thin coat of aircraft instrument oil (MIL-L-6085) to all steel parts. Thoroughly wipe off all excess oil with a clean, lint-free cloth.

NARRATIVE

5-7. CLEANING

Materials List

Solvent, Trichloroethane MIL-T-81533
Lubricating Oil Aircraft Instrument Low Volatility MIL-L-6085

CAUTION

In the following procedure, do not use cleaning solvent of any kind to clean electrical parts.

a. Clean all electrical parts as follows:

1. Clean electrical and nonmetallic parts using a stiff, nonwire brush.

2. Blow foreign particles from the part using dry, filtered low-pressure compressed air.

3. Wipe parts clean with clean, lint-free cloth.

b. Clean all other parts as follows:

WARNING

When trichloroethane is used with a vapor degreaser, the cleaning should be performed only with adequate ventilation. Avoid prolonged breathing of vapor, avoid spilling, and avoid contact with skin.

1. Clean parts using ultrasonic cleaner (UCW-200, or equivalent) or vapor liquid degreaser.

NOTE

When ultrasonic cleaner is used, use cleaning solvent approved by manufacturer. When vapor liquid degreaser is used, use trichloroethane (MIL-T-81533).

2. Dry parts with dry, filtered low-pressure, compressed air or clean, lint-free cloth.

3. Apply thin coat of aircraft instrument oil (MIL-L-6085) to all steel parts.

4. Wipe off all excess oil with a clean, lint-free cloth.

PROCEDURAL STEPS

Figure 12.5 Narrative vs. listing format.

Often they were not. The writer forgot to coordinate the wording of the labels with the wording in the instructions.

If the task you are describing is a simple one, you may be able to replace figure captions with steps in the process. For instance, in Figure 12.6 the instructions become the figure captions.

Doing Trial Runs

Testing instructions is a must, no matter how painstakingly you have prepared them. Find a group of people who don't know the procedure and whose background is similar to that of your target group. Then hand them what you've written, and watch them try to follow your instructions.

Some of the people will make mistakes you never dreamed of—that's the advantage of trying the text out.

When you are trying out your instructions, resist the temptation

Getting acquainted with your 213 projector

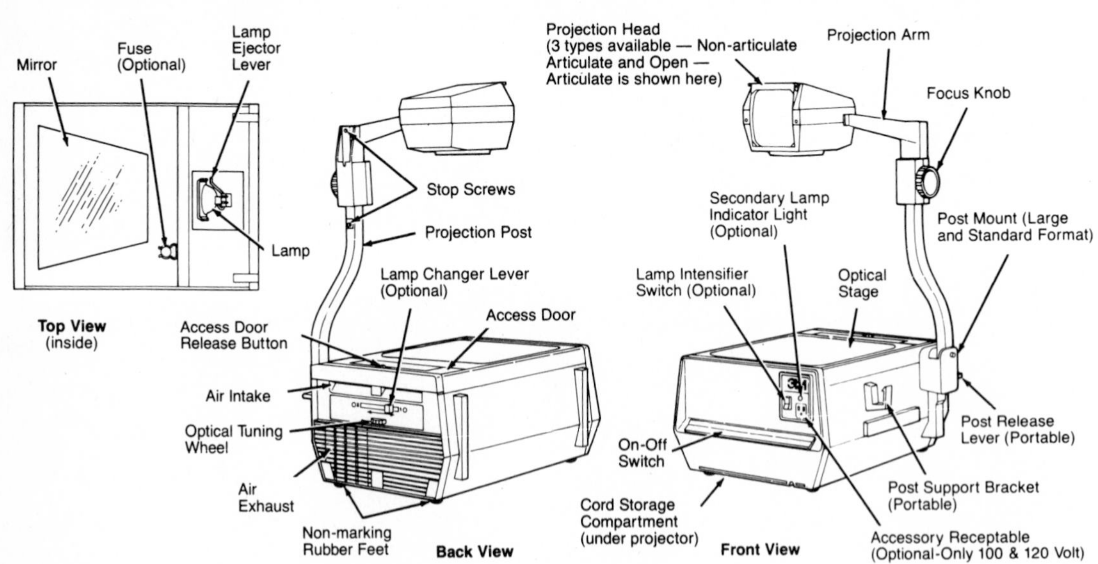

Figure 12.6 (a) Courtesy 3M.

Setup and operating procedure

1. Connect the power cord to a properly grounded outlet. (Refer to the data plate on the base of the projector for electrical requirements.)

Power Cord Slot Cover

2. Place the projector on a level surface and switch the projector on.

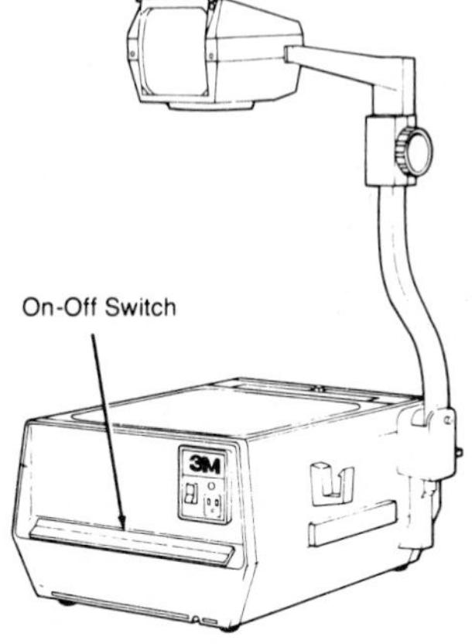

3. Adjust the image to the screen elevation by tilting the projection head.

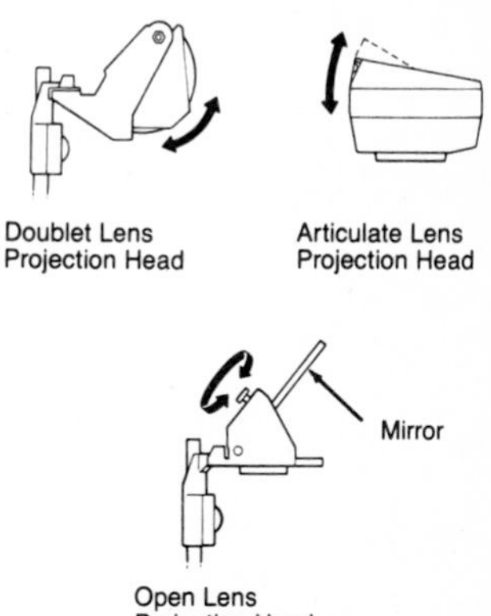

4. Place a transparency on the stage and focus the projected image by rotating the Focus Knob.

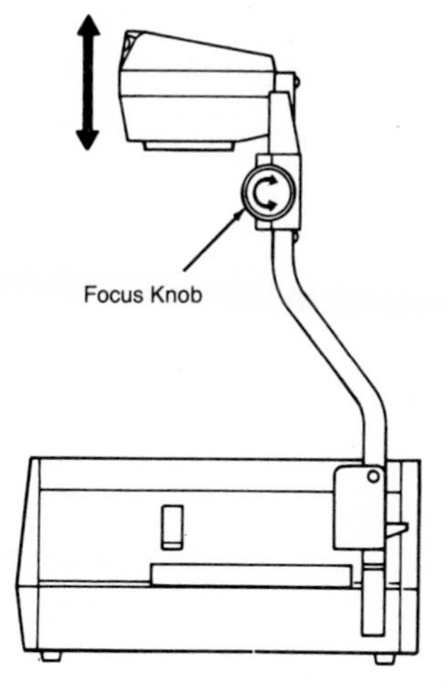

5. Rotate the Optical Tuning Wheel until the corners of the projected image are free of any red or blue corners.

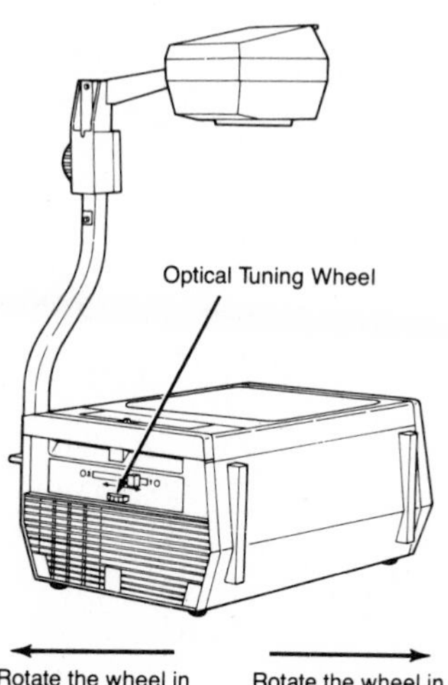

6. Correct any vertical keystoning by tilting the screen until it is perpendicular to the light beam axis.

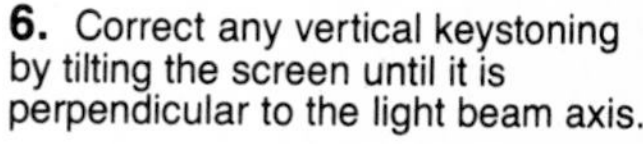

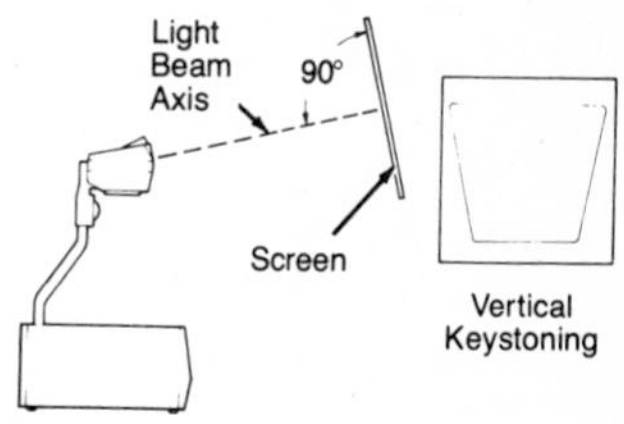

Correct any horizontal keystoning by moving the projector until it is square with the screen.

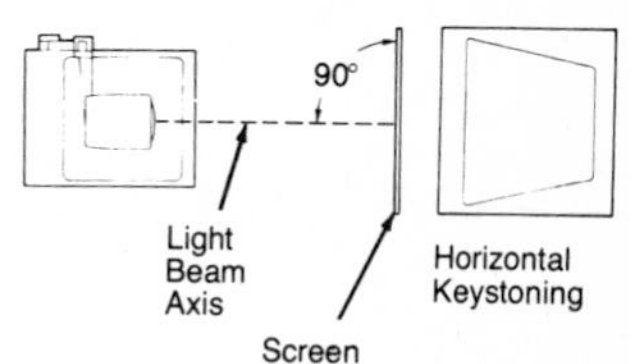

Figure 12.6 (b) Courtesy 3M.

to blame the reader. What's clear to the writer is often murky to the reader; it's the writer's job to provide the enlightenment.

But can anyone write so clearly and explicitly that each person in the targeted audience will be able to follow the instructions? Probably not, even if all of them have the background you expect. "It takes care, patience, and understanding to make the real world follow any path made clear in words," scientist and writer Philip Morrison comments insightfully. No matter how clearly you write, there will always be that one person in a hundred who is careless, who reads not what's written, but what he or she imagines *should* be written. This person is beyond your scope. Your job is to write for the other 99 who can be assumed to be unfamiliar with the task, but willing to read the instructions. Actually, the statistics may be grimmer than 1 in 100. One study suggests as a rule of thumb that printed instructions will be understood by their intended readers only two-thirds of the time.

As you revise in the light of feedback from your users, you will also want to do some editing. Look through what you've written for wordiness and errors in grammar. It's easy for such mistakes to appear if you are in the midst of working and reworking the same sentence in an effort to make it understandable. "I often don't see what I've written," one student comments. "I write and rewrite, until I'm so familiar with what I'm saying that I don't actually focus on the words. I have terrible typos and misspellings that stay in the last draft because I forget to give the paper a final word-by-word editing."

Yet this is exactly what the writer must do at the end: give the text a critical, proofreader's look, checking systematically for such errors as misplaced modification, subject-verb agreement, spelling, and wordiness.

POOR:

Rotate the screen shaft from it's verticle position to a horizontal one.

CORRECTED for grammar and spelling:

Rotate the screen shaft from its vertical position to a horizontal one.

TRIMMED to:

Rotate case to horizontal position.

POOR:

Pull screen out of screen body using screen handle. Pull out enough screen top to allow screen handle to hook into handle notch located on top of screen.

TRIMMED to:

Using screen handle, pull screen out and place handle on hook.

FOR COMPLEX CONTINGENCIES: Flowcharts and Decision Tables

Some instructions have more contingencies than others. Setting up a portable screen or overhead projector is a fairly straightforward process. But how does the writer best express the outcome of complex contingencies?

Sometimes *flowcharts* or *decision tables* are the answer. *Flowcharts* are diagrams made of symbols and connecting lines showing the step-by-step path through a procedure or system. (See Figure 12.7a.) *Decision tables* use columns to display a set of choices or outcomes for each step in a procedure or system. (See Figure 12.7b.)

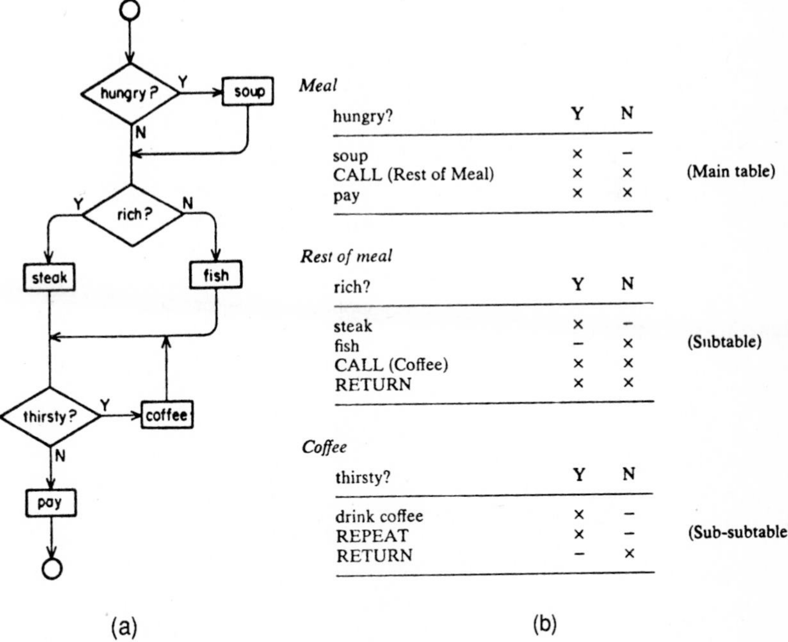

hungry?
Y
N
soup
rich?
Y
N
steak
fish
thirsty?
Y
N
coffee
pay
(a)
Meal
hungry? Y N
soup × –
CALL (Rest of Meal) × ×
pay × ×
(Main table)
Rest of meal
rich? Y N
steak × –
fish – ×
CALL (Coffee) × ×
RETURN × ×
(Subtable)
Coffee
thirsty? Y N
drink coffee × –
REPEAT × –
RETURN – ×
(Sub-subtable)
(b)

Figure 12.7 (a) Flowchart and (b) decision table notations. From M. Fitter and T. R. G. Green, "When do diagrams make good computer language?", *International Journal of Man-Machine Studies*, 1979.

With the help of flowcharts and decision tables, the reader is spared pages of "unless," "provided that," "except," "if then," "in which case," "except when." Instead, the user follows a path along which alternatives are visually distinct.

In one experiment, flowcharts gave the fewest errors among those following instructions for difficult problems. (P. Wright and F. Reid, *Written information: some alternatives to prose for expressing the outcomes of complex contingencies,* in *Journal of Applied Psychology,* 1973, Vol. 57, No. 2, 160–166.)

Flowcharts tend to be preferable when the information is highly sequential, tables when order is not critical. Fitter and Green provide an example when they contrast the uses of flowchart and decision table notation. (M. Fitter and T. R. G. Green, *When do diagrams make good computer languages?* in *International Journal of Man-Machine Studies* (1979), Vol. 11, 235–261.) Because the flowchart intertwines tests and actions, the result of the test may depend on previous actions. This is most obvious in the loop for drinking coffee (Figure 12.7). Obtaining the equivalent effect in decision tables requires the use of subtables, making the visual more difficult to follow.

A table is more effective when all the actions depend on the initial conditions of values. Fitter and Green illustrate this by modifying their flowchart so that all the actions depend solely upon knowing initial conditions. Here the flowchart becomes harder to follow than the table. "Flowcharts are better suited to highly sequential processes and decision tables to processes where order is not critical," they conclude (Figure 12.8).

One study (R. Kammann, *The Comprehensibility of Printed Instructions and the Flowchart Alternative* in *Human Factors,* 1975, 17(2), 83–191) found that *any* alternatives to prose in the "traditional bureaucratic style" were a considerable improvement in time, errors, or both—be they visual alternatives tables or flowcharts.

Tables are well suited to troubleshooting. Examples are shown in Chapter 19.

A FRAMEWORK FOR INSTRUCTIONS

Short booklets documenting a product, process, system, or procedure often are organized in a sequence drawing on the types of technical writing looked at thus far in this book.

These short booklets—called anything from memos to new product guides—are essentially miniature manuals that begin with an overview and end with instructions. Here is the typical sequence:

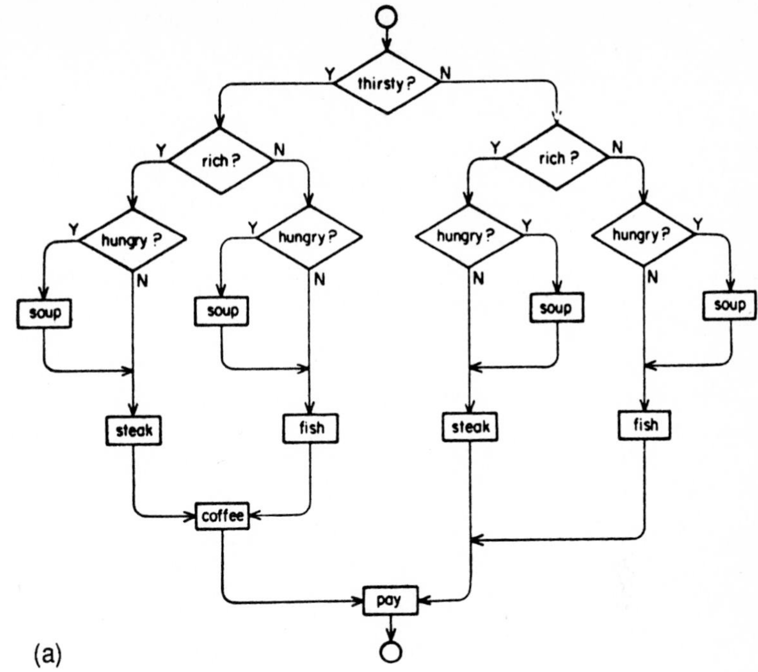

Figure 12.8 (a) Flowchart for having a meal; (b) decision table for having a meal. From M. Fitter and T. R. G. Green, "When do diagrams make good computer language?", *International Journal of Man-Machine Studies*, 1979.

Meal

hungry?	Y	Y	Y	Y	N	N	N	N
rich?	Y	Y	N	N	Y	Y	N	N
thirsty?	Y	N	Y	N	Y	N	Y	N
soup	×	×	×	×	–	–	–	–
steak	×	×	–	–	×	×	–	–
fish	–	–	×	×	–	–	×	×
coffee	×	–	×	–	×	–	×	–
pay bill	×	×	×	×	×	×	×	×

(b)

Overview (objective; in some cases, limitations, cautions, materials list.)

Definitions

Physical Description

Process Explanation

Instructions

Chapter 19 contains a full discussion of formal manuals. But you will find at work that there are many *informal* versions of the manual—such as short booklets that introduce a product, process, or service.

Instructions are the core of such booklets but the writer must set the stage beforehand, establishing the context with the necessary definitions, descriptions, and process explanations.

For instance, in this "how to" memo written by a student, you can see how the author has combined definition and description on the way to giving the instructions. The subject is how to change word processing codes to typesetting codes. The system used is Troff (Example 12.3).

SUMMING UP

• Instructions are usually written by those who know for those who don't. The instruction writer must therefore be not only methodical, but imaginative.

• Don't write over the heads of the audience. Instead, imagine what they need to know, the best order in which to tell them, the terms that need to be defined, and the visuals that might be effective. Remember: You know, but they don't. That's why they are reading the instructions.

• Orient the reader at the beginning of the instructions. This may mean a caution, a list of ingredients, a figure showing the orientation of the object, a figure that shows the parts of the whole. New users need an image of where they are heading, right from the start. There may be brief definitions and physical and process descriptions before you get to the heart of the matter—the instructions.

• Use specific language. Beware the vague phrase. Don't just say, "Be careful"; tell them *how* to be careful. Don't tell the people who are using their new car jacks for the first time to "Place jack in proper position near flat tire." Tell them what a "proper" position is. If you use technical terms, define them.

• Use a direct voice. You want the reader to have a sense of someone—preferably an intelligent, methodical person—explaining the task. Usually instructions are written in the active voice, in the imperative. You may sound a bit like a drill sergeant barking out orders, but this is probably best when the user needs to know what to do at each stage.

• You can use typography and layout to set off the imperatives dramatically, so that the readers first see what they have to do, then the details of how to do it.

Example 12.3

Typesetting an Indenture with Troff

INTRODUCTION

This memo is designed to show how to change the word processing codes of an indenture to codes that will typeset the indenture. Included are a list of definitions of the typesetting codes used for an indenture, a set of instructions for recoding, side-by-side examples of sections coded for typesetting and for word processing, and these same examples typeset and printed.

The typesetting program is Troff. Some of the codes are similar to those used in word-processing (i.e., bp, .ne). Some are slightly different (i.e., .ce) combines the .c and .ce codes and .fi is the fill command). As in the word-processing program, most Troff commands are preceded by a period (.) and appear on a separate line. However, Troff also uses commands which appear within text; these are preceded by a backslash. With these codes, the user can control the type of print (Roman, Bold, Italic), the size of print-characters, the precise positioning of the characters or lines on a page and is able to produce a document that has the appearance of being professionally printed.

When using Troff, the image on the screen will bear no resemblance to the finished document. All text is entered at the left margin and will be placed in output according to the Troff command (indent, tab, underline, etc.) and not according to how it looks on the screen. There is not, at the moment, a "proofit" program for Troff. You must print a file to see how it looks.

When you have finished recoding and are ready to print, the commands to print are:

xroff (filename)	to the Xerox 9700
hproff (filename)	to the HP Laserjet
zroff -P (# of the machine) (filename)	to the Apple Laserwriter

Example 12.3 (*continues*)

DEFINITIONS FOR TROFF FORMATTING CODES USED FOR INDENTURES

command	definition
.bp	begin new page
.ce#	center the next number of lines
\f2(text)\fl or fl (text)\fR	*change the text between commands to italics* and then change the following text back to original print
\f3(text)fl or \fB(text)fR	**change the text between commands to boldface** and then change the following text back to the original print
.fi	fill text to the margins and justify
\h'(measure)'	move horizontally the amount of the measure indicated
.in + (measure)	indent the following text by the measure indicated
.in - (measure)	cancel the indent and return to the previous margin
\l'(measure)\(ul'	draw a line the length of the measure
.ne #	need next number of lines kept together
.nf	print the following text in no-fill mode
.P	begin a new paragraph
\s#	change the size of the letter (letters or words) to # points
.sp (+ or - measure)	move vertically up or down the indicated measure
*T	this symbol signals a tab to the space set by the .ta command
.ta #R	set tab stop at the indicated number and right-justify the tabbed text
*(EM	this symbol produces a true dash; a single line rather than two dashes
.tl''''	space across the page, at left, center, and right, whatever text is between the single quotes. Text to be placed at the left side will be placed between the first and second single quote, etc.
\`\`(text)''	print small slashes and single quotes as double quotes

INSTRUCTIONS FOR CODING AN INDENTURE FOR TYPESETTING

1. Open the file that contains the start of an indenture that is coded for word processing.
2. Delete all codes at the top of the file.
3. Turn to the example labeled ''Top of File'' and type the first 20 codes into your file (.nr O 7.5P through .VM 4 4).
4. Next, replace the .f command with the .fi command for filled and justified text. Search for all .f commands and replace with .fi. Return to the first .fi command.
5. Create a blank line and type: .P Start the paragraph on a new line, flush left. (When the paragraph is printed, the first word will automatically be indents 5 spaces.) Repeat this for all paragraphs in the file, including those that begin with a letter or a number, if they are to be indented according to the paragraph style. Remove empty lines between paragraphs and return to beginning of file.
6. Change the print size for certain words or subheadings so that the initial letters are in 10p capitals and the remaining letters are in 8p capitals. This style of print, capitals in two sizes, is used (1) for subheadings under articles, (2) for the word ''Section'' under articles, and (3) for capitalized words or phrases within paragraphs. Search for these subheadings and phrases and make the print-size changes. Return to the beginning.
7. Search for all double quotes and replace with 2 short-slashes to precede quoted text and 2 single quotes to follow.
8. Replace all the centering commands (.c and .ce) with a single .ce plus a number to indicate how many of the next lines are to be centered.
9. To draw a line, use the following code: \l'(measure)\(ul'. See example. Replace all lines with this code, giving it a measure for the length of the line. The measure may be entered in inches (i), Picas (P), or points (p).
10. Remove all underline characters and replace with the command, \l'-(a measure)\(ul'.
11. Use the .ne commands when it is desirable to print several lines together.
12. Search for and remove all .ks and .ke codes.
13. If there is a Redemption Table in the document, recode the table with the commands shown in the example.

Example 12.3 (*continues*)

```
14. To indent for lists, and nested lists, use the .in +
    the number of spaces to indent. Enter a .P on the
    line before each item. Remove all indent spaces; text
    should be flush left. Remove blank lines between
    items. At the end of a list, enter .in - the number
    of spaces to return to the previous margin.
15. Recode for footnotes as follows: replace the foot-
    note with the character to be used with the footnote
    (usually *). On the next line type .FS * and hit the
    return key. Type in the footnote. Return to next line
    and enter .FE. Resume the text on the following line.
16. Replace all double dashes with the \*(EM code for a
    true dash.
17.See the examples for the coding of the final pages of
    the indenture.
```

[Examples not shown]

• Use tabular format for instructions for absolute beginners. Narrative format, which places a larger burden on the reader, is common for more sophisticated audiences.

• Present a methodical, accurate account. Some steps are harder to express in writing than others; one terrible solution is to simply leave the step out, saying to yourself something like, "Oh well, it's obvious." Don't do this. Capturing actions in a net of language is never easy; it's the writer's job to struggle with this unenviable task—not to abandon it.

• The beginner can write good instructions by bringing a fresh perspective to the task. But if you are unfamiliar with the job, learn it thoroughly before you write. Omitting or misstating steps can have serious consequences. The writer must understand the task thoroughly.

• Take advantage of headings, subheadings, bold and underline keystrokes, and generous indentations. They help the eye do its job.

• Doublecheck illustrations to make sure words used in the callouts, captions, and labels match the text. Often errors creep in when the writer revises the text, but forgets to adjust the corresponding captions and callouts.

• Edit for grammar and spelling. When you are concentrating on logic and accuracy, it's easy for typos, misspellings, and stylistic mistakes to dot the text. These are often invisible to the writer who is concentrating on getting the ideas down correctly. But once the ideas are there, proofread carefully.

- Give the instructions a trial run. Testing them is a must, no matter how painstakingly you have prepared them. People will make mistakes you never dreamed of.
- Don't blame the reader who makes the mistake. Find out how the instructions can be improved, then fix the text so that it is clearer. You won't reach everyone—there will still be the person who imagines what *should* be written, not what's actually there. But you will increase the odds of reaching more people if your instructions are crystal clear.
- Flowcharts are effective for illustrating the path of complex contingencies, especially when the information is highly sequential.
- Decision tables are useful when all actions depend on the initial conditions of values, and when order is not critical.
- Instructions may come after an extensive introduction that sets the stage. The sequence of the introduction may include an overview, definitions, a physical description, and a process explanation.

EXERCISES

1. If you are studying technical writing in a class, set up a situation that parallels the portable screen example in this chapter:
   a. Pick a task that some members of the group are familiar with—such as setting a mousetrap, tying a necktie, or calibrating a balance.
   b. Have the people who know how to do the task write the instructions.
   c. Have the people who are unfamiliar with the task try to follow these instructions. Make sure the writers don't coach the users—that is, give them hints and suggestions other than those written in the directions.

   As a group, discuss the strengths and weaknesses of the directions.
2. Photocopy a set of instructions you either like or dislike. Write an analysis of its strengths and weaknesses. For instance, how are the illustrations used? Do they provide an overview? Closeups of detail? Contrast? Are there figure references in the text? Are the instructions logical, complete, and accurate? Are steps missing? Are terms defined?

3. Write instructions for a job with which you are familiar. Some possibilities:

a. Calibrating a balance
b. Logging on to your computer system
c. Tying a bow tie
d. Making an omelette
e. Using a chalk line
f. Boning a chicken

4. Write a short (3- to 4-page) manual for a job with which you are familiar. Include

Overview

Definitions

Physical Description (when appropriate)

Process Explanation (when appropriate)

Instructions

Some possibilities:

a. Instructions for Replacing Bicycle Tires
b. Regulating a Musical Instrument
c. Developing Photographic Negatives

CORRESPONDENCE

PART V

Letters and Memos

CHAPTER 13

A telephone call is often the quickest, most direct way to find the answer to an inquiry, ask about charges, or determine a convenient time for a committee meeting. In many instances the telephone has replaced the traditional letters and memos of business correspondence.

There are still many times, though, when you will need to write letters or memos:

- When you need a record of the transaction: Even if you telephone first, a follow-up letter is the best way to document the details.
- When the message is complex and requires careful wording and organization: Ambiguities and misunderstandings often go undetected in a telephone call.

- When you are sending along another document that needs an introduction.
- When you are establishing the terms of a contract.
- When you are explaining a procedure.
- When you need to say "no." The telephone offers little help in softening or easing negative information.

Memos and letters are both forms of correspondence. Letters are usually sent outside the company or agency; memos circulate within the organization. Occasionally these forms overlap; for instance, the president of a company may send a letter, rather than a memo, to convey holiday greetings to the staff.

Cover letters are those that accompany another document; for instance, they may preface a resumé, a proposal, or a piece of sales literature. When they accompany a report or proposal, they are often called **letters of transmittal.**

Effective letters and memos depend, like other forms of technical communication, on thinking about the audience, and then casting the document so that the organization, language, and design all contribute to its readability.

CUSTOMARY FORM FOR CORRESPONDENCE

A handwritten letter of application for a job, a double-spaced business letter, a cover letter typed with a red ribbon—all of these oddities would break the customary form for correspondence, and in so doing create a negative impression on readers who expect their correspondents to follow certain traditions that have wide acceptability. Letters and memos should be typed, using the customary formats described below.

Standard Formats for Letters

Letters have a heading and an inside address, a salutation, a body and closing, and supplementary lines. They may also have a subject line. (Each part is labeled in Figure 13.1.) Both blocked (Figure 13.2) and unblocked styles are acceptable.

Heading If you are using paper with a printed letterhead, the date is the only part of the heading you'll need to add (Figure 13.3). If your paper does not have a printed letterhead, single-space address and date, leaving a margin of 1½ inches above and to the right. It's acceptable to abbreviate the name of the state; note that no comma follows between abbreviation and ZIP code (Richmond, VA 23226).

Text *continues* on page 306

Heading

Institute of Imaging Sciences
Polytechnic University
Brooklyn, New York 11201
22 February 1988

Inside Address

Ms. Yuk Lee
11 South Lincoln St.
New York, New York 10007

Subject Line

Subject: Upcoming Imaging Lecture

Salutation

Dear Ms. Lee:

Body

I am writing to invite you to the Imaging Colloquium on March 1, 1988. It will be given by John Sinclair, who is a professor in our Imaging Program. His seminar is entitled "Aerial Reconnaissance Systems."

I expect that this will be an unusual and very special presentation. As a Senior Scientist at National Laboratories, John has been concerned with images in aerial reconnaissance systems and with other high-tech applications. He has succeeded in integrating the scientific and artistic aspects of his life in a very personal philosophy.

We shall assemble in Room 372 Sinclair Hall at 4:00 pm for coffee and donuts; the seminar proper will start at 4:30 pm. Please do come if your schedule permits.

Closing and Signature

Sincerely,

Ruth Sinclair

Ruth Sinclair
Institute of Imaging Sciences

ar/ae

Figure 13.1 Unblocked style. The heading is on the right, the inside address on the left. All paragraphs are indented, and the closing is aligned with the heading.

Figure 13.2 Blocked style. Headings, inside address, and closing begin flush at the left margin. Paragraphs are not indented.

1470 Sunrise Highway
East Littleton, Virginia 23226
14 March 1988

Mr. George Davidson
Schulman & Davidson
135 South Creasy
East Littleton, Virginia 23226

Dear Mr. Davidson:

I've looked into your question of energy-related tax credits. Although energy tax credits remain available for certain tyes of alternative energy property, Congress has clearly indicated a diminishing desire to use tax credits as a means of stimulating the development of alternative energy sources. This change in legislative interest is not surprising in light of the world energy situation, which is markedly different from that prevalent when the energy tax credits were first enacted.

The tax credits available are limited to the following types of property: (1) Solar: fifteen percent in 1986, twelve percent in 1987, and ten percent in 1988; (2) Geothermal: fifteen percent in 1986, ten percent in 1987 and 1988; (3) Ocean thermal: fifteen percent in 1986, 1987, and 1988.

The credits for fuels from nonconventional sources and alcohol fuels continue to be available as under prior law.

If you have further questions, please call.

Sincerely,

Roger Lincoln

Roger Lincoln
Lincoln & Devane

rl/pd

Figure 13.3 If you are using paper with a printed letterhead, the date is the only part of the heading you'll need to supply.

1817

Harper & Row, Publishers, Inc.
10 East 53rd Street, New York, New York 10022
Cable: Harpsam Phone: 212-207-7000

December 18, 1988

Inside Address At a minimum, double-space between heading and inside address. If the body of the letter is short, add double spaces between heading and inside address until the body of the letter is centered attractively between top and bottom margins.

Salutation Double-space between last line of inside address and the salutation. End the salutation with a colon; if you are quite friendly with the recipient, a comma is acceptable.

Whether you are writing the salutation of a letter, or filling in the recipient's name in a memo, you will often be faced with a choice in honorifics (Mr., Miss, Mrs., Dr., Ms.).

If you do not know the sex of the person you are addressing, it would be wise to spend a few minutes trying to find out, rather than settling for "Dear Sir."

As increasing numbers of women join technical professions, the issue of exclusionary language has heated up, in part because of all those technical lectures at which the professor addressed the mostly male class, with its sprinkling of women, as "Gentlemen."

"Gentlemen," or its equivalent, "Dear Sir," used to be acceptable not only in classroom lectures, but in salutations and accompanying text. "When I say he, I mean she," was the informal defense. Nowadays, addressing an audience as "Gentlemen" when there are women in the group is considered to be exclusionary language. Both individuals and companies seek to avoid such sexism.

When you are casting for a letter or memo, don't assume that all engineers are male, for instance, or all food technologists female. If you are writing a form letter, include both sexes in the salutation. Alternatives such as "Dear Food Technologist," or "Dear Colleague," are acceptable for group address. So, too, is "Dear Sir or Madam." If it is a personal letter, inquire for the correct honorific in the salutation, just as you check the spelling of the name for correctness. It is one more example of a careful, professional approach to a problem. The honorific "Ms." is appropriate if that is what the subject wants. The Associated Press guidelines advise: "For women who have never been married, use Miss or Ms. . . . according to the woman's preference. For divorced women and widows, the normal practice is to use Mrs. . . . Use Miss if the woman returned to the use of her maiden name. Use Ms. if she prefers it."

(See also "Pronouns," Usage Handbook.)

Body Double-space between salutation and body. In block form, each paragraph begins at the left margin; in unblocked form, paragraphs are indented.

Closing Double-space between body and closing. In block form, the closing goes against the left margin. In unblocked form, either line the closing up with the date, or place it so that the longest line (usually your title) ends at the right margin. "Sincerely" has wide acceptability as a complimentary closing. Effusive expressions like "warmly" are inappropriate. Leave four lines between closing and typed name, to allow space for your signature. If the inclusion of your title is appropriate, add it below your name.

Supplementary Lines Place supplementary lines at the left margin, separated from the closing by at least two lines. If you are enclosing documents, say so here ("Encl." or "Enclosure: Annual Report 1988"). If you are sending copies of the letter to other readers, use "c:" or "Copy to." The abbreviation "cc," for carbon copy, is outdated in this time of high-speed copying machines and word processors. If you are including the typist's initials, they follow the initials of the author, in lowercase, separated by a slash or colon (JB/efc).

Margins Leave 1½ inches around the text. White space reduces the wall-of-print effect on your readers.

Subject Line Obligatory in memos, the subject line is optional in letters. It may be useful if you want to focus the attention of the reader immediately on the subject of the letter. For instance, you may be writing a large organization for product information. Putting this information in the subject line (Subject: Request for Specifications on IR-Z123588) allows the reader to move quickly on your request. Capitalize initial letters of important words in subject lines.

Headers If the letter extends more than a page, each subsequent page should have a header giving page number, name of receiver, and date (Figure 13.4).

Standard Format for Memos

Memos are usually sent on printed forms. Styles vary from business to business; the basic heading usually includes a "To" line, giving the name and position of addressee; a "From" line, giving name and position of writer; a "Subject" line; a "Date" line; and a "Distribution" line, if copies of the memo will be distributed to a number of people. In some companies, the writer's handwritten initials appear after the typed name in the "From" line. In others, the writer signs at the bottom of the memo. (See Figure 13.5 for a sample memo form.)

Figure 13.4 Example of a header.

page 2 Johnson 12/6/88

The results of the investigation show the waste cap shooting pond to be in a highly fractured area. Wells installed in the bedrock have intercepted water-filled fractures and show contamination by several constituents. This contaminaton is a direct result of the operation of the waste cap shooting pond.

Details appear in the attached report.

Sincerely,

Figure 13.5 Sample memo form.

To: Subject: Dr. Ralph Sinclair's
From: Donation

Dr. Ralph Sinclair intends to donate $100,000 to create an Award for Excellence. The income from the donation is to be used by the Economics Department in the following way:

Purpose: The endowment is designed to attract or retain young, bright faculty. The award is meant to supplement the awardee's regular compensation.

Criteria: The candidate must

- demonstrate excellence in research
- be under age 40
- be considered essential to the success of the Economics Department

Duration: The grant is available annually. If awarded, it can be promised for not less than one year and not more than three years. If not awarded by judgment of the committee, the income can accumulate to the following year.

Committee: The committee is to be composed of the following people:

1. Dean of Arts and Sciences
2. Executive Vice-President for Academic Affairs
3. Grantor
4. Other individuals invited at the discretion of the three above members

Reports: Research papers by the awardee are to be filed with the committee for general availabilty to the Economics Department.

LOGIC AND ORGANIZATIONAL PATTERNS IN LETTERS AND MEMOS

Letters and memos are best kept short. If you can keep a piece of correspondence to a page, do so. Your readers will appreciate this succinctness. While there will be unavoidable instances where you'll need more than a page, you will do best to imagine an outline of no more than five paragraphs, in which you follow a traditional structure called "descending order."

- In the opening paragraph, you give subject and purpose.
- In the following one, two—or at most, three—paragraphs, you develop the subject.
- In the final paragraph, you provide a strong closing with any requests for action.

Opening State your subject and purpose in the opening, not in the second or third paragraph. Concentrate on getting to the point as quickly and clearly as possible.

> Enclosed is a catalog of hardwood moldings, prepared for inclusion in the February 1987 *Architect and Interior Decoration.*
>
> I am pleased to announce the promotion of Mr. Richard Linder to Purchasing Supervisor for the Lindhaven Branch.

Letters and memos with rambling openings irritate readers who want to know the main idea at the top.

> Roundabout opening:
>
> I'm aware that your company receives a great many requests for information, some of which probably cannot be acknowledged. However, I'd appreciate your handling my request, as I need the information for a project. In my case, I need information on computer chips, specifically those for a Central Processing Unit.
>
> Focused opening:
>
> I would appreciate your sending me specifications for Central Processing Unit chips made by Hewlett Packard.

If the letter or memo is one of response, the writer should specifically link the document to prior events.

> **VAGUE:**
>
> Here is the reprint you requested.

LINKED TO PRIOR EVENTS:

Here is the information on developments in MS DOS and their applicability to our programs you requested on April 9.

VAGUE:

My staff has reviewed the activities for which you requested contract assistance back in the summer.

LINKED TO PRIOR EVENTS:

My staff has reviewed the request for contractor assistance outlined in your memo of August 12, 1988.

VAGUE:

Our agency has reviewed Environmental Impact Statements (EISs) for the proposed natural gas pipeline within the Ancona National Wildlife Refuge.

FOCUSED:

Our agency has reviewed two Environmental Impact Statements (EISs) for the proposed natural gas pipeline within the Ancona National Wildlife Refuge: EIS #2346, prepared by the Department of the Interior, U.S. Fish and Wildlife Service, dated March 1988; and EIS #2349, submitted by the Federal Energy Regulatory Commission in December 1988.

Body The body should develop the main idea stated in the opening paragraph. If, for instance, you are writing a letter of transmittal, the body contains a summary of the salient points in the enclosed document.

The enclosed catalogue includes descriptions, illustrations, and availability for panels, trim, casing, baseboards, and crowns. A price list is also attached.

If the letter accompanies a report or proposal, the body would summarize the most important topics in the report. Usually this would mean a summary of the *objective* and the *findings*.

The report covers transactions at Branch #37's automatic teller machines from January 1, 1987 to February 1, 1987. Number, type, and duration of transactions are given, as well as client background. The figures show a 4.5 percent drop in deposit activity among full-time checking account customers, but a 12 percent increase in cash withdrawals within the same group. Major gains have come not from branch customers, but from NYCE network users.

Conclusion The ending is usually a courteous invitation to keep the lines of communication open ("You can reach me at 543-

6776 if you have further questions"). It is also the place to reiterate politely but specifically what you expect or are requesting.

Ending with a courteous, upbeat statement of the specific action you want is just as important as opening with a clear statement of subject and purpose.

VAGUE CLOSING
Please get the materials to us promptly.

SPECIFIC CLOSING
We'll need to have the materials back by 19 January in order to meet the deadline for publication. Thanks very much for your cooperation.

VAGUE CLOSING
Let's get together some time soon.

FOCUSED CLOSING
Shall we meet on 21 February, if your schedule permits? I'll telephone your office during the week of 16 February to check.

VAGUE CLOSING
Let me know if any problems arise.

FOCUSED CLOSING
I'm pleased you've chosen us as the vendor. I'll telephone you in two weeks just to touch base, and to make sure the system is working smoothly.

LANGUAGE AND "LETTERESE"

Effective letters and memos are short—they get to the point quickly. A tight organizational plan like the one described in the preceding section of this chapter is one way to contain the length of letters and memos. Another way is to make the language of the letter as direct as possible.

The inflated language often found in letters and memos, sometimes called "letterese," interferes with the brief, clear delivery so important in correspondence. The involuted phrases of letterese can double the length of a document without adding in any way to either its style or substance. Letterese includes such verbal hickups as "herewith you will find" and "we will endeavor to ascertain to the fullest possible extent," such florid phrases as "predicated on the assumption that" and "pursuant to your request." Such expressions will interfere with your message; they have the same archaic ring as courtesy closings like "I remain your obedient servant."

Here is a sampling of these phrases with more direct versions given in the right-hand column.

I would appreciate your responding to the request by the earliest practicable date.	Please respond by 25 May.
If there are further matters you wish to discuss, do not hesitate to telephone me at my office.	I can be reached at 718-643-8661 if you have further questions.
It will be necessary to hold these requests in abeyance for implementation until we are able to utilize our new network.	The requests will have to wait until we have the new network up and running. or The requests will be held until the new network is functioning.
prior to	before
keep in abeyance	wait
cognizance of the fact that	know
utilize	use
endeavor to ascertain	try to find out
notwithstanding the fact that	although
despite the fact that	although
in view of the fact that	as
in the event that	if
in the likelihood that	if

THE "YOU" APPROACH

Another way to cast letters and memos effectively is to choose language that will be appealing from the reader's point of view, to use what is called the "you" approach. The invaluable "you" approach is one where the writer casts the information in the letter or memo so that the emphasis falls on the reader's needs, rather than on those of the writer. Examples 13.1a and 13.1b show the same letter cast first from the writer's point of view, and then revised so that it takes a "you" approach. When the needs of the two parties run in the same channel, the "you" approach is superfluous; however, the contrast may be substantial when there is a gulf between the goals of the two parties.

Example 13.1a Writer-centered letter.

Dear Mr. Myranes:

I am enclosing a 2-page proposal for an article on the advantages of telecommunications for consideration in Medical Suppliers. I am interested in writing such an article for your publication as Medical Suppliers reaches a broad readership. I would like to reach this readership, and feel that my background experience makes me an ideal person to contribute to your publication.

I think you will find the article timely and useful. The article contains actual examples of medical suppliers who are already offering remote order entry, inventory management, and a discount system.

Telecommunications affects almost every element in a company's value chain. My article will stress what telecommunications can do for medical suppliers. I am the author of several articles on telecommunications, all of them well received.

Dear Mr. Myranes:

Enclosed is a 2-page proposal for an article on the advantages of telecommunications for Medical Suppliers. Your readers will find the article timely and useful: many companies are missing out on the opportunities for competitive advantage that telecommunications offers. The article includes actual examples of medical suppliers who are already offering remote order entry, inventory management, and a discount system.

Telecommunications affects almost every element in a company's value chain. The article stresses what telecommunicatons can do for medical suppliers, then discusses hardware and software, and ways to smooth operations, coordinate activities, and reduce clerical expenses. The main point: If a competitive opportunity exists, set up a telecommunications system rather than be forced to do it later on the competition's terms.

Example 13.1b Letter with "you" approach.

In the second version, the author has summed up the main points in the prospective article from the point of view of the **reader.** For instance, the first version argues, "I am interested in writing such an article as your publication reaches a wide readership," presenting the writer's motives. The second version says, "Your readers will find the article timely and useful. Many companies are missing out on the opportunities for a competitive advantage," presenting benefits for the reader. This is the "you" approach. For it to work, writers must shine the spotlight on the readers' needs, not their own.

NEEDLESS NEGATIVES

Positive language is another important technique to consider when you are casting letters and memos. Often you may be able to put a positive spin on your sentences, rather than using negative phrasing that works against your own best interest. In the following examples, the same information has been cast both negatively and positively.

NEGATIVE APPROACH:
Users will be charged 40 cents for each transaction after the first three.

POSITIVE APPROACH:
The first tree transactions of each month are free. Subsequent transactions cost 40 cents each.

NEGATIVE APPROACH:
Materials in your possession should be returned within 30 days, or you will be assessed a fee of $50.

POSITIVE APPROACH:
No fine will be levied if you return materials within 30 days. After that date, you will be fined $50 in overdue charges.

NEGATIVE APPROACH:
You will not be permitted to graduate unless you return your overdue books.

POSITIVE APPROACH:
Please return overdue books within the next week at any time between 9:00 a.m. and 10:00 p.m. If you think an error has been made, speak with any of the librarians on duty, and they will be glad to doublecheck our records, as well as the shelf. If you have lost a book, we can tell you the replacement cost. On April 19, we submit a list of people who have not returned materials; if your name appears on that list, you'll be barred from receiving a diploma.

A recriminatory tone will gain no good will for the writer; avoid such finger shaking in favor of a more positive approach.

RECRIMINATORY TONE:

You have been instructed repeatedly on the procedure for vacating the building.

POSITIVE TONE:

Here's a review of the procedures for vacating the building. If you've any questions, call me at x3357.

Throughout, take the reader's point of view. "The information you requested was shipped April 5," is brief and clear, but may not address the reader's need to know when the order will arrive, or why it has been delayed. "Because of a strike by Local 42B, some of whose members operate our mailroom, we were unable to ship the circuit boards you requested until today, April 5. The boards should arrive within 48 hours of mailing. We apologize for the delay, and are glad we were able to move promptly once the strike was settled."

PAGE LAYOUT

If the memo contains detailed information, the layout should emphasize the major points. Ways to do this include subheadings, underlining and bullets. The sample memo shown in Figure 13.6, for instance, uses subheadings to distinguish main points visually.

Often, letters are quite brief; the letter of transmittal, for instance, is usually three short paragraphs. There is no need in a letter this brief to introduce subheadings to divide the text. However, if the letter is sufficiently complicated, you may want to provide some visual cues to its division. Figure 13.7 shows a letter format with typographical divisions.

Attractive page layout will add greatly to the readability of a memo. Example 13.2, for instance, is a technical memo on trademarks presented in two versions. In both cases the information is the same. In the second version, however, the design is improved:

- Main points are set off by underlining and numbers.
- Examples are set off by white space.
- Column headings are used to contrast correct and incorrect procedures.

Text *continues* on page 321

Figure 13.6 Parts of a typical memo.

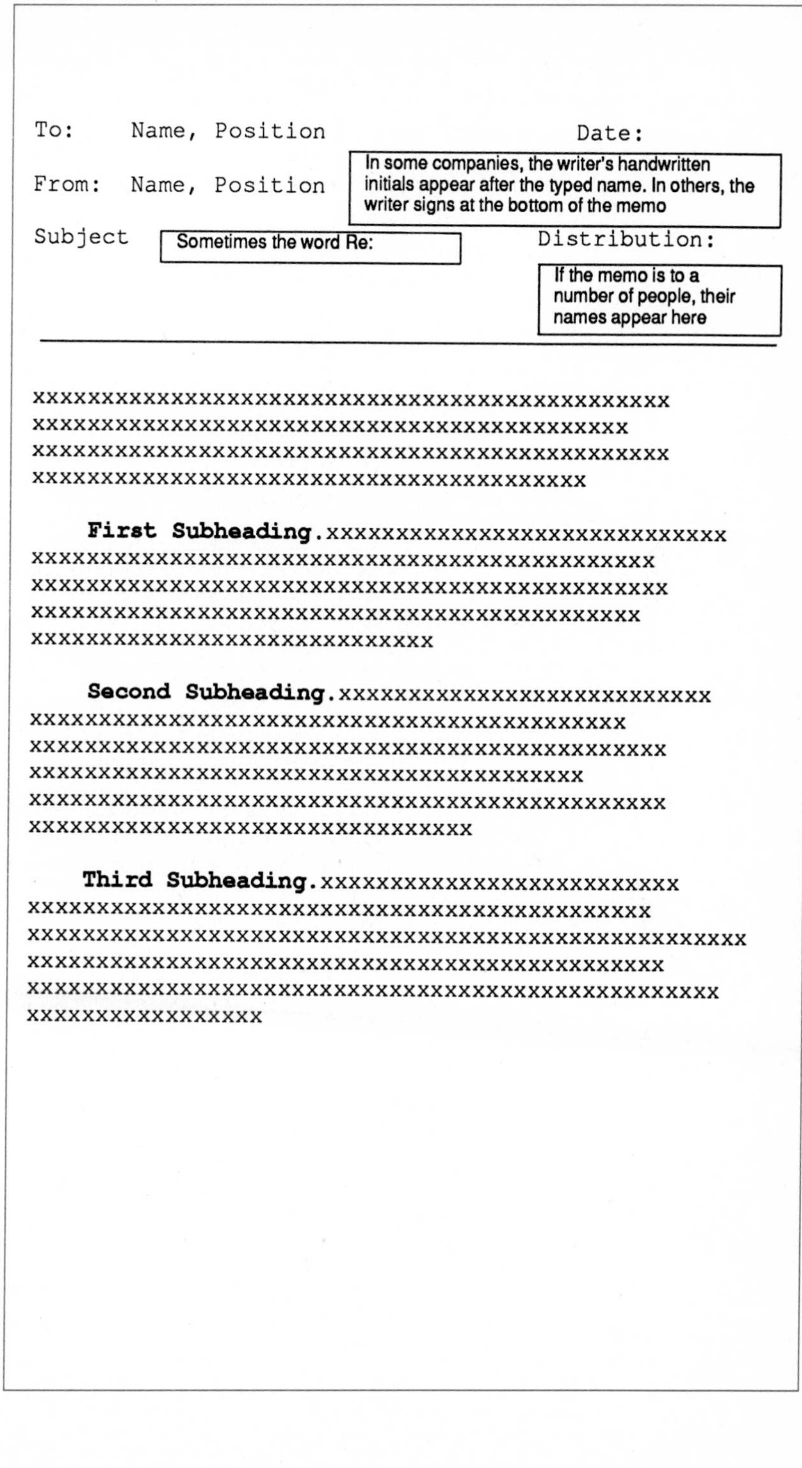

To: Name, Position

Date:

In some companies, the writer's handwritten initials appear after the typed name. In others, the writer signs at the bottom of the memo

From: Name, Position

Subject

Sometimes the word Re:

Distribution:

If the memo is to a number of people, their names appear here

---

xxxxxxxxxxxxxxxxxxxxxxxxxxxxxxxxxxxxxxxxxxxxxx
xxxxxxxxxxxxxxxxxxxxxxxxxxxxxxxxxxxxxxxxxxx
xxxxxxxxxxxxxxxxxxxxxxxxxxxxxxxxxxxxxxxxxxxxxx
xxxxxxxxxxxxxxxxxxxxxxxxxxxxxxxxxxxxxxxx

First Subheading. xxxxxxxxxxxxxxxxxxxxxxxxxxxxx
xxxxxxxxxxxxxxxxxxxxxxxxxxxxxxxxxxxxxxxxxxxxx
xxxxxxxxxxxxxxxxxxxxxxxxxxxxxxxxxxxxxxxxxxxxxx
xxxxxxxxxxxxxxxxxxxxxxxxxxxxxxxxxxxxxxxxxxxx
xxxxxxxxxxxxxxxxxxxxxxxxxxxxx

Second Subheading. xxxxxxxxxxxxxxxxxxxxxxxxxxx
xxxxxxxxxxxxxxxxxxxxxxxxxxxxxxxxxxxxxxxxxxx
xxxxxxxxxxxxxxxxxxxxxxxxxxxxxxxxxxxxxxxxxxxxxx
xxxxxxxxxxxxxxxxxxxxxxxxxxxxxxxxxxxxxxxx
xxxxxxxxxxxxxxxxxxxxxxxxxxxxxxxxxxxxxxxxxxxxxx
xxxxxxxxxxxxxxxxxxxxxxxxxxxxxxxx

Third Subheading. xxxxxxxxxxxxxxxxxxxxxxxxxx
xxxxxxxxxxxxxxxxxxxxxxxxxxxxxxxxxxxxxxxxxxxxx
xxxxxxxxxxxxxxxxxxxxxxxxxxxxxxxxxxxxxxxxxxxxxxxxxxxx
xxxxxxxxxxxxxxxxxxxxxxxxxxxxxxxxxxxxxxxxxxxxxx
xxxxxxxxxxxxxxxxxxxxxxxxxxxxxxxxxxxxxxxxxxxxxxxxxx
xxxxxxxxxxxxxxxxx

Figure 13.7 Letter with typographical divisions.

xxxxxx(date)

xxxxxxxx
xxxxxxx
xxxxxxxxxxxxxx
xxxxxxxxxxxxxx(address) Subject:xxxxxxx

xxxxxxxxx(salutation)

xxxxxxxxxxxxxxxxxxxxxxxxxxxxxxxxxxxxxxxxxxxxxxx
xxxxxxxxxxxxxxxxxxxxxxxxxxxxxxxxxxxxxxxxxxxx
xxxxxxxxxxxxxxxxxxxxxxxxxxxxxxxxxxxxxxxxxxxxxxx
xxxxxxxxxxxxxxxxxxx

1. <u>xxxxxxxxxxxxxxxxxxxxx.</u> xxxxxxxxxxxxxxxxxx
xxxxxxxxxxxxxxxxxxxxxxxxxxxxxxxxxxxxxxxxxxxxxxxxxxxx
xxxxxxxxxxxxxxxxxxxxxxxxxxxxxxxxxxxxxxxxxxxxxxxx
xxxxxxxxxxxxxxxxxxxxxxxxxxxxxxxxxxxxxxxxxxxxxx
xxxxxxxxxxxxxx

2. <u>xxxxxxxxxxxxxxxxx.</u> xxxxxxxxxxxxxxxxxxxxxx*
xxxxxxxxxxxxxxxxxxxxxxxxxxxxxxxxxxxxxxxxxxxxxxxxxx
xxxxxxxxxxxxxxxxxxxxxxxxxxx

xxxxxxxxxxxxxxxxxxxxxxxxxxxxxxxxxxxxxxxxxxxxxx
xxxxxxxxxxxxxxxxxxxxxxxxxxxxxxxxxxxxxxxxxxxxxxxxxxxx
xxxxxxxxxxxxxxxxxxxxxxxxxxxxxxxxxxxxxxxxxx

Complimentary Closing,

Signature
Name (typed)
Title (if appropriate)

Enclosure (Enclosure notation may be abbreviated Enc. or Encl. You may, when appropriate, describe the enclosure, saying, for instance, Annual Report or Maintenance Instructions.)

Copy to: (This may be abbreviated c. The abbreviation cc, for carbon copy, is becoming outdated in this time of high-speed copying machines and word processors.)

Example 13.2a Technical memo with poor design.

To: All Research Staff
From:

Date:
Re: Protection of Trademarks

Our legal department is concerned with the way we are stating our trademarks. A trademark is lost when it becomes generic--that is, when it has come to mean the product itself as distinguished from a certain brand of the product. Escalator, aspirin, cellophane and mimeograph are examples of former trademarks that are now generic. As trademarks are valuable property to the company, care should be taken to guard against their becoming generic. These procedures should be observed on all business documents, advertising literature, labels, correspondence, and displays. Visually distinguish the trademark from the text. To do this, use all capitals, or initial capitals and quotation marks. Other possibilities, depending on the context, are italics, bold-faced print, or a different color or size of print. Examples: POLYTT compact discs or "Polytt" compact discs, not Polytt compact discs.

Use a trademark notice when permitted. The registration notice r or "Reg. U.S. Pat. & TM Off." should appear directly after the mark. If the mark has not been registered, use "TM" or an asterisk and the footnote, "A trademark of Company X." Example: if registered, SANTO r or SANTO* Reg. U.S. Pat. & TM Off. If not registered, SANTO TM, or SANTO* A trademark of Company X. Reduce the possibility that the trademark be thought of as the generic product by including the common descriptive name plus the word "brand": Gore-Tex brand fabric; Band-Aid brand bandages.

Do not change the form of the word. Don't add possessives or plurals, for instance. Use "The Products of Giorgio," not "Giorgio's products"; Not "Buy several RIGHT-LOCKS at your favorite store," but "Buy RIGHT-LOCK brand key guards at your favorite store."

Example 13.2b Technical memo with improved design.

To: All Research Staff
From:

Date:
Re: Protection of Trademarks

Our legal department is concerned with the way we are stating our trademarks. A trademark is lost when it becomes generic--that is, when it has come to mean the product itself as distinguished from a certain brand of the product. Escalator, aspirin, cellophane and mimeograph are examples of former trademarks that are now generic.

As trademarks are valuable property to the company, care should be taken to guard against their becoming generic. These procedures should be observed on all business documents, advertising literature, labels, correspondence, and displays:

1. Visually distinguish the trademark from the text. To do this, use all capitals, or initial capitals and quotation marks. Other possibilities, depending on the context, are italics, bold-faced print, or a different color or size of print.

wrong	**correct**
Polytt discs	**POLYTT** discs

2. Use a trademark notice when permitted. The registration notice r or ''Reg. U.S. Pat. & TM. Off'' should appear directly after the mark. If the mark has not been registered, use ''TM'' or an asterisk and the footnote, ''A trademark of Company X.''

registered	**unregistered**
SANTO r	**SANTO** TM
SANTO*	**SANTO***
*Reg. U.S. Pat. & TM Off.	*A trademark of Company X.

3. Reduce the possibility that the trademark is thought of as the generic product by including the common descriptive name plus the word "brand."

Gore-Tex brand fabric
Band-Aid brand bandages

4. Do not change the form of the word. For instance, do not add possessive or plurals.

wrong	**correct**
Giorgio's products	The products of **Giorgio**
Buy several RIGHT-LOCKS	Buy **RIGHT-LOCK** brand key guards at your favorite store

LETTERS THAT SAY "NO"

Letters with a negative message are difficult to write in a way that guarantees the maximum good will, given the reader's predictable reaction to the sting of bad news.

One way to handle negative letters is to reverse the traditional descending order of letters, beginning instead with any good news or positive information that might soften the impact of the bad news. "Dear Mr. Sinclair: The answer is 'no' " is a poor way to begin, even if "no" is your main point. People are often more willing to listen to bad news if the information is cushioned by a reasonable explanation.

Consider, for instance, two versions of the same letter (Examples 13.3a and 13.3b). The author is a person working for a regulatory agency, the recipient an engineer in a state department of motor vehicles.

The difference in tone between the two letters is striking. Version A plunges directly into the negative news without a cushion. Version B softens the impact of the criticism that will follow by thanking the recipient for improvements and positive features in the job they have done. This builds good will, for it suggests that the writer appreciates the effort that was made, discerns the improvements, and is putting recognition before further demands.

The opening paragraphs of Version B also carefully connect the new document to past events, and announce subject and purpose.

Subject: Letter of 14 April

xxxxxxxxx(salutation):

We would like to thank you and your staff for a timely response to the concerns that we have raised about your emissions inspection and maintenance program. We were pleased to see that you have committed to a plan that generally addresses our concerns and provides for critical program improvements. The plan appears to establish a sound framework for actions to address the sticker accountability, data loss, and low failure rate problems. We are particularly pleased to see that you have already begun to implement many of the program improvements.

Notice that in connecting the present to the past, the author pro-

Example 13.3a First version of letter. The writer plunges directly into the negative news.

xxxxxxxxx (date)

xxxxxxxxx (inside address)
xxxxxxxxx
xxxxxxxx
xxxxxxxxxxxx re: Letter of 14 April 1988

xxxxxxxxxx: (salutation)

For our agency to gain a clearer understanding of how the seven key program improvements that you outlined will be implemented, we are asking that you provide us with additional information. Specifically, we would like to see a more detailed description of how some of the improvements will ultimately be enacted, an identification of the resources that will be used to carry out the improvements, and a more specific implementation schedule with interim milestones. Attachment I discusses each of our concerns in detail.

To ensure that the improvements being made have the desired effect, we suggest that you communicate with us on a regular basis.In order to meet our needs, we would like you to submit quarterly reports that describe the status of the improvements. In Attachment II we list the information that should be included as part of the quarterly reports.

Example 13.3b Second version of letter. The writer acknowledges improvements and effort before making further demands.

xxxxxxxxx (date)

xxxxxxxxxx (inside address)
xxxxxxxxxxxxx
xxxxxxxxxxx
xxxxxxxxx
xxxxxxxxxxxx re: Letter of 14 April 1988

xxxxxxxx: (salutation)

We would like to thank you and your staff for a timely response to the concerns that we have raised about your emissions inspection and maintenance program. We are pleased to see that you have committed to a plan that generally addresses our concerns and provides for critical program improvements. The plan appears to establish a sound framework for actions to address the sticker accountability, data loss, and low failure rate problems. We are particularly pleased to see that you have already begun to implement many of the program improvements.

However, for our agency to gain a clearer understanding of how the seven key program improvements that you outlined will be implemented, we are asking that you provide us with additional information. Specifically, we would like to see a more detailed description of how some of the improvements will ultimately be enacted, an identification of the resources that will be used to carry out the improvements, and a more specific implementation schedule with interim milestones. Attachment I discusses each of our concerns in detail.

To ensure that the improvements being made have the desired effect, we suggest that you communicate with us on a regular basis. In order to meet our needs, we would like you to submit quarterly reports that describe the status of the improvements. In Attachment II we list the information that should be included as part of the quarterly reports.

vides a brief summary of the past documents, rather than a vague comment.

Vague	**Nutshell Summary**
The plan appears to address many problems successfully.	The plan appears to establish a sound framework for actions to address sticker accountability, data loss, and low failure rate problems.

In the body of the letter, before stating further demands the writer gives a reason for the demands.

> For our agency to gain a clearer understanding of how the seven key program improvements that you outlined will be implemented, we are asking that you provide us with additional information. Specifically, we would like to see a more detailed description of how some of the improvements will ultimately be enacted, an identification of the resources that will be used to carry out the improvements, and a more specific implementation schedule with interim milestones. Attachment I discusses each of our concerns in detail.
>
> To ensure that the improvements being made have the desired effect, we suggest that you communicate with us on a regular basis. In order to meet our needs, we would like you to submit quarterly reports that describe the status of the improvements. In Attachment II we list the information that should be included as part of the quarterly reports.

EDITING AND PROOFREADING

Edit and proofread correspondence carefully. People make judgments about your professional abilities based on the way your memos and letters look on paper; misspellings, incorrect salutations, and grammatical errors make a bad impression.

Check the spelling and title of anyone you address. The first thing recipients will notice—and no doubt remember—is that you've misspelled their name.

Check for subject-verb agreement, pronoun antecedent, and dangling or misplaced modification. Dangling modifiers are a perennial trouble spot in memos and letters, which often begin with introductory phrases. When these phrases fail to modify the subject of the sentences that follow, they are said to dangle.

Examples

A native of Brooklyn, Sinclair's initial assignment will be to locate vendors for the staff. (The sentence says Sinclair's assignment is a native of Brooklyn.)

As the newest member of the Board, your generosity will be a beacon to others. (The sentence says your generosity is the newest member of the Board.)

(See also Usage Handbook, modification.)

SUMMING UP

- Memos and letters are both forms of correspondence. Letters are usually sent outside; memos circulate within an organization.
- Memos and letters are useful when you need a record of the information, when you are sending along another document that needs an introduction, and when you are establishing the terms of a contract.
- Cover letters accompany other documents such as product literature or resumés. When they accompany a report, they are often called "letters of transmittal."
- Correspondence is usually structured in descending order, with the most important information in the opening, the main points in the body, and the courteous closing at the end.
- You may want to delay a statement of the main point if it is unpleasant. People are often more willing to accept bad news if the news is preceded by an explanation.
- Avoid "letterese," the involuted language of letters that adds bulk without meaning.
- Try for a "you" approach in letters so that the emphasis falls on the reader's needs, not the writer's.
- Check for the correct honorific (Ms., Miss, Mr., Dr.) of the person you are addressing just as you check for spelling. This is one more example of a careful, professional approach.

Example 13.4 Example of a letter of transmittal.

15 August 1988

Dr. Ralph J. Florio
Encom Energy
2234 Linview Avenue
Lindale, Kentucky

Re: 6 June request for Vols. 1 and 2, Science and Public Policy

Dear Mr. Florio:

Thank you for your recent letter. We are enclosing the volumes you requested, as well as a brochure describing our other publicatons.

We appreciate your interest in SYNTEX publication and invite your comments on any of our books.

Sincerely yours,

Lucille Masterson
Director of Marketing

LM/spd

Enclosures

• Use subheadings, bullets, and underlining to provide division and emphasis. Page layout adds to the readability of memos and letters.

• Edit and proofread carefully. People make judgments about your professional abilities based on the way your letters and memos look on paper.

• Note: For further information on memos, see Chapter 3, organizational patterns; Chapter 16, trip memos, technical service memos.

EXERCISES

Comment on the strength or weaknesses of the following correspondence:

(LETTERHEAD)

14 March 1989

Dear Class of 1989:

Some of you have been turning in sections of your papers and others have been silent. It's up to you--if you want feedback now which doesn't affect your grade, then turn in drafts or call me with questions. If you prefer to turn in the final paper without such prior contact, you will have to

(continues)

be your own editor and critic. Remember that the paper will be graded: 1) relative to the work turned in by your peers 2) according to how well you have demonstrated an understanding of the management issues 3) last, but not least, how well you are able to convey your ideas in writen form.

Ask yourself: Have I explained the issue (problem) and why it is worth spending the time on (that is, its importance to the firm)? Have I demonstrated that I can identify and evaluate the important factors, both external to the company and internal? Have I come up with plausible, supported, and practical recommendations? Have I compiled my supporting evidence in an intelligible way? Finally, if it were someone else's paper at my work, would I find it readable? Have I written a concise Executive Summary?

Your papers are due on May 8 at the latest. I would prefer it if papers were turned in by the end of April, but recognize that some of you have had tough schedules this semester.

Sincerely,

(SIGNATURE)

22 March 1988

MEMO TO:

FROM:

SUBJECT: NEW SUN STRATEGY

This is a follow-up to our conversation of Wednesday, March 16.

Sun, as you probably know is mainly known for it's high power computers for engineers. It is a six-year old company with sales at an annual rate exceeding $1 billion. They have a new strategy. Their new hardware design is based on a microprocessor chip called a Sparc, a high-powered device that employs RISC--reduced instruction set computer techniques which make it easier and cheaper to manufacture than a conventional microprocessor. The microprocessor is the brain of a computer, manipulating data and performing calculations at the behest of the operating system.

As you may recall, Sun has announced an invitation to other manufacturers to build clones of the new Sun product--to use clones of Sun's new hardware design plus UNIX, the operating software system owned by Sun's ally, AT&T. These designs are versatile, and may be used as the basis for machines from desktops to mainframes. With such designs, corporate customers could pick and chose products from many manufacturers to build compatible computer systems to manage information. Such standardization would avoid either the present problems of multi-vendor assembly or having to utilize a single manufacturing outlet. The industry wide standards for computer hardware and software proposed by Sun have not been popular with many manufacturers who are afraid of having to compete except by proprietary technology. They are not anxious to compete the way PC companies do by constant innovating, shortening product cycles, and improving price-performance.

Sun has already started selling Sparc-based machines that reflect the price-performance advantages of it's new design. It's most costly model, priced at $40,000 provides the horsepower of comparable conventional machines costing ten times as much. Sun officials claim that the $300

(continues)

Sparc chip is cheap enough to be used as the heart of a new genre of extremely powerful personal computers that could be priced under $5,000.

The other main ingredient in Sun's strategy is the UNIX operating system. UNIX has long been used in Sun's lower-performance machines. Sun officials decided last summer that if they could persuade AT&T to help them fine-tune a special version of UNIX to run on the Sparc, the chip would stand a better chance of being adopted by other manufacturers. AT&T also wanted to refine UNIX into a more universal operating system. Finally the two companies made an alliance by which AT&T agreed to adopt the Sparc design to a new family of business computers while Sun agreed to help AT&T improve UNIX.

Sun believes that a standard computer architecture and operating system is inevitable. Sun hopes to orchestrate a revolution in the computer industry, creating the same kind of explosive growth the IBM unwittingly unleashed when it decided to use standard hardware designs and software for its PC.

All the publicity surrounding the Sparc has jarred other U.S. chip markers. Motorola has stepped up the development and promotion of its own RISC-based chip in an attempt to short-circuit the Sparc as a standard-setter.

Companies like Digital, Appolo and Hewlett-Packard have benefitted greatly from their proprietary designs, and they fear that endorsing Sun's idea will make Sun the industry's leader by default. The last thing they want, however, is to be caught at a competitve disadvantage at a time when sales of Unix-based computers are taking off. Many industry analysts believe Unix will be the fastest-growing operating system between now and the early 1990s.

We will keep an eye on these developments.

CHAPTER 14

For Job Seekers: Cover Letters and Resumés

COVER LETTERS

The cover letters you will write if you are a job applicant should follow the guidelines given in Chapter 13 for other forms of correspondence.

Organization *An opening* that states the subject and purpose, and that specifically links the letter to prior requests or events.

A body that develops the main idea.

A closing that includes a courteous request for the action you want taken.

Language *A "you" approach* that shines the light on the reader's needs, not the writer's.

Direct language that sidesteps "letterese."

A positive slant, rather than needless negatives.

Layout *An acceptable format* with generous margins, distinctive, letter-quality type, and appropriate honorific in the salutation.

Here are some examples linking these general guidelines to specific letters of job application.

Opening

Use the opening paragraph to identify yourself, the position you seek, and, when appropriate, the source of information for the job. If you are responding to an ad, name the date and the publication. If you have a personal connection to the company, and the person has no objection to your using his or her name, do so at the top of the letter.

> Dear Mr. X:
>
> John Sinclair, Associate Research Director at Willard Products, who has employed me part-time for the past two summers, suggested that I apply for the position of food technologist. I will receive my B.S. in food technology from Farleigh Dickinson in June 1988.
>
> Dear Mr. X:
>
> I would like to apply for the position of design engineer advertised in the December issue of *Collegian.* I will receive a B.S.E.E. in mechanical engineering in June from Troy State University.

You may also want to pick out your strongest points and introduce them in this paragraph, using the second paragraph for supporting details.

> Dear Ms. X:
>
> I would like to apply for the job as Residential Counselor advertised in the 24 January issue of *SUNY Life.* I will be graduated in June from SUNY Binghampton with a Bachelor of Science in Psychology, with a concentration in group work. In addition to my course work (see attached resumé), I have practical experience in counseling from working part-time for the past three years as a tutor in the University's Study Skills Center.

All three examples use an explicit expression of application such as "I would like to apply for the position" and a direct reference to the point of contact ("as advertised in," "as recommended by"). In this way, the author clearly states subject and purpose. In the third example, the author has practical experience to offer in addition to a degree. Because this is a strong point, the writer has put it at the top.

Body

The body of the text should highlight your main qualifications for the job without going line by line through your resumé. Pick out the most applicable parts of your background in relation to the job advertised. Give one or two examples. List a few of the courses that are relevant to the job application, and do so not by number ("EE 1104") but by title ("Introductory Digital Electronics"). "I have taken ____ credits in ____, including ____*(names of several appropriate courses)*____, as well as ____ hours in ____ ____*(names of several courses in secondary field of specialization)*____."

Close the body of the text with a reference to the resumé ("My enclosed resumé gives details of my education and work experience").

Throughout the letter, take a positive approach.

NEGATIVE:

Although I worked part-time throughout college, I managed to maintain a 3.1 academic average.

POSITIVE:

Throughout college I worked part-time during the semester and full-time during the summer, earning 20 percent of my expenses. My overall grade point average was 3.1.

NEGATIVES:

Although I am not a strong writer, I have a solid engineering background.

While I am not the greatest salesman, I think I can do justice to your product.

My job placement has been checkered, interrupted by the birth of two children.

My grades were sometimes not what I hoped for. Although I have a deep interest in the environmental science, I needed to work 20 hours per week, and this took away from my studies.

POSITIVES:

I have a solid engineering background, and have taken three courses in which I wrote laboratory reports or research reports.

Your product is an excellent one, and I look forward to representing you.

I have worked for two civil engineering firms where I supervised field work, prepared regulatory documents, and wrote proposals (see attached resumé).

Throughout college, I supported myself by working 20 hours per week.

Closing

The closing is a courteous invitation to keep the lines of communication open.

I hope to meet with you when you are conducting interviews at the Job Fair.

22 January 1987

Dr. L. G. Hill
Manager of Employment
Battelle Columbus Laboratories
505 King Avenue
Columbus, Ohio 43201

Dear Dr. Hill:

I am writing this letter in response to your advertisement in *Chemical and Engineering News,* 2 January, for photochemistry researchers.

I have a Ph.D. in physical chemistry and am just completing my postdoctoral tenure here at the University of Toronto, doing research in the laser-induced photophysics and photochemistry of condensed phases. I have considerable experience both with laser spectroscopy and computer modeling.

I would be glad to travel to Battelle Columbus Laboratories for an interview to discuss ways I might contribute to Battelle's research program.

Sincerely,

Example 14.1 Letter of application.

Example 14.2 Letter of application.

22 January 1987

Dr. Susan Thompson
Sinclair Research, Inc.
Stans Research Park
Lakeline Drive
Bedford, MA 01730

Dear Dr. Thompson:

This letter is in response to Sinclair Research's December advertisement in *Physics Today* for research scientists. The position matches both my academic background and research experience, for I have done both experimental and theoretical work.

While an undergraduate at Brown, I worked in the laboratory of E. B. Walford, working primarily on computer programming for spectroscopic analysis. Later, I worked with Prof. Robert Reilly doing research in the quantum scattering theory of H atom-electron collisions.

While a graduate student at MIT, I worked with George Masters on the computer modeling of atom-diatom collisions, and on experimental work in which a tunable dye laser was built to study the spectrum of gas-phase NO_2.

I have enclosed a copy of my resumé to supply you with additional information. I hope you find these credentials make me a strong candidate for the job you have advertised, and that I may have the opportunity to discuss with you the ways in which I can contribute to Sinclair Research.

Sincerely,

I look forward to discussing with you the ways I can contribute to your company.

I look forward to talking with you about the contributions I might make to the laboratory's continuing success. I would be glad to travel to Lawrence Livermore Laboratory for an interview.

Cast the information from the point of the view of the reader, using a "you" approach. Prefer "I look forward to a chance to talk with your staff about the contributions I might make to Sohio's continued growth," to "I look forward to talking with your staff about how my abilities may be just what you seek. The position seems to fit me perfectly."

"ME" APPROACH:
I have always wanted to work in the aerospace industry.

"YOU" APPROACH:
I look forward to talking with you about ways I might contribute to Grumman's outstanding work in the aerospace industry.

RESUMÉS

A resumé (from the French *résumer,* to summarize) is a summary of a job applicant's qualifications. In academic circles, the resumé goes by the name "curriculum vitae."

The resumé is an important piece of writing, for job interviewers schedule appointments based on their assessment of the resumé and accompanying cover letter.

An effective resumé depends on thinking about the audience, and then casting the information using a strong organizational pattern, clear language, and effective design.

Audience and Objective

Many beginning writers tend to plunge directly into a writing job; experienced writers, in contrast, seem to spend more time thinking about their objective, their audience, and their organizational plan. Such preliminary activities probably make sense for all forms of technical writing; one example of their usefulness is seen in the planning that precedes writing a resumé.

Begin by doing some brainstorming. Look through your records, think about the past, and start sorting the information you gather by major categories.

- educational background
- work experience
- special abilities

As you do this personal, exploratory notetaking, also do some thinking about your prospective audience—the staff members of the corporation or agency, for instance, you hope to interest.

If the company is new to you, do some research. If appropriate, look up the business in *Dun and Bradstreet* or *Standard & Poor;* check for related articles in *Business Week, Fortune,* or *Forbes.* If the company is specialized, look for articles on it in a trade publication in the field, such as *Chemical and Engineering News* or *Engineering News Record.* If it is a company that offers products or

Example 14.3

JOHN KURELIK
1415 East 14 Street
New York, New York 10007
(212) 559-0091 (home)
(718) 467-0097, Extension 237 (work)

services, telephone or write for promotional literature. Annual reports are another good source of information. A knowledge of the company will help you close in on your audience, both in casting the resumé and in preparing for an interview.

Categories in the Resumé

Identification Name, address, and telephone number—personal identification—usually form the heading for the resumé (Example 14.3). Most applicants capitalize their names. If you have two addresses—college, for instance, as well as home—give both (Example 14.4). Avoid nicknames—especially nicknames set off with quotation marks—and abbreviations, other than the capitalized abbreviation for each state.

Professional Objective This category is optional (Example 14.5). The statement of your goals may be an effective place to demonstrate your understanding of the company and how you might fill its needs. The "Objective" line may also provide useful information for the reader, particularly if the resumé is somehow separated from the cover letter. Some applicants, however, feel that an explicit objective narrows the resumé too much, and is therefore better omitted.

JOHN KURELIK

Academic Address (until June 15)	*Permanent Address*
Leeds House, Second Floor	2215 Rockingham Road
University of Virginia	Richmond, VA 23226
Charlottesville, VA 23456	
(804) 716-4284	(804) 266-2626

Example 14.4

Example 14.5

Goal: Summer position using skills achieved in aerospace engineering program.

Professional Objective

Immediate Goal. Part-time position in which my skills in engineering and computer programming can be applied to problem solving.

Long-Term Goal. Research and development in aerospace engineering.

Professional Objective: Position as computer engineer in which my knowledge of computer architecture and artificial intelligence can be used.

Education This heading usually follows after the objective, particularly for undergraduates who have more to say about their college careers than their still-to-come professional careers. However, if your work experience is quite strong, you may want to put it before education.

If you are beginning with the category of "Education," order it with the most recent information first. (This and other listings in a resumé usually follow the pattern of reverse chronology, with most recent entries at the top of the category.)

Include

- school
- major
- dates
- degree(s), either awarded or expected

You may want to include these additional items:

- class standing or grade point average
- projects completed on special assignment
- senior thesis topic or honors project
- a sequence of courses outside your major, such as in technical writing or management

If you have taken classes outside of college—an industrial short-course or class at a community college, for instance—it might be profitable to include them.

If you are working on a degree but have not completed it, state your progress positively. "Completed 36 credits toward degree" is preferable to "Dropped out after 36 credits."

Information on your high school education is unnecessary unless you have honors to mention, your school was prestigious, or you are applying for a job within two years of leaving high school. Some samples of listings for educational background appear in Example 14.6.

Honors and Awards Like the objective, this category is optional. If you do decide to list awards, don't understate them. Explain what each award represents, particularly when you think the reader may be unfamiliar with the background of the honor or prize you are listing (Example 14.7).

Special Skills Those who choose this optional category usually place it either immediately after career objective, or after education. You may want to list special skills in relation to the particular job: technical German for a job with an international firm in food technology; water chemistry for a job with the Environmental Protection Agency; knowledge of an argon laser, analytical instruments, or a programming language according to the specialties of the company. (See Example 14.8.)

Work Experience If you are strong in this category—for instance, if you are a student in a co-op program, or one who has found several part-time or summer jobs in an area closely related to your future profession—you may want to list this information before education, although education traditionally comes first.

People who are established in their professions often choose to begin the resumé with work experience, and then follow with education, particularly if their educational record is weak, or if their professional experience is more closely related to the job opportunity than is their academic background.

Whether you lead with this section or place it after education, be sure to begin with your most recent job. You should include title, place of employment, dates, and, when not self-evident from the title, a brief account of the work for which you were responsible.

You may organize the information by the dates, by the type of work you have done, or by the type of company or agency for which you have worked.

Chronological patterns are the most common ones, for they show the interviewer an applicant's growth, direction, and taking on of responsibility. (See examples at end of chapter.)

Example 14.6

EDUCATION

B.A., Business Administration (expected June 1988)
Langton Institute of Detroit

EDUCATION

Computer Science
Oswego State University
B.S. expected June 1988.

In addition to computer science sequence, I have taken fifteen credits in technical writing.

EDUCATION

B.S.M.E., June 1985. Oswego State University.

My program included graduate courses in dynamics of machines and mechanical vibrations. For my senior project, I developed a software package for the analysis of vibration in truck bodies.

EDUCATION

B.S., Electrical Engineering (to be awarded June 1988)
Oswego State University
GPA: 3.0/4.0

Areas of Proficiency: Programming in PL/1; Pascal Programming and Numerical Methods; Assembly Programming

EDUCATION

Polytechnic University 1986—present
Completed six semesters toward B.S. in Physics.

EDUCATION

University of Pennsylvania 1985-1987
26 graduate credits in biochemistry.

Example 14.7

Honors and Awards

Dean's List for six semesters, 1985-1988

Member Tau Beta Pi, engineering honor society

Honors and Awards

Dean's List 1985-1987

Winner, Society of Technical Communications Student Writing Contest. Open to all W.P.I. students, the annual award is judged by a panel of five technical writers, and carries a $500 honorarium for first prize.

Example 14.8

SPECIAL SKILLS

Translation of technical German; accurate copy editing.

If your employment history is fragmented—punctuated, for instance, by childbirth or illness—or you have only a few short-term jobs to list, organization by company may be preferable, for the name of a well-known company may add some weight to the entry (Example 14.9).

If you are an undergraduate whose work experience is completely unrelated to the job you seek, you should still list items

Employment History

TEMPLE UNIVERSITY SCHOOL OF MEDICINE (Philadelphia, PA)
Staff Technologist July 1972—December 1974

Maintained setting for thyroid function testing and thyroid cancer treatment workup.

Example 14.9

Example 14.10

Work Experience

STATE UNIVERSITY AT HILLSDALE (Hillsdale, MA). Office Assistant, Social Science Department, Summer 1986. Typing, word processing. Learned Word Perfect and DBII procedures.

separately, stressing your duties and responsibilities. While it's true that such jobs do not contribute directly to your professional training, they can show your ability to work hard and apply yourself (Example 14.10).

Some applicants use a summary of skills and qualifications at the top of the resumé before beginning a listing of work experience. This may be a useful way to organize a long list of jobs that would otherwise seem unrelated. It also provides a showcase for the highlights or strong points in the resumé (Example 14.11).

If you did several jobs within a company, you may want to list them separately to give each a distinct visual identity. If you want to emphasize each item, use bullets and underlining (Box 14.1).

Other Interests and Activities Here are some optional categories you may want to include, depending on the audience and your own strengths:

Military Status Include positive information, such as specialized technical training that may be related to the prospective job.

RACHEL SINCLAIR
3312 Sunside Avenue
Larchmont, New York 12345
914-556-7788

Summary of Accomplishments

- **Published** two articles in service journals on mass spectroscopy.
- **Designed** user manual for handheld computer.
- **Served** on editorial and policy boards of a national nonprofit organization.

Example 14.11

Box 14.1

1978-84 Research Associate

Hofstra University

- Co-edited review article published in *American Zoology* (22, 1982) on development and regeneration in single cells.
- Designed laboratory experiments on regeneration using delicate laboratory techniques.
- Performed troubleshooting, repairs, and maintenance of lab equipment.
- Managed day-to-day operations of research laboratory.

Professional Affiliations Identify the type of organization. Say "Member, National Society of Black Engineers" rather than "Member, NSBE."

Extra-Curricular Activities If you are applying directly from college, this category will be useful. A student who includes "Business Manager, School Newspaper" may find an interviewer equates that information with good organizational skills and an ability to work with others.

Community and Volunteer Work Many jobs undertaken as a volunteer for a community or religious group may show the applicant's ability to organize detail, to work creatively with a team, or to chair or supervise activities. If you are applying for an administrative job, these abilities are relevant, even though they developed outside a technical field.

Special Interests Include these if you think they are appropriate. Acquiring a complete collection of Bazooka Comics—a hobby one student proudly listed under "Special Interests"—is definitely inappropriate.

Personal Data There is no need to give height, weight, race, age, or religion. Depending on your circumstances, you may want to indicate willingness to relocate and U.S. citizenship.

References It's customary to state "References Available Upon Request." In some cases, though, applicants list names, positions, and telephone numbers, if they have the permission of the people listed, and room left on the resumé.

Design and Layout

Appearance matters in a resumé. If your typing or proofreading skills are weak, have a professional help you out, and double-check the work afterwards. Prospective employers may make judgments about your technical abilities based on sloppy headings or inaccurate spelling, no matter how sterling your professional qualifications.

Length One page is standard for entry-level jobs. If you have accumulated years of professional experience, the resumé may grow to two pages.

Layout Make sure the headings and subheadings are consistent in location, type size, and emphasis. For instance, if you list major categories to the left, all should begin at the same line position, all should be either in initial caps or all caps. If you decide to use boldface or underlining, keep the use consistent.

- Indent to show divisions within categories.
- Use bullets to emphasize items in a list.
- Leave some white space or air in the resumé. It should not present an impenetrable wall of text.

Use heavy grade paper in a conservative color such as white or ivory. The typeface should be distinct and large enough to read easily (7 points or above). Examples 14.12–14.15 show sample resumés).

EXERCISES

Write a letter of application in response to an advertisement for a job in a professional or academic journal within your field. Prepare a copy of your resumé to go with the cover letter. If you are studying in a group, exchange letters and resumés, and discuss strengths and weaknesses.

RUTH LING KO
454 16th Avenue
Rahway, New Jersey 07000
201-339-8875

PROFESSIONAL OBJECTIVE	Responsible position as a computer engineer using my knowledge of computer architecture and microprocessors.
EDUCATION	Worcester Polytechnic University B.S., Electrical Engineering to be awarded June 1988. G.P.A.: 3.75/4.0 Areas of Proficiency: Design and Analysis of Algorithms; Assembly Language Programming; Switching Circuits and Digital Systems; Engineering Circuit Analysis; Computer Architecture.
HONORS	Dean's list for all semesters. Member of Tau Beta Pi (engineering honor society).
WORK EXPERIENCE 9/87–12/87 Part-Time	Prof. Sinclair, Head, Electrical Engineering Worcester Polytechnic University. Grader for sophomore-level Engineering Circuit Analysis: responsible for grading homework and exams.
6/86–9/86 Part-Time	Contract Accounting Department, Paramount Pictures. PC system consultant; responsible for computerizing the accounting systems and teaching the staff how to use personal computers.
Summer 1984–1985	Secretary, Equitable Life Assurance Co., New York.
ACTIVITIES	Member, Chinese Student Association. Volunteer tutor, mathematics. Student Learning Center.
PERSONAL DATA	U.S. permanent resident. Will be naturalized by July 1988.

References available upon request.

Example 14.12

Example 14.13

JOHN ARNOLD

Academic Address
Worcester Polytechnic Institute
Department of Physics
Worcester, Massachusetts 02222
(617) 998-8874

Permanent Address
33 Smith Street
Amherst, Massachusetts 02111
(413) 223-0096

HIGHER EDUCATION:

Ph.D. (Physics)
Worcester Polytechnic Institute (Graduation Jan. 1987).
Dissertation: Microparticle Fluorimetery and Energy Transfer.
Advisor: Prof. Margaret Hange. Requirements completed September 1986

M.S. (Physics)
Worcester Polytechnic Institute, 1983.

B.Sc. in Applied Science (Physics and Chemistry)
Trinity College, Dublin, 1981.

EXPERIENCE:

October 1986–Present.

Post-Doctoral Research Fellow, Worcester Polytechnic Institute.
Microparticle Photophysics Laboratory, Prof. M. Hange.

Research Activities: Aerosol particle characterization using various optical techniques including fluorescence spectroscopy, elastic scattering, and intermolecular energy transfer.

September 1983–September 1986.

Research Assistant, Worcester Polytechnic Institute.

Design and construction of a fluorescence spectrometer for use in spectroscopy of single levitated aerosol particles. Investigation of electronic energy transfer in such particles.

September 1981–September 1983.

Teaching Assistant, Worcester Polytechnic Institute.

Supervision of undergraduate laboratory sessions, teaching both recitation and tutorial sessions of basic undergraduate courses and tutor for the college tutoring service.

MEMBER: The American Physical Society
The Institute of Physics (London)

GEORGE LAREDO
2335 Lilac Lane
Plainfield, New Jersey 07060
(201) 667-4099

EMPLOYMENT HISTORY

Nuclear Medicine Service
St. Jude Cancer Center, Plainfield, New Jersey

1987–Present **Laboratory Supervisor.** Maintain setting for thyroid function testing and thyroid cancer treatment workup. Responsibilities include training staff technologist, purchasing and inventory decision, and participation in ongoing research projects.

1983–1986 **Staff Technologist.** Responsibilities included quality control and quality assurance, as well as procedure documentation and benchwork.

EDUCATION

1977–1987 (part time) **St. Luke's College,** B.S., Life Science, June 1987. Minor: Physics.

2/86–5/86 **Chatham Community College** Continuing Education: Magnetic Resonance Imaging Certificate.

Example 14.14

Example 14.14 (*continues*)

LICENSES, MEMBERSHIPS

1986	Associate Member, Society of Nuclear Medicine.
1984	American Registry of Radiologic Technologists.
1980	New Jersey Board of Health, Division of Laboratories. Technician's License #1287645.

References Supplied Upon Request.

SHARON JOHNSON
1781 Edgemont Avenue
Jersey City, New Jersey 07306
(201) 998-7768

OBJECTIVE

Position in technically oriented writing or editing.

EDUCATION

B.S., Journalism/Technical Writing, expected June 1988. Drew University, Callison, New Jersey.

Major Courses: Reporting on Science and Technology; Graphics and Production Techniques; Digital Electronics; Magazine Writing; Copy editing.

EXPERIENCE

Oct. 1985–Present	Publications Assistant, University Relations, Oswego State University (part-time). Produce bi-monthly media report; write and take photographs for faculty and alumni newsletters.

Example 14.15

Summer 1985	Marketing Assistant, NEC Engineers, Inc. Wrote sections of proposals; copy edited manuals, brochures, and correspondence.
Summer 1984	Office Assistant, Department of Humanities, Oswego State. Prepared mailings, updated computer records for Freshman Placement, learned Wordstar, did general office tasks.

EXTRA-CURRICULAR ACTIVITIES

Editor, Daily Chronicle, school newspaper, 1986–87; Assistant Editor, 1985–86; Copy Editor, 1984–85.

Secretary, Pi Kappa Phi National Fraternity.

Member: Undergraduate Curriculum and Standards Committee; Volunteers for Literary.

References Available Upon Request.

PART VI

REPORTS, PROPOSALS, MANUALS, ORAL PRESENTATIONS

CHAPTER 15

Abstracts

Example 15.1 shows an abstract that appears in *IEEE Transactions on Professional Communication* before the paper it summarizes.

Example 15.2 shows abstracts from an indexing service, written for people searching the literature of a field.

Example 15.3 is an executive summary written for a more diversified audience.

These are only a few of the many types of the abstract, a popular and extremely useful form of technical writing.

The abstract provides a quick, informative glimpse of the problem, the solution, and the importance of the work.

- It is an invitation to read, telling people what to watch out for if they decide to read the paper that follows. A properly written abstract highlights the main points so that readers can scan the report more quickly and efficiently.

Research Networks, Scientific Communication, and the Personal Computer

BRYAN PFAFFENBERGER

Abstract—Personal computer-based communications media—electronic mail, bulletin boards, and computer conferencing—have great potential for integrating scholarly and scientific research networks. Research networks, or informal organizations of faculty who share an interest in a research area, are central to scholarly and scientific progress. They have been criticized, however, for their exclusion of young researchers and faculty at isolated or low-prestige institutions. Studies show that computer networking opens network access by obliterating social barriers and status distinctions. It has often been argued that, if used as a medium for research network communication, computer networking could democratize research networks. Personal computer information services designed for personal computer users, as well as personal computer-based bulletin board systems, represent the most promising avenues for research network communication owing to their low cost, flexibility, and egalitarian ethos.

Introduction

As numerous studies demonstrate, the existence of thriving networks of communicating researchers contributes to a discipline's progress. All too often, however, the private and informal nature of research network communication (such as informal gatherings at conferences) makes it difficult for would-be members to discover the networks and join them. Moreover, social barriers (erected by such factors as age, race, sex, and the academic hierarchy) may obscure or belittle contributions made by young, isolated, or female scientists. These barriers work against the very values of open access and information sharing that make research networks so vital to scientific progress.

Research shows that the social psychological effects of computer-based communications media serve to widen access, obliterate social distinctions, reduce or eliminate social barriers, and foster a strong sense of group identity. If used as a major vehicle of research network communication, therefore, these media may not only facilitate network communication; they may also rectify the networks' tendencies towards elitism and internal stratification.

Example 15.1 This prefatory abstract appears directly before the paper it summarizes. From *IEEE Transactions on Professional Communication*, Vol. PC 29, No. 1, March 1986.

29830. **Hamilton, Nina & Zimmerman, Ruth.** (U Iowa, School of Social Work, Iowa City) **Weight control: The interaction of marital power and weight loss success.** *Journal of Social Service Research,* 1985(Spr), Vol 8(3), 51–64. —Examined the relationship of weight loss maintenance to power dependence in the marital relationships of 40 25–46 yr old married, overweight women who had been in a diet program. Ss were interviewed and responded to a questionnaire on demographics, weight history, weight loss, and weight maintenance. The percent of weight loss goal maintained for 6 mo or more and loss in pounds maintained 6 mo or more were compared across categories of 9 independent variables (e.g., wife's income, who controlled the finances, time of weight onset, husband's income, who was the most supportive). Findings indicate that weight loss success was related to the wife's income and control of the finances, suggesting that having an income diminished dependence in the marriage. Having no income was a good predictor of being unsuccessful at maintaining weight loss, and higher earnings were positively associated with greater maintenance of weight loss. Results support the hypothesis that enduring weight loss is related to the power balance between spouses. (25 ref) —*C. P. Landry.*

29833. **Horowitz, Amy.** (New York Assn for the Blind, NY) **Sons and daughters as caregivers to older parents: Differences in role performance and consequences.** *Gerontologist,* 1985(Dec). Vol 25(6), 612–617. —Examined gender differences among 32 male and 99 female adult children (aged 26–69 yrs), identified as the primary caregiver to an older frail parent. Findings indicate that sons tended to become caregivers only in the absence of an available female sibling; were more likely to rely on the support of their own spouses; provided less overall assistance to their parents, especially hands-on services; and tended to have less stressful caregiving experiences independent of their involvement. It is concluded that female caregivers to the frail elderly will continue to experience high stress as a result of their caregiving activities since they provide the full range of services needed by their parents. (33 ref) —*Journal abstract.*

Example 15.2 These abstracts, written for people searching the literature of a field, are from an indexing service. (This citation is reprinted with permission of the American Psychological Association, publisher of *Psychological Abstracts* and the PsychINFO Database (Copyright © 1967–1988 by the American Psychological Association), and may not be reproduced without its prior permission.)

- It is a courtesy to those who can't, or have no need to, read an article in its entirety, but would like a nutshell description.
- It is a shuttle of information in corporations where, under the name of "executive summary" or "research highlights," it sails forth on a distribution list. In many cases, it represents all the hard work the scientist or engineer has done on a project. Often it is one of the vehicles by which the work is evaluated.

Example 15.3 This executive summary is written for a more diversified audience. (Reprinted by permission of the Harvard Business Review. Copyright © 1986 by the President and Fellows of Harvard College; all rights reserved.)

Why 'good' managers make bad ethical choices

SAUL W. GELLERMAN

Newspapers often carry stories of corporate misconduct. For decades, Manville Corporation suppressed the danger of asbestos inhalation; E. F. Hutton pleaded guilty to 2,000 counts of mail and wire fraud. How can normally honest, intelligent, and compassionate managers act in ways that seem callous, duplicitous, dishonest, and wrongheaded?

The answer is generally found in human nature—in the way ambition and duty get distorted under pressure. The border between right and wrong shifts in convenient directions, or is even ignored, though the manager has no intention of ignoring it.

The author explores the numerous rationalizations that lead to unethical behavior. He offers practical suggestions to help ensure the preservation of ethical propriety: establish clear ethical guidelines for all employees; stress formally and regularly that loyalty to the company does not excuse acts that jeopardize its good name; teach managers, "When in doubt, don't"; have company watchdogs to sniff out possible misdeeds; raise the frequency and unpredictability of audits and spot checks; when you detect a trespass, make the punishment quick, meaningful, and public. Above all, listen to your own moral voice. Chances are it's there saying "don't." Reprint 86402

This chapter has prefatory abstracts (those that come before a paper), abstracts for indexing services, abstracts for a peer audience, and executive summaries for heterogeneous audiences. In practice, many companies interchange the terms "abstract," "summary," "synopsis," and "executive summary." Still, abstracts by any name share important principles, and these will be discussed with the understanding that the word "abstract" varies with the audience—you may well find it under the name "summary" or "synopsis," depending on the conventions of the group.

Most of the time, abstracts are *short*—three to ten sentences at most. Getting them down to that size—while still preserving both information and a readable style—requires thought, practice, and revision on the part of the writer. They are yet another example of the great inverse rule of good technical writing—less work for the reader, more work for the writer.

WHY USE AN ABSTRACT TO INTRODUCE A TECHNICAL DISCUSSION?

If you write a technical paper, it is wise to include an abstract before the text. There are good reasons for this tactic, all related to the psychology of reading.

"Begin at the beginning; go on 'til you reach the end. Then stop," the Queen of Hearts advises in *Alice in Wonderland*. But this is not the way most people read technical materials. Before they read the beginning of the text, most look for clues that will tell them what they are about to read—clues such as the title, the abstract, the illustrations, and captions.

Such previewing is important. The reader uses the technique to decide whether or not to read the report itself. If the reader does decide to continue, the preview works to introduce the material, for a well-written title and abstract tell the reader what is ahead—the main points in the argument and the conclusions. This knowledge helps readers move more rapidly through the text; they have expectations, guideposts so they are not simply wandering.

The most useful device for previewing in technical writing is the abstract, for it

- gives the reader an idea of what's ahead
- reduces the reader's feeling that the report is a wall of print that must be conquered line by line until, at last, the main points are revealed

In a sense, the abstract is a lead. "Lead," sometimes spelled "lede," is a journalist's term for the short, enticing beginning that attracts the reader. There is, of course, a world of difference between news stories and technical reports. News stories must be presented in such a way that people can understand them with very little effort. Most technical writing, in contrast, requires effort on the part of the reader—such is the nature of the complex information of engineering and applied science. But the way the writer casts the information can help, rather than hinder, the reader. The abstract is one of the author's most useful aids to increase the readability of a technical document.

To see how useful an abstract is to the reader confronting technical material, consider this article from *Science*, which is reprinted here without the original abstract that accompanied it (Example 15.4). As you read the article, underline the main points in the author's argument.

Example 15.4 From *Science*, Vol. 233, 4 November 1983, pp. 523–524. Copyright 1983 by the AAAS.

Hypnotically Created Memory Among Highly Hypnotizable Subjects

In recent years, the increasing use of hypnosis to enhance the memories of victims and witnesses of crime (*1, 2*) has created controversy both within the professional hypnosis societies (3) and within the judicial system (4). The main issue concerns the alleged hypermnesic effect when hypnosis is used for memory enhancement. Data suggest that at times, this use of hypnosis may unwittingly create pseudomemories of crimes which, subsequent to hypnosis, come to be believed as true by the person hypnotized (5). Other data suggest, in addition, that hypnosis increases the frequency of both correctly and incorrectly recalled material (*6, 7*).

Hypnosis carries the implicit request to set aside critical judgment, without abandoning it completely, and to indulge in make-believe and fantasy (*8*). To the extent that a person is able to do this, such a procedure may lead to major alterations, even distortions, of perception, mood, or memory (*9*). Indeed, the person who is especially skilled at this task can be perceived as deluded in a descriptive, nonpejorative sense (*10*). Further, the fantasy of hypnosis may be so compelling and subjectively real that some investigators have described it as believed-in imaginings (*11*) and as imaginative involvement (*12*).

Given this basic characteristic of hypnosis, care is necessary when it is used in legal investigations. In common with all situations in which hypnosis is used, the setting may convey strong demand characteristics (*13*). The candidate for investigative hypnosis, a victim or witness of a crime, has generally undergone extensive routine police questioning without having provided sufficient information to furnish a positive identification of a suspect. Such a person, particularly a victim, is ordinarily highly motivated to help the police apprehend the guilty. In this context, hypnosis is usually represented as being effective in enhancing memory; some investigative hypnosis inductions represent mind as a videotape recorder and hypnosis as a means of reaching material that is stored veridically at a level not immediately available to consciousness.

In addition, such investigative hypnosis procedures virtually require fantasy; using the metaphors of televised sport, the hypnotist sometimes tells subjects that they can "zoom in," "freeze frame," and relive the events of a crime in slow motion (*1*). If an individual is asked to zoom in on an image that, in the original experience the retina could not resolve, there is no other source but fantasy for enhanced detail. This task requires the subject to see something beyond his or her capacity and is a powerful and indirect suggestion to hallucinate (*14*). Since, further, the subject ordinarily perceives the hypnotist as an expert, a process of confusing fantasy with fact may occur unwittingly and unknown to either subject or hypnotist.

In order to evaluate the extent to which an investigative use of hypnosis may have contaminated the memory of a victim or witness, Orne has proposed guidelines for conducting investigative interviews (*15*). These include videotaping all interactions (including those before and after hypnosis) between the subject and hypnotist; the hypnotist, who would be the only person with the subject during any sessions, would be required to be professionally trained and only minimally informed in writing of the events in question. This latter requirement is particularly important; for instance, if the hypnotist knows that two gunshots were fired at 4 a.m. on the night in question, it would seem natural to inquire if the subject had heard any loud noises. This seemingly benign procedure may create a pseudomemory which will persist posthypnotically and become unshakable.

The phenomenon of memory creation in highly hypnotizable individuals was first reported in the 19th century by Bernheim (*16*). He described how, during hypnosis, he suggested to a female subject that she had awakened four times during the previous night to go to the toilet and had fallen on her nose on the fourth occasion. After hypnosis, the subject insisted that the suggested events had actually occurred, despite Bernheim's insistence that she had dreamed them. Another version of this hypnotic item of pseudomemory creation has been constructed independently by Orne (*15*). No experimental study, however, has investigated this phenomenon systematically. We have evaluated memory creation among highly hypnotizable individuals, since they seem particularly vulnerable to memory distortion (*17*). Given the demand characteristics of the investigative hypnotic context, however, even witnesses and victims who are not especially responsive to hypnosis may be vulnerable also to such memory contaminants (*6, 18*). Many individuals with low susceptibility to hypnosis have imagery that is as vivid as that of highly responsive individuals (*19*).

To investigate this phenomenon, 280 subjects were screened for hypnotizability on the Harvard Group Scale of Hypnotic Susceptibility, Form A (HGSHS:A) (*20*). Subjects who seemed highly hypnotizable were subsequently screened on the more stringent Stanford Hypnotic Susceptibility Scale, Form C (SHSS:C) (*21*) to confirm their HGSHS:A scores. Of the initial 280 subjects, 27 were selected for the present study on the basis of their SHSS:C performance. The age range for the total sample of 16 females and 11 males was 21 to 48 years (mean ± standard deviation, 27.85 ± 7.26). Their HGSHS:A scores ranged from 6 to 12 (9.86 ± 1.83) and their SHSS:C scores ranged from 9 to 12 (10.74 ± 0.98) (*22*). Subjects were classified as highly hypnotizable only if they passed the posthypnotic amnesia item of the SHSS:C and most of its other 12 items (*23*).

The memory creation item used was modeled on the one described by Orne (*15*). Subjects were asked during hypnosis to choose one night of the previous week, and to describe their activities especially during the half-hour before they went to sleep; they were ascertained to have had no specific memories of awakening or of dreams occurring during the specified night. Through the use of an age-regression technique, subjects were instructed to relive the designated night and asked whether they had heard some loud noises (suggested auditory hallucination) that had awakened them. All but ten subjects reported hearing the noises, and they were encouraged to

Example 15.4 (*continues*)

describe them in detail. The regression was then terminated and the hypnosis session concluded.

Subjects were divided into two groups. For the first group of ten subjects, the suggestion was tested immediately after the hypnosis session; for the second group of 17 subjects, it was tested 7 days later. The procedures for testing the creation of a memory were otherwise identical for the two groups. A second experimenter interviewed the subjects posthypnotically in order to ascertain their perceptions of the study and to obtain their reports of what had happened during the hypnosis session. Since there was no difference between the two groups in frequency of their response to the memory creation item, the data were pooled.

Of the 27 highly hypnotizable subjects tested, 13 accepted the suggestion and stated after hypnosis that the suggested event had actually taken place on the night they had chosen, whereas 14 did not. The latter subjects stated correctly that the noises had been suggested by the hypnotist, but a few of them reached this conclusion by quite idiosyncratic means. One, for instance, decided that the event was suggested since the noise was far more vivid than any noise that he felt could occur in reality.

Of the 13 subjects who reported the suggested memory as real, six of them were unequivocal in their certainty that the suggested event had actually occurred; the remaining subjects came to this conclusion on the basis of a reconstruction of events. One subject, for example, recalled being physically startled. She stated:

> "I'm pretty sure it happened because I can remember being startled. It's the physical thing I remember. . . . I'm making an assumption that it was a noise but I was conscious of the different cars. It must have been something like that. I can remember the startle."

Even when they were told that the hypnotist had actually suggested the noises to them during hypnosis, these subjects still maintained that the noises had actually occurred. One subject stated, "I'm pretty certain I heard them. As a matter of fact, I'm pretty damned certain. I'm positive I heard these noises."

The results support Orne's contention that the memories of victims and witnesses of crime can be modified unsuspectingly through the use of hypnosis. They suggest, further, that an initially unsure witness or victim can become highly credible in court after a hypnotic memory "refreshment" procedure. Although Orne's procedural safeguards permit evaluation of the degree to which a hypnotic procedure may have inadvertently altered a person's memory, such safeguards do not prevent such memory modification from occurring. Indeed, there is no way to differentiate what actually happened during a crime from what a person recalls of it during hypnosis, other than the obtaining of independent verification of the hypnotically elicited recall (17).

The pseudomemory of some loud noises is harmless when suggested in the laboratory. In the more emotionally charged investigative situation, where motivation to please is enhanced, it assumes greater importance; a pseudomemory of a trivial event that has become inadvertently connected with the events of a crime is more likely to persist in permanent memory storage and not decay in the manner of a posthypnotic suggestion. Such "recall" could lead to a false but positive identification and to all of the legal procedures and penalties that this implies. Accordingly, the utmost caution should be exercised whenever hypnosis is used as an investigative tool.

JEAN-ROCH LAURENCE
CAMPBELL PERRY

Department of Psychology,
Concordia University,
Montreal, Quebec H3G 1M8, Canada

References and Notes

1. M. Reiser, *Handbook of Investigative Hypnosis* (Law Enforcement Hypnosis Institute, Los Angeles, 1980).
2. M. Barnes, "Hypnosis on trial" (BBC television program, London, October 1982).
3. Resolution adopted August 1979 by the International Society of Hypnosis [*Int. J. Clin. Exp. Hypn.* **27**, 453 (1979)].
4. B. Diamond, *Calif. Law Rev.* **68**, 313 (1980); D. J. Carter, *Wash. U. Law Q.* **60**, 1059 (1982).
5. C. Perry and J.-R. Laurence, *Can. Psychol.* **24**, 155 (1983).
6. J. Dywan Segalowitz and K. S. Bowers, paper presented at the 34th annual meeting of the Society for Clinical and Experimental Hypnosis, Indianapolis, October 1982.
7. M. T. Orne, affidavit to California Supreme Court on *People* v. *Shirley*, 1982 (mimeograph); J. M. Stalnaker and E. E. Riddle, *J. Gen. Psychol.* **6**, 429 (1932).
8. M. M. Gill and M. Brenman, *Hypnosis and Related States* (International Univ. Press, New York, 1959); E. R. Hilgard, *Divided Consciousness: Multiple Controls in Human Thought and Action* (Wiley, New York, 1977).
9. M. T. Orne, in *Handbook of Hypnosis and Psychosomatic Medicine*. G. D. Burrows and L. Dennerstein, Eds. (Elsevier/North-Holland, Amsterdam, 1980), pp. 29–51.
10. J. P. Sutcliffe, *J. Abnorm. Soc. Psychol,* **62**, 189 (1961).
11. T. R. Sarbin and W. C. Coe, *Hypnosis: A Social-Psychological Analysis of Influence Communication* (Holt, Rinehart & Winston, New York, 1972).
12. J. R. Hilgard, *Personality and Hypnosis: A Study of Imaginative Involvement* (Univ. of Chicago Press, Chicago, ed. 2, 1979).
13. M. T. Orne, *J. Abnorm. Soc. Psychol.* **58**, 277 (1959).
14. In a recent reported legal case, a witness of a murder provided identification during hypnosis of a youth, who was subsequently prosecuted. The case was rejected by the court because of testimony that the witness had been 270 feet away from the incident in conditions of semi-darkness: an ophthamologist testified that positive identification would not have been possible beyond 25 feet under the prevailing light conditions [*People* v. *Kempinski*, No. W80CF 352 (Circuit Court, 12th District, Will County, Illinois, 21 October 1980; unreported); (*2*)]
15. M. T. Orne, *Int. J. Clin. Exp. Hypn.* **27**, 311 (1979).
16. H. Bernheim, *Hypnosis and Suggestion in Psychotherapy* (Aronson, New York,, 1973; originally published in 1888).
17. C. Perry and J.-R. Laurence, *Int. J. Clin. Exp. Hypn.* **30**, 443 (1982).
18. G. Wagstaff, *Percept. Mot. Skills* **55**, 816 (1982).
19. J. P. Sutcliffe, C. Perry, P. W. Sheehan, *J. Abnorm. Psychol.* **76**, 279 (1970); C. Perry. *J. Pers. Soc. Psychol.* **26**, 217 (1973).
20. R. E. Shor and E. C. Orne. *The Harvard Group Scale of Hypnotic Susceptibility, Form A* (Consulting Psychologists, Palo Alto, Calif., 1962).
21. A. M. Weitzenhoffer and E. R. Hilgard. *The Stanford Hypnotic Susceptibility Scale, Form C* (Consulting Psychologists, Palo Alto, Calif., 1962).
22. The SHSS:C Kuder-Richardson reliability coefficient; r = 0.85 (*21*).
23. Hypnotizability is a stable characteristic of the individual, with from 10 to 15 percent of the population having high susceptibility, 10 to 15 percent low susceptibility, and the remaining 70 to 80 percent moderate susceptibility to various degrees [E. R. Hilgard, *Hypnotic Susceptibility* (Harcourt, Brace & World, New York, 1965); C. Perry, *Int. J. Clin. Exp. Hypn.* **25**, 125 (1977)].
24. Supported by Natural Sciences and Engineering Research Council (NSERC) of Canada grant A6361 to C.P.

Present address: Department of Psychology, University of Waterloo, Waterloo, Ontario N2L 3G1, Canada.

4 April 1983; accepted 29 July 1983

Without the abstract, the reader of this paper must pass through the introduction, the statement of the problem, the related literature, and the procedure to get a first glimpse of the results. This transfers the effort from the writer, who could have created a path, to the readers, who must hack their own way through the prose.

The abstract is that path. Ask yourself, "Would I have been able to read the article more quickly and efficiently if there were an abstract?" If the answer is yes, then the next time *you* write a paper, put an abstract at the top.

ABSTRACTS FOR PEERS

This is the classic form of the abstract, in which the information is trimmed nearly to the bone and then served up as a lean yet efficient preview for its audience.

This tight, highly focused abstract is written for others trained within a specialty. It appears either before the text or by itself, in publications like *Chem Abstracts* and *Engineering Index*.

These are the traditional parts of the abstract:

1. *Background of the problem.* This information is usually compressed into a phrase. Sometimes it is omitted.

2. *Objective.* This may appear as part of the title, or as part of the conclusions, rather than being stated separately.

3. *Procedure.* This is usually compressed to a few words or phrases, unless the method is the point of the study.

4. *Results.* These are usually given explicitly. Abstracts that calmly state "results were obtained" are irritating to the reader who would like to see the gist of the matter.

5. *Discussion: Conclusions and implications.* This information is essential; usually it is given in a highly compressed way. Writers should resist the temptation to beg the questions by stating "conclusions were drawn."

Example 15.5 is a sample of an abstract based on the article "Hypnotically Created Memory Among Highly Hypnotizable Subjects" (Example 15.4).

Example 15.5 Abstract from *Science*, Vol. 233, 4 November 1983, p. 523. Copyright 1983 by the AAAS.

> A pseudomemory of having been awakened by some loud noises during a night of the previous week was suggested to 27 hypnotizable subjects during hypnosis. Posthypnotically, 13 of them stated that the suggested event had actually occurred. This finding has implications for the investigative use of hypnosis in a legal context.

Comparing the abstract with the list of parts, one can see that many of the parts have been either compressed, or omitted.

> **Background of the problem.** This is not stated.
>
> **Objective.** This is not stated directly, but can be deduced from the title. (The article seeks to evaluate the effect of memory creation among highly hypnotizable individuals.)
>
> **Procedure.** The information is compressed until it forms the opening sentence. "A pseudomemory of having been awakened by some loud noises during a night of the previous week was suggested to 27 highly hypnotizable subjects during hypnosis."
>
> **Results.** "Posthypnotically, 13 of them stated that the suggested event had actually occurred."
>
> **Discussion.** The lengthy and important discussion in the paper is summarized in the final sentence: "This finding has implications for the investigative use of hypnosis in a legal context."

This abstract follows the traditional pattern of terseness and understatement, which is characteristic of technical and scientific writing. "The finding has implications for the investigative use of

Vague

A procedure was followed to test 27 highly hypnotizable subjects during hypnosis.

Focused

A false memory of having been awakened by loud noises during a night of the previous week was suggested to 27 highly hypnotizable subjects during hypnosis.

Vague

The suggestion worked in some cases, but not in others.

Focused

Posthypnotically, 13 of the 27 people stated the suggested event had actually occurred.

Box 15.1

hypnosis in a legal context," for instance, is deliberately understated. It is well within the tradition Crick and Watson followed, when, near the end of their monumental paper on the discovery of DNA, they commented: "It has not escaped our notice that the specific pairing we have postulated immediately suggests a possible copying mechanism for the genetic material."

ABSTRACTS FOR A BROADER AUDIENCE

The academic abstract, which is admirably short and succinct, is irreplaceable in publications for a highly uniform or homogeneous group, where it serves as an efficient previewing device.

The strings of undefined terms and intensely compressed prose, however, will not serve for a more diversified readership—for instance, for an interdisciplinary or corporate group.

Here the writer needs to broaden the information for the sake of the nonspecialist while in no way distorting it.

A good way to do this is to use the academic abstract as the basis, and then to expand it according to the audience. Examples, definitions, and information on the background and the conclusions—information so often omitted in academic abstracts—can be extremely helpful to a nonacademic audience.

The general reader comes to an abstract with a question. "What was the problem?" This is often the first item the writer needs to answer, using as few technical terms as possible. The second question to answer is, "What was the solution?" Again the language should be as clear as possible so that the reader can grasp both the problem and its solution. The final question the reader asks is, "Why is the solution significant?"

A good way to answer these questions is to expand the objective, the problem statement, and the conclusions—areas which we have seen are often either omitted or intensely compressed in the academic abstracts.

Thus, in rewriting the abstract used in the earlier example, one might add information at these key points:

> **Background.** Add: "The forensic use of hypnosis is increasing, creating controversy within the judicial system. At issue is the reliability of hypnosis: Some studies suggest hypnosis creates false memories that the hypnotized person firmly believes long after hypnosis has ended."

Why? In a few sentences, you answer the questions, "What was the problem?" "Why was it important?" This gives the reader outside the field a context for reading.

Objective. Add: "To evaluate the effect of memory creation on highly hypnotizable individuals."

Procedure. Retain the highly compressed version of the academic abstract. The general reader has little need of more detailed procedural information.

Discussion. Add: "The results of the study, which support the contention that memories can be modified through hypnosis, have important implications. The recall of such persons could lead to false positive identifications, with grave legal consequences. Accordingly, the utmost caution should be exercised whenever hypnosis is used as an investigative tool." This expanded version tells the significance of the problem.

The statement of significance—the answer to the question, "So what?"—is particularly hard to cast, given two traditions in science: the use of understatement and the desire of many workers to keep themselves and their theories modestly in the background. The result often is that writers drastically understate the implications and leave it to the reader to see the reverberations of their work. This does the author a disservice, for the general reader will often be quite unable to read between the lines.

Here are the abstracts of the same article, one for a homogeneous audience (Example 15.6a), and one for a more diversified group (Example 15.6b):

Two Contrasting Abstracts

a. Abstract for Homogeneous Audience

A pseudomemory of having been awakened by some loud noises during a night of the previous week was suggested during hypnosis to 27 hypnotizable subjects. Posthypnotically, 13 of them stated that the suggested event had actually occurred. This finding has implications for the investigative use of hypnosis in a legal context.

Example 15.6a Abstract for homogeneous audience.

Example 15.6 (*continues*)

Example 15.6b Executive summary for diversified audience.

b. Summary for Diversified Audience

The forensic use of hypnosis is increasing, creating controversy within the judicial system. At issue is the reliability of hypnosis. Some studies suggest it creates false memories; other studies suggest hypnosis increases the frequency both of incorrect and correct materials.

In this study, the researchers evaluated the effect of memory creation on highly hypnotizable individuals. They suggested, during hypnosis, to 27 highly hypnotizable subjects, the pseudomemory of having been awakened by loud noises during a night of the previous week. Posthypnotically, 13 of them stated that the suggested event had actually occurred.

The results of the study, which support the contention that memories can be modified through hypnosis, have important implications. The recall of such persons could lead to false positive identifications; accordingly, the utmost caution should be exercised whenever hypnosis is used as an investigative tool.

INDICATIVE ABSTRACTS

An indicative abstract lists the topics in the report without attempting a summary of their contents. An indicative abstract is appropriate when the information the writer is abstracting is disparate or voluminous. In Example 15.7, for instance, the author has prepared an abstract of an extensive literature review on microbial metabolism of chlorinated aromatic compounds. The abstract makes no attempt to inform the reader of the findings in each of the studies reviewed; instead, the writer indicates the major divisions and topics of the review.

WRITING STRATEGIES FOR PREPARING AN ABSTRACT

Many writers prepare the abstract last, after they have written the paper. They go through the text, underline the key points (objective, findings, discussion) and stitch the sentences together in language consistent with the original.

Afterward, they revise the language according to the audience. For an academic group, a highly compressed abstract is acceptable. If the abstract is for a more diversified group, writers struggle more

Example 15.7

Indicative Abstract

This report brings together a review of the literature of microbial metabolism of chlorinated aromatic compounds. Most attention has been given to reports of bacterial, fungal, and cyanobacterial pathways of substrate degradation where metabolites or end products have been identified. Studies reporting data of metabolites arising from incubation of the substrate with mixed cultures or environmental samples and studies that show disappearance of the compound have also been evaluated and included.

In addition to separate chapters on each class of chlorinated aromatic compounds, reviews of microbial physiology, genetics, and methods of biodegradation assessment are included. One chapter reviews biodegradation of these compounds in sealed-up processes.

The review indicated that many factors are involved in assessing the biodegradability of a compound, including the nature of the molecule, substrate concentration, environmental parameters, availability of nutrients and growth factors, and presence of degradative microorganisms.

with the language, trying to make the prose accessible to the widest possible readership. This often means the abstract grows longer. It's rare, however, to see an executive summary of more than a page.

Some authors write the abstract before they write the paper. There are many reasons for this.

- They are applying to attend a conference that makes its selection of participants based on abstracts submitted nine months in advance.
- They are abstracting ongoing work for a supervisor's report. While the final paper is still in process, an abstract of the past few months' work is required.
- They are having trouble composing the paper and use the abstract as a way to get started.

In the following example, a student read an original research report (Example 15.8) and then prepared the abstract in three steps.

1. She listed the possible parts of an abstract.
2. She filled in the information within each applicable category (Example 15.9).
3. She boiled the information down until it was clear, complete, and readable (Example 15.10).

Text *continues* on page 373

The Use of Hypnosis to Enhance Recall

The increased use of hypnosis in forensic investigation has become controversial (*1*). Although numerous case reports attest to the utility of hypnosis in enhancing the recall of the eye witness (*2*), controlled studies have produced conflicting results. Some studies have failed to demonstrate hypnotic hypermnesia, whereas those that have (*3*), have not reported errors in a systematic way nor controlled for the natural hypermnesic effects that can be achieved through repeated testing (*4*). Still others (*5*) have found that hypnotized subjects are susceptible to leading questions. Although scientists are wary of the reliability of forensic hypnosis, police investigators are lobbying to sanction its use in criminal investigation and the judiciary is seeking evidence on which to base legal decisions. The relation between hypnosis and memory enhancement needs to be clarified.

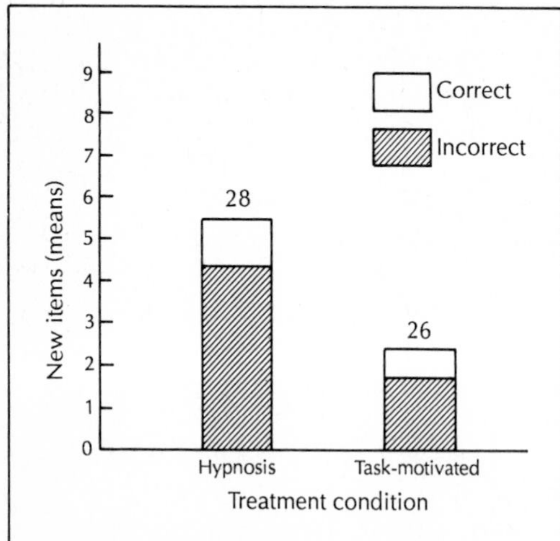

Fig. 1. New items presented as memories by subjects after hypnotic or task-motivating suggestions to enhance recall. All items were designated by subjects as true memories. The number of subjects in each group is shown above each bar.

We now report that any pressure to enhance recall beyond the initial attempt may increase the number of items recalled but increase the number of errors as well. The use of hypnosis exaggerates this process, particularly for those with hypnotic ability. When hypnotized, the highly hypnotizable subjects recalled twice as many new items as controls but made three times as many new errors.

Fifty-four subjects were selected on the basis of their hypnotic ability as measured by a group adaptation of the Stanford C Scale of Hypnotic Susceptibility (SHSS:C) (*6*). Subjects with low susceptibility had SHSS:C scores from 1 to 6, and those with high susceptibility, from 7 to 12. All subjects were presented with a series of 60 slides of simple black-and-white line drawings of common objects (*7*), presented at a rate of 3½ seconds per slide. They were then given a recall sheet and requested to write the name of a line drawing in each of the 60 blank spaces provided for this purpose, indicating as well which items represented memories and which were just guesses. This forced recall procedure is standard in hypermnesia studies (*8*). Subjects were initially given three trials in the laboratory with 3-minute rest periods between trials.

Subjects were then instructed that during the next week they were to recall as many of the line drawings as they could once each day, and to write their recollections on the take-home recall sheets provided. They were asked to deposit each recall sheet in a convenient dropbox daily for 6 days. Altogether, subjects completed nine trials over a period of 7 days before their second laboratory session.

The mean number of items recalled on the first trial was 30. By trial 9 the cumulative

Example 15.8 Original paper used for the abstract. From *Science*, Vol. 222, 14 October 1983, pp. 184–185. Copyright 1983 by the AAAS.

mean had risen to 38 items—an increase of 27 percent. The number of errors increased as well, from an average of less than one error on the first trial to an average of four errors by the ninth. Most subjects approached asymptotic levels of output by about trial 7, 4 days after a single viewing of the stimuli.

The next step was to see whether hypnotic suggestions for increased recall would enable subjects to retrieve more information after asymptotic recall had been reached. During this second laboratory session, subjects were told to relax and focus all their attention on the slides they had seen the week before. They did so either while hypnotized (hypnosis condition) or without hypnosis (task-motivated condition). Before this session subjects did not know which condition they would be in, and the experimenters were unaware of subjects' hypnotic ability. Consistent with these precautions, independent sample *t*-tests indicated no difference between high and low susceptible subjects in the cumulative number of correct items retrieved over the week before treatment [$t(26) = 0.49$] or for the cumulative errors retrieved prior to treatment [$t(26) = 0.14$].

Figure 1 illustrates the number of items reported on the treatment trials that had never been reported as memories before. Subjects in the hypnosis group reported over twice as many new items (both correct and incorrect) as subjects in the task-motivating condition did. The correct information retrieved by subjects in both conditions remained proportional to this shift in total output. Those higher in hypnotic ability in the hypnosis condition were primarily responsible for the increase in output, and hypnotic suggestion was no more potent than task motivating suggestion for those lower in hypnotic ability (Fig. 2).

A two-way analysis of variance based on the total increase in items indicates a significant main effect for condition [$F(1, 50) = 5.63$, $P < 0.03$] and a significant interaction of condition with hypnotic susceptibility [$F(1, 50) = 4.31$, $P < 0.05$]. When just the correct information was considered, the interaction between condition and hypnotic ability was significant as well [$F(1, 50) = 4.95$, $P < 0.05$]. Using new errors as the dependent measure yielded a significant main effect for condition [$F(1, 50) = 5.38$, $P = 0.03$], but the interaction in this case was not statistically significant [$F(1, 50) = 3.10$, $P = 0.08$]. Even though hypnotizable subjects in the hypnosis condition showed a statistically significant increase in accurate recall, this increase was small in absolute terms. No subject in even this most responsive group retrieved more than five new correct items (mean = 1.40), and six of them failed to produce any new correct information at all. The cost of correctly recalling these few items was considerable, since it was accom-

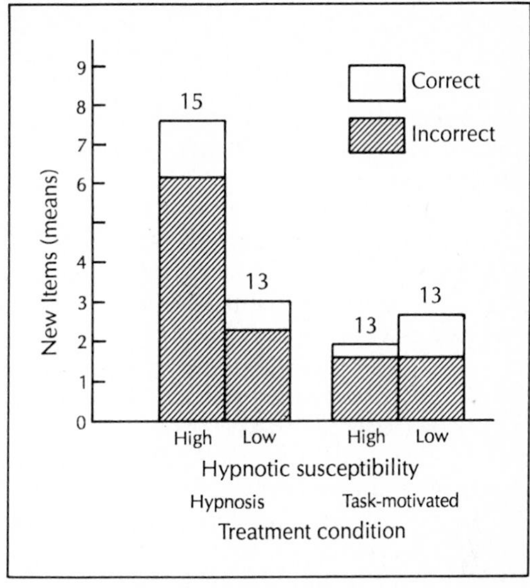

Fig. 2. New items presented as memories by subjects of high and low susceptibility to hypnosis after hypnotic or task-motivating suggestions to enhance recall. The number of subjects in each group is shown above each bar.

panied by almost three times as many errors as were made by subjects in any other condition. We have replicated this pattern of results on a new sample of 56 subjects (*9*).

The probability of correctly recalling new items under hypnosis seems directly related to the number of items a subject is willing to report as memories, a finding that could be interpreted as being due to a shift in report criterion. That is, the increase in correct recall may not represent increased sensitivity to memory traces, but may instead result from less caution by subjects in what they are willing to report as memories. This criterion shift could be attributed to various demand characteristics, social cues, and expectations engendered by the hypnotic situation.

Another possible explanation for the effect of hypnosis on memory depends less on a shift in the report criterion than on the frequency with which the individual's criterion for memorial judgment is subjectively met. Hypnosis may heighten the sense of recognition associated with even falsely recalled items, in effect "fooling" a central processor or editor responsible for memorial judgments (*10*). It may be that one of the criteria upon which this sense of recognition is based is the vividness with which the subject is able to envision those items generated as possible memories during recall attempts. If hypnosis enhances the vividness of mental imagery (*11*), perhaps the vividness with which the subject is able to envision these possibilities becomes compelling. Under these circumstances, the editor could mistake vividly imaged possibilities for memories of the stimuli; the enhanced vividness could lead to a false sense of recognition and hence the inflated output as well as the surprising certainty that subjects have about their hypnotically enhanced recall (*12*).

The role that affect may play in the relationship between hypnosis and memory has not been explored in this investigation, but may be relevant to the use of hypnosis in forensic settings. Nonetheless, our observations of hypnotically enhanced recall should give pause to those advocating the use of hypnosis in situations in which the veridicality of information is of prime concern.

JANE DYWAN
KENNETH BOWERS

Department of Psychology,
University of Waterloo,
Waterloo, Ontario N2L 3G1

References and Notes

1. C. Perry and J. Laurence, *Int. J. Clin. Exp. Hypn.* **30**, 443 (1982); B. L. Diamond, *Calif. Law Rev.* **68**, 313 (1980); H. Spiegel, *Ann. N.Y. Acad. Sci.* **347**, 73 (1980).
2. W. Kroger and R. G. Doucé, *Int. J. Clin. Exp. Hypn.* **27**, 358 (1979); T. S. Worthington, *ibid.*, p. 402; D. S. Carter, *Wash. Law Q.* **60**, 1059 (1982).
3. F. A. DePiano and H. C. Salzberg, *Int. J. Clin. Exp. Hypn.* **29**, 383 (1981); T. P. Dhanens and R. M. Lundy, *ibid.* **23**, 68 (1975).
4. The phenomenon was first described by W. Brown [*J. Exp. Psychol.* **6**, 337 (1923)] and brought into prominence again by M. H. Erdelyi and J. Becker [*Cognit. Psychol.* **6**, 159 (1974)].
5. M. Zelig and W. B. Beidleman, *Int. J. Clin. Exp. Hypn.* **29**, 401 (1981): W. H. Putnam, *ibid.* **27**, 437 (1979).
6. A. M. Weitzenhoffer and E. R. Hilgard, *Stanford Hypnotic Susceptibility Scale, Form C* (Consulting Psychologists' Press, Palo Alto, Calif., 1962). Test-retest reliability is 0.85 ($N = 307$) [E. R. Hilgard, *Hypnotic Susceptibility* (Harcourt, Brace & World, New York, 1965)].
7. The slides were selected to be high on image and name agreement from a set of pictures that had been normalized on a number of dimensions relevant to memory studies [J. G. Snodgrass and M. Vanderwart, *J. Exp. Psychol. Hum. Learn. Mem.* **6**, 174 (1980)]. This selection procedure ensured against ambiguity in scoring.
8. The hypermnesic function has been replicated in numerous studies that used repeated recall over the course of single laboratory session [for example, A. D. Yarmey, *Bull. Psychonom. Soc.* **8**, 115 (1976) and over the course of a week [M. H. Erdelyi and J. Kleinbard, *J. Exp. Psychol.: Hum. Learn. Mem.* **4**, 275 (1978)].

9. All subjects were hypnotized, with those unresponsive to hypnotic suggestion serving as controls. The susceptible subjects reported significantly more information in response to hypnotic suggestion than the unsusceptible subjects [$t(27) = 2.68$, $P < 0.01$]. However, only 10 percent of the new information reported by either group was accurate. While hypnosis produced a small but significant increase in new information [$t(27) = 1.90$, $P < 0.05$], it also produced a significant increase in new errors [$t(27) = 2.67$, $P < 0.01$]. Since this replication examined specific hypotheses, one-tailed tests were used.
10. M. K. Johnson and C. L. Raye, *Psychol. Rev.* **88**, 67 (1981).
11. Experimental evidence on this point is controversial since most studies base their results solely on self-rating scales. H. J. Crawford, using self-rating scales and an objective performance criterion, reported a shift in the vividness of visual imagery during hypnosis for those in the upper range of hypnotic ability [paper presented at the National Meeting of the Society of Clinical and Experimental Hypnosis, Denver, Colo., October 1979].
12. Pseudomemories, originally developed in hypnosis, may come to be accepted by the subject as his actual recall of the original events; they are then remembered with great subjective certainty and reported with conviction [M. T. Orne, *Int. J. Clin. Exp. Hypn.* **27**, 331 (1979)].
13. We thank S. Vandermeulen and D. Chansonneuve for assistance in collecting and scoring the data; S. Segalowitz, J.-R. Laurence, E. Woody, and M. P. Bryden for their advice and editorial comments; and all members of the University of Waterloo Hypnosis Research Team for their suggestions and encouragement. Supported in part by Social Sciences and Humanities Research Council grant 037-8195.

21 March 1983; revised 21 July 1983

The Relation Between Hypnosis and Memory Recall

Overview

There is a growing interest on the part of police investigators and the judiciary in the use of hypnosis on eye-witnesses to increase their recall. Scientists are attempting to determine the effect of hypnosis in increasing accurate recall. Results of studies up to now have been inconclusive. A recent study by the Department of Psychology at the University of Waterloo, Ontario, presents convincing evidence that while hypnosis can improve the number of items remembered, the number of wrong recalls will increase significantly. In addition, improvement in recall is directly related to the subject's susceptibility to hypnosis.

Example 15.9 Writing an abstract. Student first summarized sections of the paper before tightening the information into an abstract.

Example 15.9 (*continues*)

Procedure

The study was conducted in two parts. In a first session, 59 subjects viewed 60 slides of common objects. They were then tested on their ability to recall the images in 9 attempts over a seven day period. For the second session the group was divided, one-half received hypnosis and the others served as a control group. They repeated the procedure for session one. All subjects had been rated on their hypnotic receptivity, and both groups in the second session contained subjects with high and low ratings.

Results

On the first trial, the mean number of items recalled rose from 30 (first attempt) to 38 (9th attempt), an increase of 27%. However, the errors rose from an average of 1 error to 4 errors, a 300% increase. It was further noted that the responses of the high hypnosis susceptible subjects was no different in number of correct recall than the low susceptible subjects.

When the group was divided in session two, the subjects in the hypnosis group reported over twice as many new items (both correct and incorrect) as the control group. The increase was due mainly to the responses of those higher in hypnotic ability. Even among these subjects, however, the increase in correct recall was small in absolute terms, (no subject retrieved more than 5 new correct items and 6 failed to produce any new correct information), and their error rate went up threefold. Within the group that was hypnotized, the subjects rated lower in hypnotic ability showed no greater increase in recall than the control group.

This test of the hypnosis factor on memory recall has been replicated with a sample of 56 new subjects, yielding the same pattern of results.

Conclusions & Implications

The results of the study by the Department of Psychology at Waterloo University on the relation of hypnosis to memory recall indicated that only subjects highly responsive to hypnosis are able to increase their recall and the additional recall will have a high percentage of errors. Hypnosis of eye-witnesses in conjunction with police investigations or court proceedings should be used with caution and may produce incorrect information.

Example 15.10 Final version of the abstract.

> Abstract: The forensic use of hypnosis is increasing. A hypermnesic procedure was used in an experiment that calls this practice into question. Subjects tried for a week to recall 60 previously presented pictures. They were then either hypnotized or not and encouraged to recall even more pictures. Most of the newly recalled material was incorrect, especially for highly hypnotizable subjects in the hypnosis condition. Such errors in recall can have profound implications for forensic investigations.

You may find that writing an abstract pays benefits, not only for the reader, but for the writer. In returning to the original paper to hunt for key sentences for the abstract—lead or topic sentences that summarize entire sections—you may find you've forgotten to write such sentences.

The abstract thus provides a second chance to tighten and focus the original manuscript, for as you look for statements of key points, you may find that in your immersion in detail, you have forgotten to state them explicitly.

SUMMING UP

- The abstract provides a quick, informative glimpse of the problem, the solution, and the importance of the work.
- It is a courtesy to those who can't or don't have time to read an article in its entirety, but would like a nutshell description.
- It is a small shuttle of information when it sails forth on corporate distribution lists. In many cases it stands for all the hard work the writer has done on a project.
- Prefatory abstracts come before the reports they summarize.
- Abstracts in indexing services are reprinted independent of the original paper.
- Abstracts may be written for peers or for more heterogeneous groups. Those written for management are sometimes called "executive summaries."
- Many companies interchange the terms "abstract," "summary," "synopsis," and "executive summary."

• Abstracts are short. Getting them down to size—while preserving both information and a readable style—requires thought, practice, and revision.

• A well-written abstract functions as a "lead," showing the reader the path the report will take.

• Writers who delay a clear statement of the problem, the solution, and the significance until the concluding section of a report force the reader to go through the entire paper. Instead, this information should be packaged in the abstract.

• Abstracts that blandly state "results were found" and "conclusions were drawn," without summarizing the results or the conclusions, are particularly ineffective.

• With the audience in mind, cast the abstract so the reader will quickly grasp the gist of what you've done. The reader seeks the answers to the questions, "What?" and "So What?" The answers lie in a clear statement of the objective, procedure, results, conclusion, and implications.

• For peers, the introductory information may be curt, with only the most compressed of discussions on the theoretical background. This compressed style, often a kind of professional shorthand, is appropriate for a homogeneous group steeped in the problem and familiar with the vocabulary and terse style of the field.

• Undefined terms, intensely compressed prose, and understated conclusions won't serve for a more diversified readership that can't read between the lines. For this interdisciplinary audience, the academic model can be adapted by expanding the problem statement, the objective, and the discussion—areas that are usually highly compressed in academic abstracts.

• The abstract should be written so that the generalist reading it will understand these questions: "What was the problem?" "What was the solution?" "Why is the solution significant?" The language should be as clear as possible so that the reader who is not a specialist in the field can grasp both the problem and its solution.

• The statement of significance is particularly hard to cast, given two traditions in science: the use of understatement and the desire of many workers to keep themselves and their theories modestly in the background. The result often is that the writer drastically understates the implications, leaving it to the reader to see the reverberations of the work. This does the author a disservice, for the general reader will often be quite unable to read between the lines.

• Many writers prepare the abstract last, after they have written the paper. In returning to hunt for key sentences for the abstract—lead or topic sentences that summarize entire sections—they often find they've forgotten to write such sentences. The abstract thus provides the writer with a second chance to tighten and focus the manuscript.

EXERCISES

1. Example 15.11 is a set of abstracts written for "The Relationship Between Hypnosis and Memory Recall" (Example 15.8). Discuss the effectiveness of these sample abstracts. Are the shorter ones successful in presenting the salient points in the article? How about the longer ones? Is longer necessarily better?

The Use of Hypnosis to Enhance Recall

Abstract: Police investigators and the judiciary are seeking substantial evidence to justify the use of hypnosis as a tool for corrective actions.

Abstract: Hypnosis to enhance memory recall has been evaluated for its use in forensic investigations. Findings have led to controversy and are inconclusive.

Abstract: The relationship between hypnosis and accurate enhanced recall is investigated in a controlled study. Subjects are tested on recall on a standard set of images. Statistical analysis shows that hypnosis did increase recall but also increased incorrect recall.

Abstract: The effectiveness of hypnosis to enhance memory was analyzed among 54 subjects. The subjects were selected on the basis of their hypnotic ability. All subjects were first shown a series of 60 black-and-white slides of common objects, and given recall sheets to write down their recollections

Example 15.11

Example 15.11 (*continues*)

to find their asymptotic levels. Next these same subjects were divided randomly; one group was under hypnosis while the other was not. Then both groups were presented with a series of slides again. The study shows that the increase in accurate recall in hypnotized subjects was actually small in absolute terms. Hypnosis may heighten the sense of recognition, but has also led to mistaking vivid and false images as memories.

Abstract: A study was conducted using 54 subjects, graded according to hypnotic susceptibility in order to determine the effect of hypnosis and task motivation on enhanced recall. Memory recall was tabulated during 2 laboratory sessions, consisting of 9 trials each, for viewing 60 items by slide presentation while hypnotized or without hypnosis. Results indicated not only is there a significant relationship between the level of recall and the condition of hypnotic susceptibility, but in the number of recalled items and increases in error reporting as well.

In order to determine if hypnosis enhances recall, groups of subjects with high and low hypnotic susceptibility were tested for recall both with and without hypnosis. Although the results showed a small but significant increase in recall under hypnosis, there was a large concomitant increase in the number of errors made.

Hypnosis was evaluated as a memory enhancing device for potential forensic applicability. A blind crossover design, determining the ability to recall 60 common objects showed that hypnotically stimulated subjects recalled more items while making three times as many new errors as control groups. While the hypnotic stimulus seemed to reflect a breakdown in inhibition or willingness to report, no evidence was shown for reporting accuracy.

2. Collect five or six abstracts or executive summaries of articles within a professional field. Analyze the types of information the authors include (objective, procedure, results, conclusions). Do the abstracts vary in language and level of detail according to the audience they address? Explain.

3. Write an abstract of "Migration and Western Europe: The Old World Turning New" (Example 15.12). Design the abstract for a wide readership of varied academic backgrounds.

Migration and Western Europe: The Old World Turning New

GÖRAN THERBORN

Post–World War II migrations have transformed Western Europe to an extent and a depth which Europeans—citizens, politicians, official statisticians, scholars—are still only beginning to cope with. Currently there are more foreign-born residents of Sweden (7.8% in 1985) or of the United Kingdom (7% in 1983) than of the United States (6.0%, 1981–1985) (*1*). The proportion of resident aliens at the end of 1984 or 1985 was 9.1% in Belgium, 8.1% in France, 14.4% in Switzerland, and 7.1% in West Germany (*2*).

Recent migrations have changed the position of Europe in the world and inter-European relations as well as the internal structure of Northern and Central European societies. Of old, at least since the beginning of the conquest of the Americas, Europe was a continent of emigration. Between 1850 and 1960 it has been estimated that about 55 million people, equivalent to about 18% of the Western European population in 1910, left Western Europe for other continents (*3*). However, after the end of World War II, Western Europe became a region of immigration.

If we disregard the forced migration of Germans from Eastern Europe into West Germany in the aftermath of the defeat of the Nazis—an exodus of massive proportions, landing 8 million people in the Federal Republic by 1950, 16% of the total population of the country—the historical turn took place in two stages. The first one occurred in the 1950s, resulting in a migration surplus in Western Europe (including Greece) of almost 500,000 for the decade (*4*). But this first wave of net immigration had a rather special character that might be thought of as temporary. It was heavily dominated by East Germans moving into West Germany, 3 million between 1950 and 1960 (*5*).

The second phase, however, showed decisively that fundamental structural changes were taking place. Counting in decades, the key period runs from 1964 to 1973. Then migration for all the countries of the area (including Greece) taken together showed a surplus of 2,314,000 (*6*). In France and Germany the demographic impact of immigration was quite dramatic: 37% of French population growth between 1964 and 1973 was due to immigration, and at its recent height, in 1970, net immigration meant a population increase of 0.35%. At its peak in Germany, in 1968–69, net immigration each year added 0.9% to the total population, and for the years 1964 to 1973, immigration accounted for 90% of population growth (*6*). As a yardstick for comparison, take a figure from U.S. immigration at its peak. In 1913, net arrivals of immigrants from overseas corresponded to 0.9% of the then American population (*7*), the same as net migration into West Germany in 1968 and 1969.

The turn of Western Europe from a people-exporting to a people-importing area was the product of two migratory changes. One, and the more important one, was the opening of immigration routes from outside Western Europe. By the mid-1980s, there were from major

Example 15.12 From *Science*, Vol. 237, 4 September 1987, pp. 1182–1188. Copyright 1987 by the AAAS.

Example 15.12 (*continues*)

ethnic groups about 6.7 million registered non-Western European resident aliens and ex-colonial immigrants in the Western European countries of significant gross immigration, about 2.2% of the total population (8). This opening was above anything else a reversal of the old colonial relationship of European settlement. Modernizing social changes combined with little or truncated development, after as well as before independence, turned the old colonizing countries into ex-colonial labor markets. Of the 6.7 million mentioned, 3.9 million are ex-colonials. (Returned European settlers are not counted here.)

The second process involved has been a redirection of Western European emigration from intercontinental to intracontinental migration. Finnish emigration was redirected already after World War II to Sweden. But for Greece, Italy, Portugal, and Spain the turn came about 1960. The change was mainly an end of mass emigration to Latin America, due to economic stagnation there. Both processes, then, imply an epochal shift in the relationship of Europe to what is now called the Third World (9).

National Trajectories of Migration

Western Europe is not a political unit, nor until the 1960s was it an area of migration. Therefore, we had better distinguish the national histories of international migration (Table 1).

All countries of Western Europe, except France, have been emigration countries. Since 1815, Western European migration has had six periods of nationally unevenly distributed trend breaks (the beginning of overseas emigration uncounted): the late 19th-century boom, the aftermath of World War I, the onset of the Depression, the aftermath of World War II, the final phase of the post-World War II boom, and the onset of the 1973 oil crisis. But ceasing to be a country of out-migration does not necessarily mean becoming an immigration country, here defined as having a sustained net immigration equal to or exceeding one promille of the resident population. Denmark, Norway, and the United Kingdom—in the latter case in spite of a substantial gross immigration—have not become immigration countries in this sense. And recent Finnish, Greek, Italian, and Portuguese—and Irish in the 1970s—immigration is almost exclusively or overwhelmingly the return of former emigrant nationals.

The Historical Sociology and Politics of the Demand for Foreign Labor

European emigration was basically supply-determined by the push of demography, closed opportunities, and persecution, and by the pull of available land, higher wages, and more freedom (*10*). The major waves of Western European immigration during the past century, on the other hand, have been strongly demand-determined (except for West Germany until about 1960). Demand has initiated, directed, and controlled immigration, and in this century demand has been politically organized by, or at least under, the auspices of immigration governments. The master form has been a bilateral treaty between a recruiting and a recruitment country (*11*). Demand has been for labor, which takes us to the functioning of labor markets. We shall see that this demand is not simply and directly related to an overall scarcity of native labor, but is also largely due to domestic social divisions, both outside and in the labor market. The popular concept of a "dual labor market" appears too crude to capture the complexity and variety of these divisions. To demonstrate this, we shall briefly go through the most important cases of turns toward massive immigration.

Switzerland in the 1890s. Switzerland was an old out-migration country, mainly export-

Table 1. Periods of significant net migration into or from Western European states, 1815 to 1984 (*42*). Significant net migration is defined as at least one per thousand population. Longer periods may include odd years with a migratory balance or with an opposite sign. Except for the times since 1973, periods shorter than 5 years are not noted, nor are the years of the two World Wars.

Country	Emigration	Immigration
Austria*	1922–26, 1949–61	1965–73
Belgium	1847–66	1922–33, 1945–75
Denmark	1866–1914, 1926–30	
Finland†	1860s†, 1945–70	1981–
France		1830–37, 1875–86, 1920–31, 1948–73
Germany‡	1848–96	1947–1973
Greece§	1950–74	1975–
Ireland‖	1832–1970, 1982–	1971–78
Italy	1860s–1971	1982–
Netherlands¶	1880–1913, 1946–57	1964–81
Norway	1849–1930	
Portugal**	–1973	1981–
Spain**	1888–96, 1904–14, 1919–25, 1949–1974	
Sweden	1830s, 1848–1930	1945–70, 1974–
Switzerland‡‡	1870s–1891, 1919–25, 1975–78	1892–1913, 1928–31, 1945–1973
United Kingdom‡‡	1901–31	
England and Wales§§	1881–1926	1838–49, 1931–39, 1959–64

*The Austrian Republic from 1920 only. †Before World War II Finland has a pattern of migration oscillating strongly from year to year. ‡From 1842, territory as of 1871–1937, after World War II the Federal Republic only. §Covered from 1920 only. ¶From 1910, territory of the Republic only. ¶Data from 1841. **From 1850 onward. ††From 1872. ‡‡Covered from 1901 only. §§From 1838 to 1975.

ing mercenaries but also civilians. As emigration units the majority of the confederated Swiss cantons, in 1827, concluded agreements of free settlement with the kingdoms of France and Piedmont-Savoy, 21 years before there was free settlement among the Swiss cantons themselves. The post-1848 federal government then concluded free settlement treaties with a large number of countries. Here was the legal framework for the sudden turn in the 1890s (*12*). The change involved a decrease, by 40% compared to the previous decade, of Swiss emigration, but the latter still remained high, 1.7% of the native population from 1888 to 1900, about the same from 1900 to 1910 (*13*). The new demand for labor came in part from the engineering industry but, above all, from the surge of Alpine railway construction and from the construction industry in general. The Swiss were little interested in those jobs, in part because strong handicraft traditions and a widespread ownership and co-ownership of land had impeded proletarianization, in part because many had the option of emigrating to a higher end of the labor market as craftsmen and skilled service personnel (*13*, p. 182).

After World War I in Switzerland, France, and Belgium. The postwar migration policies of these countries were established during the war. Switzerland, as a small neutral country anxious to keep out of the conflicts of her neighbors, closed her borders to keep draft evaders and refugees out. A quasi-insurrectionary general strike by a largely foreign radical working class frightened the ruling class, and a postwar policy of diminishing economic dependence upon foreign workers ensued (*12*, p. 198). Therefore, Switzerland fell out from

Example 15.12 (*continues*)

the other immigration countries in this period, for a time becoming an out-migration country again.

In France, on the other hand, as part of mobilization for the war effort, the state took charge of the recruitment of foreign workers, mainly Spanish, Portuguese, and Greek. After the war, a special labor importation corporation was set up, in close cooperation with the state, by the big employers of the Mine and Metallurgy Industries Association. The main objective was to recruit Italians and Poles for the mining and steel industry, for which few French workers could be found in spite of considerable paternalist efforts (14). Before the war, Belgians had provided a substantial part of mining and steel work in northern France, as well as at home, but after it there was in Belgium a scarcity of native labor for this kind of work. Belgium too resorted to an organized recruitment of Poles, Italians, Czechs, and other Eastern Europeans (15). In both countries there was also a substantial nonorganized immigration of Eastern European Jews and of Russian refugees.

After World War II. The special German case apart, the largest postwar immigration effort was proposed by the French. French immigration policy came out of a very interesting political and scientific debate. One camp in the early postwar discussion, represented by, among others, the great demographer Alfred Sauvy and the future Foreign Minister Robert Schumann, argued the necessity of a massive immigration in order to achieve the necessary reconstruction of the country. The Planning Commission, headed by Jean Monnet, on the other hand, saw mass immigration as only one option. The alternative, also deemed possible, was raising the productivity of the French economy. The populationists won politically, and a state body was set up to organize immigration. The immigration of Germans, Spanish, Portuguese, North Africans, and sub-Saharan Africans was deemed undesirable, which mainly left Italy as a recruitment area. Due to the malfunctioning of the French postwar currency and for other reasons, France turned out much less attractive to Italian immigrants than planned. Instead came the first massive Algerian immigration, made possible by the proximity of and the legally open entry from this then part of the French empire (14, p. 10).

The Belgian government also turned to Italy, signing an agreement with the latter in 1946 to recruit 50,000 Italian workers in exchange for the delivery of 3 million tons of coal (15, p. 312).

Switzerland resorted to her pre–World War I policy of recruiting foreign, mainly Italian, labor for expanding the heavy manufacturing and construction industries. But this time it was an immigration corseted in a draconian framework of legal and police control, established during the Depression of the 1930s. Whereas the French immigrationists planned for the settlement of families of desirable immigrant workers, their assimilation, and even for the "francisation" of their names, the Swiss created a class apart of an "alien work force" (12, p. 200). Internationally competitive wages made this policy successful.

The peak of the postwar boom. The largest part of this immigration wave was directed by the West Germans, and the Federal Republic also had the most extensive debate about migration policy. The debate about and the planning for non-German immigration had already began in 1954 and 1955. There was at this time still not full employment, but the drying up in the near future of the Eastern European reservoir could be foreseen and, most directly, there were difficulties in recruiting laborers to agriculture. The reasons for labor import were set out in the October 1955 issue of the business journal *Industriekurier (16)*:

> —native labor reserves can only be utilized through regional mobility which is limited by available housing, whereas alien labor can be housed in barracks

(which also happened in the first period of non-German mass immigration);

—an increased entry of women onto the labor market is undesirable for reasons of family policy;

—an accelerated technical rationalization faces limits on the capital market;

—a prolongation of working time is politically unfeasible.

Employers also stressed the "total" mobility of foreign workers as a major advantage. Government spokesmen emphasized the general labor scarcity to be foreseen after the drying up of the East German reserves and the specific advantages to be derived from the price-stabilizing effects of the high savings ratio of foreign workers and from their contributions to paying for German pensions (*5*, p. 195).

A labor import agreement, for a time a fairly marginal one, was made with Italy in 1955, followed by ones with Greece and Spain in 1960, with Turkey in 1961, with Portugal in 1964, and with Yugoslavia in 1968. As in Switzerland, the import of labor into West Germany was legally framed by very restrictive decrees dating from the 1930s, narrowly circumscribing the permits of aliens and fully controlling their movements, a framework from which European Economic Community nationals (Italians) were exempted (*17*). Remarkably enough this piece of German legislation from January 1933, which had been brought into Austria by the Nazi Anschluss, was still in force, governing the recruitment of foreign workers into Austria, which gathered momentum toward the end of the 1960s (*18*). But the guest worker legislation of the Depression was not capable of mastering the tight labor market and the more humane sociopolitical situation of the 1960s and early 1970s. By 1978, 23.7% of the alien population of West Germany was under 16 years of age, in Switzerland 27.2% was under 16—about the same proportion of foreign children as in the more official immigration countries of France (25.2% in 1975) and Sweden (30.3% in 1976) (*19*).

Competition in the Italian labor market, from the Germans, the Swiss, and North Italian employers, led the French government in the early 1960s to issue a new series of immigration or free migration agreements, with Spain in 1961, with Morocco, Tunisia, Portugal, Mali, and Mauretania in 1963, with Senegal in 1964, and with Yugoslavia and Turkey in 1965. The peace agreement with the Algerians in 1962 also included a clause of free migration (*20*).

Manufacturing and Construction

The bulk of immigrant laborers are manufacturing workers; manufacturing ranges from 43% of foreign employment in Switzerland to 54% in West Germany with France in between (*21*). The last Western European wave of immigration coincides with the last expansion of manufacturing employment in the immigration countries in a context of full employment, with a registered unemployment below 2% by around 1960. For the two major net immigration countries of this period, manufacturing employment reaches its all-time high in 1974 in France and in 1970 in Germany. In Austria this occurred in 1982 and in Belgium, Sweden, and Switzerland about 1965. As a proportion of total employment, manufacturing in France and Germany peaked in 1973–74 and in 1970, respectively (*22*).

By the end of 1985, aliens accounted for 10.7% of all employment in the West German manufacturing industry and for 30.5% in the Swiss (*21*). Their share of manual work is much higher, for example, 27% of all assembly and assistance workers in engineering, 25% of chemical workers, and 24% of metal manufacturing workers in Germany (*23*).

There is one major industrial branch with a remarkable variation of immigrant employ-

Example 15.12 (*continues*)

ment—construction. In France and Switzerland, this is an important part of alien employment. In France in 1979, construction work employed 29% of foreign workers, which constituted about 20% of the employment in the industry (*24*). In Switzerland, at the end of 1985, construction workers represented 12.5% of the foreign labor force, and aliens made up 38% of the construction work force (*25*). In Britain, the Irish are traditionally concentrated in construction, with 30% of the Irish work force in Britain in 1966, and 8.6% non-British workers, slightly above the relatively even sector-distributed foreign labor force (*26*). In Germany in 1985, only 10% of construction workers were foreign, down from 12.5% about 1970, and only 9% of employed foreigners are to be found in construction (*27*). But the real contrast is Sweden, where (at the end of 1986) construction constitutes no more than 3.6% of foreign employment and where aliens account for only 3% of the work volume carried out (*28*). The wage structure is correspondingly different. Putting average wages in manufacturing at 100, in 1982 construction wages were 98.6 in France, 101 in Switzerland, 102 in Britain, 105.5 in Germany, and 116.8 in Sweden (*29*). This is most likely rather the cause of different sources of labor supply than the effect of foreign imports. Construction has been a high wage industry in Sweden for some time, with entry through a well-established apprentice system and with very strong unions. Construction shows more clearly than most branches—although the few Irish in English coal mining also provides a good contrast to mining in Belgium and France—that the native attractiveness of jobs is partly variable, with particular regimes of labor.

Immigration into Particular Regimes of Labor

Demographic developments and general labor market conditions provide only a broad, usually low-working causal basis for the new European pattern of migration. The concrete demand for foreign labor has been determined by conjunctions of specific, historically evolved national labor regimes. These regimes involve the industrial and property structure, technological patterns, the position of women in the labor market, the position of labor in the domestic job markets, and internal migration patterns. The most relevant aspects of the historical evolution of these labor regimes are experiences of non-regulated in-migration and national population politics.

The little proletarianized Swiss labor regime faced great problems in meeting the sudden demand for masses of unskilled labor at the time of the difficult railway construction in the late 19th century, a demand which easily could be met by foreign labor because of pre-existing migration treaties made for opposite purposes. This made labor by aliens part of the Swiss way of life, of the Swiss labor regime, but the size of the alien work force remained a political option. The organized labor import into Belgium after both World Wars and into France after World War I was too specific to be explained by the hecatomb of the war, and it seems rather to have speeded up a process, begun before the war, of substituting foreign for reluctant native labor. This acceleration was directly made possible by state experiences of labor mobilization during the war and, possibly, the prospects of increased indigenous labor elsewhere. Concern with a low birth rate and slow-growing population had been a largely militarily motivated national concern in France since its first defeat at German hands in 1871. After World War II a peopling concept of reconstruction came to prevail over a productivist one.

The "German miracle" had obviously, and unexpectedly, benefited from the large influx from the East, and a continuous import of labor had become part of the West German labor regime. Well before the domestic labor force was fully employed, the ruling forces of the country were planning for a widening of the areas of foreign supply. On a much smaller scale, Swedish industry had become accustomed to Nordic immigration after the war, which also made it easy to throw the recruitment net wider in the 1960s.

The extensive postwar industrialization required more factory labor, which was not that easily attracted in the prosperous 1960s, although Norway and Denmark managed it without making significant use of foreign labor, and Britain without active recruitment. The late emigration countries, Austria and the Netherlands, on the other hand, found it most convenient to import labor, in spite of an extremely low, till the early 1970s stably so, female labor force participation in the Netherlands. In Austria, the import of labor was accompanied by a decline of female labor force participation in the 1960s and 1970s (*30*).

The Crisis, New Labor Regimes, and Foreign Labor

The 1973 crisis brought a structural transformation of the world economy. It immediately altered the labor market situation and, with the exception of Sweden, Norway, Austria, and Switzerland, European governments soon abdicated from attempting to maintain full employment (*31*). Although the size of alien populations has continued to rise, through family reunions mainly, the number of foreign workers has gone down.

The most drastic change is to be found in Austria. Between 1973 and 1984 the number of employed foreign workers in Austria declined by 38.7% at the same time as the number of native workers and employees increased by 27.4% (*32*). Austrian full employment policy is more nativist than the better known Swiss one. Foreign employment in Switzerland was 11.9% less in 1985 than in 1973 after a partial recovery since 1977, whereas native employment was 0.4% less (*33*).

Changes in West Germany have also been dramatic. Between the end of 1985 and 1973, the number of employed alien wage and salary earners has been reduced by 39%, whereas the number of German workers went down by 5.1% (*34*). Aliens in the labor force have also decreased in Belgium, the Netherlands, and Sweden, by 13 to 15% since the peak. But their number has risen in France, by 30.4% between 1973 and 1983 (there was a decline from 1977 to 1981), and by 23.9% in the United Kingdom; these declines were accompanied by increase in the native labor force of 6.3 and 3.6%, respectively (*35*).

National patterns continue to vary. Active, although voluntary, return migration policies have been pursued by Germany, most effectively, France, the Netherlands, and Belgium. But more important has been the government control of short-term labor contracts, decisive in Austria and Switzerland, significant also in Germany. The Swedish reduction is completely spontaneous, mainly due to a narrowing of the income and labor market gap between Sweden and Finland (*36*). Migrant workers have been affected both by exported and by host country unemployment. Global French and British unemployment rates are higher than the German but have never been as high as that among foreign workers in Germany, 13.9% in 1985 and 21% in the four northernmost German states (*23*, p. 31).

An extreme case is the Dutch situation. In early 1987 the unemployment rates among West Indians and the Moluccans (both ex-colonial immigrants) in Holland was 45%, among Turks and Moroccans (recruited labor) 40% (*37*).

The labor regime is changing. The heavy, unskilled manufacturing jobs, for which immigrants were recruited, are irreversibly declining. Many of the latter are, or are considered by employers, incapable of learning the new automated forms of production (*38*). There are and there will probably be some more outlets in the catering industry and in

Example 15.12 (*continues*)

personal services. But the main tendency of European labor markets seems to be that the new labor regime will require more formal qualifications and will offer fewer jobs without formal qualifications. The children of immigrants receive much less post-elementary education than natives. The former made up 7.9% of the total German school population, but accounted for only 2.7% of the apprenticeships in 1983 (*39*). In this way, the unemployment of tomorrow is being produced today.

The New World as a Ghetto?

Western Europe has indeed become a New World, in the sense of multi-ethnic and multicultural societies shaped significantly by immigration. In spite of restrictive labor market policies and of official encouragement of return migration by continental European countries, the resident alien population of the Western European immigration countries increased by 2.1 million (naturalizations uncounted) between 1985 and 1973 (*40*). Even countries of recent major emigration, such as Greece and Italy, are getting immigrant ethnic minorities (*8*). The rich cultural potential of this new influx has been very little tapped so far. European social sciences and humanities, for example, have, on the whole, expressed little interest in and produced little new knowledge about the countries of out-migration, about their relations to Western Europe, and about the bearing of immigrant experience upon psychology and psycholinguistics. Current migration research tends to stick very closely to immediate policy issues. (As such, it usually plays a supportive role, which is far from unimportant.) With a few odd exceptions here and there, the esthetic impact has been confined to English and French ex-colonial literature.

The prospects for a large number of recent immigrants into Western Europe and for their children seem bleak. They arrived in the final stage of a labor regime of extensive industrialization, which has now ended. The young, vigorous workers recruited from outside in the 1960s and early 1970s facilitated the mobility of the native Central and Northern Europeans into more pleasant jobs, increased the profitability of the industrial expansion, and contributed heavily to paying for the enormous growth of the welfare states, from which old-age natives benefited most (*41*). The migrants and their families also gained, from work and higher wages, perhaps also from wider cultural experiences. But now, many migrants are worn out, their labor is no longer wanted, and they are uprooted from their countries of origin, while confronted with rising racism in their new home countries. Their children are often already strongly disadvantaged in the schooling process.

Options are still open, but the current predominant tendency—not without its counterpoints—is that the Old World turned New is getting the worst of both worlds, the underclass ghettoes of the New while keeping the traditional cultural closure of the Old.

References and Notes

1. *Statistisk Årsbok 1987* (Statistiska Centralbyran, Stockholm, 1986), tables 15 and 30; Central Statistical Office, *Social Trends 16* (Her Majesty's Stationery Office, London 1986), p. 24; G. Borjas and M. Tienda. *Science* 235, 645 (1987), table 1.
2. The percentages were lower in Austria, 3.6; Denmark 2.1; the Netherlands 3.9; and Norway 2.4 [M. Frey, in *Die "neue" Auslanderpolitik in Europa*, H. Körner and U. Mehrländer, Eds. (Verlag Neue Gesellschaft, Bonn, 1986), p. 17. French and Swiss data are from the end of 1985; data for the others from 1986. Austria had at the same time an alien population of 3.6%, Denmark 2.1%, the Netherlands 3.9%, and Norway 2.4%.
3. Calculations from W. Woodruff, *Impact of Western Man* (Macmillan, London, 1966), pp. 104 and 106.
4. Including Malta, 486,000; 440,000 excluding. Calculations from the U.N. Economic Commission for Europe, *Labour Supply and Migration*

in Europe (United Nations, New York, 1979), tables 11.8 and 11.9.

5. U. Herbert, *Geschichte der Auslanderbeschaftigung in Deutschland 1880 bis 1980* (Dietz Nachf, Berlin, 1986), p. 181.
6. Calculations from the Organization for Economic Cooperation and Development (OECD), *Labour Force Statistics 1964–1984* (OECD, Paris, 1986), country tables 1. Malta is not included. The data source is national population statistics.
7. *Historical Statistics of the United States* (U.S. Bureau of the Census, Washington, DC, 1975), series A6–8 and C296–301.
8. There are about 1.9 million Turks, 1.8 million North Africans, 1.2 million from the Indian subcontinent, 0.9 million Yugoslavs, 0.7 million Caribbeans (in France uncounted), and 0.2 million Africans. Smaller contingents, like North and Latin Americans, are not included. Host countries covered are Austria, Belgium, France, Germany, the Netherlands, Sweden, Switzerland, and the United Kingdom. Calculations from OECD, "Continuous reporting system on migration," *SOPEMI 1986* (OECD, Paris, 1986), p. 49, and population figures from OECD labor force statistics; Central Statistical Office, in (*1*); Sociaal en Cultureel Planbureau, *Sociaal en Cultureel Rapport 1986* (Staatsuitgeverij, The Hague, 1986), p. 615. Since the mid-1970s there has also been a certain non–Western European immigration into Greece and Italy (D. Papademetriou, in *International Labor Migration in Europe*, R. Krane, Ed. (Praeger, New York, 1979), p. 193; *SOPEMI 1986* (OECD, Paris, 1986), pp. 33–34.
9. T. Lianos, in *The Politics of Migration Policies*, D. Kubat, Ed. (Center for Migration Studies, New York, 1979), p. 210; M. B. Rocha Trinidade, *ibid.*, pp. 220–221; D. Demarco, in *Les Migrations Internationelles de la Fin du XVIIIe Siècle à Nos Jours* (Editions de CNRS, Paris, 1980), p. 598.
10. A broad and subtle analysis of European emigration in the North Atlantic economy is given by B. Thomas, *Migration and Economic Growth* (Cambridge Univ. Press, Cambridge, ed. 2, 1973). One of the most successful econometric accounts of the supply side of emigration is provided by T. Moe [thesis, Stanford University, Stanford, CA (1970)].
11. J. Salt, in *Migration in Post-War Europe*, J. Salt and H. Clout, Eds. (Oxford Univ. Press, Oxford, 1976), p. 98.
12. H. J. Hoffmann-Nowotny and M. Killias, "Switzerland," in *The Politics of Migration Policies*, D. Kubat, Ed. (Center for Migration Studies, New York, 1979), p. 194.
13. Calculations from E. Gruner in *Les Migrations Internationelles de la Fin du XVIIIe Siècle à Nos Jours* (Editions de CNRS, Paris, 1980), p. 189.
14. G. Tapinos, *L'immigration étrangère en France* [Presses Universitaire de France (PUF), Paris, 1975], p. 6; G. Noiriel, *Longwy. Immigrés et prolétaires 1880–1980* (PUF, Paris, 1984), pp. 152 and 164.
15. J. Stengers, in *Les Migrations Internationelles de la Fin du XVIIIe Siècle à Nos Jours* (Editions de CNRS, Paris, 1983), p. 310.
16. *Industriekurier* **4**, 10 (October 1955). The account here is based on Herbert (5, p. 196).
17. Decrees of 23 January 1933 and 22 August 1938 (5, pp. 199 and 260).
18. E. Gehmacher, *The Politics of Migration Policies*, D. Kubat, Ed. (Center for Migration Studies, New York, 1979), p. 96.
19. A Pollain-Widar, *Special Problems Posed by Second Generation Migrants* (Council of Europe, Strasbourg, 1980), p. 7.
20. The French lost out to the Germans also on the Spanish and Turkish labor markets [Tapinos (*14*, p. 106) and Salt (*11*, p. 89)].
21. Calculations from *Statistisches Jahrbuch der Schweiz 1986* (Berne, 1986), pp. 97 and 341; *Statistisches Jahrburh für die Bundesrepublic Deutschland 1986* (Wiesbaden, 1986), p. 105; *Données Sociales* (INSEE, Paris, ed. 4, 1981), p. 92. The French figures refer to the 1970s, the others to 1985. Manufacturing work occupied 60% of the foreigners in Sweden in 1970, but an extensive employment of immigrant women in the public social sector, largely as hospital auxiliaries, has led to a situation where manufacturing and public services each account for a good third of immigrant employment [E. Wadensjö, *Immigration och Samhällsekonomi* (Studentlitteratur, Lund, 1973), p. 158; *Arbetsmarknaden 1970–1983* (Statistiska Centralbyrån, Stockholm, 1984), p. 101].
22. Data from OECD (6, country table III) and P. Flora *et al.*, *State, Economy, and Society in Western Europe 1815–1975* (Campus, Frankfurt, 1987) vol. 2, chap. 7.
23. H. Werner, *Arbeitsmarktsanalyse 1985 anhand ausgewählter Bestand und Bewegungsdaten* (Institut für Arbeitsmarkt- und Berufsforschung, Nürnberg, 1986), unpublished work report, p. 36. French data appear more irregularly, but, for example, by the end of the 1970s, 26% of the

employment in the French auto industry was alien [M. Goubet and J.-L. Roucolle, *Population et société française 1945–1981* (Sirey, Paris, 1980), p. 25.]

24. See *Données Sociales* (INSEE, Paris, 1981), p. 92, and OECD (*6*, p. 225).
25. Calculated from *Statistisches Jahrbuch der Schweiz* (*27*, pp. 97 and 341).
26. Calculated from S. Castles and G. Kosack, *Immigrant Workers and Class Structures in Western Europe* (Oxford Univ. Press, London, 1973), table 15.
27. Calculated from *Statistisches Jahrbuch für die Bundesrepublik* (*27*, p. 105).
28. Calculated from *Statistiska Meddelanden Am 11 8701* (Statistiska Centralbyrån, Stockholm, 1987), pp. 36 and 40.
29. Calculated from Swedish Employers' Confederation. *Wages and Total Labour Costs for Workers, International Survey 1972–1982* (SAF, Stockholm, 1984), pp. 38 and 72.
30. OECD, *Historical Statistics 1960–1983* (OECD, Paris, 1985), p. 34.
31. See G. Therborn, *Why Some Peoples Are More Unemployed Than Others* (Verso, London, 1986).
32. Calculations from Frey (*2*, p. 20) and OECD (*6*, pp. 158–159).
33. Calculations from *Statistisches Jahrbuch* (*27*, pp. 91 and 341). Swiss reduction of foreign employment was somewhat more drastic than the Austrian, 24.7% from 1973 to 1977 as against 23.2% from 1973 to 1976 in Austria.
34. Calculations from Frey (*2*, p. 20), *Statistisches Jahrbuch* (*27*, p. 105), and OECD (*6*, p. 240).
35. Calculations from Frey (*2*, p. 20), and OECD (*6*, pp. 220–221 and 444–445). The Austrian, German, and Swiss figures refer to employment, the other are labor force data, including the unemployed.
36. See H. Korner and U. Mehrlander, Eds., *Die "neue" Auslanderpolitik in Europe* (Verlag, Neue Gesellschaft, Bonn, 1986).
37. Internal report of the Dutch Ministry for Social Affairs, presented in the VPRO television program on 1 February 1987.
38. A series of Dutch examples of this new exclusion process was given in *NRC Handelsblad*, 10 June 1987, Supplement, p. 3.
39. U. Mehrländer, in *Guests Come to Stay*, R. Rogers, Ed. (Westview, Boulder, CO, 1985), pp. 169 and 174.
40. Frey (*2*, p. 17) and *SOPEMI* (*8*, p. 70).
41. Estimates for Sweden have indicated a substantial transfer of incomes from immigrants to natives by the public sector. It culminated around 1970 but still occurs [J. Ekberg, *Inkomsteffekter av invandring* (Acta Wexionensia, Växjö, 1983).
42. Except for the period before 1950 in Greece, Portugal, and Spain, the data come from official population statistics: P. Flora *et al. State, Economy, and Society in Western Europe 1815–1975* (Campus, Frankfurt, 1987), vol. 2, chap. 6.1; OECD, *Labour Force Statistics 1964–1984* (OECD, Paris, 1986), national table 1; from 1964 the end-year data of the OECD have been used. For Greece, Portugal, and Spain, population statistics for 1950 to 1963 are taken from U.N. Economic Commission for Europe, *Labour Supply and Migration in Europe* (United Nations, New York, 1979), table 11.9; thereafter data are from OECD. For the period before 1950, the data derive from the registration of emigrants, related to the population [W. Woodruff, *Impact of Western Man* (St. Martin's, New York, 1966), p. 106; B. R. Mitchell, *European Historical Statistics 1750–1970* (Macmillan, London, 1975), tables B1 and B9].

Short Reports

CHAPTER 16

> **"First we had breakfast at the motel. Then we went over to see about renting a car. Then . . ."**
>
> **—excerpt from a rambling trip report**

Many technical people find that the reports they write are short—one- or two-page nutshell accounts of progress, of a trip taken, of a set of data.

Although such reports are short, they are by no means easy to do, for typically the writer must find a way to sum up or epitomize much detail using little space. One solution is simply to write "see attached documents" on a cover page and dispatch both documents and covering note to the recipient. The problem with this expedient is that the writer abandons any attempt to organize or focus the data, leaving this job entirely to the reader. At the other extreme are narratives that contain the information, but in an unfocused, rambling, or disjointed way.

In between these two extremes are various kinds of short reports that present technical detail effectively. Although each is cast differently, according to purpose and audience, effective short reports share common qualities. *The most important quality is organization.*

Short reports should be done in such a way that the writer's *organizational scheme* is apparent to the reader throughout.

Readers should be able to

- quickly understand the nature of the problem and its significance
- quickly be able to find the parts of the report that are essential reading for them
- easily see the pattern within which the writer has fitted details

Some people think that organization is saved for the grand occasions in life—full-scale reports, formal proposals. This is not so. Tight organization applies as well to the humble trip report as to the mighty proposal for capital funds. The discursive, poorly structured trip report that begins, "We had breakfast at Bargain Inn, and then . . ." is the despair of anyone who has to wade through it.

AUDIENCE, LOGIC, LANGUAGE, LAYOUT

These are some of the occasions that call for a short report:

- progress reports of an individual or a group
- monthly or interim status reports
- meeting reports that summarize a technical discussion
- trip reports, on occasions such as technical exhibitions or conferences
- test reports summarizing data collected by a service group
- confirmation reports cementing oral agreements
- justifications for new equipment, new procedures, or changes in equipment or procedure

Considerations of audience, language, logic, and layout are the same for short reports as for any other form of technical writing.

Audience The writer must first think about who the primary and secondary readers are likely to be; and the best strategies for reaching these readers.

Logic and organizational patterns Short reports are structured so that results and conclusions flow logically from the objective and procedure.

- *Summary or overview* What was the objective? What is the essence of the problem under discussion? Why do we need to know? Sometimes this information is presented solely in an "Objective" line; in other cases a separate paragraph serves. You may need to give the limits or scope of the report, too.
- *Body* The type of development depends on the data you are presenting. Common patterns are listing, chronology, problem-solution, and cause-effect.
- *Ending* Short reports wrap up with conclusions, implications, and recommendations; sometimes there are attachments such as tables or illustrations.

Language Transitional phrases and strong lead sentences matter for continuity. Use of detail, comparison, and contrast help send home the point. In the examples that follow, the tone ranges from conversational to formal, depending on the audience.

Design and Layout There are many striking ways to organize short reports so that they are more attractive on the page. In the following examples, first-, second-, and third-level headings are deployed effectively, as are indentation, bullets, informal tables, underlining, and boldface type.

FOUR EXAMPLES OF SHORT REPORTS

Following are four examples of short reports (Examples 16.1–16.4). An analysis of audience, language, logic, and page design is given for each example.

1. *Technical Confirmation* This brief report confirms a job assignment.
2. *Trip Report* This spare memo reports succinctly on a trade fair.
3. *Preliminary Report of an Investigation* In this short, tightly organized report, the results and conclusions flow logically from the objectives.
4. *Test Report* In this report, the author tested plastic sheeting to determine its tensile heat distortion temperature.

Example 16.1

Date: Copies to:
To:
From:
Subject: Analyses for paint thinner and paint

This memo will confirm our discussion of the two separate projects you would like the laboratories to undertake.

- The first is the analysis of competitive domestic paint thinners.
- The second is the investigation of paint-odor problems.

A description of each project and outline of our proposed approach follows.

1. Competitive Paint Thinner Analysis

You will supply us with samples of domestic paint thinner for which you wish to know the major solvents and the relative amounts of those solvents. We will analyze the solvents by capillary gas chromatography/mass spectrometry after first removing the insoluble portions of the formulations. Relative amounts of each solvent will be determined chromatographically by comparing with standard mixtures of the solvents.

We estimate that each sample will take us up to two hours to prepare, analyze, calculate, and report. We understand that you currently expect to send us 10–15 samples.

2. Paint Odor

(a) Final Product Odor

You will supply us with paint judged to have objectionable odor. We will analyze the headspace over these paints using a Purge & Trap Sampler and capillary gas chromatograph, and attempt to correlate the chromatographic data with subjective odor-level evaluation. Once we have agreed on the peak or peaks to be used to describe odor level objectively, we will determine odor levels instrumentally on production items.

The analysis of each paint sample should take approximately one hour. The number of samples is not known at this time.

(b) Qualification of Raw Materials

In order to determine which lots of raw materials, especially N-propanol, used in preparing paint contain contaminants that might result in objectionable odor, you will send us samples for analysis. We will chromatographically determine if there are any extraneous peaks in the samples and whether such contaminants are present in high or low concentrations.

Each of these analyses should take approximately one hour. The number of samples is not known at this time.

In order for us to begin working on these two projects, we will need written authorization to charge you. You will be charged only for the actual number of hours spent. If you like, when authorizing charges you can specify that total charges not exceed a certain limit.

We will begin work as soon as we have the samples and authorization. Results will be reported both by telephone and memo report.

(signature)

Analysis of Example 16.1

Audience:

Primary:

The customer, who has some technical background and a general understanding of the sorts of analytical tests she wants performed.

The writer's supervisor, who approves all confirmations issued by his division. His concern is with the contractual obligation (Can the lab deliver what it is promising?) and the price (Will the lab make a profit on the transaction?).

Secondary:

The customer's supervisor, who, presumably, wants an assurance that the job will be done promptly, precisely, and economically.

Logic:

Conditions of the work are spelled out systematically in problem-solution format.

Organizational Pattern:

Overview.
Body: Division into parts.
Conclusion.

Language:

The sentences are short and clear. The paragraphs are unified. The diction is spare. The tone is friendly without being effusive, and suggests competence. The author uses first person and addresses the reader directly. Verbs are active.

Design and Layout:

In the overview, key points are set off by bullets, and key words are underlined.

In the body, first- and second-level headings are used. Second-level paragraphs are indented.

Example 16.2

To: Date:
From:
Organization Visited: CompuFair
Hinton Center Hotel
Boulder, Colorado
Date of Visit: 22 May 1987
Technical Subject Matter: MOS transistors; High-Voltage ICs

I attended two sessions at CompuFair. The first session was on MOS transistors, the second on high-voltage ICs.

- **Discrete power MOS transistors.** Circuit applications and failure modes were discussed. The most significant feature of these discrete circuit topologies was the use of one or more high-voltage devices in a source follower or cascaded configuration.
- **High-voltage ICs.** The ICs had integrated high-voltage devices, but not high current-handling capability. Also presented was an IC with an LT device integrated into an oxide isolation process.

Most of this information has been presented in larger exhibits. In future, I suggest we skip this meeting.

Analysis of Example 16.2

Audience:
Primary
One supervisor, whose technical background is nearly identical to that of writer.

Organizational Pattern:
Overview.
Body: Division into parts.
Conclusion.

Language:
The tone is first-person informal ("In future, I suggest we skip this meeting") but informed.

Design and Layout:
Division shown through boldfaced subheadings.

Author: Copies to:
Date:
Subject: Computer Modeling of Norman, Oklahoma, Pipeline

Objectives:

1. Develop a computer model of the Norman, Oklahoma, pipeline system that produces a graph of Norman's throughput up to the limit of the pumps and motor;
2. Determine the factors prohibiting the further increase in throughput.

Findings:

Results
1. A computer model of the Norman pipeline has been developed.
2. For the proposed line and pump configuration, the maximum throughput is 5750 std. bbl/day.

Results are listed in Tables 2 and 3 and plotted in Figures 1 and 2. The pump horsepower requirement is listed in Table 2 and pipeline discharge pressure as a function of flow is plotted in Figure 3.

Example 16.3

Example 16.3 *(continues)*

Conclusions

1. The pump is the factor prohibiting throughput increase.
2. The booster pump by itself is capable of moving 2250 bbl/day, but if the booster pump is operated alone, the system will be pump-limited.

Discussion:

The modeling of the Norman pipeline was divided into two segments.

Physical Properties. Physical properties of the oil were developed. A table of enthalpy, entropy, compressibility, density, liquid volume percent, and bubble points, as a function of temperature and pressure was prepared. (See Table 1.) The data are required input for the program used.

Pipeline Hydraulics. The pipeline hydraulics were simulated with the computer program, PDR7, to determine the maximum throughput as well as the prohibiting factors for further throughput increase. Crude leaves the stabilizer at 85°F, 12 psia. The booster pump moves it at a speed of 820 rpm through an exchanger that cools the fluid to about 25°F. The mainline pump transports the crude at 1780 rpm. All pipeline, pump curves, and pressure drop information assumed in the study can be found in the Appendix.

NOTE: Tables and figures not shown in this example.

Analysis of Example 16.3

Audience:

Primary:

Ten readers in management positions within the large company. They are all technically trained, but in various disciplines.

The author attempts to address them by keeping the heart of the report lean and streamlined, and putting the tables and supporting data in the appendixes.

Secondary:

Five or six readers in parallel technical positions who will read the sections that interest them, including the appendixes.

Logic:

This is an interesting and creative adaptation of the classic problem, "What should I do when my subject is complex and I have many readers with many different backgrounds?"

The author rejects simplifying or watering down the subject; instead, he casts it in such a way that people can overview and select appropriate sections according to their needs.

In this case, the author has

- ordered the elements so that the main points are at the top (results and conclusions first, discussion afterwards)
- used appendixes for supporting information, leaving the body of the report free for a tight narrative presentation

Organizational Pattern:

The author adapts classical report order: objective, abstract, procedure, results, conclusions, implications.

Development of Detail:

The supporting materials are placed at the end, rather than occupying the front of the report. In this way, the spotlight is on what's important for the primary readers—the findings.

Appendixes are used for supporting information. This permits the author to tighten the body of text.

Language:

Address is formal (for instance, note use of passive) to suit the varied expectations of the readership.

Design and Layout:

Listing and indentation of objectives.
Use of second- and third-level headings.
Third-level headings on same line as text.

Analysis of Example 16.4

Audience:

Person who requested test. This reader has a technical background similar to that of the writer's.

Language:

As the audience is extremely familiar with the subject, the author's language is shorn of any explanatory information: acronyms, for instance, are not defined; conclusions are stated boldly, as the subject is familiar to the reader. No background is needed, nor given. Sentences are often cast in the passive so that emphasis falls on the test ("Test pieces were cut").

Copies: Requested by:
Project No.: Tested by:
Date: Reported by:
Subject: Testing of Formplex Sheets A & B

Objective:

To determine the standard HDT, Tensile Strength and Modulus, Notched Izod and Brabender Torque at 235°C Formplex Sheet A (clear) and Sheet B (yellow).

Method:

Two sheets of plastic were received from Formplex. The first material, labeled "A", was a clear, colorless plastic. The second material, labeled "B", was a clear, yellow plastic. Test pieces were cut out from each material for physical tests. A sample of Plexiglas for HDT comparison was included.

Results:

		Material & Values		
Type of Test	*Units*	*A*	*B*	*Plexiglas*
264 psi HDT	°C	71.5	75	100.5
Tensile Strength	psi × 10^3	12.1	12.6	—
Tensile Modulus	psi × 10^5	6.06	5.61	—
Notched Izod	ft. lbs./in.	0.54	0.42	—
Brabender Torque @ 235°C		No fusion	No fusion	—
Melt Index @ 200°C, 27½ lbs.	gms/10 min.	No flow	No flow	—

Conclusions:

The HDT of both materials is near that of Sudoy 100 resin and the tensile properties and Notched Izod are similar to those of ANS base resins.

Example 16.4

A CLOSER LOOK: The Role of Organization

It's crucial to organize your short reports so that detail is introduced within a pattern. In the following section, the same sort of writing assignment—a trip report—is organized four different ways; each way permits the orderly introduction of detail.

Listing Formats

This first example starts with a set of notes taken by an analytical chemist who attended a conference. She subsequently turned the notes into a trip memo that was distributed to people working in the chemical analysis group, the separations group, the laser research group, and the library.

Here are the notes:

Notes taken at the Pittsburgh Conference, March 10-12, 1988.

1. Dionex has a new electrochemical detector which is used along with the conductivity detector. This must be placed in a high conductivity area--between the separator and the suppressor. There is the option of using a silver or a platinum working electrode, depending upon whether sensitivity or selectivity is desired. With the silver electrode, .1 ppm bromide can be determined in the presence of 1000 ppm chloride. With the platinum electrode, both mono- and divalent cations can be

determined on the same set of columns, avoiding the necessity of dedicating one set of columns to divalent cations alone.

2. Dionex also has new fast run separator columns which give greater sensitivity and resolution without changing flow rates or eluents. They also have an "Anion Standard-Nitrate Column" which shifts the nitrate peak to very late in the run, permitting analysis of earlier eluting ions in the presence of high concentrations of nitrates. A year or so ago, when this column was first discovered, it was shelved for lack of applications, but now a few are out for experimentation.

3. Other detectors of interest.

a. Tracor Instruments has a photo-conductivity detector. It is a dual conductivity cell used to subtract out the background caused by the eluent. At Allied Chemical Co., they have used a variable wavelength UV-Vis detector in series with the usual conductivity detector. This aids in peak identification and can simplify complex samples.

To eliminate interferences caused by broadening of bands as weakly ionic or neutral organic compounds progress through the suppressor column, the UV-Vis detector can be placed before the suppressor.

In this position, though, there is a higher wavelength cut-off due to eluent interference. By placing the detector after the suppressor, compounds that absorb in the range of 190-210 nm can be determined. Anions such as iodate and bromate can be determined and the water dip interference can be eliminated with this detector.

In casting the notes as a trip report, the writer

- composed a lead or overview statement that summarized and focused the information that followed.
- listed the supporting details, giving the significance for each item on the list (the data and conclusions)

Her final trip memo is shown in Example 16.5. The overview and simple listing pattern give coherence to the detail that follows.

A second example of a simple listing format, reinforced by effective page layout, is shown in Example 16.6.

Purpose/Discussion Format

The more effort the writer puts into crafting a highly organized piece of prose, the less effort the reader has to exert.

Trip Report
Pittsburgh Conference
March 10–12, 1988

From: Susan Held

I found that the most valuable papers at the Pittsburgh Conference were those on Ion Chromatography. There were presentations of methods of analysis both using and not using suppressor columns following the separator.

Those who did not use suppressors used low conducting eluent solutions and felt that it was an advantage not to have "down time" caused by the necessity of regenerating the column.

Several different detectors were discussed:

1. *Tracor Instruments* has a Photo-Conductivity Detector. It is a dual conductivity cell used to subtract out the background caused by the eluent.

2. At *Allied Chemical,* they have used a variable wavelength UV-Vis detector in series with the usual conductivity detector. This aids in peak identification and can simplify complex samples. To eliminate interferences caused by broadening of bands as weakly ionic or neutral organic compounds progress through the suppressor column, the UV-Vis detector can be placed before the suppressor.

In this position, though, there is a higher wavelength cutoff due to eluent interference. By placing the detector after the suppressor, compounds that absorb in the range of 190-210 nm can be determined. Anions such as iodate and bromate can be determined and the water dip interference can be eliminated with this detector.

3. *Dionex* has a new electrochemical detector which is used along with the conductivity detector. This must be placed in a high conductivity area—between the separator and the suppressor. There is the option of using a silver or a platinum working electrode, depending upon whether sensitivity or selectivity is desired. With the silver electrode, .1 ppm bromide can be determined in the presence of 10,000 ppm chloride. With the platinum electrode, both mono- and divalent cations can be determined on the same set of columns, avoiding the necessity of dedicating one set of columns to divalent cations alone.

Example 16.5
Simple organizational pattern: Listing format.

AMERICAN VACUUM SOCIETY 30th NATIONAL SYMPOSIUM
NOVEMBER 1–4, 1987, Boston, Mass.

Filed by: A. A. Alvarez
Purpose: 1. Short Course, "Sputter Deposition and Ion Beam Processes"
2. Short Course, "Technology of Surface Preparation"

1. Sputter Deposition and Ion Beam Processes

This course is designed for those interested in fabricating thin film devices. Two days of lectures were devoted specifically to ion beam, conventional flow discharge, and magnetron sputtering. For each process, Thornton's model was used to discuss the effects of substrate temperature and vacuum chamber pressure upon the thin film properties. The mechanisms involved in film growth were explained in great detail. Also discussed to a lesser extent was reactive sputtering.

At the end of the course, magnetron sputtering was reviewed and a portable demonstration system was used to explain both cylindrical post and hollow cathode magnetron sputtering.

2. Technology of Surface Preparation

This course was intended to address the preparation, handling, and storage of substrates prior to use, although the main topic of discussion was substrate handling. Explained were the substrate surface conditions which will determine the film properties of nucleation and growth. Discussion included topics such as adhesion, characterization, surface conditioning, and the physical, electrical, and chemical properties which may be desired of a clean substrate. One important item was brought up concerning glass substrates. It was stated that most plate glass is formed by floating the molten glass material on top of a pool of molten tin. This would leave tin oxide on the "bottom" surface of the glass. It was suggested the preetching of the glass substrates would be the most effective way of removing the tin oxide.

Example 16.6
Simple organizational pattern: Listing format.

Example 16.7 shows a trip report that the writer has spent time organizing. Thus:

- The reader sees the organization of the piece from the beginning. For instance, the writer divides and distinguishes between the seminars and the exhibits. The distinction is made both in the language, in explicit lead sentences, and in the layout, where boldface and italics are used to divide the text.
- It is easy to understand what the author thinks is worthwhile. Instead of a simple listing, there is an *evaluation* of the list.

TRIP MEMO

25th BLUE MOUNTAIN CONFERENCE
RALEIGH, NORTH CAROLINA

Filed by: R. Cannova

Purpose of Trip

The principal objectives of this trip were to note new developments in spectroscopy by attending papers presented at the symposia, and to visit exhibits displaying new analytical instrumentation.

Symposia

The conference was jointly sponsored by the Blue Mountain Section Society for Applied Spectroscopy and the Blue Mountain Chromatography Discussion Group. It was divided into thirteen separate symposia in which over 300 papers and posters were presented. Symposia I attended and found useful included Atomic Spectroscopy, Chromatography, Computer Applications, EPR Spectroscopy, Ion Chromatography, Mass Spectroscopy, NMR Spectroscopy, and Raman Spectroscopy. Those sessions I found most interesting are discussed below.

Future Applications of Computers in R & D
by R. D. Wang, RFD Company, Cleveland

Dr. Wang explained at some length the extensive systems of computers which are in use at RFD, especially at their Richmond R & D facility. Their applications include all the normal ones, such as data retrieval with automatic archiving, and electronic mail. (It was interesting to note that their electronic mail was deliberately set up as an informal communication system, where style, punctuation, and format are ex-

Example 16.7
Purpose/Discussion format

pected to be imperfect. I think this freedom may provide a high dividend when it comes to informal communication.)

They also have many specialized and unusual applications. Several involve the customization of computer hardware and software to generate unique systems required for a particular application. Particularly interesting was the control of a small robot by a personal computer to automate an NMR spectrometer. The instrument was already fitted with a computer from the manufacturer for data handling. The system perfected is a truly automated instrument with all four subsystems (NMR, data computer, robot, and robot computer) working together.

Additionally, RFD has actually manufactured several laboratory instruments from combinations of existing ones, building all their own control modules. Probably the most memorable RFD system described by Dr. Wang involves the use of a video camera as part of a computer system. Image signal is digitized with the "as viewed" screen being divided by a matrix. The color, hue, and color intensity are measured at each sector of the matrix. All these features combine to allow detailed pattern recognition and quantification of diaper rash and other site-specific phenomena.

Languages, Operating Systems, and Networks
Dr. Carmilla Bernstein, Worcester Polytechnic Institute

Bernstein chose to deviate from her proposed discussion of computer communications. Instead, in her usual deliberate way, she responded to Mr. Wang, amplifying certain points and taking exception to others. She began by developing and embellishing one point: There is no longer a question of whether or not computers will be used extensively in the laboratory, there is only the question of how far behind some laboratories will become before they're forced to admit it. To punctuate her point, she cited GM as an example. They are undertaking an expenditure of $7.5 M to put a small computer on every desk of every professional.

Exhibits

The most interesting new piece of equipment I saw at the exhibits was the Metrohm 636 Titroprocessor. Built entirely around a microprocessor, the 636 has the capabilities of a fully automated titration system from sample preparation, through actual titration and calculation of results, to final data handling and storage.

It can be programmed for eighteen different test procedures using up to three directly connected burets. Two additional burets can be programmed in conjunction with the automatic sample handlers. The largest sample handler has a transport capacity of forty-four 250 ml beakers.

Abstract/Conclusions/Recommendations Format

In the next example (Example 16.8), the author puts the abstract, conclusions, and recommendations at the top. This is an immensely useful device for the reader, for it permits a preview. Notice that the author not only summarizes the conference, but includes an evaluation. He gives not only the "What?", but the "So What?"

TRIP MEMO
Short Course on Coal Characteristics
and Coal Utilization, Ohio State University
November 1–5, 1987

Filed by: W. Hoi 20 November 1987

ABSTRACT

This course dealt with the various characteristics of coal and their effects on processes for coal utilization. Presentations were made by faculty members from several universities, and in particular from the College of Earth and Mineral Science of Ohio State University. Topics included coal petrography, pyrolysis and conversion, combustion, and analytical methods for coal conversion products.

CONCLUSIONS AND RECOMMENDATIONS

Although only 24 persons attended (as opposed to over 50 last year), it was by no means a reflection on the quality or relevance of the course. In general, the presentations were excellent, as were the personal interactions. Attendance was worthwhile and is highly recommended. Emphasis was placed on the existing need to continue research in the development of conversion processes. Although work in these areas has slowed down due to the economy, it was stressed that current research is essential to the future production of non-petroleum hydrocarbons. Attention was also given to the immediate need for improved methods of handling and transportation of coal.

DISCUSSION

Richard Greenfield, Los Alamos Energy Technology Center, has investigated characteristics of low-rank coals that are important in coal utilization. He discussed liquefaction reactivity's

Example 16.8
Abstract/Conclusions/Recommendations format.

dependence on lignite oxygen functionality. His group has also spent time studying the problem of slagging and fouling associated with high sodium levels in low-rank coals.

Jane Kovald of Worcester Polytechnic Institute discussed pyrolysis and the effects of reaction conditions (temperature, particle size, pressure) on product yields and distributions. Rapid heating pyrolysis does result in larger volatile yields. Kovald believes that this is an effect of the experimental techniques used to achieve the faster heating and not the heating rate itself. Essentially, it is the pyrolysis temperature and not the heating rate that results in increased volatile yields.

Harry Roth from Ohio State spoke on direct liquefaction mechanisms and processes. He emphasized the importance of the solvent as a means of solid coal transportation within the system and as a significant hydrogen source for coal conversion. Roth also mentioned the new two-state process that Chevron recently developed. In this process coal is dissolved and dispersed in the first stage. Catalysis occurs in the second stage prior to solid removal. Supposedly, this decreases the amount of time during which first-stage products can undergo condensation reactions. These condensed products are not easily converted to fuel products. The result is a better product yield.

SUMMING UP

The key ingredient in short reports is organization.

• Short reports should be structured in such a way that the writer's organizational scheme is apparent to the reader.

Readers should be able to

- quickly understand the nature of the problem and its significance
- quickly be able to find the parts of the report that are essential reading for them
- easily see the pattern within which the writer has fitted details

• Short reports should be structured so that the results and conclusions flow logically from the objective. The report is usually divided into a summary or overview, which gives the objective; the body, which develops the details; and the wrap-up, which combines conclusions, implications, and recommendations. Sometimes there are attachments—such as tables or illustrations—particularly if the writer wants to keep the body of the report lean.

First-, second-, and third-level headings can be used to organize short reports so they are more attractive on the page. Indentations, bullets, informal tables, underlining, and boldface type serve that purpose, too.

If your subject is complex, but you have many readers whose backgrounds are varied, try ordering elements so the main points are at the top for the generalists, with the details organized below so that non-specialists can get the gist of the report without reading the supporting technical detail.

EXERCISES

1. Assume that, as part of your work for an agency, you have combined data from a mail survey with economic census data to develop a profile of businesses owned by women.

Once the data are tabulated, you prepare a report of no more than two pages. In it you summarize the findings, telling

- what characteristics of the owners are revealed (typical age, race, and educational background, for instance)
- what characteristics of the majority of their businesses are demonstrated

Prepare a brief, readable report discussing the findings as reflected in Tables 1 through 9.

Notes: the term "firm" in the survey refers to sole proprietorships, partnerships, and corporations. The term "business receipts" includes the gross value of all products sold, services rendered, or other receipts from customers during the year. "Employees" includes all full-time and part-time employees on the payroll, but excludes proprietors, partners, or owners who work in the business, and family members or others who work in the business but are not subject to withholding taxes.

2. Do a site-visit at your school or place of employment. Write a brief report on one of these subjects:

- special facilities for the handicapped
- special facilities for the faculty or, if at a business, senior management

Table 1: Age and Marital Status of Women Business Owners, by Receipts Size, Employment Size, Geographic Division and Legal Form of Organization: 1977

Receipts size, employment size, geographic division, and legal form of organization	Firms (number)	Owner's age as of December 31, 1977 (percent)					Owner's marital status (percent)			
		Less than 25 years	25 to 34 years	35 to 44 years	45 to 54 years	55 years or more	Never married	Married	Divorced or separated	Widowed
All industries	701,957	5.3	17.0	17.3	18.8	41.7	18.8	27.0	26.6	27.6
RECEIPTS SIZE										
Less than $5,000	298,159	7.7	20.0	13.6	13.7	44.9	24.5	15.5	30.2	29.8
$5,000 to $9,999	101,621	6.7	13.6	20.2	21.6	37.8	21.5	19.5	33.2	25.8
$10,000 to $24,999	109,330	3.1	20.8	21.8	17.2	37.2	19.4	28.1	27.6	24.9
$25,000 to $49,999	64,310	3.1	13.8	16.9	23.5	42.7	13.0	35.5	20.8	30.7
$50,000 to $99,999	52,278	1.4	12.1	22.0	26.4	38.1	8.2	40.6	26.0	25.2
$100,000 to $199,999	36,440	1.6	10.5	18.3	27.8	41.8	4.7	61.9	9.5	23.9
$200,000 to $499,999	26,127	.4	12.7	16.2	28.1	42.6	6.9	61.9	5.2	26.0
$500,000 to $999,999	8,106	.6	7.2	20.7	30.2	41.3	2.7	67.3	1.7	28.4
$1,000,000 or more	5,586	1.3	5.5	14.5	28.9	49.8	4.0	64.2	5.5	26.4
EMPLOYMENT SIZE[1]										
No employees	534,224	6.3	18.6	17.3	16.5	41.3	21.7	20.8	29.8	27.7
1 to 4 employees	118,017	1.7	13.5	17.2	24.1	43.6	9.9	44.9	17.6	27.6
5 to 9 employees	28,653	.3	7.1	17.9	31.8	42.9	6.9	58.7	9.2	25.2
10 to 19 employees	13,611	.6	4.9	17.9	37.3	39.4	2.2	54.8	14.6	28.4
20 to 49 employees	5,770	.4	3.7	14.3	35.3	46.3	2.4	66.2	5.7	25.7
50 to 99 employees	1,245	.1	1.3	20.1	41.8	36.7	.7	66.6	1.7	31.0
100 or more employees	437	3.6	4.8	14.4	34.7	42.5	3.5	67.1	8.8	20.6
GEOGRAPHIC DIVISION										
New England	37,680	9.3	19.9	22.9	17.9	30.0	21.1	27.6	34.7	16.7
Middle Atlantic	112,582	1.9	16.5	17.8	22.3	41.5	22.4	28.0	20.4	29.3
East North Central	114,453	6.2	16.0	20.7	17.8	39.3	18.4	30.3	20.7	30.6
West North Central	54,589	5.7	17.8	12.4	11.4	52.7	16.8	24.7	20.9	37.6
South Atlantic	106,196	4.3	15.3	16.3	18.4	45.7	16.1	32.3	26.4	25.2
East South Central	37,408	7.2	12.1	10.8	17.2	52.7	19.7	30.1	16.8	33.5
West South Central	70,656	5.3	13.4	15.6	21.9	43.8	16.8	20.1	29.5	33.6
Mountain	38,006	3.5	17.6	13.6	21.9	43.5	15.3	31.8	28.7	24.1
Pacific	130,387	6.7	21.6	18.6	18.5	34.6	20.4	20.2	39.1	20.3
LEGAL FORM OF ORGANIZATION										
Sole proprietorship	531,856	5.9	16.6	15.2	17.3	45.1	22.0	13.0	32.6	32.3
Partnership	111,430	3.4	20.3	23.2	21.3	31.8	10.4	72.6	6.5	10.4
Corporation	58,671	2.5	14.6	26.8	29.7	26.4	3.1	78.7	5.1	13.1

[1]Includes both full- and part-time employees.

3. Write a brief trip report on one of these topics:

- a trade show or fair within your technical field
- a museum of science, technology, or business
- a special exhibition on science, technology, or business

4. Write a brief report on your progress in technical writing.

Table 2: Race and Ethnicity of Women Business Owners, by Industry Division and Major Group: 1977

Industry	Firms (number)	Race of owner (percent)			Ethnicity of owner (percent)	
		White	Black	Other	Spanish/ Hispanic	Not Spanish/ Hispanic
All industries	701,957	94.1	3.8	2.0	2.6	97.4
Construction	21,129	96.9	1.7	1.4	2.2	97.8
Special trade contractors	14,409	96.7	2.0	1.3	2.2	97.8
Other contract construction	6,720	97.3	1.0	1.7	2.1	97.9
Manufacturing	18,914	96.9	1.0	2.0	2.9	97.1
Apparel and other textile products	2,725	90.3	.4	9.2	8.9	91.1
Printing and publishing	4,914	98.5	1.5	-	2.0	98.0
Stone, clay, and glass products	2,166	97.2	1.3	1.6	1.3	98.7
Miscellaneous manufacturing industries	2,388	99.0	.4	.6	.3	99.7
Other manufacturing	6,721	97.7	1.1	1.2	2.5	97.5
Transportation and public utilities	11,874	94.2	4.1	1.7	2.0	98.0
Trucking and warehousing	4,804	96.7	1.6	1.6	1.7	98.3
Transportation services	3,233	95.1	2.6	2.3	2.4	97.6
Other transporation and public utilities	3,837	90.3	8.5	1.2	1.9	98.1
Wholesale trade	16,133	97.7	1.3	1.0	1.0	99.0
Durable goods	7,446	98.3	1.0	.7	1.1	98.9
Nondurable goods	8,687	97.2	1.5	1.3	.8	99.2
Retail trade	211,723	93.9	3.7	2.4	3.0	97.0
Building materials and garden supplies	5,145	99.3	.7	-	1.6	98.4
General merchandise stores	3,770	95.4	2.0	2.6	3.8	96.2
Food stores	21,309	91.7	6.0	2.3	2.9	97.1
Automotive dealers and service stations	8,186	98.2	1.4	.4	4.1	95.9
Apparel and accessory stores	16,716	96.0	3.0	1.1	2.0	98.0
Furniture and home furnishings stores	8,949	97.9	.9	1.2	2.2	97.8
Eating and drinking places	39,415	89.9	5.6	4.5	2.5	97.5
Miscellaneous retail	108,233	94.4	3.3	2.3	3.3	96.7
Finance, insurance, and real estate	66,257	97.9	1.3	.9	1.4	98.6
Insurance agents, brokers, and service	8,596	97.1	.7	2.2	1.1	98.9
Real estate	55,093	97.9	1.4	.7	1.4	98.6
Other finance, insurance, and real estate	2,568	98.7	.9	.4	1.2	98.8
Selected services	316,031	92.8	5.1	2.1	2.5	97.5
Hotels and other lodging places	12,590	93.9	6.1	(Z)	1.6	98.4
Personal services	95,202	86.9	9.3	3.9	1.3	98.7
Business services	47,436	97.6	1.6	.8	1.9	98.1
Automotive repair, services, and garages	4,636	94.0	5.0	1.0	3.1	96.9
Miscellaneous repair services	4,670	96.1	2.3	1.6	4.2	95.8
Motion pictures	1,439	99.3	-	.7	.7	99.3
Amusement and recreation services	16,763	94.6	3.8	1.6	4.5	95.5
Health services	44,762	90.4	6.1	3.5	4.3	95.7
Legal services	6,405	94.5	3.7	1.9	1.8	98.2
Educational services	23,148	96.9	2.4	.8	1.5	98.5
Social services	2,780	89.0	9.7	1.2	.5	99.5
Miscellaneous services	56,200	97.9	1.6	.5	3.8	96.2
Other industries	12,087	96.4	.6	3.0	2.2	97.8
Not classified by industry	27,809	95.9	2.0	2.0	6.1	93.9

Table 3: Highest Level of Formal Schooling Attended by Women Business Owners: 1977

Owner's highest level of formal schooling (percent)							
Less than 9 years	9 to 11 years	12 years or high school equivalent	Junior or community college	Vocational or technical school	4 year college or university	Graduate or professional school	Other
4.2	3.9	17.3	9.5	14.0	28.2	22.6	.2

Table 4: Participation of Women Owners in Management of Their Businesses, by Receipts Size, Employment Size, Geographic Division, and Legal Form of Organization: 1977

Receipts size, employment size, geographic division, and legal form of organization	Firms (number)	Owner active in making decisions (percent)		Average hours per week owner spent managing the business (percent)					
		Yes	No	Less than 10 hours	10 to 19 hours	20 to 29 hours	30 to 39 hours	40 to 49 hours	50 hours or more
All industries	701,957	84.1	15.9	27.3	12.2	9.7	11.3	18.1	21.3
RECEIPTS SIZE									
Less than $5,000	298,159	84.8	15.2	38.0	19.6	13.2	9.0	10.7	9.5
$5,000 to $9,999	101,621	85.9	14.1	15.2	9.3	11.8	18.2	24.3	21.2
$10,000 to $24,999	109,330	88.1	11.9	14.1	8.5	7.6	15.1	25.8	28.9
$25,000 to $49,999	64,310	87.8	12.2	19.3	6.6	6.6	9.7	22.7	35.2
$50,000 to $99,999	52,278	85.5	14.5	19.0	5.3	3.5	9.9	21.0	41.3
$100,000 to $199,999	36,440	78.7	21.3	26.1	6.6	3.2	8.0	22.2	34.0
$200,000 to $499,999	26,127	62.8	37.2	45.7	3.0	7.5	10.9	16.8	16.2
$500,000 to $999,999	8,106	66.6	33.4	35.9	4.3	10.9	11.9	20.2	16.8
$1,000,000 or more	5,586	47.0	53.0	59.0	4.1	5.2	7.6	12.4	11.8
EMPLOYMENT SIZE[1]									
No employees	534,224	85.6	14.4	27.2	14.5	10.9	11.7	16.3	19.4
1 to 4 employees	118,017	83.0	17.0	23.9	5.2	5.2	10.2	25.8	29.7
5 to 9 employees	28,653	74.8	25.2	33.0	5.0	7.1	10.0	20.4	24.5
10 to 19 employees	13,611	70.1	29.9	35.3	3.6	4.6	9.1	20.3	27.0
20 to 49 employees	5,770	57.7	42.3	46.0	3.5	5.4	8.8	20.6	15.7
50 to 99 employees	1,245	72.1	27.9	30.7	3.5	15.3	18.4	13.2	18.7
100 or more employees	437	47.5	52.5	58.6	3.8	2.7	9.9	10.6	14.4
GEOGRAPHIC DIVISION									
New England	37,680	83.1	16.9	29.8	12.4	7.6	11.9	17.2	21.0
Middle Atlantic	112,582	80.5	19.5	34.9	12.3	10.2	9.7	17.4	15.5
East North Central	114,453	84.6	15.4	26.1	13.9	10.9	11.9	13.5	23.8
West North Central	54,589	87.2	12.8	27.7	8.7	7.7	12.6	19.3	23.9
South Atlantic	106,196	84.3	15.7	27.6	11.0	11.0	12.9	16.7	20.9
East South Central	37,408	85.3	14.7	27.5	11.5	7.8	10.1	23.9	19.1
West South Central	70,656	83.8	16.2	23.6	13.2	6.1	9.6	22.2	25.4
Mountain	38,006	91.6	8.4	20.0	9.8	9.5	11.1	20.2	29.4
Pacific	130,387	83.2	16.8	24.5	13.5	10.7	11.8	19.5	19.9
LEGAL FORM OF ORGANIZATION									
Sole proprietorship	531,856	86.7	13.3	24.4	13.7	10.5	11.4	17.9	22.0
Partnership	111,430	79.8	20.2	32.9	8.5	6.8	12.7	18.5	20.6
Corporation	58,671	70.2	29.8	40.6	7.2	7.9	8.5	18.6	17.2

[1]Includes both full- and part-time employees.

Table 5: Percent of Income Women Owners Derive From Their Businesses and How They Acquired Their Businesses, by Receipts Size, Employment Size, Geographic Division, and Legal Form of Organization: 1977

Receipts size, employment size, geographic division, and legal form of organization	Firms (number)	Income owner derives from the business (percent)							How owner acquired the business (percent)			
		None	Less than 10 percent	10 to 24 percent	25 to 49 percent	50 to 74 percent	75 to 99 percent	100 percent	Original founder	Purch-ased	Inher-ited	Other
All industries	701,957	11.2	21.0	10.3	8.6	8.1	12.9	27.8	70.5	20.2	8.2	1.1
RECEIPTS SIZE												
Less than $5,000	298,159	15.1	37.4	13.5	9.4	5.8	5.3	13.5	85.6	9.9	3.3	1.3
$5,000 to $9,999	101,621	7.5	13.1	10.4	8.6	12.0	14.6	33.8	74.0	17.5	5.8	2.7
$10,000 to $24,999	109,330	11.5	9.0	6.6	7.1	6.7	18.6	40.6	68.6	23.1	7.9	.4
$25,000 to $49,999	64,310	11.6	9.6	8.2	9.4	8.1	16.6	36.4	56.4	30.5	12.7	.4
$50,000 to $99,999	52,278	6.2	6.1	7.0	8.6	9.9	19.8	42.3	44.8	37.6	16.9	.6
$100,000 to $199,999	36,440	4.1	8.3	8.9	6.9	13.5	20.1	38.1	39.4	40.5	19.6	.5
$200,000 to $499,999	26,127	7.7	6.8	5.8	10.3	11.7	22.0	35.7	44.5	34.6	20.9	(Z)
$500,000 to $999,999	8,106	6.6	4.0	4.8	5.7	7.6	35.3	35.9	31.6	44.3	24.1	.1
$1,000,000 or more	5,586	5.8	6.0	5.9	8.6	13.7	31.7	28.2	39.3	24.4	35.9	.4
EMPLOYMENT SIZE[1]												
No employees	534,224	12.4	24.6	11.0	8.6	7.2	10.9	25.3	78.0	14.9	5.8	1.3
1 to 4 employees	118,017	8.0	9.0	8.9	9.1	10.4	17.4	37.2	46.5	38.0	14.8	.8
5 to 9 employees	28,653	4.7	9.2	5.6	10.3	12.9	22.8	34.5	35.1	44.8	19.9	.3
10 to 19 employees	13,611	5.2	3.8	5.7	5.6	12.3	26.3	41.1	31.3	40.2	28.4	.1
20 to 49 employees	5,770	5.6	7.5	10.6	4.8	15.6	35.4	20.6	42.1	33.2	24.7	-
50 to 99 employees	1,245	4.6	1.5	1.6	2.4	20.2	34.1	35.6	20.0	51.7	28.2	.1
100 or more employees	437	6.0	6.0	4.2	6.0	19.2	34.7	24.0	45.4	24.5	30.1	-
GEOGRAPHIC DIVISION												
New England	37,680	12.7	19.0	14.0	4.4	5.9	16.8	27.2	74.2	16.9	8.1	.7
Middle Atlantic	112,582	12.6	23.2	9.8	12.2	6.5	11.8	24.0	74.3	15.6	9.3	.7
East North Central	114,453	12.4	23.7	7.5	6.4	5.6	13.1	31.4	67.1	22.0	9.8	1.0
West North Central	54,589	11.8	17.9	10.1	11.6	9.2	13.0	26.5	61.7	28.1	8.5	1.7
South Atlantic	106,196	13.0	21.5	9.2	9.3	9.2	10.7	27.1	71.7	18.2	8.9	1.3
East South Central	37,408	10.5	19.7	15.8	7.7	10.6	11.5	24.1	65.6	20.7	10.4	3.3
West South Central	70,656	12.1	22.3	10.2	9.9	10.4	16.5	18.5	67.8	20.0	11.7	.4
Mountain	38,006	7.0	21.9	11.7	5.9	11.6	10.2	31.6	67.0	26.6	6.0	.4
Pacific	130,387	8.1	17.3	11.3	7.6	8.5	13.9	33.4	75.5	19.6	3.6	1.2
LEGAL FORM OF ORGANIZATION												
Sole proprietorship	531,856	9.6	23.0	10.7	8.4	8.0	13.0	27.3	74.5	17.1	7.1	1.3
Partnership	111,430	16.3	15.2	9.6	10.3	8.7	11.6	28.3	56.8	32.4	10.3	.4
Corporation	58,671	17.0	12.3	8.4	7.8	8.7	14.5	31.3	56.4	27.5	15.9	.2

[1]Includes both full- and part-time employees.

Table 6: Selected Characteristics of Businesses Owned by Women: 1977

Business operated from residence (percent)		Year business was founded (percent)				Percent of business owned by women (percent)		
Yes	No	Before 1960	1960 to 1969	1970 to 1975	1976 to 1977	50 to 74 percent	75 to 99 percent	100 percent
46.6	53.4	30.8	18.7	26.3	24.2	16.7	.7	82.6

Table 7: Full-Time Employees in Businesses Owned by Women: 1977

Number of full-time employees in businesses owned by women as of March 12: (percent)											
1977						1978					
None	1 to 4	5 to 9	10 to 19	20 to 49	50 or more	None	1 to 4	5 to 9	10 to 19	20 to 49	50 or more
71.4	21.0	4.3	1.7	1.1	.6	72.6	19.3	4.6	1.9	.9	.7

Table 8: Part-Time Employees in Businesses Owned by Women: 1977

Number of part-time employees in businesses owned by women as of March 12: (percent)											
1977						1978					
None	1 to 4	5 to 9	10 to 19	20 to 49	50 or more	None	1 to 4	5 to 9	10 to 19	20 to 49	50 or more
76.2	19.4	2.3	1.0	.5	.6	76.3	18.8	2.7	.9	.6	.7

Table 9: Selected Characteristics of Women Who Owned Businesses During 1977, by Owner's Marital Status

Item	Owner's marital status (percent)				
	Total	Never married	Married	Divorced or separated	Widowed
How owner acquired the business	100.0	18.8	27.0	26.6	27.6
Original founder	70.5	15.6	17.3	20.2	17.5
Bought the business	20.2	2.2	7.5	5.5	5.0
Inherited the business	8.2	.9	2.0	.5	4.9
Other	1.1	.2	.3	.5	.2
Major source of financing and/or capital used by women to become owners	100.0	18.8	27.0	26.6	27.6
Family or friends	19.5	4.9	6.1	5.0	3.5
Individual or joint savings	63.6	11.8	13.8	17.9	20.1
Government program	.5	-	.2	.2	.1
Commercial bank loan	14.7	1.9	6.3	3.1	3.4
Other	1.7	.2	.6	.4	.5
Amount of capital, including financing, women invested in their businesses to become owners	100.0	18.8	27.0	26.6	27.6
None	28.3	6.1	4.4	7.4	10.3
$1 to $9,999	53.6	11.2	13.3	15.7	13.4
$10,000 to $24,999	10.4	1.1	5.1	2.1	2.1
$25,000 to $49,999	4.4	.2	2.2	.9	1.0
$50,000 to $99,999	2.1	.1	1.2	.2	.6
$100,000 or more	1.1	.1	.7	.2	.1
Percent of income owner derives from the business	100.0	18.8	27.0	26.6	27.6
None	11.2	1.3	4.1	2.8	2.9
Less than 10 percent	21.0	3.8	4.6	6.2	6.4
10 to 24 percent	10.3	1.7	2.6	2.4	3.6
25 to 49 percent	8.6	1.2	2.3	2.0	3.1
50 to 74 percent	8.1	1.6	2.3	1.8	2.4
75 to 99 percent	12.9	2.3	2.5	3.7	4.4
100 percent	27.8	7.0	8.3	7.5	4.9
Number of years owner had full- or part-time paid employment	100.0	18.8	27.0	26.6	27.6
None	11.2	1.6	2.5	2.3	4.8
Less than 2 years	2.3	.6	1.1	.3	.3
2 to 5 years	10.2	3.2	3.2	2.1	1.8
6 to 9 years	12.0	3.1	4.5	2.6	1.9
10 to 19 years	27.8	4.3	8.0	10.2	5.3
20 years or more	36.5	6.0	7.8	9.1	13.5
Number of years owner worked in a managerial capacity	100.0	18.8	27.0	26.6	27.6
None	16.0	3.6	3.6	4.1	4.7
Less than 2 years	7.4	1.6	2.5	2.1	1.2
2 to 5 years	22.1	5.8	6.4	7.2	2.8
6 to 9 years	15.9	2.4	5.2	4.9	3.4
10 to 19 years	18.9	2.6	5.2	4.8	6.3
20 years or more	19.7	2.9	4.1	3.5	9.1

CHAPTER 17

Formal Reports

The technical report is a historical form as venerable as the sonnet. In fact, it is much older. The format was developed in antiquity for forensic uses. The classical rhetorical scheme included an introduction; an exposition presenting the circumstances defining the issue; a proposition that the remainder of the report established as true, valid, or probable; a division of parts; a short outline for the audience of main points to be discussed; a proof; a refutation of anticipated objections; and a summary. Revived in the West during the Renaissance, the report survives today in virtually the same logical order it had assumed by 300 B.C.

REPORTS FOR DIFFERENT AUDIENCES

The audience for reports ranges from a narrow band of technical experts to a varied public readership.

For Government Agencies

These are reports prepared for the technical experts at sponsoring agencies such as the Environmental Protection Agency, the Department of Defense, or the National Aeronautics and Space Administration.

The National Technical Information Service, established to help readers locate technical reports done either by agencies or under contract, lists about 70,000 technical reports each year.

Figures 17.1a and 17.1b show the title pages of a government report. Figure 17.1a is the general title page; Figure 17.1b the technical report standard title page. Figure 17.2 shows a table of contents.

For Journal Publication

Some research done in American laboratories is proprietary—that is, it is privileged information belonging only to the company and those to whom the company makes disclosures. Results appear only within confidential reports that are circulated exclusively within the corporation.

Most research, though, is not confidential; if the work is acceptable, it appears in a specialized journal. Example 17.1 shows the opening of a published research report.

For Internal Corporate Use

At virtually every level within a company, reports are written to document work.

These are among the many types of corporate reports:

• *Service Reports.* A person, team, or even department is asked to perform a service—inspect a transformer, analyze a tax problem, troubleshoot a production snag. When the job is done, the team writes a report.

• *Research Report.* In addition to periodic reports, researchers file a final document in which they report their findings. If the research is proprietary, the report remains in-house. If the researchers are permitted to do so, and their work is competitive, they may seek to publish the work in a refereed journal.

• *Literature Review.* Because technical fields change so rapidly, periodic reports on the literature are essential. Today these are usually done with the aid of a computer search. The writer sorts, evaluates, and presents the information in a review.

• *Topical Reports.* Workers present a review or analysis of a topic on which they are working that has commercial potential.

Computer Science
and Technology

NBS Special Publication 500-147

Guidance on Requirements Analysis for Office Automation Systems

(Update: NBS SP 500-72)

Lynne S. Rosenthal
Elizabeth G. Parker
Ted Landberg
Shirley Ward Watkins

Center for Programming Science and Technology
Institute for Computer Sciences and Technology
National Bureau of Standards
Gaithersburg, MD 20899

U.S. DEPARTMENT OF COMMERCE
Malcolm Baldrige, Secretary

National Bureau of Standards
Ernest Ambler, Director

Issued March 1987

Figure 17.1a General title page for government report. Source: U.S. Department of Commerce, National Bureau of Standards.

U.S. DEPT. OF COMM.
BIBLIOGRAPHIC DATA SHEET *(See instructions)*

1. PUBLICATION OR REPORT NO.	2. Performing Organ. Report No.	3. Publication Date
NBS/SP-500/147		March 1987

4. TITLE AND SUBTITLE
Computer Science and Technology:
Guidance on Requirements Analysis for Office Automation Systems
(Update: NBS SP 500-72)

5. AUTHOR(S)
Lynne S. Rosenthal, E. Parker, T. Landberg, S. W. Watkins

6. PERFORMING ORGANIZATION *(If joint or other than NBS, see instructions)*
NATIONAL BUREAU OF STANDARDS
DEPARTMENT OF COMMERCE

7. Contract/Grant No.

8. Type of Report & Period Covered
Final

9. SPONSORING ORGANIZATION NAME AND COMPLETE ADDRESS *(Street, City, State, ZIP)*
Same as item 6.

10. SUPPLEMENTARY NOTES
Library of Congress Catalog Card Number:87-619807

11. ABSTRACT
This report is designed to help managers maximize the benefits to be achieved through the application of office automation technologies. It presents a systematic planning method which will guide the manager to technology solutions which can improve the quality, efficiency, or effectiveness of an organization's products or services. Planning for office automation is accomplished through a requirements analysis study. The study can be initiated in response to installing office automation systems in a non-automated office or acquiring additional systems for an existing automated environment. This report provides guidance in the overall process of determining requirements for office automation systems.

12. KEY WORDS *(Six to twelve entries; alphabetical order; capitalize only proper names; and separate key words by semicolons)*
Benefit cost analysis, computer based office systems, office automation, post implementation audit, requirements analysis, system design model

13. AVAILABILITY
[XX] Unlimited
[ ] For Official Distribution. Do Not Release to NTIS
[XX] Order From Superintendent of Documents, U.S. Government Printing Office, Washington, D.C. 20402.
[ ] Order From National Technical Information Service (NTIS), Springfield, VA. 22161

14. NO. OF PRINTED PAGES
105

15. Price

Figure 17.1b Technical title page for government report. Source: U. S. Department of Commerce, National Bureau of Standards.

TABLE OF CONTENTS

Figure 17.2 Table of contents from government report. Source: *Microcomputers in Transit: A Hardware Handbook*, Final Report, July 1984, Urban Mass Transportation Administration.

Squeeze Film Forces in a Magnetic Shaft Suspension System

Viscous squeeze film forces were studied as they pertain to a magnetic shaft suspension system (magnetic bearing). A nonrotating magnetically suspended shaft generates hydrodynamic fluid forces as it oscillates radially within a close fitting concentric cylinder. Experimental results from a scaled model verify the theoretical analysis. The study includes the finite and infinite length cases (1/4<L/D<8) for large diameter to clearance ratios (D/c>50). The viscous squeeze film forces increase the phase margin in the magnetic bearing closed-loop control systems used in state-of-the-art cryogenic refrigerators.

L. D. KNOX Philips Laboratories, North American Philips Corp., Briarcliff Manor, N.Y. 10510

Introduction

The interest in squeeze film forces between concentric cylinders arose with the realization of a magnetic bearing used in state-of-the-art cryogenic refrigerators. A magnetic bearing is comprised of four electromagnets and four position transducers both evenly spaced around the circumference of the shaft. A closed-loop control system acts to minimize the radial position error by energizing the appropriate electromagnets. Two of these bearings radially position the shaft within a close fitting concentric cylinder (25 μm radial clearance) to form a clearance seal to reduce axial leakage through the bearing. The stability of the radial closed-loop control system was found to depend upon the clearance seal dimensions. Thus an analysis was performed to quantify the effect of the clearance seal upon the closed-loop control system.

Prior to this analysis, despite similar work performed on rotating systems [4, 7, 8, 13], the state-of-the-art of squeeze film forces between non-rotating concentric cylindrical elements was limited to analytical work. The solution of the infinitely long bearing is a text book problem stemming from Reynold's equation which is a special case of the Navier-Stokes equation [10]. A numerical solution was obtained for the finite length case in 1961 by D. F. Hays of General Motors [6]. A. Cameron later presented a first order approximation to the results of this analysis. The need remains, however, to verify these models experimentally and to apply the results to the specific case of interest.

Theoretical Analysis

Figure 1 shows how squeeze film forces act to retard relative radial motion between the shaft and cylinder. As the shaft approaches the cylinder wall, the displaced fluid must travel circumferentially around the shaft. The circumferential pressure gradient required to force the fluid through the annulus acts over the surface area of the shaft to retard its motion. For the geometries, fluid properties, and frequencies of interest, the flow is purely viscous. Hence the fluid forces can be modeled as a damping element in the magnetic bearing control system model. This readily explains the sensitivity of the closed-loop stability upon the clearance geometry since it directly affects the magnitude of the system damping.

Example 17.1 Excerpt from journal report written by corporate scientist.
Source: *Journal of Tribology*, Vol. 106, October 1984.

These highly focused reports on a service or topic, on literature, or on research, are a staple in most departments that offer technical services. In many instances, the report *is* the service.

For the Public

Occasionally the audience is broader than a journal's readership or the names on a corporate distribution list. For example, the Rogers Report on the Challenger (*Report of the Presidential Commission on the Space Shuttle Challenger Accident*) was written not only for technical experts, but for the public. So, too, was "Wind Machines," prepared originally for the National Science Foundation and then distributed widely through the U.S. Government Printing Office (Figure 17.3).

IS THERE A "BASIC REPORT FORMAT"?

No single, all-purpose "basic report format" exists, with one iron-clad list of components and sequence. Given the variations in audience needs and expectations, you would be unlikely to find a single format suitable to a universe of readers.

While no one absolute form exists, the writer can often take advantage of standard components or elements of the report, adjusting these parts according to the audience. Because any one of these standard elements may be crucial when you are building an effective report, each element is discussed separately in this chapter. That doesn't mean you will need every part, every time. For instance, the list of terms becomes very important in a report for the general public; it is unnecessary in a proprietary report where the readers are deeply involved in the project. Another example is the letter of transmittal, which is appropriate when the report is mailed to an outside reader, but inappropriate when the report circulates internally within a company.

The emphasis in a report may fall anywhere, for emphasis shifts according to the writer's purpose and the reader's needs. Example 17.2, for instance, gives the U.S. Environmental Protection Agency (EPA) guidelines for reports written on tests of pesticide residue in crops. The people who write these reports are chemists, biologists, or other technical professionals who work for manufacturers of pesticides. The readers are staff at the EPA who monitor any traces of pesticide in crops in the light of potential danger to the public.

While the parts of this report assume a traditional order, the emphasis is in the "Materials/Methods" section. It is by far the most

Text continues on page 423

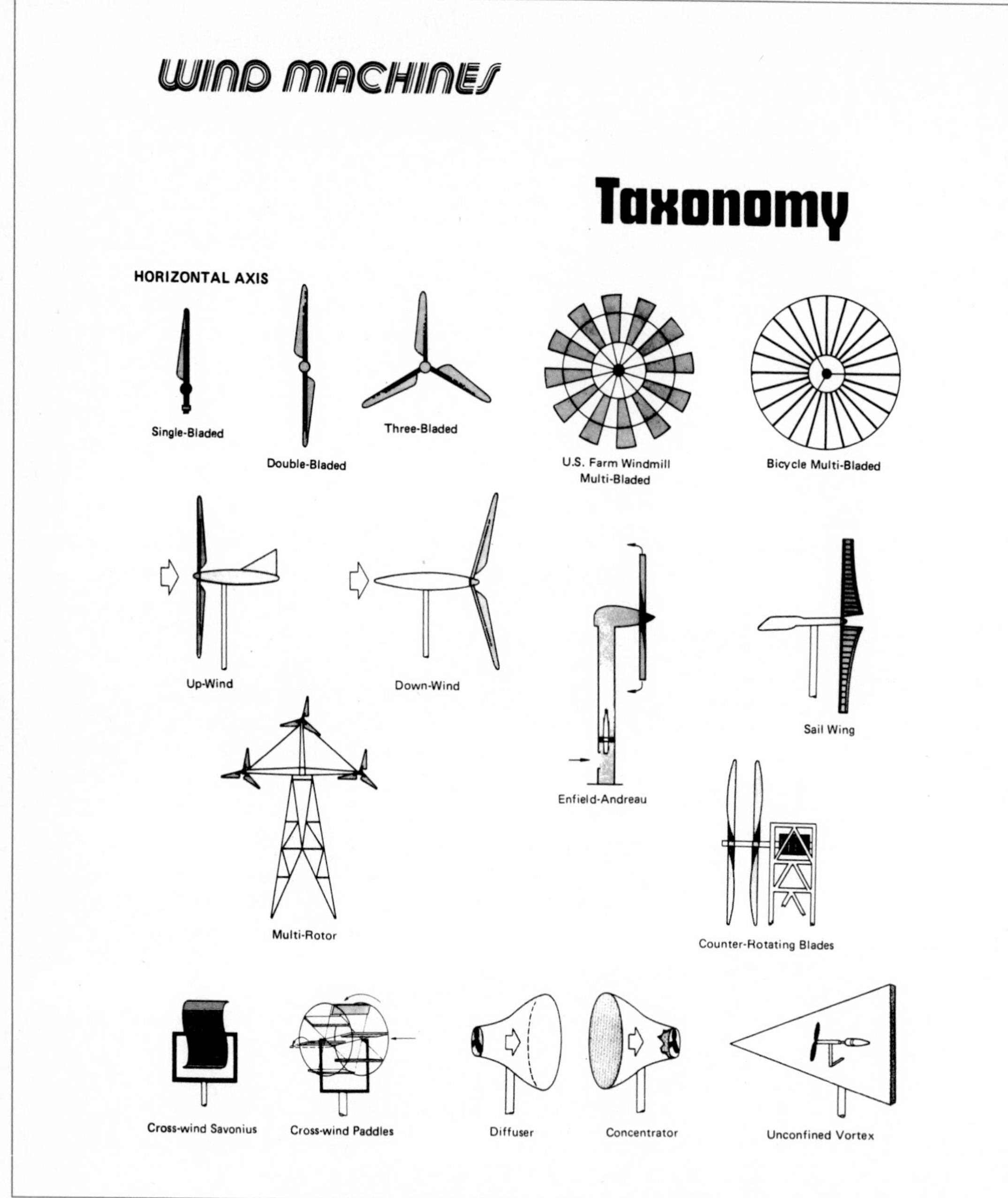

Figure 17.3 Excerpt from a report, "Wind Machines" (October 1975), prepared for the National Science Foundation and then distributed to the public through the U.S. Government Printing Office.

Example 17.2 Guidelines for technical reports written for a specialized audience. Note emphasis on the "Materials/Methods" section. In "Analytical Methods, Magnitude of the Residue: Crop Field Trials and Storage Stability Study," U.S. Environmental Protection Agency, Report PB86–248192.

GUIDELINES

TITLE/COVER PAGE

Title page and additional documentation requirements (i.e., requirements for data submission and procedures for claims of confidentiality of data) if relevant to the study report should precede the content of the study formatted below. These currently proposed requirements are described in 49 FR (188) 37596 (9/26/84).

TABLE OF CONTENTS

I. SUMMARY/INTRODUCTION

A. *Scope* (suitable matrices) and source of method (e.g. Pesticide Analytical Manual, company reports, etc.); and

B. *Principles* of the analytical procedure, including identification of the chemical species determined and the limits of detection and sensitivity.

II. MATERIALS/METHODS

A. *Equipment* (list and describe);

B. *Reagents and standards* (list and describe source and preparation);

C. *Analytical procedure* (detail in a stepwise fashion, with special emphasis on reagents or procedural steps requiring special precautions to avoid safety or health hazards):

1. Preparation of sample;

2. Extraction [demonstrate efficiency, if relevant (e.g., dry crop substrates, "bound" residues, etc.)];

3. Fortification, if applicable (i.e., during method validation runs];

4. Clean-up; and

5. Derivatization (if any).

D. *Instrumentation* [to include information on]:

1. Description [e.g., make/model, type/specificity of detector(s), column(s) (packing materials, size), carrier gas(es), etc.];

Example 17.2 (*continues*)

2. Operating conditions [e.g., flow rate(s), temperature(s), voltage, etc.]; and

3. Calibration procedures.

E. *Interferences*(s) (describe tests):

1. Sample matrices;

2. Other pesticides;

3. Solvents; and

4. Labware.

F. *Confirmatory techniques* (describe);

G. *Time required for analysis* (to carry a sample/set completely through the analytical procedure, including the determinative step);

H. *Modifications or potential problems*, if any in the analytical method(s) (detail circumstances and corrective action to be taken);

I. *Methods of Calculation* (describe in a stepwise fashion):

1. Calibration factors; and

2. Analyte in sample.

J. *Other* (any and all additional information the petitioner considers appropriate and relevant to provide a complete and thorough description of residue analytical methodology and the means of calculating the residue results).

III. RESULTS/DISCUSSION (describe expected performance of method)

A. *Accuracy* (expected mean and range of recoveries);

B. *Precision;*

C. *Limits of detection and quantification* (provide definition);

D. *Ruggedness testing*, if performed; and

E. *Limitations.*

```
 IV. CONCLUSION

     Discuss applicability of analytical procedure for
     measuring specific test compound(s) in various
     test substrate(s), ready availability of equip-
     ment, interference(s), etc.

  V. CERTIFICATION

     Certification of authenticity by the study direc-
     tor (including signature, typed name, title, af-
     filiation, address, telephone number, date).

 VI. TABLES/FIGURES

VII. REFERENCES

VIII. APPENDIX(ES)

     A. Representative chromatograms, spectra, etc. of
        reagent blanks, solvent blanks, reference
        standards, controls, field samples, fortified
        samples, etc. (cross-referenced to individual
        field trial study reports);

     B. Reprints of published and unpublished litera-
        ture, company reports, letters, analytical
        methodology, etc., cited (or used) by the
        petitioner/registrant (unless physically lo-
        cated elsewhere in the overall data report, in
        which case cross-referencing will suffice);

     C. Other (any relevant material not fitting in
        any of the other sections of this report).
```

detailed of the sections. Why? The EPA seeks reliable analytical methodologies for data gathered on pesticide residue. To insure reliability, they need a description of the method that is sufficiently detailed to allow replication in the EPA's laboratory (should that be necessary). This goal is reflected in the structure of the report. While the methodology section is lengthy, other parts of the report are shorter than they might have otherwise been, if the report were written for a different audience or purpose. For instance, there is no separate abstract. Instead, the abstract and the introduction are combined in Part I as a "Summary/Introduction" that gives the *scope* of the study and the *principles* of the procedure.

LOGIC, LANGUAGE, AND LAYOUT

Like any other piece of technical writing, the report should be readable for its audience. This readability will be reflected in the logic and organizational pattern, the language, and the layout:

• *Logic and organizational patterns.* Establish the main points before introducing detail. To do this effectively, start with an abstract that summarizes the findings. This will give the reader the necessary scaffold. Use the introduction to sum up the problem and its significance. Divide the text into major supporting sections. Define all key terms. Use examples and comparisons to bring abstractions to life. Begin each section with a summary or encapsulating statement that previews the organization of the section.

• *Language.* Address the reader directly, avoiding nominalizations, passives, redundancy, inflated diction, and unnecessarily specialized vocabulary. Check for grammatical correctness in modification, subject-verb agreement, pronoun antecedent, and other trouble spots. Keep the body of the report as lean as possible by placing supporting materials in the appendixes. Note that in Example 17.2 the appendixes are used for supporting materials such as chromatograms, spectra, reprints of published literature, and letters.

• *Layout.* Use headings and subheadings to divide and distinguish main points. Take advantage of margins, tabular layout, boldface, and italics to divide and distinguish the text. Use tables and figures to display data compactly and vividly.

ELEMENTS IN THE REPORT

Reports divide into three parts:

Front matter
Introduces problem, significance
Gives scope, limits
Summarizes related work in field
Concludes with roadmap or guide to contents that follow

Body
Presents major arguments and supporting evidence
Analyzes evidence supporting arguments
Draws conclusions and implications

Back Matter
Supports body with references, attachments, glossary

Within these three major divisions, the writer may choose subdivisions, depending on the audience. Many formal reports follow this sequence, with adjustments made for the audience and nature of material:

- Front Matter
  - Letter of Transmittal or Distribution List
  - Title
  - Table of Contents
  - List of figures/illustrations
  - List of tables/charts
  - Abstract
- Body
  - Introduction
    - Statement of Purpose
    - Conditions under which work was done
    - Background
    - Limitations
    - Acknowledgement of cooperation
    - Summary of previous work in field
    - Notes on most important prior publications
  - Glossary (may also appear at end of Back Matter)
  - Methods and Materials
  - Procedure
  - Results
  - Discussion
  - Conclusions
  - Recommendations, Implications
- Back Matter
  - References
  - Bibliography
  - Appendixes
  - (Glossary)

ANATOMY OF KEY ELEMENTS

Letter of Transmittal

See Chapters 10, 13, and 14 for examples and discussion.

Title

"How long, oh Lord?" is the question in *Ecclesiastes*. People have asked the same question about titles, as they look at spectacular 17- and 18-word specimens.

Box 17.1

FOCUSING THE TITLE

Notes on New Operation →

Comments on New Operation →

New Operation: Selective Posterior Rhizotomy →

Selective Posterior Rhizotomy : Neurosurgical technique to reduce spasticity in youths with cerebral palsy.

Actually, long titles are one case where prolixity is not the villain. There are sensible reasons for longer titles. The writer needs to get the key words into the title for indexing and retrieval; if the writer studies poly(2,3-dimethylbutadiene), then "poly(2,3-dimethylbutadiene)" must be in the title.

Indeed, considerable thought should go into the title, to shape it into a useful preview of the report's contents. The sensible author will try to include as much information about the objective, the findings, and the implications as possible.

Many writers have trouble with understatement: They claim too little in the title. A spare "Comments on FT-IRs," for instance, strikes an elegant but ineffective note, for it fails to epitomize the findings. There is also the opposite problem—claiming too much, something the scrupulous will avoid.

Box 17.2

FOCUSING HEADINGS

Use headings that summarize or preview the text, rather than merely labeling parts.

Procedure → Use of Carbolic Acid with Compound Fractures.

The Table of Contents

There are many standard formats, organized either by size and style of type, or numerically, or both.

Figure 17.4 shows a table of contents in which the divisions are emphasized by boldfaced, slightly larger type. Figure 17.5 shows a table of contents in which the divisions are set off both by capitals and lower case headings, and numbering.

The Executive Summary (also known as the Executive Abstract, Management Summary, Management Synopsis, Synopsis, Summary)

This is a prefatory summary that has grown in popularity as more and more people join the distribution list for a technical report.

Many of the readers of a report lack a strong technical background in the subject, yet they make vital decisions based on the report. Therefore this broader version of the engineering abstract has grown increasingly important. It is one of the chief ways the writer has to adjust for audience background.

The executive summary, within a few paragraphs, should tell what was done, and why; what the results were; and why they matter.

Many people visualize the executive summary as a three-tiered section divided into these parts:

FINDINGS
RECOMMENDATIONS/IMPLICATIONS
(commercial, technical, scientific)
BACKGROUND

In Example 17.3 all four tiers fit on one page. (For other examples, see Chapter 15.)

The Abstract

The abstract should give the results of the work. Often the abstract is printed independent of the text, and is the only basis for another reader's decision whether to send for the original. It is imperative that it make clear the author's objective and findings. (See also Chapter 15.)

Will you need both an abstract and an executive summary? That depends on your audience. Usually an abstract is sufficient, but in companies where the readership is highly diversified, you will want to include an executive summary that reduces the technical detail

Figure 17.4 Table of contents. Divisions are emphasized by boldfaced, slightly larger type. Courtesy Aldus.

Contents

Figure 17.5 Table of contents. Divisions are emphasized by capital and lower-case headings, and by numbers.

CONTENTS

Example 17.3

EXECUTIVE SUMMARY

Visit of Dr. C. Alair 26 August 1988

Purpose

Dr. Alair was asked to consult on the fire and explosion hazards involved in the high-temperature, high-pressure operation of the benchscale methanation reactor.

Background

The benchscale methanaton reactor was designed to collect kinetic data to support the Wylie Gasification modeling effort. The reactor will be used to react flammable gases at temperatures and pressures that have never been studied here, and which are at the limits of the materials. The consultation was requested to ensure the safe operation of the reactor.

Conclusions

1. All gas mixtures in the reactor system will be combustible.
2. Gas mixtures will not pose an explosion threat if the leak rate is less than or equal to the normal flow rates in the system. However, CO gas can cause asphyxiation even at the low flow rates.
3. In the event of a reactor explosion, the shock wave will not be dangerous, but the fragments blown off the reactor will.

Recommendations

1. A CO monitor should be put in the work area to warn personnel in the event of a leak.
2. Flow limiters should be used on the feed tanks to control the flow if there is a leak.
3. Shielding should be built of 1/4" steel to surround the reactor and extend two feet above it and one foot below it to deflect fragments in the event of an explosion.

and specialized vocabulary, stressing instead the applicability of the findings.

Body of the Report

The Introduction. In a few paragraphs, the introduction should provide an overview and necessary background by first orienting the reader and then connecting the new work to older work in the same area. (Example 17.4).

The introduction usually includes a short statement of the objective, any limiting conditions, and references to related work in the field. It answers the questions of background: What problem was addressed? Why was the problem significant? Who else has worked on this problem? Sometimes it ends with a summary statement and a roadmap indicating the scope or division of the paper that follows.

Some introductions include definitions of terms or a glossary. In Figure 17.6, the glossary appears as part of the front matter. Glossaries may also appear at the end of the report.

The order of the introduction—with its background to the problem, significance of the problem, related literature, and summary statement or roadmap—may seem modern. Actually, it is a classically rooted part of the report. Figure 17.7 shows an introduction from a famous nineteenth-century report of an investigation.

Procedural Information

A "methods" section is obligatory in reports of experiments, unless the details are published elsewhere, in which case clear references are adequate. If either materials or procedure is new, however, the writer's goal should be to describe the new in such detail that another researcher can duplicate the experiment solely by reading the text. (See Example 17.2 for typical guidelines for the procedure section of an analytical report.)

You might use a procedure section in a troubleshooting report where you are describing how you went about solving a problem, or in a feasibility study, where you show the steps you took to evaluate the data. Many types of formal reports, however, are outside the realm of experiments, troubleshooting, or feasibility studies, and require no procedural discussion. Reports, for instance, that depend on interviews and published sources will require neither a "methods and materials" nor a "procedure" section.

The Use of Hypnosis to Enhance Recall

What is the problem?

Why is it significant?

Who else has studied this problem?

The introduction often ends with a summary statement of the findings

The increased use of hypnosis in forensic investigation has become controversial (1). Although numerous case reports attest to the utility of hypnosis in enhancing the recall of the eye witness (2). controlled studies have produced conflicting results. Some studies have failed to demonstrate hypnotic hypermnesia, whereas those that have (3) have not reported errors in a systematic way nor controlled for the natural hypermnesic effects that can be achieved through repeated testing (4). Still others (5) have found that hypnotized subjects are susceptible to leading questions. Although scientists are wary of the reliability of forensic hypnosis, police investigators are lobbying to sanction its use in criminal investigation and the judiciary is seeking evidence on which to base legal decisions. The relation between hypnosis and memory enhancement needs to be clarified.

We now report that any pressure to enhance recall beyond the initial attempt may increase the number of items recalled, but increase the number of errors as well. The use of hypnosis exaggerates this process, particularly for those with hypnotic ability. When hypnotized, the highly hypnotizable subjects recalled twice as many new items as controls but made three times as many new errors.

Example 17.4 Sample introduction. Source: *Science*, Vol. 222, 14 October 1983, p. 184. Copyright 1983 by the AAAS.

Results

The division and analysis of the results depend on the nature of the material. If, for instance, you are presenting the results of an experimental study, you may present the data in tables, and pick out the most striking or important items for discussion. If you are presenting a feasibility study analyzing two separate telephone systems for possible installation, you might discuss the capabilities of System A, then System B in the narrative, and support the discussion with a tabular display of contrasting capabilities of the two systems. If you are preparing a final report on a statewide program to monitor inspections of automobile inspection stations, you may discuss the results county by county, following the geographical divisions established by the state. If your report is a comparative study of two possible printing systems, you might present a comparative, feature-by-feature analysis of the two systems, using the text to discuss the

Figure 17.6 This list of definitions appears in the front matter of a technical report. Glossaries may also appear at the end of a report. Source: "A Review and Evaluation of the Langley Research Center's Scientific and Technical Information Program," NASA Technical Memorandum 83269, April 1982.

The results of the study were reported without regard to economic consideration. For this reason, before implementing changes based on the results, cost/benefit analyses should be performed to ensure that reader benefits will outweigh the cost to the producer.

The study spanned the period from December 1980 to November 1981.

DEFINITION OF TERMS

Academic organizations. Academic organizations were interpreted as institutions of higher education or consortia of these institutions.

Back matter. Back matter of a technical report was defined as the section immediately following the body or text of a technical report. Supplemental materials such as appendixes, index, references, and bibliography appear in this section.

Body or text. The body or text of a technical report was defined as the section immediately following the front matter. The development of the central theme of the report including the introduction; the investigative, analytical, or theoretical material; the description of the research; the results and discussion; and the conclusions, appears in this section.

Congeniality. Congeniality was used to describe the subjective impression that typographic arrangements convey, or as commonly stated, "judging a book by its cover." Factors such as reader preference and appropriateness of a type face determine the congeniality of printed material (Zachrisson, 1965).

Current practice and usage. Current practice and usage were defined as the methods or modes for report format, organization, and presentation employed by the selected technical reports surveyed and analyzed as part of this study.

Diagrams. Diagrams were defined as the group of visual representations which are primarily concerned with displaying relationships of contingency (computer flow charts and verbal algorithms), mechanism (schematic and circuit diagrams), and subordination (management organization charts and charts of biological orderings).

Experimental/theoretical literature. Experimental/theoretical literature was defined as periodicals, literature, reports, and research containing empirical results specifically concerned with or related to the organization, the language, and the presentation components of the technical report.

Figure 17.6 (*continues*)

Figures. The term figures in this report was used to mean the entire family of visual representations other than the tables. Major classifications are charts, graphs, diagrams, and illustrations.

Front matter. Front matter of a technical report was defined as the section immediately preceding the body or text. Included in this section are the forword, preface, and contents. This section is related only to the writing of the technical report itself and is not essential to the subject matter.

Government organizations. Government organizations were defined as state and federal agencies engaged in research, regulatory, or service functions.

GLOSSARY

ABDEN	Aerospace, Basic Research, Defense, Energy, and Engineering
AERO	Aerospace
ANSI	Amercian National Standards Institute
APA	American Psychological Association
BRS	Bibliographic Retrieval Services
Chicago	*Chicago Manual of Style*
COSATI	Committee on Scientific and Technical Information, U.S. Federal Council on Science and Technology
DoD	Department of Defense
DTIC	Defense Technical Information Center
ERIC	Educational Resources Information Center
GBC	General Binding Corporation
GPO	Government Printing Office (U.S.)
LaRC	Langly Research Center
LISA	Library and Information Science Abstracts
NACA	National Advisory Committee for Aeronautics
NAS	National Academy of Sciences
NASA	National Aeronautics and Space Administration
NMI	NASA Management Instruction
NTIS	National Technical Information Service
R&D	Research and Development
RECON	Remote Console Interactive Computer System
SATCOM	Committee on Scientific and Technical Communication, National Academy of Sciences National Academy of Engineering
SDC	System Development Corporation
SR	Scientific Report

Figure 17.7 The order of the introduction has classical roots. This nineteenth century report by Joseph Lister begins with a statement of the problem and its significance, and ends with a summary statement of the findings.

ON THE ANTISEPTIC PRINCIPLE IN THE PRACTICE OF SURGERY

In the course of an extended investigation into the nature of inflammation, and the healthy and morbid conditions of the blood in relation to it, I arrived, several years ago, at the conclusion that the essential cause of suppuration in wounds is decomposition, brought about by the influence of the atmosphere upon blood or serum retained within them, and, in the case of contused wounds, upon portions of tissue destroyed by the violence of the injury.

To prevent the occurrence of suppuration, with all its attendant risks, was an object manifestly desirable; but till lately apparently unattainable, since it seemed hopeless to attempt to exclude the oxygen, which was universally regarded as the agent by which putrefaction was effected. But when it had been shown by the researches of Pasteur that the septic property of the atmosphere depended, not on the oxygen or any gaseous constituent, but on minute organisms suspended in it, which owed their energy to their vitality, it occurred to me that the decomposition in the injured part might be avoided without excluding the air, by applying as a dressing some material capable of destroying the life of the floating particles.

Upon this principle I have based a practice of which I will now attempt to give a short account.

The material which I have employed is carbolic or phenic acid, a volatile organic compound which appears to exercise a peculiarly destructive influence upon low life forms of life, and hence is the most powerful antiseptic with which we are at present acquainted.

main advantages and disadvantages, and the tables to provide detailed specifications.

If you are presenting a report on an experiment, the data should, of course, be presented in their rawest form. If this can't be done, the manipulations on the data should be described clearly so that the reader can recover original values. Those who work from notebooks should indicate the page references, should the need arise to trace the work from notebook to final published version.

Illustrations Supporting Data

Graphs These are excellent to show relationships and trends, especially if followed by tables that give the data. Of course, it is also possible to make graphs large enough to show the actual values, although this is often impractical.

Variables should be clearly distinguished on the graph if the author seeks to show a comparison.

Make sure the lettering is three times the size you want it to be in its final appearance. Graphs are usually reduced to fit available space. The smaller letters (such as the units of measurements on the ordinate and abscissa) often virtually disappear from view unless the author is careful.

Figure 17.8 shows a graph in which the variables are clearly distinguished. The labeling, however, is somewhat small.

Tables These are usually a great aid to both reader and writer—by far the best way to display numerical data.

Data are poorly displayed when they are simply set out line by line in the narrative. A table, by contrast, will allow you to present the same information concisely and vividly.

Note that tables use titles, rather than captions. Number titles sequentially, and cite each on first reference.

Line Drawings Indispensable for depictions of laboratory equip-

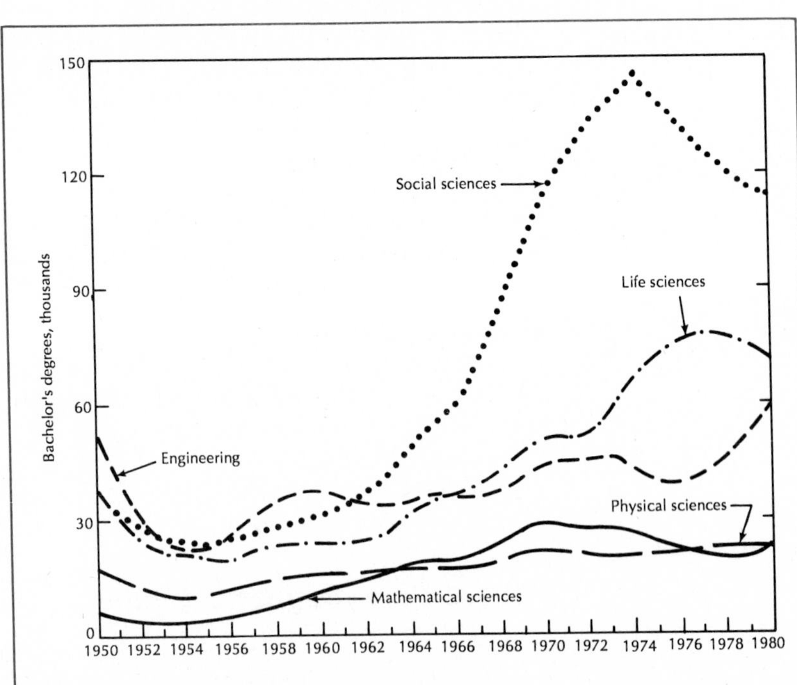

Figure 17.8 Variables are distinguished clearly in the graph, although the labeling is somewhat small.

ment, they are a great aid to the reader, although often expensive to prepare.

Photographs These are often of poor quality, and therefore ineffective. To reproduce well, photographs must be black and white, high-contrast studio quality. Unless such photographs are available, the author may do better to consider line drawings, especially for the description of equipment.

(For a more detailed discussion of visuals in support of data, see Chapter 5.)

TECHNICAL WORD PROCESSING

Technical word processing is beginning to make a substantial difference in the mathematical materials of scientists, engineers, and technical professionals. Technical word processors are those that handle mathematical symbols, as well as other special symbols and diagrams.

Technical word processing permits those who work with equations to sit down to a keyboard and produce these symbols with proportionally spaced text, built-up equations, and other elegant touches formerly the exclusive province of the typesetter.

There are, as of this writing (1988), two ways of handling mathematical materials using technical word processing.

Markup languages TEX and UNIX's EQN/TROFF are examples of markup languages begun on mainframes and now available on personal computers. Markup languages are those in which the user explains the format; for example, instead of simply typing an equation with superscripts, the users says which items are superscripted. An example of TEX markup language (a) and output (b) is shown in Figure 17.9.

What-you-see-is-what-you-get (WYSIWYG) languages WYSIWYG is possible in business word processing, but it becomes far more complicated when the chore involves proportional fonts. The reality for technical word processing is usually a compromise—one, for instance, in which the user puts the superscript on the line above the equation, or uses a markup command, and then sees the result on the screen.

In one WYSIWYG system, Murray Sargent III has developed a linear equation notation. To obtain

$$\frac{\alpha_2}{\beta_2^3 + \gamma^3}$$

(a)

```
$$\leftline{$\displaystyle W^{3\beta}_{m_in_in_2}{p_1,p_2 =
U^{3\beta}_{m_in_1}{p_1,p_2}
+ \int_0^\infty {dp_ep_3^2 \over 8\pi^3}
\sum_n\sum-M\sum-{zalph_2}\sum_{beta_2}\sum_{n^\prime
\sum_{n^{\prime\prime\prime}}(-1)^$}$$
$$\times\left ({U^{33}_{m_in_1}{p_1,p_2} \over p_3^2 -
k^2}\right)z_{3m_in_1}j_n(p_3,p_2)\cdotp
a^{\alpha_23}_{mn(m_1-m)n^{\prime}n_2}
a^{\beta_2\beta}_{-mn(-m_1+m)n^{\prime\prime}n_2}
W^{\alpha_2\beta_2}_{(m_1-m)n^{\prime}n^{\prime\prime}}
(p_3,p_2).\eqno(110)$$
```

(b)
$$W^{3\beta}_{m_1n_1n_2}(p_1,p_2) = U^{3\beta}_{m_1n_1}(p_1,p_2) + \int_0^\infty \frac{dp_3p_3^2}{8\pi^3} \sum_n \sum_m \sum_{\alpha 2} \sum_{\beta 2} \sum_{n'} \sum_{n''} (-1)^m$$

$$\times \left(\frac{U^{33}_{m_1n_1}(p_1,p_2)}{p_3^2 - k^2}\right) z_{3m_1n_1} h_n(p_3,p_2) \cdot a^{\alpha 23}_{mn(m_1-m)n'n_2} \alpha^{\beta 2\beta}_{-mn(m_1+m)n''n_2} W^{\alpha 2\beta 2}_{(m_1-m)n'n''}(p_3,p_2). \quad (110)$$

Figure 17.9 An examples of TEX markup language (a) and output (b). Source: "Technical Wordprocessors for the IBM PC and Compatibles. A report by the PC Technical Group of the IBM PC Users Group of the Boston Computer Society," 1985.

you type

$$\alpha_2/(\beta_2^3 + \gamma^3)$$

The program measures the size of the numerator and denominator, centers one over the other, and prints. Sargent's system has a graphic preview, so that the user can look at built-up equations. ("Technical Word Processing," Murray Sargent III, *Optics News*, June 1985.)

The American Physical Society accepts electronic manuscripts. It uses TROFF/NROFF with EQN as its typesetting system. A booklet from the APS gives detailed instructions for the preparation of electronic manuscripts.

DISPLAY OF EQUATIONS

Short equations are treated as part of the text; equations that are set off, however, are numbered and referred to as "displayed."

There are some procedures for mathematical display you may find useful in preparing reports:

Punctuation Usually punctuation is dropped *after* a displayed equation, but retained before it.

According to Snell's law,

$$n_1 \sin \theta_1 = n_2 \sin \theta_2$$

Displayed Equations on the Same Line These are usually separated by space, but no punctuation. If the equations are joined by a conjunction (*or, and*) they are also set off by space.

$$r = \sqrt{x^2 + y^2} \qquad \text{and} \qquad \theta = \tan^{-1} \frac{y}{x}$$

Breaking Equations Break them before the operations sign. The sign appears at the beginning of the new line.

$$\begin{aligned} y = v + z = &\sqrt[3]{\{-\tfrac{1}{2} r + \sqrt{(\tfrac{1}{27} q^3 + \tfrac{1}{4} r^2)}\}} \\ &+ \sqrt[3]{\{-\tfrac{1}{2} r - \sqrt{(\tfrac{1}{27} q^3 + \tfrac{1}{4} r^2)}\}} \end{aligned}$$

DOCUMENTING REFERENCES

If your report draws on published materials or interviews, you will need to document your sources. There are four times in the report that typically call for citations:

The *introduction* often includes reference to others in the field who are working on or have worked on the problem.

The *procedure* often includes references to others whose work contributed to the methodology, or to earlier works by the author.

The *results* may include references to sources of equations, arguments, or outside data.

The *conclusions* sometimes include references to other workers who have reached similar or conflicting conclusions.

Which form of citations should you use? If you are writing a report that you wish to publish in a journal, you should follow the style sheet of publication you have in mind. You'll find that there are considerable variations in style between, for instance, the American Medical Association's *Manual for Authors and Editors,* and the American Chemical Society's *ACS Style Guide*. Here is a list of the most popular specialized guides.

American Chemical Society. *The ACS Style Guide:* A Manual for Authors and Editors. Washington, DC: American Chemical Society, 1986.

American Institute of Physics. *Style Manual*. New York: American Institute of Physics, 1978.

American Mathematical Society. *Manual for Authors of Mathematical Papers, 5th ed*. Providence, RI: American Mathematical Society, 1980.

American Psychological Association. *Publication Manual of the American Psychological Association*, 3rd ed., Washington, D.C.: American Psychological Association, 1983.

Council of Biology Editors, Committee on Form and Style. *CBE Style Manual*, 4th ed. Washington, DC: American Institute of Biological Sciences, 1978.

Engineers Joint Council, Committee on Engineering Society Editors. *Recommended Practice for Style of References in Engineering Publications*. New York: Engineers Joint Council, 1977.

Citation formats in these guidelines are somewhat specialized, with many small variations in style idiosyncratic to each professional field. For instance, style guides for physicists and chemists often dispense entirely with the title of the article in bibliographic references. Instead, the journal name, volume, date, and page suffice to identify the article.

Two widely accepted styles for acknowledging sources are the **author and date system** (e.g., Smith 1988) or a **numbering** system ([1] (1)[1]). Both systems follow with **endnotes** (placed at the end of the report) rather than **footnotes** (placed at the bottom of the page). Either system has wide acceptability both in business and among government agencies.

Author and Date System

In-Text Citation The author and date system is widely used for reports in the life sciences, social sciences, and business. To use it for a citation, give the author's name and publication date in the text within parentheses.

xxxxxxxxxxxxxxx(Garetz and Lombardi 1983)xxxxxxxxxxxx

Give page numbers or other divisions, such as section number, after the date. Separate volume and page number with a colon.

(Mitcham 1988, 121–25)

(Robinett 1987, sec. 24.5)

(Mitcham 1987, 3:26)

Box 17.3

AUTHOR AND DATE SYSTEM

Example of In-Text Citation

Many studies of game behavior are based on the explicit premise that cooperation, competition, and individualism can explain the choices of most laboratory subjects (e.g., Kuhlman & Marshello 1975; Kuhlman & Wimberly 1976).

Example from Reference List

Kuhlman, D. M. & Marshello, A. J. F. (1975). Individual differences in game motivation as moderators of preprogramming strategy in Prisoner's Dilemma. *Journal of Personality and Social Psychology, 32,* 922–931.

If the author's name appears in the sentence, give the year in parentheses.

> According to Morrison (1987) SEED is used as an optical switch.

For two authors, list both (Thomas & Thomas 1986); for three authors, list all three initially (Thomas, Thomas & Morrison 1986) and then abbreviate (Thomas et al. 1986).

Place page numbers directly after quoted material.

> Weitzenbaum (1986) comments that "the computer is a solution in search of a problem." (p. 122)

If the author has more than one work in the same year listed in the reference, use an a, b to distinguish items.

> Fitter (1986a) comments on a weakness in the study.

If no author is listed, use the name of the sponsoring agency, institution, corporation, or group.

> The study (Environmental Protection Agency 1987)

If you are citing a source that is not included on your reference list, give the details within parentheses:

Mr. Sinclair (telephone interview 1 August 1986)

References All references follow after the report in a list called **References** or **References Cited.** Type the list in alphabetical order, double-spaced, on a separate piece of paper. If there is no author's name, use the first significant word in the title.

1. The first line of each entry goes against the margin (flush left); succeeding lines in an entry are indented three spaces.
2. Capitalize only the first letter in the names of journal articles.
3. Underline or italicize the name of the journal and capitalize major words.
4. Underline or italicize the volume numbers.
5. When you cite a book, underline or italicize the title and capitalize only the first letter.

Numbering Systems

In-Text Citation Most publications in science, engineering, and medicine use a numbering system. To follow this system, you assign numbers to each reference and then place the numbers in the body of the report. As three slight variations in the number system exist (depending on the professional field), you should check guidelines if you are preparing a report in strict accordance with, say, the American Chemical Society style sheet. However, in most cases any one of the three variations is entirely correct. Simply be consistent within the report.

- **Parentheses or brackets.** Either may be used to enclose the reference.

When results do appear in print, authors often obscure them by overusing the passive voice, hedging, ducking responsibility, using impersonal language, and using lengthy or disrhythmic sentences. [2, 10]

The hypothesis that as much as 80 percent of cancer could be due to environmental factors was based on geographic differences in cancer rates and studies of migrants. (136)

- **On the line or above the line.** Reference numbers may be on the line or above the line (superscript).

> Claims for enclaves of longevity in which people are said to live into the 120s and 130s have been refuted by scientific scrutiny. In human beings, the documented life span is approximately 115.[2]
>
> We have made impressive gains in extending life expectancy at birth—from 45 years in 1900 to 71 years for men and 78 for women in 1983.[3] As a result, increasing numbers of Americans are reaching the 8th, 9th, and 10th decades of life.[2, 3]
>
> Several reports describe the value of magnetic resonance imaging (MRI) in visualizing the lesions of multiple sclerosis. [1, 3, 5, 8–10]

- **Ordered alphabetically or as cited.** Follow either method:

1. Order the references alphabetically by author or, if there is no author, first significant work in title. Then number the list and use these numbers for in-text citation.

> Although computer-based communications are impersonal, they appear to foster greater group intimacy and attachment than noncomputer media. [5, 14]

2. Number the references as they occur in the text. If you cite the reference again, use the same number.

> Claims for enclaves of longevity in which people are said to live into the 120s and 130s have been refuted by scientific scrutiny. In human beings, the documented life span is approximately 115.[2]
>
> We have made impressive gains in extending life expectancy at birth—from 45 years in 1900 to 71 years for men and 78 for women in 1983.[3] As a result, increasing numbers of Americans are reaching the 8th, 9th, and 10th decades of life.[2, 3]

Endnotes In-text citations are followed by a list called "References," "Works Cited," "Sources," or "Endnotes." Type the list double-spaced on a separate page.

Box 17.4

ENDNOTE FORM

1. G. Herzberg, *Spectra of Diatomic Molecules* (New York: Van Nostrand Reinhold, 1950), 23.

BIBLIOGRAPHIC FORM

Herzberg, G. *Spectra of Diatomic Molecules.* New York: Van Nostrand Reinhold, 1950.

These samples of endnote style are based on those in *The Chicago Manual of Style*. Here are representative examples:

Books

One Author

1. Richard K. Beardsley, *Village Japan* (Chicago: University of Chicago Press, 1959), 102.

Later Reference

2. Beardsley, 105.

Two Authors

3. John P. Dean and Alex Rosen, *A Manual of Intergroup Relations* (Chicago: University of Chicago Press, 1959), 303–04.

Later Reference

4. Dean and Rosen, 307.

Three or More Authors

5. Richard K. Beardsley, John W. Hall, and Robert E. Ward, *Village Japan* (Chicago: University of Chicago Press, 1959), 303–04.

Journals

6. Alan Grob, "Tennyson's The Lotus Eaters: Two Versions of Art," *Modern Philology 62(1964): 110.*

Popular Magazine

7. E. W. Caspari and E. E. Marshak, "The Rise and Fall of Lysenko," *Science*, 16 July 1965, pp. 275–78.

Edition

8. John W. Hazard, *The Soviet System of Government*, 4th ed. rev. (Chicago: University of Chicago Press, 1968) 25.

Newspaper
9. *New York Times,* 11 August 1965, p. 3.

Reference Books
10. *Encyclopaedia Britannica,* 11th ed., s.v. "Prayers for the Dead."
11. *Webster's New International Dictionary,* 2nd ed., s.v. "epistrophe."

If you wish to include a separate bibliography in addition to endnotes, use a format like the one for notes, except that

- the authors' names are reversed, as the bibliography is arranged in alphabetical order
- periods, rather than commas, separate the three elements author, work, and publication
- specific references such as pages, sections, and equations are omitted

DESIGN AND PAGE LAYOUT: Standardizing Report Format

Consistency of typographical elements is important in reports. Here the author has varied the style of subheadings, rather than making them consistent.

<u>**Introduction**</u>
Background to the Problem
METHODS

Consistent versions might look like this:

Introduction	<u>Introduction</u>
Background to the Problem	<u>Background to the Problem</u>
Methods	<u>Methods</u>

Those interested in developing their own guidelines may find the following checklist useful, particularly in the preparation of longer reports and manuals. Here consistency in format becomes very important, especially when the work is divided among team members, each with his or her own style. If you are using a word processor, you'll need to give the line position for headers, footers, hanging paragraphs, second- and third-level headings, and other indented items in the text. (Definitions and examples of typographical elements are in Chapter 5.)

REPORT FORMAT

TITLE

All Caps
Initial Caps Only
Center
Double-space Before;
Single-space After

Underscored
Boldface
Double-space Before & After
Other ______________

SECTION HEADINGS

On Line by Self
Begin Flush Left
Begin Pos. #_____
All Caps
Initial Caps
Double-space Before & After

Underscore
Boldface

FIRST-LEVEL SUBHEADINGS

On Line by Self
Begin Pos. #_____
Initial Caps

Boldface
Underscore

SECOND LEVEL SUBHEADINGS

On Line by Self
Text follows on same line
Begin Pos. #_____

All Caps
Initial Caps

LISTS

bullets

Space between bullet and text _____
Begin Pos. #_____

numbers

Parenthesis (1) (2) (3)
Begin Pos. #_____

Periods 1. 2. 3.

punctuation

for all lists, including simple one
for all lists with punctuation within
for any item with a verb

HANGING TEXT

If a list continues past one line, begin typing following line pos.
#______

BODY

Type size.
- narrative ______
- quoted material ______
- examples ______
- figures
  - callout ______
  - captions ______
  - legends ______

Pitch ______
Line Length ______
Margins
- Left Margin ______
- Right Margin ______
  - Justified left right
  - Ragged Right

Tab sets ______
Space between paragraph ______
Page numbers
- by section
- consecutively for entire report
- lower case roman for front matter
- header. Position ______.
- footer. Position ______.

Figures
- Figures may be placed sideways
  may not
- Rotate and reduce figures if necessary

EXAMPLES OF FORMAL REPORTS

One adaptation of the formal report is the feasibility report, in which the writer uses a set of criteria to weigh various solutions to a problem.

The level of engineering detail in a feasibility report is usually quite high. For instance, in *Principles of Water Resources Planning*, the author cites these activities needed to prepare a feasibility report for planning an urban flood control project (Example 17.5).

Example 17.5 The level of engineering detail in feasibility reports is usually high. Shown is a list of activities needed to plan one such report. From Alvin S. Goodman, *Principles of Water Resources Planning*, Prentice-Hall, Inc. 1984.

- Management and coordination
- Analysis of basic data —maps, aerial photos, stream flow, etc.
- Determination of needs for flood control
  —delineation of area affected by floods
  —determination of floodplain characteristics
  —forecast of future activities in affected area
  —estimates of existing and future flood damages
- Consideration of alternative ways of meeting needs
  —upstream reservoir
  —local protective works for urban area
  —nonstructural measures
- Studies for reservoir
  —selection of site
  —selection of capacity
  —selection of type of dam and spillway
  —layout of structures
  —analysis of foundations of structures
  —development of construction plan
  —cost estimates of structures
  —layout and cost estimates of access roads, bridges, communication facilities, construction camp, etc.
  —identification and estimates of requirements for lands, relocations, easements, etc.
  —consideration of reservoir for multipurpose use with pertinent analyses of layouts, capacities, costs, etc.
- Studies for local protective works—levees, walls, river shaping and paving, interior pumping stations
- Studies of nonstructural measures—land use controls, flood warning systems, flood proofing, etc.
- Formulation of optional combination of structural and nonstructural components for flood control project
- Economic analyses
- Financial analyses
- Assessments of environmental impacts—ecological, archeological, historical, geological, air and water quality, land sedimentation and erosion, etc.
- Sociological impact assessment
- Public information and participation programs
- Feasibility Report preparation

The planning activites shown by the list above constitute the work needed to prepare a *feasibility* report. The level of engineering detail for such a report is higher than for a *preliminary* report, but lower than for the design of a project.

The feasibility report for this project would include the parts listed in Example 17.6.

An example of the feasibility study process, which may be quite complicated, is shown in this chart prepared by the EPA for guidance on feasibility studies prepared under the Comprehensive Environmental Response, Compensation and Liability Act of 1980 (CERCLA) (Figure 17.10).

Determination of Need The process usually begins with a description of the problem and the need for a solution. Alternatives are proposed, with the understanding that each will be exposed to detailed technical and cost analysis.

Determination of Criteria The next step is usually to establish standards against which the alternative solutions will be measured. These criteria are usually formulated in relation to

- **Effectiveness.** How does each alternative alleviate specified problems?
- **Efficiency.** How cost-effective is each alternative for relieving the specified problems?

The feasibility report should include the following:

- Descriptions and analyses of the data
- Confirmaton of construction feasibility based on additional field and laboratory investigations, studies of project arrangements and indiviual project features, and analysis of construction methods (sources of construction materials, access to the project site, diversion of water during construction, etc.)
- Final recommendations for arrangement of project works, preliminary plans and other analyses to determine the principal qualities of construction, a reliable cost estimate, and discussions of the design criteria
- Construction schedule showing the timing and costs of project features
- Economic analyses of the project
- Financial analyses projecting the year-by-year costs, revenues, and subsidies for the project
- Plans for financing construction, and for managing the construction and operation of the project
- Institutional and legal requirements
- Assessments of the environmental and social impacts of construction and operation, and other impact studies if required

Example 17.6 Source: Alvin S. Goodman, *Principles of Water Resources Planning*, Prentice-Hall, 1984.

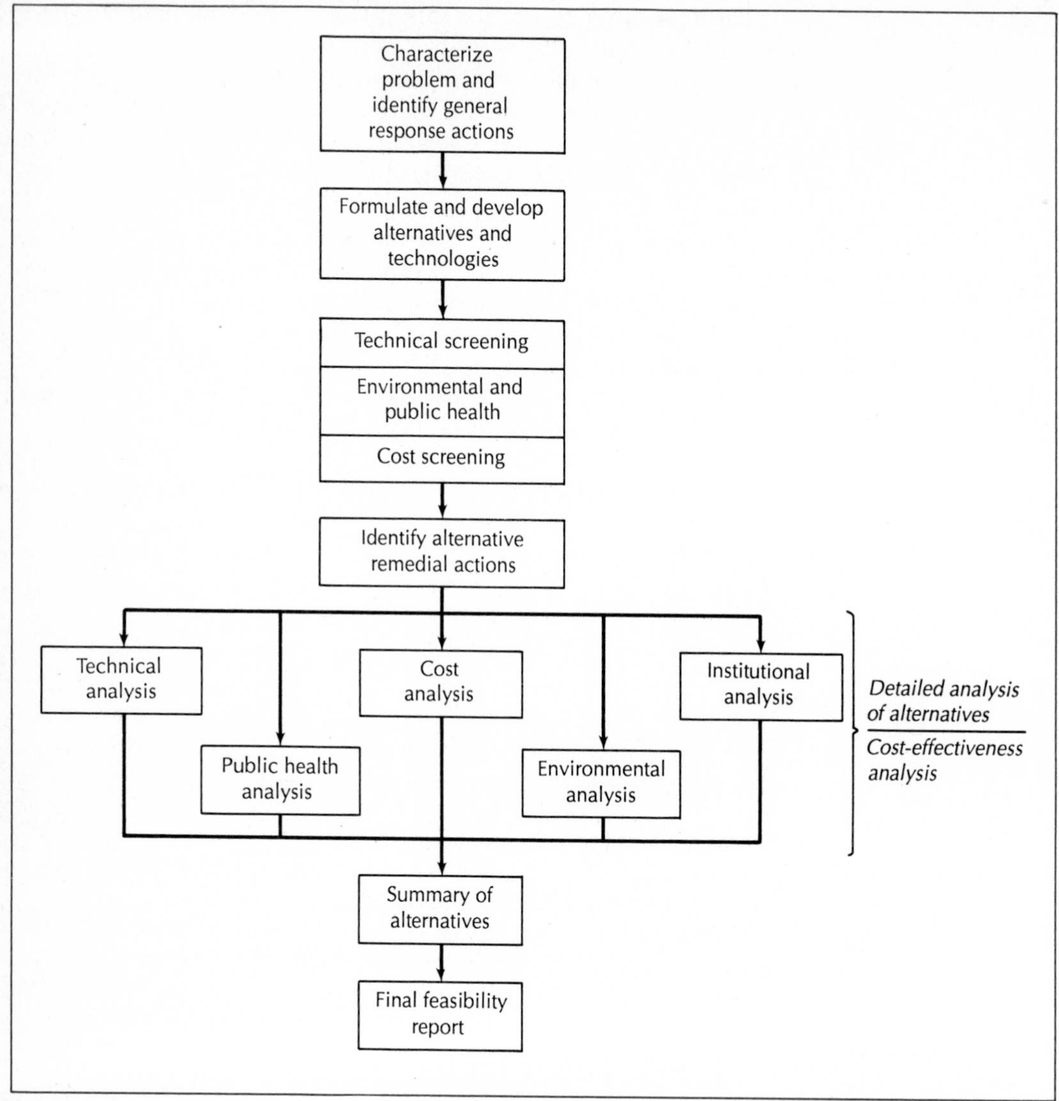

Figure 17.10 Source: "Guidance on Feasibility Studies Under CERCLA," U.S. Environmental Protection Agency, EPA 540G85003, June 1985.

Technical Analysis This includes specifications, estimated costs, advantages, and disadvantages for each.

Financial Analysis This includes a comparison between alternatives.

Conclusions and Recommendations Sometimes the authors draw conclusions; at other times, the writers simply present alternative solutions, without endorsing any one of them.

The final report usually includes these parts:

Title page

Table of contents

Lists of illustrations, maps, figures

Abstract

Glossary of Terms

List of Criteria

Introduction, including the purpose of the report, a statement of the problem addressed, and a statement of the scope or limits of the report

Data, organized by alternative solutions

Conclusions

Recommendations

References

Appendixes

A Sample Feasibility Report In this instance, the writer was preparing a feasibility report on the installation of printers for a large law office. The office had already determined there was a need; the existing copying system was inefficient, often failing to keep up with demand. The possibility of buying several kinds of printers was discussed, and the writer set to work.

The criteria established in this instance, the technical and financial standards against which each alternative would be measured, were

longevity

durability

productivity

versatility

health and safety

The writer then measured each alternative against the criteria, doing both a technical analysis—specifications, costs, advantages, disadvantages—and financial analysis. In this case, the author drew conclusions and made recommendations.

Some pages from this feasibility report appear in the following example (Example 17.7).

Example 17.7 Student-prepared feasibility study.

FEASIBILITY STUDY
SELECTION OF DESKTOP PRINTERS
for use of INCO OFFICE PERSONNEL

Abstract

I conducted research on the five basic types of desktop printers: dot matrix, daisywheel/thimble, ink-jet, thermal transfer, and laser. Only the laser technology offered the speed and the excellent print quality that is necessary for a large legal organization. I recommend that we purchase 49 Hewlett-Packard LaserJets at a ratio of one to three terminal stations. I further recommend that a printer with real typesetting capabilities, the Apple LaserWriter, be purchased to experiment with typesetting documents that are currently sent out for printing.

The Need to Update the INCO Printing System

INCO has been expanding every year and, with expansion, the amount of printed material generated by the firm has multiplied. The present printing system, which includes two Xerox 9700 printers, has become inadequate.

At times, the queue for printing causes extensive delays. There can also be a considerable delay until the job is delivered from the copy center on the 38th floor to one of our seven floors of offices. The situation is aggravated when one of the 9700 printers shuts down, which happens on the average of once a week.

In recent years, great strides have been made in desktop printer technology. Improvements in speed, clarity of print, and variety of print style make desktop printers a possible alternative to the 9700s. Their accessibility would be a great convenience.

In conducting research on desktop printers, I first interviewed INCO's Office Manager, Mr. Robert Hendel, to establish criteria, the capabilities in a printer necessary for the operations of a large law firm. Next, I looked at recent periodicals for information on the latest technology and the reviews of different models. Lastly, I spoke with salesmen, at two computer salesrooms located in the downtown area, familiar with supplying and servicing printers for legal and financial firms.

Criteria for Legal Printing

Following are the criteria for a printer that would serve our needs.

• High Quality of Print--typewriter or better

• Print Versatility--a variety of fonts (print type), points (size), and style (bold, italic, sub- and super-script, underlining)

• Speed--Mr. Hendel suggested that printing capacity be estimated based on a secretary's busiest week (c. 150 printed pages a week). If each printer networked with three terminals, it would need to produce 450 pages per week.

• Compatibility with IBM PC

• Networking capability

• Reliability--for high volume work

• Reasonable equipment cost--comparable to competitive machines

• Cost and ease of maintenance

Results

Research on desktop printers turned up five types of printers based on different technology: dot-matrix, daisywheel/thimble, ink-jet, thermal transfer, and laser. While dot-matrix machines outnumber all other types of printers, they are not suitable for legal work. Some models can produce letter-quality or near letter-quality print, but they do so by overstriking each word several times, an action that greatly reduces printing speed. The thermal transfer machines are based on a new technology, not fully tested for reliability. Further, their use of special materials--coated paper, transfer sheets, and ribbons--suggests costly and time-consuming maintenance. The ink-jet technology produces poor-quality text and also requires special paper, ribbons, and ink. It is a superior technology for color work but not a good choice for legal requirements.

The print quality is about equal for the NEC Spinwriter and the more expensive Hewlett-Packard LaserJet, with the LaserJet a bit clearer. Each machine can accept different type, point, and style elements and can network with three or more computer terminals. There are some significant differences, however. The Spinwriter is very noisy and needs a sound cover if it is to operate in a network of terminals. The added price of this, plus several other pieces of equipment necessary to its operation in this kind of office setting, bring the cost of machinery close to the cost of the LaserJet. In addition, the Spinwriter has more accessories to replace for maintenance and the cost of operation is more than the LaserJet, ($.045-.05 per page to $.03 per page). The most serious drawback to the Spinwriter, however, is its slow speed. At top speed, its output is 55 cps as compared with 300 cps for the LaserJet. Speed is a crucial criterion for law firm operations. INCO's needs, estimated by Mr. Hendel at 125 cps, are clearly out of the range of

Example 17.7 (*continues*)

the Spinwriter yet well within a LaserJet's capability of 300 cps (with leftover capacity for firm expansion).

Within the laser printer category, Apple has recently marketed the LaserWriter. This is a truly remarkable machine with the most powerful computer Apple has ever sold. It has 8 times more memory than most desktop computers. It is really a typesetting machine with a built-in software that can print four typefaces, in a variety of styles (bold, italic, etc.), and in different point sizes. It produces any form of written or graphic communication and the quality of the print is publication-quality. The LaserWriter is similar in size to the LaserJet and shares the same ease of maintenance with one, easily replaced, toner cartridge. One pays for the power, however, with a price almost double that of the LaserJet.

Recommendations

On the basis of its speed, clarity of print, acceptable variety of print style, price, and room for further INCO expansion, I recommended that INCO purchase 49 Hewlett-Packard LaserJets, one for every three IBM PCs.

In addition, I recommended that INCO purchase one Apple LaserWriter to experiment with typesetting the documents that are currently being printed by professional printers. If the firm can produce typesetting equal to that of a professional layout, the savings in time and cost, given the volume of INCO's operations, would be considerable. The Bond Department alone has estimated that a professional printer's fee to a client of $75,000 for typesetting and printing a 28-page prospectus (which includes overtime due to the time pressures associated with meeting an SEC filing date) could be done for $9,000 with the convenience of a computerized, laser typesetting system. The potential savings to INCO clients would be considerable.

NOTE: Attachments not shown in this example.

Example 17.8 is another example of a formal report, also written by a student. The report on moderately priced bicycles is cast as a buyer's guide.

Example 17.8 Student-prepared report.

A BEGINNER'S GUIDE TO BUYING A MODERATELY PRICED BICYCLE

TABLE OF CONTENTS

Example 17.8 (*continues*)

LIST OF TABLES

ABSTRACT

This report, derived from the current literature as well as from interviews with a retailer and competitive cyclists, teaches novice bicyclists how to purchase a moderately priced bicycle. While racers and long distance cyclists should focus their attention on ten-speed bicycles, commuters and back-road cyclists in need of a more rugged bike should focus on mountain bicycles. The report provides techniques for judging models of ten-speed and mountain bikes according to a checklist of criteria that includes the quality of test rides and components. Some components, such as the bicycle frame, should be judged by examining them. Other components that do not show obvious indicators of quality, such as derailleurs and cranks, are easiest to judge according to name brands. By preparing novice cyclists to make purchases that provide the best possible value and match their needs, the information in this manual can help novices maximize their cycling enjoyment. It can also introduce beginners to the inner workings of their machines.

INTRODUCTION

Since the exercise craze and energy shortage first fueled the bicycle revival in the early 1970s, myriad bicycle brands and models have flooded the market. The untrained eye may have difficulty in recognizing the differences between these bicycles. This manual teaches novice riders how to judge the quality of a bicycle's construction and components when purchasing a moderately priced bicycle, costing approximately $200 to $400.

PROCEDURE

The contents of this report were derived from the current literature about bicycle purchasing, from interviews with bicycle retailers and competitive cyclists, and from the author's own inspections and test rides of some of this year's bicycle models.

MAKING A SELECTION

Selecting the right bicycle first requires narrowing your focus either to ten-speed bikes or mountain bikes, the two bicycle types that currently dominate the market. Next, you must select a model within the desired type, and finally determine the right size machine.

Choosing Between Bicycle Types

Mountain bikes and ten-speed bikes are wholly different from one another in function and form (Table 1).

TEN-SPEED BICYCLES

Ten-speed bikes are used for racing and long-distance touring. Their lean, lithe design includes the following features (Berman 1987):

- Skinny, bald tires that minimize friction. Friction's slowing effect is an enemy of the speed demon.
- Racing, drop-type handlebars that require the rider to crouch with the back at a 45 degree angle to the ground. Although the position may seem awkward, it encourages the most efficient use of muscular energy, particularly during long distance rides. It also cuts down on speed-hampering wind resistance, thereby promoting greater cycling efficiency.

Example 17.8 (continues)

TABLE 1. SUMMARY OF KEY DIFFERENCES BETWEEN MOUNTAIN BIKES AND TEN-SPEED BIKES

	MOUNTAIN BIKES	TEN-SPEED BIKES
PURPOSE	Riding on rugged terrain.	Speed racing and distance touring.
TIRES	Fat, heavily treaded balloon tires.	Skinny, bald tires.
HANDLEBARS	Wide, straight handlebars for an upright position.	Curled handlebars for a crouching position.
GEARS	Extra low gears for low speed power needed for climbing mountains.	High gears for high-speed riding.
FRAME	Strong, sturdy, resilient.	Light and streamlined. Sensitive to shocks.
WEIGHT	Heavy. Usually about 35 pounds.	Light. Usually about 25 pounds.
BRAKES	Heavy weight, but impressive control under all conditions.	Light weight, but less control under adverse conditions.
GROUND CLEARANCE	High.	Not particularly high.
RIDING SPEED	Relatively slow.	Fast.

- Gears that support high-speed riding.
- Lightweight brakes. Be aware that these brakes sacrifice quality in favor of light weight; they are more likely than heavier brakes to fail under adverse conditions such as rain. (So great is the racer's desire to reduce weight that most ten-speed bikes used in professional races don't have any brakes at all.)
- A light, slender, streamlined frame, usually weighing about 25 pounds.
- An ability to accelerate swiftly and to reach high speeds easily. This characteristic is made possible by the ten-speed bicycle's light weight, skinny tires, and high gears.

MOUNTAIN BICYCLES

The mountain bicycle is a sensible choice for those who are frustrated by the fragility of the ten-speed bicycle (Feldman 1987). In fact, mountain bicycles, which were unheard of only just a few years ago, now claim the fastest-growing share of the adult bicycle market.

Mountain bikes can handle the pounding doled out by rough terrain, such as hiking and horseback riding trails, and city streets lined with potholes and curbs. Their strong and resilient design includes the following features:

- Fat, balloon tires with pronounced treads which help the bike withstand abuse from obstacles, provide traction over treacherous terrain, and make the ride feel less punishing.
- Wide and straight handlebars that force the rider to sit up and thus gain a better view of hazards such as jaywalkers, rocks, branches, and holes. The upright position also gives the cyclist more control, an imperative on daredevil slalom courses.
- Extra low gears supporting low-speed power needed for climbing mountains.
- Relatively heavyweight brakes that give the cyclist impressive control under all conditions and terrains such as wet, winding, unpaved steep trails.
- A sturdy, heavy frame. This type of frame absorbs jolts instead of passing them on to riders. Therefore, mountain bikes offer a smoother ride than ten-speed frames. A sturdy, heavy frame is also less likely to sustain damage from shocks.
- A hefty weight, averaging between 25 and 35 pounds. The mountain bike's heavy weight--a product of its fat tires, sturdy frame, and heavier brakes--gives the cyclist greater stability.
- High ground clearance. The distance between the pedals and the ground is high, to protect the cyclist's feet from debris kicked up by the bicycle.
- Lever at the seat post for quick adjustment of seat height without getting off the bike. Climbing a hill requires a high seat so that the cyclist can extend his/her leg fully, and thus pedal more efficiently. Descending mountain trails requires a low seat for better control.

Choosing a Model

Once you decide whether to buy a mountain bike or a ten-speed, it's time to settle on a model. Since bicycle models usually become obsolete by their first birthday, a review of this year's ten-speed and mountain bikes would have only short-

Example 17.8 (*continues*)

term value (Wheeler 1987). Of more use to the bicycle purchaser is an ability to discern quality by judging a bike's components and workmanship and by test-riding.

Remember to keep your expectations for a moderately priced bicycle realistic. You will probably have to sacrifice some less valued hallmarks of quality for others that you consider more important. But take heart. Since many bicycle parts are interchangeable, you may be able to request nonstandard parts to bring the price of an otherwise unaffordable bicycle within reach.

COMPONENTS AND WORKMANSHIP

You can easily judge the quality of some bicycle parts just by examining them. In other cases, inexperienced purchasers must rely on manufacturers' reputations. Fortunately, quality manufacturers usually have enough confidence to display their name prominently on their products.

The Frame: The frame is a tubular construction, welded or brazed together to form one structural unit. Other bicycle parts are bought from specialized suppliers and mounted to the frame in the manufacturer's plant. As the bicycle's backbone, the frame absorbs much of the stress and strain of a ride and literally holds all the pieces together. How and what a frame is made of has more bearing on a bicycle's strength and longevity than the frame manufacturer's name (Wheeler 1987).

The most common types of frames are seamed tubing frames and seamless chrome alloy steel tubing frames (Van der Plas 1985). In general, a lighter frame usually costs more.

Less expensive bicycles usually feature seamed tubing frames (Consumer Guide 1980). They are constructed by wrapping a flat strip of steel with rollers into a tubular shape. The sides of the metal sheet are then welded to form a tube. But since the junctures between each tube and other parts of the bicycle—the most stress-prone areas—are not specially re-enforced, even typical riding habits can eventually lead to failure of a seamed tubing frame.

Chrome alloy steel tubing frames are more expensive than seamed tubing frames, but also considerably more durable (Consumer Reports 1985). Such frames are often double-butted, which means that both ends of each tube have been thickened for special re-enforcement. Fortunately, the outside diameter of a double-butted tube is not increased, so double-butting does not weigh the bike down. Double-butting may not be readily apparent, but most manufacturers proudly advertise this feature. As the price of the bike increases, so does the number of double-butted joints in the frame.

Another hallmark of quality is lugging (Coles and Glenn 1973). Lugs are small, reinforcing, metal sleeves that are placed at the juncture between tubes prior to welding.

<u>The Power Train</u>: The power train is the bicycle's engine. It conveys the energy that you put into the pedals to the bike's back wheel. While the front section of the power train consists of the pedals, the cranks, the front derailleur, and the front sprockets or chainwheels, the back section consists of the rear sprocket cluster and the rear derailleur. The bicycle chain connects the front section of the power train to the back section.

When you move the appropriate gearshift levers near the handlebars, derailleurs shift the chain between chainwheels on each sprocket cluster. The standard ten-speed racing bike has at least five chainwheels clustered near the rear-wheel hub, and at least two chainwheels on the front cluster near the pedals. Mountain bikes have either five or six chainwheels at the rear and three chainwheels near the pedals. This arrangement yields from 15 to 18 gears (<u>Consumer Reports</u> 1986).

While a larger chainwheel on the rear sprocket cluster puts the gear into a lower gear, a smaller chainwheel puts the bike into a higher gear. Intermediate gears are set by shifting between front chainwheels.

A bicycle in high gear operates by a principle similar to that driving an egg beater. One rotation of an egg beater's handle makes each beater revolve about five times. Likewise, when a bicycle is in high gear, a single turn of the crank makes the bicycle wheel revolve many times. In other words, the bicycle wheel rotates much faster than the crank. This arrangement gives you the greatest distance traveled for each revolution of the pedals. In contrast, a bicycle in a lower gear is easier to pedal, but moves less distance per full pedal revolution.

In order to invest muscle power most efficiently and ward off fatigue, cyclists must pedal at a steady cadence without bearing down hard on the pedals (<u>Consumer Reports</u> 1985). This requires downshifting when pedaling uphill, thereby sacrificing distance in order to make pedaling easier. When pedaling downhill or on flat terrain, cyclists upshift for distance.

<u>Gear Ranges</u>: Bicycle gearing levels indicate the distance that one full revolution of the pedals moves the bike. Multiplying the number given for a bicycle gearing level by three tells you

Example 17.8 (*continues*)

how many inches one full revolution moves the bike. The high-speed racing demands put on ten-speed bicycles require that they have a range of gearing levels from about 33, at the low end, to 100 or more at the high end (Wheeler 1987). In contrast, mountain bikes, which are more likely to encounter rough, slow-going terrain, need more low gear power. The usefulness of very high fast gears on a mountain bike is also limited by the slowing effect of the rider's upright position. Therefore, mountain bikes should have a lower gear range than ten-speed bikes; from the low- to mid-20s to the 80s is reasonable.

Note also that mountain bikes should have the gearshift levers on the handlebars rather than on the handlebar stem. That way, your thumbs can shift without letting go of the handlebars, a definite safety plus when trying to maintain your balance on treacherous terrain.

Cranks: Campagnolo, T. A. Criterium, Stronglight, and Williams all make first-rate, strong aluminum cranks. Nervar steel cottered cranks or Stronglight 3-pin steel cranks also inspire confidence (Coles and Glenn 1973).

Derailleurs: Huret, Shimano, Simplex, Prestige, Huret-Alvit and Campagnolo all manufacture error-free derailleurs that shift quickly, easily and smoothly (Consumer Reports 1980).

Pedals: Although Campagnolo pedals lead the pack, Lyotard makes a reliable and more affordable alternative. Be sure that pedals feel sturdy underfoot, and can be dismantled for cleaning and packing (Consumer Reports 1986).

Brakes: Squeezing the handbrake levers presses rubber brake pads against the wheel rims, thereby stopping the bike. For mountain bikes, look for the new cantilever style of caliper brakes that come with heavy-duty motorcycle-type hand levers. This type of brake system makes it particularly easy to apply and hold the desired amount of braking force on each wheel.

Until recently, center-pull caliper brakes, particularly those made by Mafac and Weinmann and by Campagnolo, were the favorite for ten-speed bikes. However, new varieties of side-pull brakes produced by Campagnolo and Universal, distinguished from center-pull brakes by a slight difference in design, now also have a faithful following. The most dependable center-pull and side-pull brakes are made of aluminum alloy (Feldman 1987). Beware of inexpensive side-pulls that grip the rims unevenly, causing poor stopping ability that may endanger the cyclist.

Some ten-speed bikes feature easy-to-reach auxiliary brake levers that parallel the horizontal section of the handlebars. Despite their extra weight, they are a valuable safety feature.

Wheels: A bicycle wheel consists of hub, spokes, rim, and tire. The wheel size is measured as the diameter of the inflated tire. While mountain bikes usually have 26-inch wheels and the conventional wired-on tires, ten-speed bikes usually have 27-inch wheels and tubular racing tires, which are each cemented to the special rim.

The rear wheel of both mountain and ten-speed bikes should have 36 spokes, and the front wheel should have at least 32 spokes. A wheel with fewer than the standard number of spokes is probably weak, and thus a potential source of trouble under pressure. All spokes should be double-butted, easily identified by their thickened ends.

Security conscience city riders should look for cam-lock levers, which enable you to free the wheels from the frames, without tools, for safekeeping. Note that this feature could be a hazard on bikes ridden in the mountains; the cam lever could snag on a branch and flip open, thereby accidentally releasing the front wheel.

Rim: Stick with aluminum rims. Fine aluminum particles that rub off of aluminum rims and become embedded in the brake pads provide good braking action in the rain. The pearly gray finish of aluminum rims is easily distinguished from the mirrorlike, chrome-plated finish of less expensive, less durable, and heavier steel rims.

Hub: The hub consists of a set of ball bearings mounted on a central axis, which is fixed in the bicycle's frame. The very finest hubs are made by Campagnolo, but Normandy, Simplex, and Cinelli also make satisfactory hubs.

TEST RIDING

Even a bike that adheres to all of these recommendations and sports the finest name brands will get you nowhere if you find it uncomfortable. The only way to judge your compatibility with a bicycle is to test-ride it (Consumer Report 1986).

Select a bicycle that is easy to pedal and handle—predictable and responsive rather than cumbersome and sluggish. Expect a gentle squeeze on the brake levers to result in smooth and uniform braking that stops the bike within 12-15 feet of your command, without throwing you over the handlebars. Test wet braking by having a companion direct a fine spray at the

Example 17.8 (*continues*)

wheels just before and during braking. Gears should click into place quickly, crisply, and precisely with just a slight tap on the gear lever. Give a low ranking to bikes that demand excessive pedaling effort.

Here are some special recommendations for each type of bike:

Ten-Speed Bikes: Test ten-speed bikes on straightaways, up hills, and steep winding roads. Look for a model that moves smoothly and precisely in response to the slightest commands from your hands and feet. Gears should shift crisply right after you flick the lever. In addition, the bike should be relatively easy to pedal up hills, stay under control when gliding downhills, and accelerate briskly. Give a lower rating to bikes that oversteer at low speeds, or are reluctant to turn at high speeds. Beware of rear derailleurs that overshoot one of the rear cluster's middle chainwheels.

Mountain Bikes: Make sure that the mountain bike lives up to its "all terrain" promise by testing it both on and off the road. If possible, run the bike over obstacles such as ruts, rocks, steep inclines, and soft sandy areas, as well as through slalom maneuvers at various speeds. Pay particular attention to turning and braking performances. Look for a model that feels surefooted and nimble, and avoid those that demand excessive steering corrections or seem to wander aimlessly under adversity.

COMPARING MODELS

Once you have narrowed your selection to a few models, systematically compare their virtues and drawbacks. This can be done quickly and efficiently via the checklist provided in Table 2. Each numbered column under the "BICYCLE" heading provides an opportunity to record one model's performance for the "CRITERIA." Therefore, each row should give you a comparison of how the four models fare for a single criterion.

Selecting the Right Size Bicycle

A bicycle must fit properly to be safe. Indeed, a National Safety study indicates that riding a bicycle that is too large leads to a loss of control that increases by five times the chances of an accident (Coles and Glenn 1973).

The size of the bicycle frame is usually measured as the total height of the seat tube from the center of the bottom bracket to the top of the seat tube.

TABLE 2. CHECKLIST FOR COMPARING BICYCLE MODELS

CRITERIA	**BICYCLE MODEL**			
	1	2	3	4
PRICE				
FRAME				
Seamed tubing or seamless chrome alloy steel construction?				
Double-butting?				
Lugs?				
POWER TRAIN				
Brand of cranks				
Brand of derailleurs				
Brand of pedals				
Gear range				
WHEELS				
Size				
Number of spokes				
Are spokes double-butted?				
Brand of hubs				
Aluminum rims?				
WEIGHT				
TEST RIDE				
Comfort				
Breaking action: • Wet • Dry				
Steering responsiveness				
Gear shifting				
BRAKES				
Type				
Brand				

Table 2 (continues)

Check the fit of a bicycle by straddling the top tube with both feet flat on the floor. At least 1/2 inch of clearance between you and the top bar is required for a proper fit. More than two inches of clearance indicates that you should consider a larger size. And if the bar is uncomfortably close, consider a smaller size.

If an assembled bike is not available to try out, measure the length of your leg inseam. Subtracting ten inches from the figure will give you the proper size.

GLOSSARY

Derailleur: A mechanism that, upon the cyclist's command, forces the bicycle chain from one sprocket to another, thereby causing the bicycle to shift gears.

Derailleur Gearing System: The most common type of gearing system found in bicycles today. It consists of front and rear derailleurs, front and rear sprockets and shift levers.

Double-butting: A process by which both ends of a metal tube, such as the spokes in a bicycle wheel or the metal tubes in a bicycle frame, are reinforced with extra thickening for structural strength.

Frame: A tubular assembly that forms the bicycle's backbone. Additional components, such as the wheels and power train, are attached to the frame.

Gearing Level: The number given for a gearing level is equal to one-third the distance in inches that one full crank of the pedals will move the bicycle. Low gearing levels sacrifice distance for pedaling ease. Therefore, they are usually used for pedaling uphill. In contrast, high gearing levels sacrifice pedaling ease for distance. Therefore, they are usually used for pedaling downhill and on flat terrain.

Gearing Range: The range of gearing levels over which a bicycle is capable of traveling.

Lug: A metal sleeve that is welded onto the junctures between the metal tubes in a bicycle frame for added structural strength.

Mountain Bicycle: A bicycle with flat handlebars, fat high-pressure tires, usually 15-speed derailleur gearing, and a strong, sturdy frame. It is particularly suitable for use on unsurfaced trails and on poorly surfaced roads and streets.

Power Train: A set of components that transmits the rider's effort to the rear wheel. It consists of pedals, cranks, a front and a back sprocket cluster, a front and a back derailleur, and a chain.

Sprocket Cluster: A set of stacked chainwheels. Moving the bicycle chain from the largest to the smallest rear chainwheel moves the bike into progressively higher gearing levels. In addition, for each shift at the rear, the chain can also be shifted between front chainwheels to put the bicycle into intermediate gears.

Ten-Speed Bicycle: A bicycle with drop handlebars, derailleur gearing, narrow tires and a sleek, light-weight frame. It is particularly suitable for racing and long-distance touring.

Wheel: A bicycle wheel consists of a hub, a network of spokes, and a rim on which the tire is mounted.

REFERENCES

"All Terrain Bicycles," Consumer Reports, Feb. 1986, 97–100.

Berman, Martha. competitive cyclist. Personal communication (July 7, 1987).

"Bicycle Buyer's Guide 1980." Consumer Guide, June, 1980, 1–20.

Coles, Clarence and Glenn, Harold. Glenn's complete bicycle manual. New York, NY: Crown Publishers, Inc. 1973.

Feldman, Peter. competitive cyclist. Personal communication (July 3, 1987).

"Touring bicycles." Consumer Reports, November 1985, 651–655.

Van der Plas, Rob. Bicycle repair book. San Francisco, CA: Bicycle Books Publishing Inc., 1985.

Wheeler, Charles, bicycle salesman. Interview 10 July 1987.

SUMMING UP

• There is no one all-purpose "basic report format." There are however, standard elements the writer can work with, adjusting them according to the audience.

• Reports are divided into three parts: front matter, body, and back matter.

• Considerable thought should go into the title, to shape it into a useful preview of the report contents.

• Major divisions in the table of contents can be shown through size and style of type, numerically, or through a combination of these techniques.

• Try for headings that summarize or preview the text, rather than merely labeling the parts.

• The executive summary should tell within a few paragraphs what was done and why it matters. Often it is cast for a very broad audience that ranges up and down the corporate distribution.

• The abstract should give the findings and their implications.

• The introduction should give an overview and necessary background. It answers the questions, "What was the problem?" "Why was the problem significant?" "Who else has worked on this problem?" Sometimes it ends with a summary statement and a roadmap indicating the scope or division of the paper that follows.

• Present data in their rawest form. If graphs are used to illustrate the data, variables and units of measurements should be clearly distinguishable.

• Technical word processing permits those who work with equations to sit down to a keyboard and produce proportionally spaced text, built-up equations, and other elegant touches that were formerly the exclusive province of the typesetter. Most programs are a compromise between markup languages, where the user explains the format, and what-you-see-is-what-you-get languages, where the user enters a markup command, but then sees the result on the screen.

• If you set off equations from the text, number them.

• Consistency of typographical elements is important in reports. A set of guidelines may be useful as a checklist when you are preparing the final version of the report.

• The feasibility report is a variation on the formal report, in which the writer uses a set of criteria to weigh various solutions to a problem. The level of detail in a feasibility report is usually quite high. Alternative solutions are evaluated both technically and financially. Sometimes the authors draw conclusions; at other times the writers simply present alternative solutions, without endorsing one of them.

EXERCISES

1. Assume your company is exploring the possibility of installing an exercise room in response to employee requests. The company has 800 employees. You are asked to prepare a brief research report on exercise rooms: What are some of the typical pieces of equipment popular in exercise rooms? How much space would be needed? Are there ventilation requirements? What would approximate costs be? Prepare this report. Include a memo instead of a letter of transmittal, as the report will be distributed within the company. Use an informative title. Make sure your abstract includes all salient points. Allow three weeks to prepare the report, as you will need to

- telephone or write manufacturers for product literature.
- visit exercise rooms for some first-person observation
- interview people who are involved in managing exercise rooms

2. Your company is interested in the advantages and disadvantages of offering early retirement benefits to its senior employees. Such an offering would reduce mandatory layoffs that would hit pensionless younger workers the hardest. Prepare a topical report concentrating on a literature review of advantages and disadvantages of early retirement plans. The readership is a group of managers well acquainted with the terminology of retirement. The index of the *Wall Street Journal,* as well as other business indexes (see Chapter 7), will be a good place to get started. Be sure to include an abstract with your final report, as well as endnotes documenting your sources. Include a listing of any tables or graphs in your table of contents.

3. Your business is considering the advantages of a telecommunications system, having learned that telecommunications might produce a competitive edge in such areas as remote order entry, inventory management, and discount systems. Prepare a research report on the competitive advantages of a telecommunications system for a specific business field such as medical supplies.

Use a memo instead of a letter of transmittal, as the report will be read by management within your company.

4. You are asked to evaluate the relative merits of for-profit and nonprofit hospital corporations. Prepare this report for a community board that seeks to know which of the alternatives might provide better benefits for patients.

5. Prepare a literature review on a recent advance in the technology of your field for an audience outside your immediate field. Imagine your readers to be people interested in recent advances, but without your identical technical background. Make sure, in other words, that your report is readable for people trained outside your specialty. Some possibilities:

- an advance in medical imaging
- an advance in the technology of airline safety
- an advance in heart pacemakers
- an advance in the laying of deep-water pipelines

Include an executive abstract instead of an abstract. Make sure that illustrations are captioned and listed in the table of contents. Indicate sources in the endnotes. Use informative headings and subheadings.

6. Hundreds of pigeons have chosen the roof of your company's building as their home. Their noise and droppings are creating problems both for staff members and for customers. Prepare a report for management reviewing possible solutions.

7. Your company is concerned about corporate espionage, as several disquieting instances of spying have recently been reported in the business press. You are asked to write a brief topical report (10 pages maximum) on the subject, concentrating on recent events. How serious is the problem nationally? What are some of the solutions that have been proposed? Use business indexes in the library as part of your search.

8. The community group to which you belong inherits a substantial bequest from a member of the group. You are asked to prepare a report on leading industrial indexes (Dow Jones industrial index, Standard & Poor's 500-stock index, and Value Line index), so the group can better understand the nature of its inheritance. Write the report in such a way that people who are literate, but unfamiliar with the stock market, can still follow your points. Be sure to explain how each of the indexes works.

9. Select two research publications in a technical field of interest to you. Study the instructions to the author in each publication. Such instructions usually appear either in each issue, or annually. If they do not appear in the publication, they may be obtained by writing to the editorial offices at the address given on the masthead of the publication. Then study five or six issues of both publications. Analyze the structure of the articles. See if you can identify the parts of the reports; analyze how these parts correspond to standard report format. Prepare a report on the two publications, concentrating on how they approach their audience and on their choice of language, organizational patterns, and visual display. Use examples from both publications.

10. Prepare a feasibility report on a technical subject. Allow at least three weeks for the assignment. You'll need to

- establish the problem, through interview and reading
- define criteria in relation to the problem
- investigate alternative solutions
- compare and contrast these alternatives
- draw conclusions and make recommendations

The subject you choose should be one in which you can develop a substantial technical understanding of the problem.

11. A small business approaches you and asks your consulting firm for help in selecting a computer and printers. The busines is a wholesale jewelry firm that needs to keep track of a complicated inventory. The firm produces its own catalogue, and sends mailings regularly to retailers. The company, run by two partners, employs one secretary, a bookkeeper, and five sales representatives. The partners have neither the time nor the technical abilities to investigate computers and printers for the small business market. Prepare a report for them. They do not know the language of the computer industry, so use as few technical terms as possible, and define those that you do use. Include an abstract that summarizes your findings. Use informative subheadings to divide the text. Include simple tables that show differences in capability among models; include examples of typeface.

Proposals

CHAPTER 18

• Joan Smith is a scientist who works for a chemical company. To keep her laboratory up to date and competitive, she regularly writes proposals for new equipment. She doesn't always get what she requests, though. Management in her company reads not only her proposals, but dozens of others, selecting only a few; the winners are usually the ones who argue their cases most persuasively.

• Harry Classon is a college senior enrolled for a thesis project. Before doing this thesis, however, he must write a proposal for the work—a proposal members of the thesis committee find acceptable. Like Joan Smith, he finds that his advancement depends on his ability to formulate and deliver a cogent argument.

These are two examples among many that illustrate the importance of the technical proposal. A proposal is a suggestion, a "put-

Box 18.1

TYPICAL OUTLINE FOR A PROPOSAL

Front Matter

Letter of Transmittal
Title Page
Table of Contents
List of Figures and Illustrations
Summary

Introduction

Problem	What problem does the project address? Explain why this problem is significant.
Objectives	What are your specific goals? If you have short-term and long-term objectives, discuss both. Be sure to relate your goals to the stated goals of the funding source.
Advantages	What gains will your project bring? Why do these gains matter? State gains from the reader's point of view.

Body

Technical Plan

Description	What will you do? How? When? What outcomes do you anticipate? This is the place to hit your home runs. Pick your approach and sell it hard. If you are proposing research, relate it to the past and present research that provides its theoretical or methodological framework. Identify any lacks in this framework that your proposal will help to fill. Be sure to discuss any ongoing or recently completed work in your field that is conceptually or empirically linked to your work. In this way, you place your work in a scholarly context and show how it will help build the foundation of knowl-

edge in a particular discipline. If you are proposing the solution to a business problem, show the benefits—such as reduced costs, improvement in morale, safety, or prestige to the institution.

Management Plan

Personnel	What makes you (and your staff) the right people for the job? Don't include resumés in the body of the text. Cross reference them and attach in the appendix.
Facilities	What facilities and equipment do you have? What needs to be bought?
Budget Plan	Use both figures and an accompanying narrative unless the funding source supplies a prescribed form.
Evaluation	How will you measure the success of your undertaking?

Conclusion

Upbeat restatement of the significance of the problem and the proposed benefits. Discuss any implications that may arise from the work.

ting forth" of an idea. Whether you work in sales or marketing, research or product development, in an independent engineering consulting firm or within a vast corporate structure, you'll find proposals are an important part of the job.

Proposals answer these questions:

- What is the proposed service, product, idea, or plan?
- What is the problem or need it addresses? Why is this problem significant?
- What are the benefits of your solution?
- How will you do the work? When? What will it cost?
- How will you measure the effectiveness of your work?

Proposals have many functions in common with reports; for instance, both documents usually define a problem and its solution. A proposal, however, is speculative: You tell how you'll carry out your plan—the costs, the timetable, the facilities, the staff—and list expected outcomes. In the report, in contrast, you give the actual outcomes, not projections.

By their nature, proposals must be persuasive documents, telling not only what you want to do, but why you are the best person to do it, the person with the solution that most clearly benefits the reader. Once you have argued the need for your solution, you then present a carefully structured plan as to how you'll bring it about. The details of the plan may be technical, but the arguments themselves must be persuasive ones, arguments that sell you, your idea, your competence to do a job.

"I hate to sell," one engineer in the throes of writing a proposal commented. "That's why I became an engineer. I figured that was the one safe area where I would not have to promote myself." Yet engineers, scientists, and technical professionals must often bid in a crowded field for jobs, contracts, budget allocations. The route to these objects is the proposal.

Proposals may take the form of a letter or a 30-page dossier. Leonardo da Vinci's letter to his prospective patron, Ludovico Sforza, for instance, is an outstanding example, just the sort of forceful document one would expect a genius like da Vinci to supply. (Figure 18.1).

Longer proposals are often divided into a **technical section,**

Most Illustrious Lord, having now sufficiently seen and considered the proofs of all those who count themselves masters and inventors of instruments of war, and finding that their invention and use of the said instruments does not differ in any respect from those in common practice, I am emboldened without prejudice to anyone else to put myself in communication with your Excellency, in order to acquaint you with my secrets, thereafter offering myself at your pleasure effectually to demonstrate at any convenient time all those matters which are in part briefly recorded below.

1. I have plans for bridges, very light and strong and suitable for carrying very easily, with which to pursue and at times defeat the enemy; and others solid and indestructible by fire or assault, easy and convenient to carry away and place in position. And plans for burning and destroying those of the enemy.

2. When a place is besieged I know how to cut off water from the trenches, and how to construct an infinite number of bridges, mantlets, scaling ladders and other instruments which have to do with the same enterprise.

Figure 18.1 Leonardo da Vinci wrote this direct, clear proposal to his patron, Ludovico Sforza.

3. Also if a place cannot be reduced by the method of bombardment, either through the height of its glacis or the strength of its position, I have plans for destroying every fortress or other stronghold unless it has been founded upon rock.

4. I have also plans for making cannon, very convenient and easy of transport, with which to hurl small stones in the manner almost of hail, causing great terror to the enemy from their smoke, and great loss and confusion.

5. Also I have ways of arriving at a certain fixed spot by caverns and secret winding passages, made without any noise even though it may be necessary to pass underneath trenches or a river.

6. Also I can make armoured cars, safe and unassailable, which will enter the seried ranks of the enemy with their artillery, and there is no company of men at arms so great that they will not break it. And behind these the infantry will be able to follow quite unharmed and without any opposition.

7. Also, if need shall arise, I can make cannon, mortars, and light ordnance, of very beautiful and useful shapes, quite different from those in common use.

8. Where it is not possible to employ cannon, I can supply catapults,angonels, *trabocchi*, and other engines of wonderful efficacy not in general use. In short, as the variety of circumstances shall necessitate, I can supply an infinite number of different engines of attack and defence.

9. And if it should happen that the management was at sea, I have plans for constructing many engines most suitable for attack or defence, and ships which can resist the fire of all the heaviest cannon, and power and smoke.

10. In time of peace I believe that I can give you as complete satisfaction as anyone else in architecture in the construction of buildings both public and private, and in conducting water from one place to another.

Also I can execute sculpture in marble, bronze or clay and also painting, on which my work will stand comparison with that of anyone else whoever he may be.

Moreover, I would undertake the work of the bronze horse, which shall endure with mortal glory and eternal honour the auspicious memory of the Prince your father and of the illustrious house of Sforza.

And if any of the aforesaid things should seem impossible or impractical to anyone, I offer myself as ready to make trial of them in your park or in whatever place shall please your Excellency, to whom I commend myself with all possilble humility.

which answers the questions of what and when; a **management section,** which makes the case for your group, explaining why you have the background, facilities, track record, and staff to do the job; and a **budget,** which tells how much the job will cost.

The proposal may be **unsolicited**—you write it yourself to sell a book, get a contract, interest an agency in a project you want to run; or **solicited**—in answer to a request for proposals (RFP) from, for instance, a source like the Environmental Protection Agency.

In this chapter there are examples of different proposals; the writing varies according to the type of work proposed and the expectations and background of the audience. Despite adjustments made for the nature of the task and the needs of the audience, it's safe to say that in general, proposals are hybrid documents: They combine the technical rigor of a problem-solving report with the political astuteness of a sales document.

SHORT NARRATIVE PROPOSALS

Whether short or long, a proposal must answer essential questions: What is the project? Why is the project significant? How will we benefit from funding this proposal? Why are you the best person to do the job? How will you do the work? What will it cost?

The following example shows how one writer answered these questions. This example is a short, narrative proposal of five pages. Unlike longer, more formal proposals, it has no extensive **subproposals** within it—separate management sections, for instance, or a separate extensive technical discussion.

This narrative proposal was written in response to a request from the National Science Foundation (NSF). Some government agencies, such as the NSF, announce a competition in two stages: First the applicants are asked to write preliminary proposals, and then a fraction are asked to write fullscale proposals, based on the reviewers' reactions to the preliminary documents. The example below was in response to a request for preliminary proposals.

In the proposal the writer was supposed to cover these areas, all of which were requested in the guidelines issued by the NSF:

- objectives
- expected outcome
- significance of work
- credentials of investigator
- proposed method
- time table
- budget
- plans for evaluation

To show how an actual proposal corresponds to these general guidelines, Example 18.1 has been indented slightly. In the margin are comments connecting the requests in the NSF guidelines to the writer's actual solutions in the proposal.

The writer attached his vita (academic resumé) to the proposal, as well as other supporting documents, and forwarded the proposal with a letter of transmittal. The funding agency sent the proposal out for review to three experts in the field. The writer was later encouraged to write a fullscale proposal.

PARTS OF A FORMAL PROPOSAL

If your proposal is for research funds or extensive contract work (such as that advertised in Figure 18.2), you'll need to prepare a formal proposal in accordance with a set of guidelines distributed by the funding source. (See Box 18.1 at beginning of chapter for a typical outline.)

Each of the parts in a representative set of guidelines for proposals is discussed below. You may not need each section when you are preparing a proposal, because categories and their order vary within agencies and companies. Regardless of variations in format, however, your reviewers are certain to ask questions such as these as they evaluate your proposal:

- Why is the project valuable?
- What makes you qualified to do this job?

Example 18.1 Preliminary proposal.

ETHICAL ISSUES IN MILITARY RESEARCH

Author states objective and expected outcome

This proposal is for an eighteen-month project to explore the ethical issues in military research. It would culminate in a conference co-sponsored by the Philosophy & Technology Studies Center and the Hastings Academy of Sciences. Papers and discussion from the conference would be published as a volume in the <u>Proceedings</u> of the Hastings Academy.

Author relates preliminary proposal to funding source

<u>Background</u>

In the Fall of 1985, a Hastings Academy committee which is involved in planning for the "High Technology and Diplomacy" conference and is chaired by Prof. Jaon Silver (Anderson University) began to consider the possibility of a further discussion of ethical and moral issues surrounding high technology military research. At the same time, the Philosophy & Technology Studies

Example 18.1 (*continues*)

Center began independently to formulate research plans in this area. When the two efforts became aware of each other, Prof. John Hemp (Director, Philosophy & Technology Studies Center) was invited to attend a meeting of the Silver committee.

The outcome of this November meeting was an agreement to work together to organize a conference on ''Ethical Issues in Military Research'' for the spring of 1989 which would build on and complement the ''High Technology and Diplomacy'' conference scheduled for the spring of 1987.

Focus

The purpose of the "Ethical Issues in Military Research" conference would not be to take a partisan stand or debate once again the pros and cons of the Strategic Defense Initiative (SDI). Instead, the aim would be to bring scientists and engineers representing both sides of the political-moral spectrum in the current SDI debate together with philosophers and others who could help place the debate in a larger context by considering

Author argues strongly for why this proposal addresses a significant problem

- the historical background of the current debate,
- the sociological aspects of the debate,
- the kinds of ethical arguments employed by both sides in the debate, and
- the moral options open to persons on both sides of the debate.

In the course of exploring these four aspects of the larger context, the following kinds of specific questions would be addressed: What are the strengths and weaknesses of various ethical frameworks -- utilitarian, ontological, natural law -- in dealing with the ethical issues of military research? Are some ethical frameworks more often employed to justify and others to criticize military research? Are there ethical issues in this area which are not adequately addressed by traditional ethical theories? What is the relation of just war theory to military research? Are (and should) religious beliefs (be) involved in making moral evaluations about military research?

Does the political system within which decisions about research priorities are made significantly affect the moral responsibilities of the individual researcher? Does the social status of the scientist or engineer alter moral responsibilities or obligations? Historically, what have been the moral imperatives associated with scientific research? What have various moral philosophers (and scientists) had to say about the ethical issues of military research?

Schedule

During the Fall semester, 1988, the Director of the Philosophy & Technology Studies Center would work to set up an organizing committee, and would work with the committee to identify issues and individuals for the conference. Invitations would be made to ten persons to prepare special presentations for the conference.

The three-day conference itself would be held in the spring of 1989, and would include major presentations as well as workshop discussions. The conference would be well publicized and at least the major presentations open to the public for a modest registration fee. (It might prove desirable to have the workshop sessions restricted.)

The summer and fall of 1989 would be devoted to preparing the proceedings for publication.

Budget	
A. Senior personnel	
Director, Phil & Tech Studies Center	
(one course release per semester)	$19000
Fringe benefits (30%)	5700
B. Indirect costs (71% of A)	17537
C. Honoraria for invited speakers	
(ten speakers at $500 each)	5000
D. Other personnel	
Hastings Academy of Sciences staff	
Administrative	7500
Editorial	7500
E. Other direct costs	
Duplicating	500
Communications (postage, telephone)	500
Distribution of proceedings	3000
Travel for invited speakers ($200 per person)	2000
Accommodations for speakers ($75 per diem)	2250
Conference site expenses	10000
Equipment rental	500
Auxiliary personnel services	
Receptionists	500
Projectionists	400
Recording and transcription	1200
Printing of programs and abstracts	4000
Mailing of programs	1000
Conference promotion	2000
SUBTOTAL	27850
TOTAL	$90087

In regard to the above budget, line items A and B are derived from the university. Line B is the overhead percentage negotiated with the federal government on the basis of the most recent audit. Line items D and E are derived from the Hastings Academy of Sciences. In both cases, the institutions involved are willing to cost-share 25%.

Figure 18.2 An advertised request for bids and proposals.

BIDS & PROPOSALS

THE PORT AUTHORITY OF NY & NJ

Sealed bids for the following proposals will be received at the Office of the Manager, Purchase & Supply Services Division, The Port Authority of New York & New Jersey, One World Trade Center, Room 82S, New York, New York 10048 until 11:00 A.M. on the date indicated below at which time and place said proposals will be opened and read.

Bid documents are mailed upon request by telephoning (212) 775-8994 between the hours of 9:30 A.M. to 11:30 A.M. and 1:30 P.M. to 3:30 P.M.

PROPOSAL #6060/52
TITLE: CAFETERIA PLASTIC PLATES
BIDS DUE: WEDNESDAY, MARCH 2, 1988

PROPOSAL #6061/02
TITLE: SUPPLY ONLY CUSTOM MADE PASS MACHINE TUFTED 100% WOOLCARPET FOR WTC.
BIDS DUE: THURSDAY, MARCH 10, 1988

PROPOSAL #6063/44
TITLE: POWER MODULES AS MFGR. BY TEXAS INSTRUMENTS, JOHNSEN CITY, TENNESSEE
BIDS DUE: WEDNESDAY, MARCH 2, 1988

PROPOSAL #6064/44
TITLE: "SKYCAP" WINTER JACKETS/ TROUSERS - SUMMER TROUSERS PER P.A. SPECIFICATIONS
BIDS DUE: WEDNESDAY, MARCH 4, 1988

- How is the proposal different from what we have already?
- How thorough and realistic is the plan?
- How cost effective is the project?
- Is the budget appropriate for the scope of the proposed activities?

Cover or Title Page

The cover page may give the title of the proposal, the agency or group it solicits, the number of the grant or competition it pertains to, the authors, and the date. If the proposal is a bid to do a job, writers include the dollar amount of the bid, unless the bid is supposed to be sealed.

A cover page supplied by the National Science Foundation appears in Figure 18.3. Figure 18.4 shows a sample title page made by the proposer of a short, narrative proposal.

Table of Contents

The table of contents lists the major divisions in the proposal and the page where each begins. Most funding agencies prefer that the writer keep the body of the report lean. In Figure 18.5, for example, a proposal to the National Science Foundation, the Project Description itself is 11 pages. Supporting information is placed in attachments.

COVER SHEET FOR PROPOSALS TO THE NATIONAL SCIENCE FOUNDATION

FOR CONSIDERATION BY NSF ORGANIZATIONAL UNIT (Indicate the most specific unit known, i.e. program, division, etc.)	PROGRAM ANNOUNCEMENT/SOLICITATION NO./CLOSING DATE

SUBMITTING INSTITUTION CODE (if known)	FOR RENEWAL ☐ CONTINUING AWARD ☐ ACCOMPLISHMENT BASED RENEWAL ☐ REQUEST, LIST PREVIOUS AWARD NO.:	IS THIS PROPOSAL BEING SUBMITTED TO ANOTHER FEDERAL AGENCY? Yes___ No___; IF YES, LIST ACRONYM(S)

NAME OF SUBMITTING ORGANIZATION TO WHICH AWARD SHOULD BE MADE (INCLUDE BRANCH/CAMPUS/OTHER COMPONENTS)

ADDRESS OF ORGANIZATION (INCLUDE ZIP CODE)

IS SUBMITTING ORGANIZATION: ☐ For-Profit Organization; ☐ Small Business; ☐ Minority Business; ☐ Woman-Owned Business

TITLE OF PROPOSED PROJECT

REQUESTED AMOUNT	PROPOSED DURATION	DESIRED STARTING DATE

CHECK APPROPRIATE BOX(ES) IF THIS PROPOSAL INCLUDES ANY OF THE ITEMS LISTED BELOW:

- ☐ Animal Welfare
- ☐ Endangered Species
- ☐ Human Subjects
- ☐ Marine Mammal Protection
- ☐ Pollution Control
- ☐ National Environmental Policy Act
- ☐ Research Involving Recombinant DNA Molecules
- ☐ Historical Sites
- ☐ Interdisciplinary
- ☐ International Cooperative Activity
- ☐ Research Opportunity Award
- ☐ Facilitation Award for Handicapped
- ☐ Proprietary and Privileged Information

PI/PD DEPARTMENT	PI/PD ORGANIZATION	PI/PD PHONE NO. & ELECTRONIC MAIL

PI/PD NAME/TITLE	SOCIAL SECURITY NO.*	HIGHEST DEGREE & YEAR	SIGNATURE
ADDITIONAL PI/PD (TYPED)			
ADDITIONAL PI/PD (TYPED)			
ADDITIONAL PI/PD (TYPED)			
ADDITIONAL PI/PD (TYPED)			

For NSF Use:

AUTHORIZED ORGANIZATIONAL REP.	SIGNATURE	DATE	TELEPHONE NO.
NAME/TITLE (TYPED)			
OTHER ENDORSEMENT (optional)			
NAME/TITLE (TYPED)			

Figure 18.3 The sponsoring agency often supplies a title or cover page for the proposal. Courtesy National Science Foundation.

Figure 18.4 The author of a preliminary proposal provided this cover page.

Preliminary Proposal

ETHICS AND VALUES IN SCIENCE AND TECHNOLOGY

TO:
Directorate for Biological, Behavioral & Social Sciences
National Science Foundation
Washington, DC 20550

FROM:
Philosophy & Technology Studies Center

PROJECT TITLE:
Ethical Issues in Military Research

ESTIMATED BUDGET:
$90,087

TABLE OF CONTENTS

Figure 18.5 Table of contents for a proposal. Note that the body of the proposal is a trim 11 pages. The supporting information is in the attachments.

Project Summary

The abstract, or summary, must sell the proposer's idea. It is neither the place for language readers won't grasp, nor for marginal technical details. Resist the desire to echo the table of contents: "Included in this proposal are an introduction, a discussion of the problem, a proposed procedure, and anticipated outcomes." Instead, summarize what you're proposing, how you'll go about doing the job, and how your idea matches the guidelines or requirements. The summary should persuade the reader that you are the person to do this job—that you have the personnel, expertise, facilities, and savvy to do the work properly. If your audience is composed of technical experts, you may use technical terms. If the audience is managerial, cast the abstract so that people will understand the problem, the solution, the method, and the significance, without getting mired in information they have neither the background nor inclination to understand.

Introduction

The introduction should state the problem and your solution to this problem. Make your approach clear—emphasize what you plan to do, how you'll carry out your goals, and why your plan meets the customer's needs. Do this in as brief a space as possible. If, for instance, your total proposal is 10 pages, the introduction should consume no more than a page.

The introduction is extremely important; proposals often stand or fail based on the persuasiveness of the problem statement. These are some of the questions you need to think about before you write the introduction.

- What is the purpose of the proposal? If the proposal is unsolicited, or if you have modified the customer's request, you'll need to spend more time answering this question than you would were the proposal in response to a specific request.
- What is the origin of the proposal? Is it in response to an invitation? If you are responding to a request, this answer can be handled in a phrase. If you are the originator—that is, if the proposal is unsolicited—you will need to think about how to connect your proposal to the customer's needs.
- What is the scope of the proposal? What are its objectives?
- What is the problem you are addressing? Why is this problem significant? Of what value is the proposed work? The value must be stated from the reader's point of view.
- What are the expected outcomes?

Technical Plan

The technical discussion is the heart of the proposal, the place "to hit all your home runs," according to one company's guidelines to its proposal writers, which advises writers to "present one approach and sell it hard. Don't leave the reader in doubt as to what you are proposing and why it is the best approach." If you are proposing a solution that differs from the one specified in the RFP, clarify why you've decided on a different solution, and why your solution is technically superior.

The technical discussion covers both the experimental and theoretical methods you will use. Depending on your subject, you may include a discussion of the analytical method, the experimental methods, the test instrumentation, reliability, quality control, special treatment of properties, materials, or processes, design approach, or special techniques. You may have to account for aspects of the electrical or mechanical design, for restraints on the design, or for tradeoffs you have made.

- Preface the technical discussion with a brief summary of its contents.
- Use subheads to divide the text according to the main points.
- Make the discussion sufficiently thorough to show your expertise.
- To streamline the text, place as many of the supporting documents as possible in the appendixes.

Management Plan

What you are going to do goes in the technical discussion, but *how* you'll do it goes in the management section. This means that the management sections should include the plan, the schedule, the facilities you'll use, and the personnel who will be assigned.

Make clear what you'll do, how long it will take, and who will be responsible for each part of the work. Most clients appreciate a chart or time line showing when to expect delivery of each part of the goods (Figure 18.6). Describe facilities and people briefly in the narrative. Enclose resumés and standard information on the facilities (these and other standard handouts, or fixed sections, of a proposal or report are often called "boilerplates") in the appendixes.

If you are preparing a larger proposal, the project director will probably work collaboratively with marketing people and budget people to develop strong Management and Budget sections.

```
                         ESTIMATED MANPOWER LOADING SUMMARY

                             Months after receipt of contract        Approximate Manhours
                            0 2 4 6 8 10 12 14 16 18 20 22 24      Scientific
Major Task                                                           Personnel        Tech

Material Growth               ^_______________________^                300            1300

Mask Making                 ^___^              ^_____^                 107             183

Input Optimization               ^_________^                           900            1200

Output Optimization               ^________^                           700             900

3-Terminal Device Studies        ^________ ^       ^_______^          1300            1200

Signal Process Optimization                  ^___^                     200             100

Input-Output Integration                         ^_________^           800            1000

Hybrid Integration                                    ^_____^          100             200

Microwave Evaluation             ^__________________________^         1200            1200

Reports                                    ^___^        ^___^          302
                                                                     ______          ______
                                                            Total     5909            7283
```

Figure 18.6 Many funding sources require a milestone chart in the proposal.

Budget Plan

Lay out costs both in the budget and in an accompanying narrative that relates expenses both to program goals and to a timetable. The budget is almost always prepared in response to guidelines supplied by the funding source. If you have no guidelines, use an informal model such as the one given in Example 18.1.

For models of letters of transmittal, see Chapters 10, 13, and 14. For models of listings of tables and figures, see Chapter 17.

THE PROCESS OF WRITING A PROPOSAL

There are many proposals whose formats fall somewhere between short narratives and longer, formal presentations. *All* require that the writer state the problem, argue the significance of the solution, and spell out the benefits—for instance: cost reduction, safety, gains in the environment, savings in energy, improvement in production rates, gain in morale, or a new product.

Beyond these universals, there's no fixed recipe for writing a winning proposal. The process usually begins with the writer or writers exploring the problem and building bridges to the contracting agency, client, or source. Since the proposal is a sales document, the writer needs to think not only about technical problems, but about any political problems along the way. What are the idiosyncrasies of the funding source? What do the grants officers like and dislike? The answers to these questions matter, for as a writer you need to persuade the client that your group understands the problem and has the know-how to solve it. This may be as much a political process as a technical one. Writers must feel their way carefully, establishing a connection to the funding source.

Proposal writing is another example of technical writing that must be reader-centered to do the job. The writer must imagine the reader and then aim directly for that person. How do you gauge this person's point of view? The answer depends on the background, expectation, and needs of the reader, whether that person is a manager at work, an expert doing reviews for a national agency, or a grants officer at a foundation who does not share your technical background.

Some steps in reader analysis are straightforward. If, for instance, you are applying to a national agency, you might begin researching its interests by reading the annual reports or other publications of the agency. Often you will be able to interview staff members and find out what sorts of work the group encourages. You may be able to initiate a relationship with the funding source, finding out which of your projects might be viewed more favorably. You may be able to find out how sophisticated the technical background is of the prospective reviewer. If you then discover you will be writing for generalists, the entire focus and language of the proposal should be altered to accommodate their understanding.

A proposal's chances improve dramatically if it is tailored to the readers' interests and background knowledge. Readers generally look for three basic qualities in all proposals: significance of the problem the writer addresses, the quality of the proposed solution, and the track record of the people who will be running the project. Beyond that, each agency has its own quirks; a grant seeker who understands these quirks probably will get further than one who doesn't. If the audience is broad rather than specialized, the abstract, introduction, and conclusion should be purged of specialized terms. If you expect the review to be read by a panel of peers, however, it's essential to include all recent, related research work; if you accidentally omit important work done by one of the reviewers, you will decrease your chances for a respectful reading.

To appeal to readers who confront a waiting stack of proposals as though it were a tidal wave, writers will be wise to make the text as tight as possible. Authors who simply echo the wording in the guidelines for the competition not only lengthen their proposals, but also suggest they have failed to master their own subject. Similarly, padding of the proposal with extraneous text will reflect poorly on the writer. The proposal that uses 200 words in the background to discuss how the school was founded 100 years ago, and then throws in a list of the Board of Trustees, wastes the time of its reviewers.

Authors should proofread carefully for inconsistencies in style and grammar. Stylistic and grammatical lapses can prejudice reviewers who may reason that if the prose is flawed, then the ideas, too, are probably weak.

SUMMING UP

- Proposals are sales documents, and to do their job—get money, equipment, time, staff—they must sell the ideas of the proposer.
- Proposals may take the form of a letter or a 30-page dossier. They may be brief narratives, or longer, multi-part documents with separate sections for the introduction, technical discussion, management plan, evaluation, and budget.

• Proposals must answer certain implicit questions: Why should we spend this money? Why should we spend this money on *you*?

• To answer these questions successfully, the writer must build a tight argument for the significance of the problem, the benefits of the solution, and the qualifications that make the proposer the person for the job.

EXERCISES

1. Identify a problem in your school or place of work, such as difficulties in the computer lab, discrepancies in language abilities of teaching assistants, poor organization of the parking lot, or the lack of a snack bar in a particular building. Write a short proposal addressing this problem. Include a summary, as well as a letter of transmittal. If you are writing the proposal to address a problem at school, identify an official—a president or dean, for example—to whom you might realistically address the proposal. Imagine this person as you are writing. Decide what arguments would be most appealing if, say, the dean were your reader: Student retention? Improvement in morale?

To get help with the budget, make appointments with officers in the school who can help you calculate costs such as electricity, heating, or fringe benefits.

2. Prepare a proposal for your final report for this class. Include an introduction stating the problem and its significance. Discuss the proposed outcome—the main divisions of the topic you expect to explore. Give a timetable for the work.

3. Prepare a proposal on behalf of an organization to which you belong, such as a sports team, choir, computer users, or ham radio operators club. You may apply to a central organization for funds for a project, for instance, or to a city, state or federal agency. You may apply to a foundation for help with a program, or for equipment or services.

CHAPTER 19

Manuals

Manuals are guidebooks that explain how to use a product, process, or system. To do a good job, the manual must be accurate, complete, and readable for its audience. This requires

- **research,** lots of it—understanding what you are describing by reading specifications, talking to production staff, interviewing, listing, noting, studying
- **thinking** about the audience
- **organizing** the material logically to guide the reader
- **choosing** language that is as free of jargon as possible
- **creating** a design and illustrations that give directions their strongest possible visual impact
- **revising** after you've listened, written, and tried out the text
- **editing** for readability, consistency, and style

This chapter looks at ways to handle these imperatives of audience, logic, language, and visual display using examples from effective manuals.

AUDIENCE ANALYSIS

At one time, design engineers were famous for their abrupt dismissal of the technical writer. "Never mind the manual," they said. "It's the product that sells." To them, the manual was a piece of window dressing, the manual's author a nuisance foisted on them by interfering management.

"It's the product that sells," may have been true in the days of the Eastman Kodak slogan, "You push the button, we do the rest." But nowadays there are a lot of buttons to push on products, and people are increasingly aware that the manual can either make or break sales. In the personal computer market, for instance, stories abound of how products stand or fall on the strength of their user manuals.

There are many audiences for manuals, including

- educated people with no specialized technical background—the folks, for instance, who buy and use a new word processing program
- trained technical specialists—the people, for instance, who repair an argon-ion laser
- clerical workers—people who follow a manual to enter, keystroke by keystroke, the data that are the lifeblood of businesses from brokerage houses to advertising agencies

How do you know which of these multitudes are your readers? If you are working as a contractor for a government agency, a lot of your questions will be forestalled, if not answered, by detailed specification sheets that describe the audience you are required to address. Outside of contract work, the question of identifying the audience is more problematic. Some managers and technical publications directors admit they use "educated guesses" when deciding what their audience knows and needs to know. Others hire marketing analysts to help get the answers. Analysts build dossiers and user profiles, they take surveys and polls, they interview, they prod the writers and artists to keep the user in mind from beginning to end.

If you are writing a manual, you have to imagine the user when deciding every element, from order of detail to visuals: Who is going to use it? What do they know already? What do they need to know? Find out about your audiences, and keep them in mind throughout the process. They will color every decision you make.

ORGANIZATIONAL PATTERNS

Overviews, abstracts, roadmaps—all the previewing devices used in the early chapters come into play when writing a manual.

It's easy to see why if you think about the different ways people use a manual. Most readers give the manual a serious look when they first get the item or system it documents; later the manual becomes a reference book to flip open when the user needs help. It is rarely read cover to cover.

Previews to tell them what's coming, and why they need to know it. Provide these

- at the beginning of the manual
- at the beginning of each chapter, section, or module

Wrap-ups that summarize what you've said

- at the end of each chapter, section or module; readers use these for review and quick reference, when the memory of the chapter has faded
- at the end of the text, in the form of glossaries and an index

Within, text is organized by what the user needs. For beginning users of software applications, for instance, the text is often broken into a "tutorial"—a short, guided tour of what the program can do—and "reference sections"—detailed explanations of all options.

In a technical manual, the conventional divisions are

- overview
- description
- principles of operation
- testing and troubleshooting
- assembly/disassembly
- maintenance
- packaging and preservation
- illustrated parts breakdown

PREVIEWING DEVICES

Table of Contents

Since many people look here first, try to make the titles and first-level headings as informative as possible, postponing technical terms to second-level headings when you can. Keep your eye on the *major* and *minor* categories, building the table of contents until it shares the directness and clarity of a good roadmap.

Use type that reflects the major/minor division. You might use larger and darker type for the major headings, and lighter type for the minor ones. You can use all capitals for the major points, initial capitals and indentation for the minor ones.

Keeping the language consistent (parallel) can also help make a table of contents more readable. Mixing phrases and sentences creates a jarring effect.

The use of direct questions helps liven a table of contents, as do verbs that tell people what they will be doing—"Using Tabs" instead of "The Use of Tabs."

Entry in Table of Contents that fails to reflect major and minor divisions

WORD PROCESSING CODES FOR AN INDENTURE

CODES FOR TYPESETTING THE INDENTURE
INSTRUCTIONS FOR CODING AN INDENTURE FOR TYPESETTING

PRINTOUTS OF CODED EXAMPLES

Entry in Table of Contents where type size and use of indentation emphasize major and minor divisions

CHANGING INDENTURE CODES FROM WORD PROCESSING TO TYPESETTING

Using Word Processing

Adapting to Type Setting

Printing Out Coded Examples

Practice Exercises

Box 19.1

Box 19.2

MIXED TITLES IN TABLE OF CONTENTS

Editing the layout

Selecting text and graphics

How do tabs work?

You can change the look of the type.

PARALLEL TITLES IN TABLE OF CONTENTS

Editing the layout

Selecting text and graphics

Using tabs

Changing the look of the type

Chapter Table of Contents

A table of contents before each chapter or module permits you to show second-level headings (Figures 19.1 and 19.2).

Sometimes the table of contents that begins the chapter is printed on heavier stock, or color-coded stock, particularly if the manual is bound looseleaf or spiral.

The introductory page for this section of a manual on electronic mail, pictured in Figure 19.3, is printed on heavy, brightly colored paper. The page, which functions as a chapter overview, summarizes the six steps that are the basis for the chapter on "Message Types."

Introduction

The introduction is the place to tell users what the manual will do for them, what it won't do, and what equipment they need to have on hand.

Apple II's proDOS User's Manual has a preface, "About This Manual," that gives a roadmap. (Figure 19.4.)

A roadmap is a division into parts. The Apple II roadmap explains the division of the manual and summarizes each of the three parts.

Text *continues* on page 500

Contents

Figure 19.1 Excerpt from table of contents for a manual. Courtesy Texas Instruments.

2 Getting Started

Figure 19.2 A chapter-level table of contents permits the author to show second-level headings. Courtesy Texas Instruments.

Sending a Message

Each time you connect to EasyLink there are six steps you need to complete to send any message.

1. Dial EasyLink

2. Connect to EasyLink

ID?

3. Type Customer Identification ↵

PTS

4. Type Address for Message + ↵

GA

5. Type the Text of the Message ↵

6. Send Message by Typing LLLL ↵

Accepted Message Number

The following chapter describes each of EasyLink's messages and shows you samples detailing the basic steps required.

Message Types

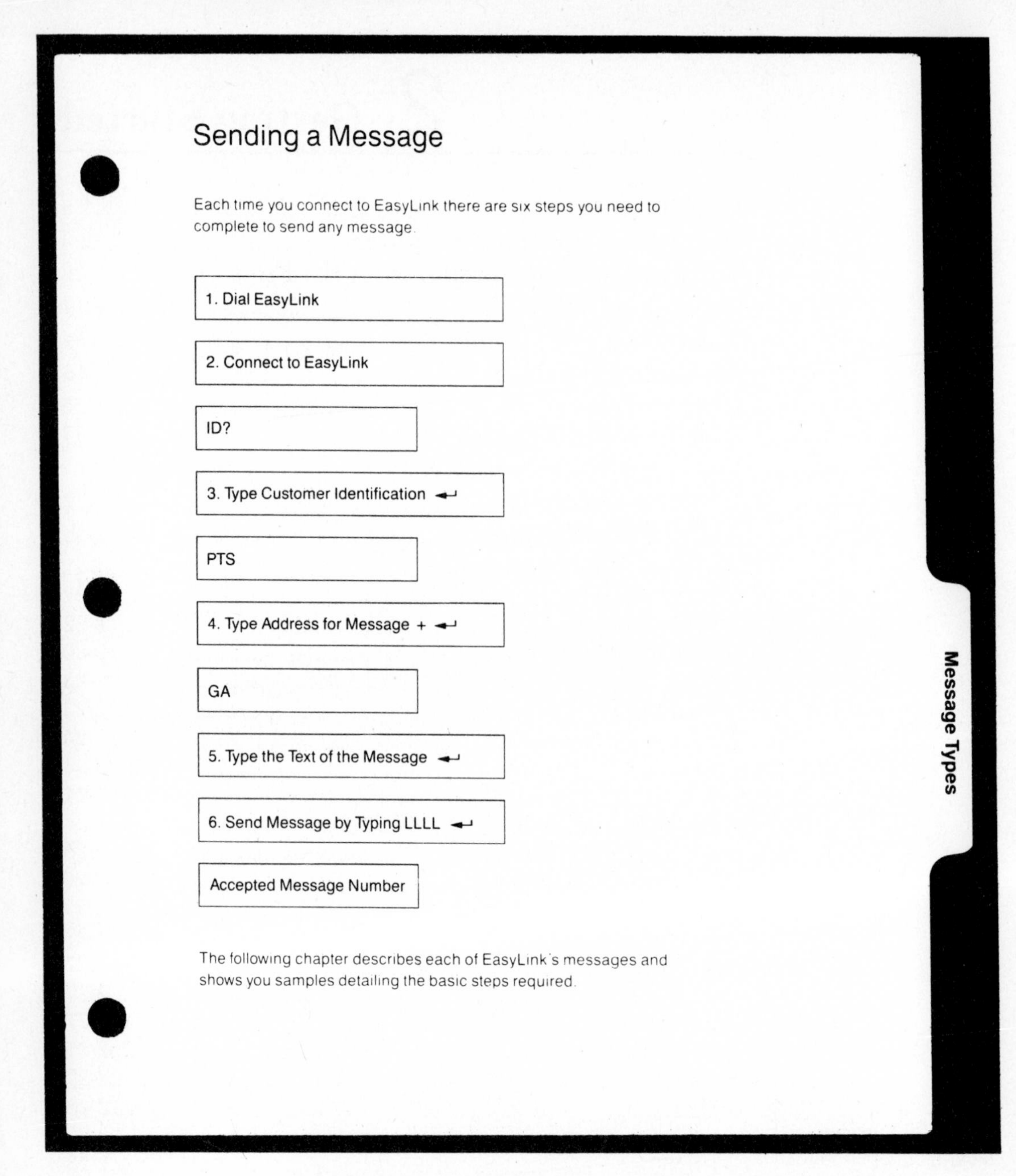

Figure 19.3 Prefatory page, printed in heavy stock, summarizes main points in the chapter it introduces. Courtesy Western Union *Easy Link Training Manual*.

How to Use This Manual

This manual is divided into three parts. The first part is a brief introduction to the *User's Disk*. Because the *User's Disk* is fairly self-explanatory, this may be the only part of the manual you will need to read.

The second part discusses the ProDOS Filer in detail. If you are not familiar with how a hierarchical file structure works and how to use pathnames, you may want to read Chapters 2 and 4. Chapters 3 and 5 tell you how to use the volume and file commands. These chapters are written so that you may use them just for reference. But don't hesitate to work through the chapters, trying out every command. Chapter 6 is about the default assumptions used by the Filer.

The third part explains how to use the DOS-ProDOS Conversion Program included on the *User's Disk.* This program allows you to convert your DOS 3.3 files to ProDOS files and vice versa.

Practice Makes Perfect

Reading a recipe for whole wheat bread is not the same as mixing the ingredients, kneading the dough, and baking the bread. Similarly, reading about how to copy a file is not the same as actually copying one. Try out each command as you read about it. That's the best way to find out if you understand the material.

Summary Sections

Some people like to race through manuals; others like to take it slow and steady. The skimmers among you will appreciate the summaries at the end of each chapter. Each summary is a digest of all the new commands and vocabulary presented in that chapter.

You may use the summaries as a quick reference or to test your mastery of the material you've just covered. If any of the commands are still hazy, it's a clue to go back and review before tackling new material. If you're an old hand with computers and disk operating systems, the summaries may be all the instruction you need.

Figure 19.4 The introduction is the place to tell users how the manual is organized and what it will do for them. Notice the informal diction, the "you" address and the contractions, all of which help establish a direct voice. Courtesy of Apple Computer, Inc., *proDos, User's Manual,* 1986.

Notice the "you" address and the contractions, both of which work to establish a direct voice. Sentences are short and loose, rather than long and inverted. Diction is informal ("don't hesitate," "an old hand with computers").

Many manuals have a preview section to orient the reader. Here, for example, is a "Before You Begin" from the opening of a manual.

BEFORE YOU BEGIN

The voice is direct: the reader is addressed as "you." Some of the chapter headings are direct questions ("Are you ready?") to establish a direct, "spoken" tone.

Chapter Overviews

In the same way that a report is prefaced by an abstract giving the gist of the findings, each chapter in the manual should have a precis of what will follow, telling the readers what the chapter will be about, and why this information matters.

LANGUAGE AND VOICE

Avoiding jargon If you are writing for the beginner, you have to deal immediately with the problem of jargon, that specialized or technical language of a trade, profession, class, or fellowship.

Some technical terms are inevitable. Use them, providing definitions both on the spot the first time, and in the glossary.

Being friendly "User-friendly" does not mean breathing in the reader's face. Prose that is too friendly can take the form of breathless exclamations—"Congratulations! You're going to be delighted with this new microplexer!"—or condescension: "Don't be afraid of the manual. It won't bite you!"

Address your readers directly, but don't buttonhole them or become patronizing. Talking to people as clearly as possible doesn't mean grabbing them by the lapels.

The "you" form of address makes sense in a manual. It's a way to talk to readers without entering a thicket of passive constructions. If you do use direct address, though, be careful to stay with it; don't switch back and forth. Figure 19.5 is an "Introduction to Unix." The writer switches from "you" to "one" to "he."

Figure 19.5 "You" address makes sense in a manual, helping the writer avoid a thicket of passive constructions. Don't switch, though, from "you" to "he" to "one," as the writer has done in this example.

PREFACE

THE UNIX operating system provides many features you can use to develop both programs and documents. This mini-manual helps one learn the basic functions needed to get started. After the beginner is on his way, he'll find the detailed program guide useful.

The language of a manual is most effective when it helps bring the reader closer. Contractions, active verbs, direct address, and questions are all devices that help create the illusion that the writer is speaking directly to the reader.

When you couple accurate, methodical presentation with direct language, you can create the sense of the writer as an informative, helpful guide. (For a more extensive discussion of voice in technical writing, see Chapter 4.)

You want the reader to hear a friendly, competent voice, the voice of a person who has thought about the problems and has the wit and organizational skill to present solutions.

This voice can be established in the preface and introductory sections of the manual, and carried throughout the reference sections.

VISUALS AND DESIGN

Skilled manual writers look for opportunities to show instead of tell, or, as one editor put it, to "say a page in a figure."

White space, attractive page layout, effective print specifications—all these elements of design can help the reader move through the text rapidly and efficiently.

Visuals Line drawings are popular in manuals, for they permit great flexibility and control in closeups and level of detail.

Photographs are popular for introductory shots. "Photos show this product must really exist. They build confidence. People are reassured that we really have an item to sell," joked one technical publications manager. In the age of vaporware (software that is promised, but never appears), such concern is reasonable. Usually photographs are used for control panels, or other overviews. They are followed by either line drawings, which permit closeups and selective views, or graphs and tables. (For a more extensive discussion of visual display, see Chapter 5.)

Troubleshooting Sections have their own conventions, all of them related to government specifications that developed to guarantee consistency in manuals written for the armed forces. Many government specifications for technical manuals require that troubleshooting sections be in the form of logic trees or diagrams. Outside the world of government contracts, people adopt these standards for their own uses.

Format for Warnings Given the subjects many manuals address, **warnings** (usually protecting people) and **cautions** (usually shielding equipment) have a special and important place.

• Put them *before,* not after, the hazardous step.

• Keep them whole. If they begin at the bottom of the page, don't break them in two. Instead, rearrange the text so the warning or caution is whole.

• Set them off dramatically with some combination of boldfaced capital letters, border, typesize, and color. Traditionally, red is used for warning, orange for caution. An acceptable form for government manuals is shown in Figure 19.6.

Illustrated Parts Breakdown Both government and nongovernment manuals follow similar formats for illustrated parts break-

WARNING

CAUSTIC CHEMICALS IN BATTERIES

• Use rubber gloves and apron to avoid severe burns.

• If chemicals get on your skin, clothes, or equipment, wash immediately with water.

• If chemicals get in your eyes, wash them with plenty of water and get medical help immediately.

Figure 19.6 The word "warning" is set in large bold type. Each alternative is separated to add distinct visual identity. The entire warning is boxed. From *Military Handbook 360386.*

downs. Usually the callout goes directly on the illustration, connected by arrows or leader lines (Figure 19.7). Arrows should not zig-zag on route to the object.

If the callout is complex, the parts are indexed and called out separately in a legend or listing (Figure 19.8). In Figure 19.9, instructions for disassembly are set beside an indexed illustration. The index reduces the clutter in the illustration.

Exploded Views These are drawings that show the unit disassembled, but with all the parts positioned in correct relationship to one another. Exploded views are common in illustrated parts breakdowns, as well as in the materials used in the manual to illustrate installation, removal, disassembly, assembly, repair, and replacement.

Figures 19.10 and 19.11 illustrate good and bad exploded views. All numbers used to index the callout should be outside the boundaries of the parts illustrated. In Figure 19.10, some of the numbers are set within the parts, and are hard to locate, such as 6, 13, 19, 20, and 24. Figure 19.11 shows an exploded view in which the indexed numbers are easier to follow. It also has a valuable locator view, showing how the part relates to the larger unit.

Measured Quantities Quantities such as length, time, and amount are usually expressed in numerals, regardless of the value. Thus, "6 years old, 25 hours later, 1.24 inches, 0.25 inch.

Unit modifiers are usually hyphenated. Thus, 5-turn rotation; 5-percent reduction.

Maintenance Instructions For the maintenance sections in manuals, government specifications suggest formats like Figure 19.12 when the illustration requires only a short written accompaniment. The page is broken into two columns, with the procedure presented in tabular layout on the left, while the illustration occupies the right column.

Standard Format You need to specify line position, type size, and typestyle for these elements:

title
first-, second-, and third-level headings
warnings
tables
figure captions
running text
running heads

Text *continues* on page 510

NAVAIR 01-XXXXX

006 00

Page 3

WARNING

Dry cleaning solvent, P-D-680, used to clean parts is potentially dangerous to personnel and property. Avoid repeated and prolonged skin contact. Do not use near open flame or excessive heat.

g. Carefully clean all residual oil and preservative compound from top surface of arbor shaft, suspension rod, and indicator bushing. For best indication, thoroughly degrease these surfaces, using clean cloth moistened with solvent (P-D-680, Type II).

h. Using two small pieces of paper, or equivalent, as shims (approximately 0.003 inch thick) under indicator bushing, reinstall the indicator bushing on to suspension rod and tighten setscrew.

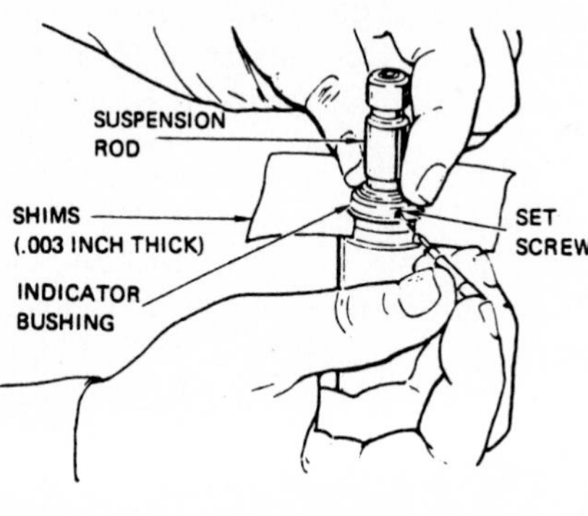

Step h

i. Remove shims and check that the indicator moves freely, without any evidence of binding, over the black indicator disk incorporated into the arbor end.

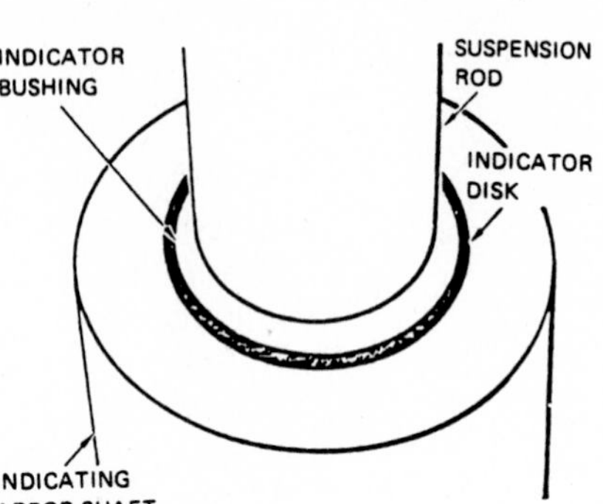

Step i

j. If indicator arbor flops or oscillates from one side to the other (or inadvertent loss of dash-pot oil), remove button-head hex socket screw (5/64-inch hex key wrench) at 11-inch scale position. Fill arbor to hole level with mineral oil (MIL-L-15018, Grade 5150). Reinstall buttonhead socket screw.

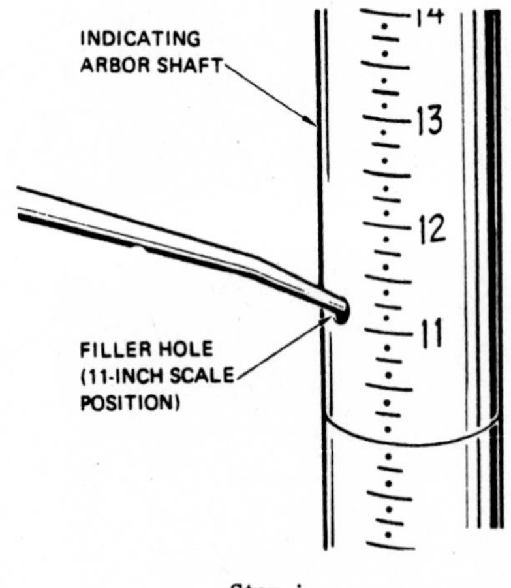

Step j

Figure 19.7 Direct Callout. In this page from a navy manual, names of parts are placed directly on the figure and connected to the illustration with arrows. Note the boxed warning.

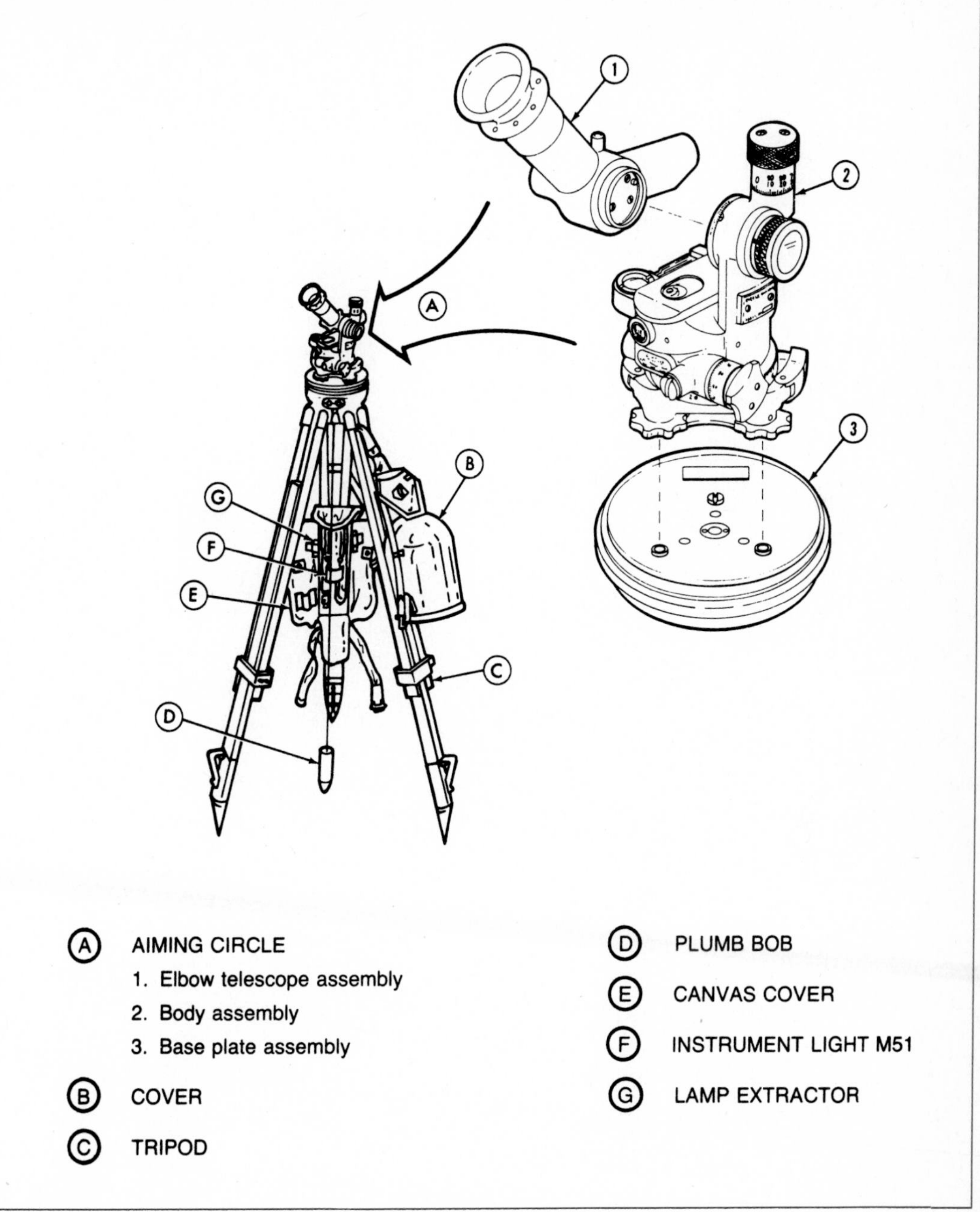

Figure 19.8 Indexed Callout. Here the names of parts are indexed (A, B, C, 1, 2, 3) and called out below in a separate legend.

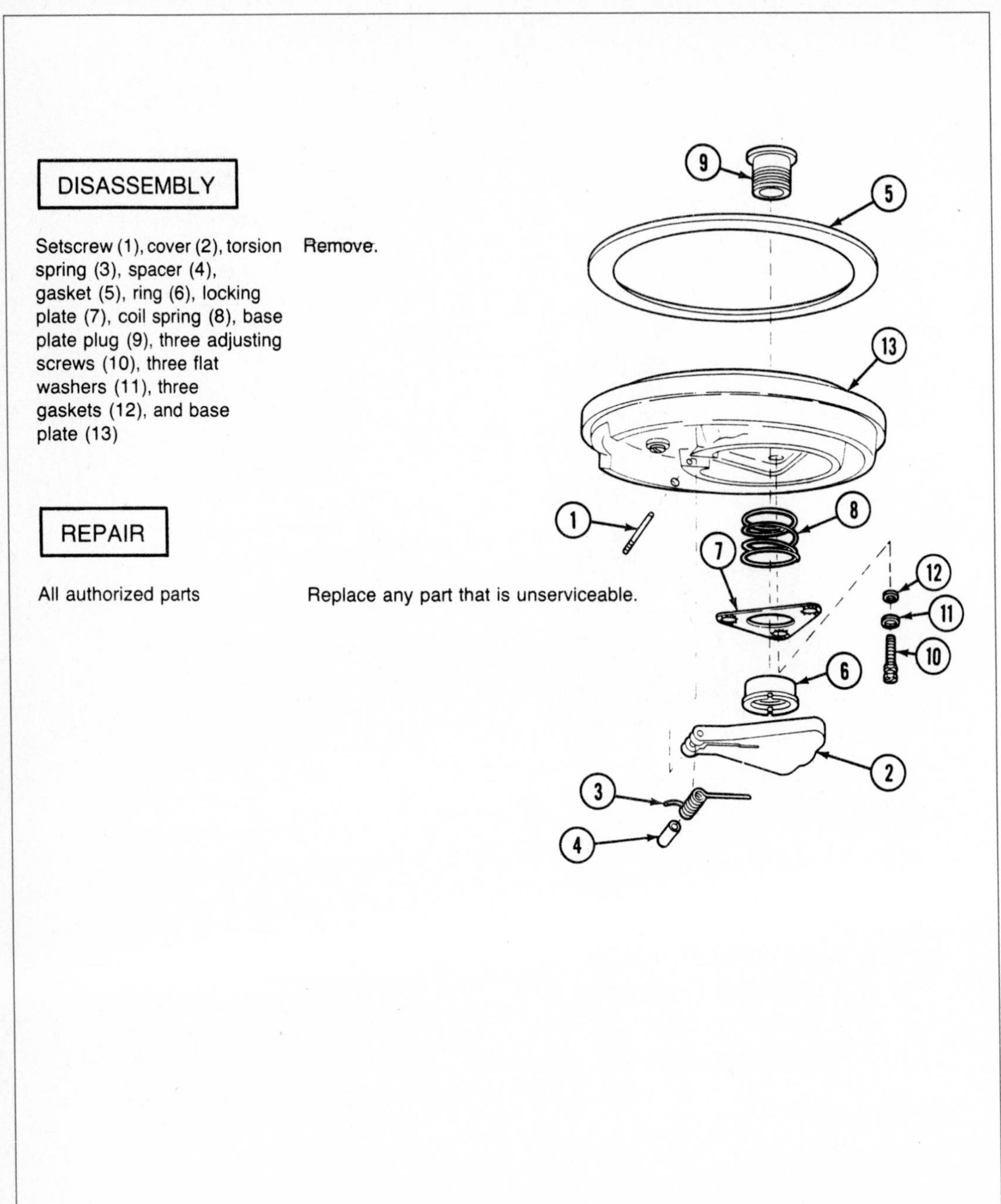

Figure 19.9 Indexed callout reduces clutter in this illustration from a government manual.

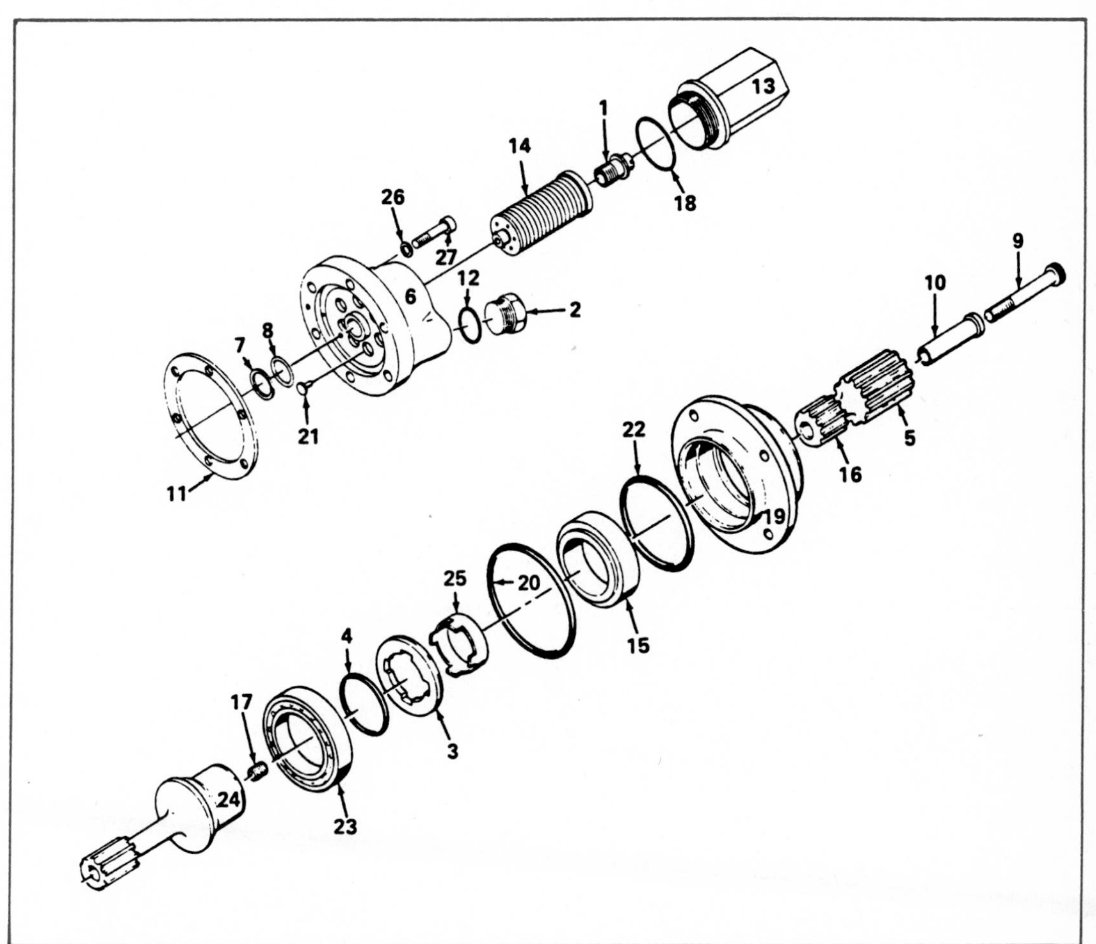

Figure 19.10 Exploded View. Poor Index Placement. Some numbers (6, 13, 19, 20, 24) set within parts are hard to locate. From *NAVAIR Manual 00-25-700*.

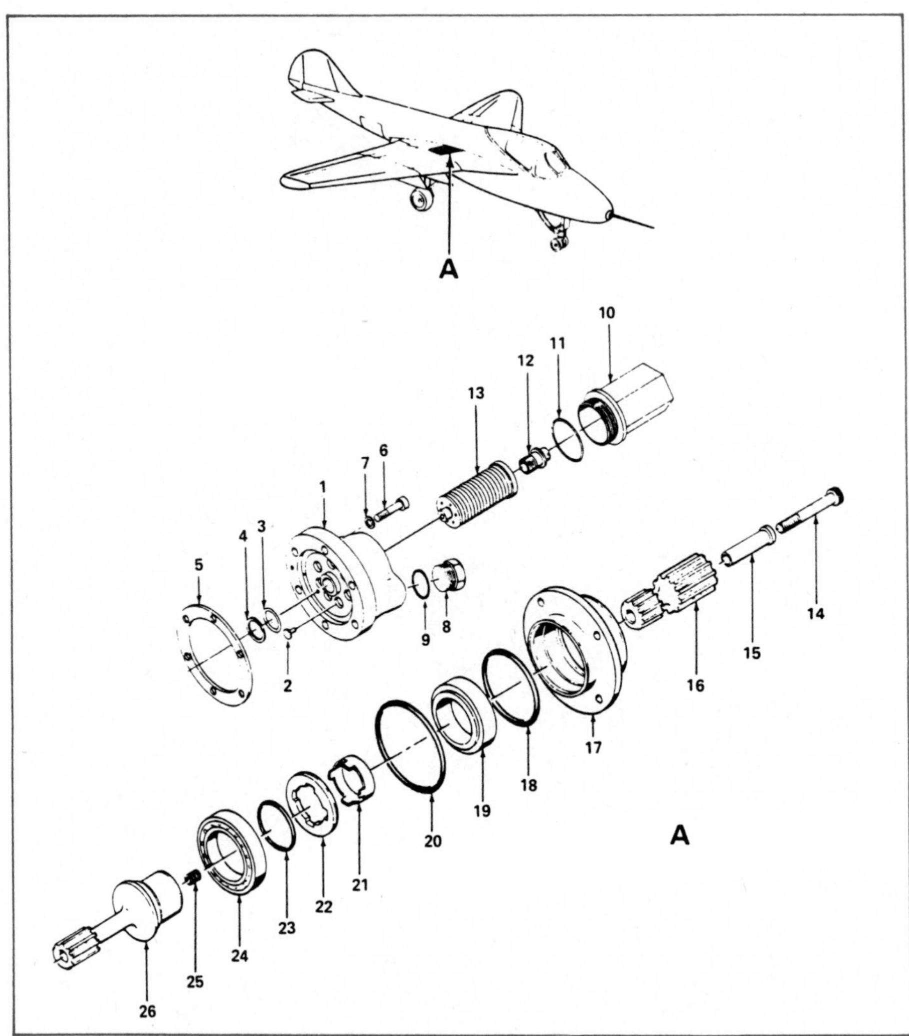

Figure 19.11 Exploded View. Improved Index Placement. Note valuable locator (Label A), which shows relationship of parts to larger unit. From *NAVAIR Manual 00-25-700.*

NAVAIR 01-XXXXX

115 00
Page 2

Step 1. Removing Oil Cooler Door.

a. Remove nut, washers, and bolt securing actuating rod to oil cooler door.

b. Remove six bolts (three on each side, one through top, two through bottom) securing oil cooler door hinge fitting to oil cooler duct.

c. Remove oil cooler door and hinge fitting.

Step 2. Removing Oil Cooler Door Operating Mechanism.

a. Remove oil cooler door. (Refer to step 1.)

b. Remove screws securing access panel to duct assembly; remove access panel.

Note

When removing access panel, exercise care to prevent damage to oil cooler door actuating rod seal.

c. Disconnect oil cooler door cable at main landing gear door operating linkage by removing nut, washer, and bolt.

d. Remove bolts and washers securing oil cooler door operating mechanism to wing leading edge over boom beam.

e. Position operating mechanism to gain access to pulley guard pin and cable; remove pin and cable.

f. Leave cable installed in wing; remove oil cooler door operating mechanism. Replace cable if damaged.

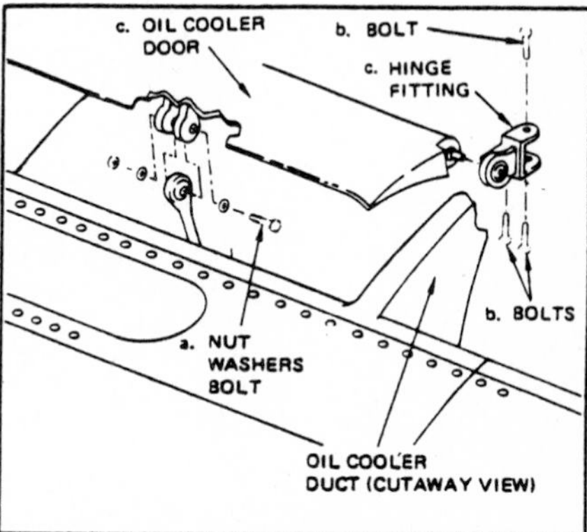

Step 1

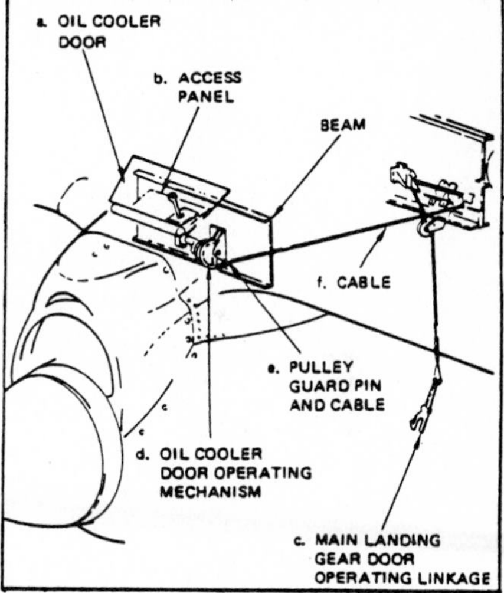

Step 2

Figure 19.12 Maintenance Instructions. In this format from a navy manual, the procedure is in the left column, the accompanying illustrations to the right.

You need to decide on position and typestyle for

bulleted lists

numbered lists

subscript and superscript

mathematical formulas

callouts

callout arrows

Most of these elements are the same as those you need to specify for reports. (See standardized format sheets, Chapter 17.)

If you are working alone on a project, you may think format standards are frills that need concern only the typist who is working on the final version of your report or manual. Actually, the consistency of typographical elements—the formal decisions you make about where to place bullets or secondary headings, for instance, or which type to use in the table of contents—makes a big impact on the reader. When these elements are consistent or parallel throughout the text, the manual has a pleasing coordination. By the same token, a mixture of typefaces is as jarring as a mixture of nonparallel grammatical structures, as unsettling as reading a phrase such as "She likes singing and to dance."

Making the manual's style consistent becomes even harder when there is more than one person working on a section. That's when the format sheet is particularly handy; it is a place to specify and regularize all the details of typography and layout. Will you use second-level headings? Will you begin the paragraph on the same line as the second-level heading, or drop to the next line? Will you use end punctuation for lists? The format sheet will help you make decisions about these matters, and then act consistently on your decisions.

In some companies, an editor is assigned toward the end of the manual's production to comb the copy and make sure it is consistent.

Earl Clifford, head of technical publications at Timeplex, comments. "Our technical writers are translators. They translate from the engineering level to the user's level. We have a writing team of 5–6 people, and each person inevitably formats and writes somewhat differently. That's fine in the opening stages, because the writers are focusing on the information—on getting the text down accurately, completely, and understandably. We have to do a 500-page manual in five or six months, so the content is first. Then, once the content is correct, we have production editors who take the manual over, and make the style more consistent."

EDITING: Two Kinds of Tables for a Troubleshooting Guide

Alan Gaynor is a technical writer who works for a manufacturer of statistical multiplexers that permit a number of ports to share a data link efficiently. The statistical multiplexers are used in networks to connect computer ports, terminals, printers, and other devices throughout a site. A supervisory terminal provides network control from a single place.

Gaynor was preparing a revision of the troubleshooting guide, the set of tables users consult when there is a problem. Figure 19.13 shows two pages of a table from the original guide. The column on the far left lists problems. (Since the problem can be listed more than once in the guide, the user checks the entire table.) The next four columns show the status of the indicators on the control panel—off, normal, alarm, flashing. The last two columns show probable cause and correct action.

The tables were both clear and correct, but there was room for improvement. First, the tables made no attempt to list mutually exclusive categories. Instead, the user had to look through the whole table. Second, the language was sometimes vague, as in "various conditions."

Gaynor and his associates wanted to produce a section of the report in which the graphical display made the text easier to follow.

The revision of the table is a dramatic improvement (Figure 19.14).

When the user looks at the troubleshooting guide, the configuration of the panel is shown. In miniature, the heads of the table provide an analogue of the physical panels of the device the user is troubleshooting.

The problem list runs across the top. Subordinated beneath it is the panel with its indicators. Thus for the problem "No data flow on all ports," the user looks at the top columns and sees two possible displays. In one, all the indicators are off (represented by the blank circles in the "normal," "alarm," and other indicator columns). In the second major division, some circles have an "F" within them. The note explains that they are flashing.

The probable causes and corrective actions are beneath, set off by boldface and bullets.

A doublescored vertical line separates the last column, "no or faulty data flow," from the first two columns.

This is an ingenious solution. The graphical display seeks to do two important things: help the viewer to compare different data and show the data graphically at several levels of subordination.

Text *continues* on page 517

Problem	Indicators: Main Module	Indicators: Data Link Module (NM/QSM only)	Indicators: Expander Modules	Indicators: Power Supply	Probable Cause	Corrective Action
No data flow on all ports.	All off.	All off.	All off.	All off.	Loss of power	Check power cord is connected and power circuit breaker is ON.
					Faulty Power Supply	Replace Power Supply Module.
	Continually resets: ALARM and NORMAL indicators flash alternately.	Continually resets: ALARM and NORMAL indicators flash alternately.	Continually resets: ALARM and NORMAL indicators flash alternately.	NORMAL indicator lit.	Defective Main Module	Replace Main Module.
No or faulty data flow on some or all ports	Various conditions	Various conditions	Various conditions	ALARM indicator lit.	Faulty Power Supply	Replace Power Supply Module.
No data flow on all ports on a data link	CARRIER FAIL and LINK ERROR indicators lit. OVERLOAD indicator may be lit.	CARRIER FAIL and LINK ERROR indicators lit. NORMAL indicator off.	NORMAL indicator lit.	NORMAL indicator lit.	Problem on data link(s).	Enter LINK command or SY2 mode to isolate problem to 1 or more data links.
					Faulty cabling between NSM and modem on data link.	Check cabling.
					Faulty telephone line or modem on data link.	• If using an AIM modem on an SM, perform data link modem loopbacks from supervisory terminal. • If troubleshooting an NM/QSM perform modem diagnostics per modem manual.
No or faulty data flow on all ports on a data link.	NORMAL and LINK ERROR indicators lit. OVERLOAD and ALARM indicators may be lit.	LINK ERROR indicator lit. ALARM indicator may be lit.	NORMAL indicator lit.	NORMAL indicator lit.	Problem on data link(s)	Enter LINK command or SY2 mode to isolate problem to 1 or more data links.
					Faulty cabling between NSM and modem on data link.	Check cabling at both local and remote units.
					Faulty telephone line or modem on data link.	• If using an AIM modem on an SM, perform data link modem loopbacks from supervisory terminal. • If troubleshooting an NM/QSM, perform modem diagnostics per modem manual.

Figure 19.13 First Version of Trouble Shooting Guide. Courtesy Timeplex, Inc.

Problem	Indicators: Main Module	Data Link Module (NM/QSM only)	Expander Modules	Power Supply	Probable Cause	Corrective Action
(Cont'd)					Faulty remote NSM.	Perform diagnostics at remote unit.
					Faulty local or remote Main Module or Data Link Module.	Perform STATUS command for local and remote units.
	NORMAL, CARRIER FAIL, and LINK ERROR indicators lit; and OVERLOAD indicator may be lit.	LINK ERROR and CARRIER FAIL indicators lit.	NORMAL indicator lit.	NORMAL indicator lit.	Faulty cabling between NSM and modem.	Check cabling.
					Faulty telephone line or modem.	• If using an AIM modem on an SM, perform data link modem loopbacks from supervisory terminal. • If troubleshooting an SM with non-AIM modem or an NM/QSM, perform modem diagnostics per modem manual.
	LINK ERROR and/or CARRIER FAIL indicators lit.	NORMAL indicator off, various conditions otherwise.	NORMAL indicator lit.	NORMAL indicator lit.	Faulty Data Link Module.	Replace Data Link Module.
No or faulty data flow on some or all ports.	NORMAL indicator lit.	NORMAL indicator lit.	NORMAL indicator lit.	NORMAL indicator lit.	Faulty data base or incorrect routing of port.	• Reprogram routing as required. • Display hardware and firmware status report at supervisory terminal by a STATUS command. Check for data base error. • If resetting the system is possible, initiate SY1 and SY6 tests at front panel. Check for data base error. • If resetting the system is possible, initiate SY5 calculate checksum test to correct data base checksum error.
No or faulty data flow on some or all ports and ALARM indicator lit on Main Module.	NORMAL and ALARM indicators lit.	NORMAL indicator lit.	NORMAL indicator lit. (ALARM indicator may be lit.)	NORMAL indicator lit.	Incorrect programming or data base checksum error.	1. Check all programming. Reprogram if necessary. 2. Display hardware and firmware status by STATUS command or SY6 mode. If data base error: • If resetting the unit is possible, enter GO,R and STATUS commands or SY1 and SY6 modes. Check for data base error.

Figure 19.13 (*continued*)

Problem: No data flow on all ports

Indicators / Module	Normal	Alarm	Link Error	Carrier Fail	
Data Link	○	○	○	○	
			Carrier Fail	Link Error	Over Load
Main	○	○	○	○	○
Expander	○	○	Problem No. 1		
Power Supply	○	○			

- **Probable Cause:** Loss of power at affected Multiplexer

  Corrective Action: Check power cord is connected and power circuit breaker is ON. Check fuse on 8-port Multiplexers.

- **Probable Cause:** Faulty Power Supply at affected Multiplexer

  Corrective Action: Replace Power Supply Module.

Problem: No data flow on all ports

Indicators / Module	Normal	Alarm	Link Error	Carrier Fail	
Data Link	Ⓕ	Ⓕ	Ⓕ	○	
			Carrier Fail	Link Error	Over Load
Main	Ⓕ	Ⓕ	○	Ⓕ	○
Expander	Ⓕ	Ⓕ	Problem No. 2		
Power Supply	●	○			

NOTE

NORMAL and ALARM indicators flash alternately; Main, Data Link, and Expander Modules continually reset.

- **Probable Cause:** Defective Main Module at affected Multiplexer

  Corrective Action: Replace Main Module.

Problem: No or faulty data flow on all ports on a data link

Indicators / Module	Normal	Alarm	Link Error	Carrier Fail	
Data Link	○	‡	●	●	
			Carrier Fail	Link Error	Over Load
Main	‡	‡	●	●	◐
Expander	●	○	Problem No. 3		
Power Supply	●	○			

- **Probable Cause:** Problem on data link(s)

  Corrective Action: Enter [ee]LINK command or SY2 mode to isolate problem to 1 or more data links between affected Multiplexers.

- **Probable Cause:** Faulty cabling between Multiplexer(s) and modem(s) on affected data link

  Corrective Action: Check cabling between Multiplexer(s) and modem(s) on affected data link.

- **Probable Cause:** Faulty Data Link Module at affected Multiplexer

  Corrective Action: Replace Data Link Module.

- **Probable Cause:** Faulty telephone line or modem on affected data link

  Corrective Action:

  - If using an AIM modem on a single/dual link Multiplexer, perform Loopback Troubleshooting Procedure for Failure of All Ports on Data Link. (See Figure 3.)
  - If troubleshooting a single/dual link Multiplexer with non-AIM modem or a Quad Multiplexer, perform modem diagnostics per modem manual and Loopback Troubleshooting Procedure for Failure of All Ports on Data Link. (See Figure 3.)

Figure 19.14 Revised Version of Trouble Shooting Guide. Courtesy Timeplex, Inc.

Problem: No or faulty data flow on all ports on a data link (Cont'd)

Indicators / Module	Normal	Alarm	Link Error	Carrier Fail	
Data Link	●	⊗	●	⊗	
			Carrier Fail	Link Error	Over Load
Main	●	◐	⊗	●	◐
Expander	●	○	Problem No. 4		
Power Supply	●	○			

- **Probable Cause:** Problem on data link(s)

  Corrective Action: Enter [See$]LINK command or SY2 mode to isolate problem to 1 or more data links.
- **Probable Cause:** Faulty cabling between Multiplexer(s) and modem(s) on affected data link

  Corrective Action: Check cabling at both affected Multiplexers.
- **Probable Cause:** Faulty Main Module or Data Link Module at affected Multiplexers

  Corrective Action: Perform [See$]STATUS command for affected Multiplexers Perform appropriate corrective action.
- **Probable Cause:** Faulty telephone line or modem on affected data link

  Corrective Action:
  - If using an AIM modem on a single/dual link Multiplexer, perform Loopback Troubleshooting Procedure for Failure of All Ports on Data Link. (See Figure 3.)
  - If troubleshooting a single/dual link Multiplexer with non-AIM modem or Quad Multiplexer, perform modem diagnostics per modem manual and Loopback Troubleshooting Procedure for Failure of All Ports on Data Link. (See Figure 3.)

Indicators / Module	Normal	Alarm	Link Error	Carrier Fail	
Data Link	⊗	○	●	●	
			Carrier Fail	Link Error	Over Load
Main	●	◐	●	●	◐
Expander	●	○	Problem No. 5		
Power Supply	●	○			

- **Probable Cause:** Faulty cabling between Multiplexer and modem on affected data link

  Corrective Action: Check cabling.
- **Probable Cause:** Faulty telephone line or modem on affected data link

  Corrective Action:
  - If using an AIM modem on a single/dual link Multiplexer, perform Loopback Troubleshooting Procedure for Failure of All Ports on Data Link. (See Figure 3.)
  - If troubleshooting a single/dual link Multiplexer with non-AIM modem or a Quad Multiplexer, perform modem diagnostics per modem manual and Loopback Troubleshooting Procedure for Failure of All Ports on Data Link. (See Figure 3.)

Problem: No or faulty data flow on ALL ports of an Expander Module

Indicators / Module	Normal	Alarm	Link Error	Carrier Fail	
Data Link	●	○	○	○	
			Carrier Fail	Link Error	Over Load
Main	●	◐	○	○	○
Expander	◐	◐	Problem No. 6		
Power Supply	●	○			

- **Probable Cause:** Expander Module not properly seated at affected Multiplexer

  Corrective Action: At affected Multiplexer, push Expander Module into connector.

Figure 19.14 *(continued)*

Problem	No or faulty data flow on ALL ports of an Expander Module (Cont'd)				
Indicators / Module	Normal	Alarm	Link Error	Carrier Fail	
Data Link	●	○	○	○	
			Carrier Fail	Link Error	Over Load
Main	●	●	○	○	○
Expander	●	○	Problem No. 7		
Power Supply	●	○			

- **Probable Cause:** Data base error at affected Multiplexer

  Corrective Action:

  - Check all programming at affected Multiplexer. Reprogram if necessary.
  - Display hardware and firmware status report of affected Multiplexer at supervisory terminal by a [See$]STATUS command.
    Check for data base error.
  - If resetting the affected Multiplexer is possible, initiate SY1 and SY6 tests at affected Multiplexer front panel.
    Check for data base error.
  - If resetting the affected Multiplexer is possible, initiate SY5 calculate checksum test at affected Multiplexer to correct data base checksum error.

Problem	No or faulty data flow on 1 or more ports of an Expander Module				
Indicators / Module	Normal	Alarm	Link Error	Carrier Fail	
Data Link	●	○	○	○	
			Carrier Fail	Link Error	Over Load
Main	●	●	○	○	◐
Expander	●	●	Problem No. 8		
Power Supply	●	○			

- **Probable Cause:** Faulty Expander Module at affected Multiplexer

  Corrective Action: Replace Expander Module.

Problem	No or faulty data flow on 1 or more ports				
Indicators / Module	Normal	Alarm	Link Error	Carrier Fail	
Data Link	‡	‡	‡	‡	
			Carrier Fail	Link Error	Over Load
Main	‡	‡	‡	‡	‡
Expander	‡	‡	Problem No. 9		
Power Supply	○	●Ⓕ			

- **Probable Cause:** Electrical short on Main, Data Link, Expander Module, or motherboard at affected Multiplexer

  Corrective Action: Remove all modules except Power Supply Module. If Power Supply Module NORMAL indicator lights and ALARM indicator goes off, reinsert Main, Data Link, and Expander Modules 1 at a time. If Power Supply Module indicators show problem after reinsertion of any module, module may be faulty. Remove questionable module and reinsert in another position (slot).

  - If module is OK in new position (slot), call field service and report faulty position (slot).
  - If module causes indicators on Power Supply Module to show problem, replace faulty module per Corrective Maintenance Manual.

- **Probable Cause:** Faulty Power Supply at affected Multiplexer

  Corrective Action: Replace Power Supply Module.

Figure 19.14 *(continued)*

EXERCISES

1. *Individual Assignment:* Technical manuals for the armed services are prepared in accordance with one or more military specifications. See Figure 19.15 for a recent list of specifications.

One of the publications that gives guidelines for technical manuals is the Naval Air Systems Command's NAVAIRN. Using these guidelines, a student prepared the following parts of a manual for an airbrush (see Example 19.1):

Testing and Troubleshooting

Maintenance Instructions

Operating Instructions

The student's manual appears on pp. 519–526. Included are sections on testing and troubleshooting, maintenance, and operation.

Do a similar project.

- Choose a set of guidelines such as the one for NAVAIRN.
- Write a manual for a product, following these guidelines.

2. *Team Assignment:* You are as likely to find a technical writer at a brokerage house, documenting the paths of commission payments, as in a naval engineering firm, working on an installation manual for a sensing device.

Manuals are often written by a team. Sometimes the engineers or programmers who designed the product are the authors; sometimes technical writers and editors—joined, on occasion, by marketing consultants—work as a group with engineering or programming staff to produce the manual.

The manual may be split up into jobs, some done by the writing group, others done by the production group. Sometimes there is a separate editing group. If you are using this book in a classroom, divide into similar groups and prepare a manual.

3. *Analysis:* Find a manual you think is either effective or ineffective. Write a brief (1- or 2-page) analysis of the techniques used in the manual.

TECHNICAL MANUAL SPECIFICATIONS

MIL-M-5288	Manuals, Technical: Cargo Loading and Offloading; Preparation of
MIL-M-7298	Manual, Technical: Commercial Equipment
MIL-M-008910	Manual, Technical: Illustrated Parts Breakdown, Preparation of
MIL-M-9854	Manuals, Technical: Structural Repair (For Aircraft)
MIL-M-23305	Manual, Technical Operation: Maintenance and Overhaul Instructions with Illustrated Parts Breakdown (For Aircraft Launching and Recovery Equipment), Preparation of
MIL-M-23618	Manuals, Technical: Periodic Maintenance Requirements, Preparation of
*MIL-M-23695	Manuals, Technical: Test Equipment, Calibration Procedures; Preparation of
MIL-M-38784	Manuals, Technical: General Requirements for Preparation of
MIL-M-38793	Manuals, Technical: Calibration Procedures, Preparation of
MIL-H-46855	Human Engineering Requirements for Military Systems, Equipment, and Facilities
MIL-M-81043	Lubrication Charts For Air Weapons Systems, Preparation of
MIL-M-81203	Manuals, Technical: In-Process Reviews, Validation and Verification, Support of
MIL-M-81218	Manuals, Technical: Aircraft Engine Intermediate and Depot Maintenance, Preparation of
MIL-M-81928	Manuals, Technical: Aircraft and Aeronautical Equipment Maintenance, Preparation of (Work Package Concept)
MIL-M-81929	Manuals, Technical: Illustrated Parts Breakdown, Preparation of
MIL-M-85025	Manuals, NATOPS Flight: Requirements for Preparation of
MIL-M-85041	Manuals, Technical: Conventional Microfilm Compatible
MIL-C-81222	Checklists, Flightcrew Pocket, Preparation of
MIL-M-81260	Manuals, Technical: Aircraft Maintenance
MIL-M-81310	Manuals, Technical: Airborne Weapons/Stores Loading (Conventional and Nuclear)
MIL-M-81700	Manual, Technical: Airborne Armament Equipment
MIL-M-81701	Manuals, Technical: Airborne Missiles and Guided Weapons, Preparation of
MIL-M-81702	Manual, Technical: General Airborne Weapons (Conventional)
MIL-M-81715	Manuals, Technical: Ship Weapon Installations
MIL-M-81748	Manuals, Technical: Rapid Action Changes, Requirements For Preparation of
MIL-M-81754	Manuals, Technical: Weapon System Technical Documentation List, Preparation of
MIL-M-81792	Manuals, Technical: Loading and Transport of Nuclear Weapons in Cargo Aircraft, Preparation of
MIL-M-81834	Manuals, Technical: Aircraft Tactical, Requirements for Preparation of
MIL-M-81901	Manual, Technical: Aircraft Fire Fighting and Rescue; Requirements for Preparation of Information for Inclusion in
MIL-M-81919	Manuals, Technical: Support Equipment, Preparation of
MIL-M-81927	Manuals, Technical: General Style and Format of (Work Package Concept)
MIL-M-85337	Manuals, Technical: Quality Assurance Program; Requirements for
*AR-75	Manuals, Technical: General Style and Format (Microfilm Compatible)
*AR-76	Manuals, Technical: Aircraft, Equipment and Component Maintenance Preparation of (Microfilm Compatible)
*AR-78	Manuals, Technical: Illustrated Parts Breakdown; Preparation of (Microfilm Compatible)

*Superseded Specifications – Listed because certain manuals in current use are still being maintained in accordance with these documents.

Figure 19.15

SHOP REPAIRABLE ASSEMBLY MAINTENANCE
TESTING AND TROUBLESHOOTING
BADGER 150 DUAL-ACTION, INTERNAL-MIX AIRBRUSH

Reference Material

Badger 100 Series Instruction Book (10-78) . . . . BA 319
Airbrushing for Fine & Commercial Artists . . . . ISBN 0-442-27508-0
Paschal, R. New York, Van Nostrand Reinhold. 1981.

Alphabetical Index

TESTING AND TROUBLESHOOTING

DESCRIPTION. This work package describes the most common problems faced in using the airbrush. The coverage reviews the problems and offers solutions.

BASIC CLASSES OF PROBLEMS. A dirty airbrush causes most of the problems met in airbrush operation. Poor adjustment of the moving parts also causes problems. The user almost always finds the source of the problem by direct inspection. The user can clean dirty parts or change adjustments directly, without seeking outside help.

TROUBLESHOOTING TABLE. Scan the problems listed in Table 1. Look for the probable cause and take corrective action.

Example 19.1 Student-prepared manual for an airbrush.

Table 1. Testing and Troubleshooting

PROBLEMS	QUESTIONS AND PROBABLE CAUSES	SOLUTIONS
1. Bubbles form in paint jar during operation.	Is the spray regulator turned out too far?	Inspect the regulator. If needed, turn the regulator in (clockwise) one or two turns.
	Is air leaking back into the airbrush and down the siphon?	Check the head washer. It should be firmly in place at the rear of the head assembly. If the washer is firm but bubbles still form during operation, apply softened beeswax all around the washer to restore the seal.
2. Paint does not spray during triggering.	Is the vent open in the jar lid? If the vent has been covered by accident, a vacuum forms in the paint jar.	Check the vent. If the pinhole is covered or clogged, clear or clean it.
3. Spray cannot be shut off.	Is the tip clean? The tip may be clogged with dried paint. This prevents the needle from closing off the paint passage.	Check the tip. Remove any bits of dried paint.
4. Restricted spray.	Is the spray regulator screwed into the head assembly too far?	Check the regulator. If needed, open it up one or two turns. If the problem persists, read problem 6 in this Table.
5. Crooked direction of spray.	Is the tip of the needle bent? This will prevent fine-line delivery of spray.	Check the tip. A bent tip can be fixed with gentle use of pliers. Carefully straighten the tip. Then sand the tip on emery sandpaper.

Example 19.1 *(continues)*

Table 1. Testing and Troubleshooting (continued)

PROBLEMS	QUESTIONS AND PROBABLE CAUSES	SOLUTIONS
6. Irregular air supply. Symptoms: restricted spray or a grainy spray pattern.	Are all air chambers tightly sealed? A small air leak in the air-supply system can cause this.	Check the airhose connections. They should be tight. Check the hose for leaks as follows: a. Look for worn fabric or a tear in the braiding. b. Keeping the ends clear, dip the hose in a basin of water. Look for bubbles. If the hose is problem-free, check the compressor as follows: a. Look for a dirty filter. Clean or replace it if needed. b. Look for a worn or missing gasket. Replace it if needed.

Example 19.1 (*continues*)

SHOP REPAIRABLE ASSEMBLY MAINTENANCE
MAINTENANCE INSTRUCTIONS
BADGER 150 DUAL-ACTION, INTERNAL-MIX AIRBRUSH

Reference Material

Badger 100 Series Instruction Book (10-78) . . . . BA 319
Airbrushing for Fine & Commercial Artists . . . . ISBN 0-442-27508-0
Paschal, R. New York, Van Nostrand Reinhold. 1981.

Alphabetical Index

DESCRIPTION.

ITEMS TO CLEAN. This work package describes how to maintain the 150 airbrush. It covers the following topics:

a. Cleaning the Color Passage.
b. Cleaning the Needle.
c. Cleaning the Spray Regulator.
d. Removing Dried Paint.

MAINTENANCE TABLE. Follow the cleaning steps presented in Table 2 to ensure a clean tool, ready for use at all times. Maintaining an airbrush requires inspection and cleaning only. No lubrication or other kinds of care are needed.

Example 19.1 (*continues*)

Table 2. Cleaning Steps.

ITEM TO BE CLEANED	CLEANING STEPS
1. Color Passage	1. Fill siphon with cleaning agent that matches the type of color medium. For example, water thins acrylic paints. 2. Insert the jar with the cleaning agent into the force fit holder. 3. Use disposable paper towels or rags to spray into. 4. Depress airbrush trigger and draw back on it all the way. This sprays the cleaner through the paint passage at high volume. 5. When the fluid sprays clear, the color passage is clean.
2. Needle	1. Spray the cleaning agent through the airbrush as though cleaning the color passage. Do this until the spray is clear. 2. Remove the handle that houses the back of the needle. 3. Remove the needle from the assembly. Make sure the needle adjustment screw and needle chuck are attached. 4. Wipe the needle clean with a soft cloth dampened with cleaning agent. 5. Replace the needle in the assembly gently. 6. Spray the cleaning agent through the airbrush once more. 7. Remove the jar from the force-fit holder.

Example 19.1 *(continues)*

Table 2. Cleaning Steps (continued)

ITEM TO BE CLEANED	CLEANING STEPS
2. Needle (cont.)	8. Keep spraying until the cleaner is used up.
3. Spray Regulator	1. Remove the head assembly using the wrench that comes with the airbrush. Turn counterclockwise. 2. Soak the spray regulator in a jar of cleaner for 1-2 hours, or until it is needed again. 3. Replace the spray regulator.
4. Parts Caked with Dried Paint	CAUTION Never soak the entire airbrush. Soak only the parts through which paint flows. 1. Soak the part on which deposits of dried paint have formed. The deposits should float off. 2. If soaking fails, send the airbrush to the manufacturer or to a cleaning service for sonic cleaning.

Example 19.1 *(continues)*

OPERATING INSTRUCTIONS
BADGER 150 DUAL-ACTION, INTERNAL-MIX AIRBRUSH

Reference Material

Badger 100 Series Instruction Book (10-78) . . . . BA 319
Airbrushing for Fine & Commercial Artists . . . . ISBN 0-442-27508-0
Paschal, Robert. New York, Van Nostrand Rinehold. 1981.

Alphabetical Index

1. OPERATING INSTRUCTIONS.

2. DESCRIPTION. This work package explains how to use the 150 airbrush. Figure 1 is a cutaway of this dual-action, internal-mix model.

3. MIXING PAINT. Mix any color combination desired and load the paint jar as follows:

CAUTION

Paints must be compatible. For example, mix enamels only with other enamels and enamel solvents. Mix lacquers only with other lacquers and lacquer thinners.

a. Use correct thinner to soften the paint consistency. Mix paint and thinner in a separate jar to be used for mixing and thinning only.

Example 19.1 *(continues)*

b. Read carefully the proportions of thinner to paint to get the best mix.
c. Test the mixture with a sable brush. When the paint is thin enough, it will tint the brush, but not color it solidly.
d. Pour the paint from the mixing jar to the paint jar.
e. Attach the paint jar by pushing the force-fit cylinder onto the receiving end just behind the head assembly.

4. CONNECTING THE AIR SUPPLY. Attach the airbrush to the air supply as follows:

a. Attach the airbrush to the hose by gently turning the airbrush clockwise onto the hose fitting.
b. Attach the other end of the hose to the brass compressor fitting.
c. Plug the compressor cable into the foot pedal receiver.
d. Plug the foot pedal cable into the wall outlet.

5. HOLDING THE AIRBRUSH. To get the most ease of movement during operation, follow these steps:

a. Hold the airbrush as if it were a pen or pencil. Place the thumb and middle finger on each side of the shell and place the index finger on top of the trigger.
b. Drape the airhose over the forearm. This will help to avoid dragging it across the work.

6. TRIGGERING THE AIRBRUSH. The dual-action trigger changes the volume of paint sprayed as follows:

a. Depress the trigger. This action releases air through the airbrush.
b. Gently draw back the trigger. This action releases paint through the airbrush. Draw the trigger back more to boost the volume of fluid released. Move the trigger forward to decrease the volume of fluid.
c. If desired, turn the trigger—adjusting screw clockwise to move the trigger toward the handle. This adjustment presets the volume of spray.

7. DIRECTING THE SPRAY. Keep the airbrush at a right angle to the work. Move your entire arm smoothly to get an even pattern of spray.

CHAPTER 20

Aloud: Oral Technical Reports

> **So now you know what I'm going to talk about. The next question is, will you *understand* what I'm going to tell you? Everybody who comes to a scientific lecture knows they are not going to understand it, but maybe the lecturer has a nice, colored tie to look at.**
>
> **Richard Feynman, *QED*.**

> **I could somehow see nothing all around me but the paper, and I felt as if my body was gone, and only my head left.**
>
> **Charles Darwin, on his feelings before he spoke at the Geological Society.**

There are technical people whose talks are legendary—Richard Feynman's series of filmed physics lectures given at Cornell in 1964 and later published as *The Character of Physical Law;* Philip Morrison on subjects from time to termites. Michael Faraday's lectures, such as "The Chemical History of a Candle," are earlier examples of the same clarity in explication, as are Darwin's. These hypnotic speakers have given lectures later remembered by people in the audience virtually word for word.

The famous are not the only ones who get up before groups and talk. Most technical and scientific professionals spend a surprising amount of time giving oral reports. These are some typical instances.

- **Orientation and training talks.** Talks for new staff members include introductions to equipment and procedures, whether in the laboratory or in production, in quality control or in design.
- **Professional presentations at conferences and meetings.** This is the classic way scientists and technical staff share the news of their discoveries.
- **Project analyses, status reports, new products, new processes, proposals.** The work may be presented orally to colleagues or supervisors, to sales and marketing staff, or to officials from a regulatory agency.

What are the important ingredients when you are giving a report aloud? Some of the techniques are the same as those for writing: There is no replacement for audience analysis, logical presentation, clear language, and effective visual display, whether you are talking or writing.

Accurate, thorough preparation is as essential for speaking as it is for writing. Few people speak well without doing their homework. Everyone has a favorite horror story—the speaker who wandered from the announced point, never to return, or the speaker who decided to "warm up" the audience with a series of jokes that ranged from stale to offensive.

While some of the ingredients that make for a good talk are the same as those for good writing, others are different. These include the "platform skills," those in-person items such as eye contact, voice, bearing, poise, and dress; the ability to interact with and respond to the audience; the timing and integration of verbal and visual information; and handling nervousness.

Some of the key ingredients in a successful speech are nonverbal: The poise, the energy and the *brio* that characterize the liveliest speakers cannot be added to a talk like sauce to a steak. These attributes will come naturally out of the preparation and effort you put into the talk. With effort, you can develop the appearance of effortlessness that is at the heart of poise. With the assurance that the details of the talk have been mastered, your interest and competence will have a chance to shine through. There may even be the opportunity to display some of the wit and authority that are the prerogatives of any knowledgeable speaker.

STEPS IN PREPARATION

Decide on a Goal and Response

The classic division of talks is into those that inform, those that persuade, and those that entertain. Most technical talks inform; some both inform and persuade. Either type of speech may entertain in a subtle way—through the clarity and assurance of the argument.

Here are some examples of the distinction between informative and persuasive titles:

INFORMATIVE:

How to Use Lotus 1-2-3, Version 2.0

Understanding Three Basic Principles of This Machine

Use of Nuclear Magnetic Resonance (NMR) Techniques

PERSUASIVE:

Lotus 1-2-3 Appears to Be the Best Software Program of its Kind on the Market.

Why You Should Buy This Machine

Reasons to Consider Switching to NMR

IN A SENTENCE OR TWO, SUMMARIZE THE MAIN IDEA YOU INTEND TO MAKE, BASED ON YOUR PURPOSE AND DESIRED RESPONSE.

Example:

Purpose:
Introduce Principles of Use for Lotus 1-2-3

Response:
At the end, people should be able to edit a cell's contents.

Main Idea:
Using Lotus 1-2-3, it is easy to edit a cell's contents.

Box 20.1

COMBINATION:

Applications of the NMR to Our Laboratory (What it is, and why it would be useful)

Lotus 1-2-3 and the User's Needs

A good talk has a purpose the speaker can articulate in a sentence or two. This purpose helps the speaker construct a logical presentation, for the content flows from the purpose.

Beginning speakers often set out to accomplish the impossible in a speech—cramming into their presentations eight or nine major points, without giving the examples or clarifications the audience needs to grasp the major theme.

Instead, choose one central point, an idea you want the audience to leave with after your talk is done.

Develop an Outline That Considers How Much Time You Have To Speak

If you plan on using the usual rhetorical devices to expand and illustrate your thesis—examples, analogies, comparisons, contrasts—you will find yourself against a sharp limit: the amount of time you have.

Given your rate of talking and the slot alloted for your talk, you are likely to be limited in what you can say—unless you are like the man in the famous fast-talking commercial clocked at an impressive 400 words per minute.

One obvious solution is to practice aloud, and develop a sense as to how long it is going to take you to say the words. But even before that, consider time limits as you develop your outline. Each of your supporting points should be buttressed with examples or explanations, not simply announced. The more general the audience background, the more examples, explanations, and contrasts you will want to weave into the talk, and therefore the fewer supporting points you will be able to make.

All of the talk should be subsumed under the main-idea thesis or statement—the idea you want the audience to leave with.

Main Idea

I. Point One
Support

II. Point Two
Support

III. Point Three
Support

TIPS FROM A NOBEL WINNER

In a charming discussion of how to present a scientific paper in *Advice to a Young Scientist* (Harper & Row, 1979), Sir Peter Medawar, a Nobel Prize Laureate, includes the following:

- "People with anything to say can usually say it briefly; only a speaker with nothing to say goes on and on as if he were laying down a smoke screen."
- "Under no circumstances whatsoever should a paper be read from a script. It is hard to overestimate the dismay and resentment of an audience that has to put up with a paper read hurriedly in an even monotone."
- "A lecturer can be a bore . . . because he goes into quite unnecessary details about matters of technique. . . . If it is important to know and if the audience wants to know the order in which the speaker dissolved the various ingredients of his nutritive culture medium, he will be asked immediately after the lecture, or privately later on."
- "Even the most experienced speakers feel nervous before a talk, and it is very right that they should do so, for it is a sign of anxiety to do well. . . . Audiences respond better to evidence that the speaker has been at pains to prepare whatever he has to say."
- "If people *do* sleep in their lectures, speakers should try to get some comfort from the thought that no sleep is so deeply refreshing as that which, during lectures, Morpheus invites us so insistently to enjoy."

Box 20.2

Tight, clear presentations where the speaker makes the point, explains the implications, and then answers questions, are far preferable to lengthy speeches. The successful speaker is one who delivers the message *before* the audience's attention span is depleted. Plan to fit your argument within a tolerable time limit. How long is long enough? Attention often begins to wander after 20 minutes. If you are slotted for a longer time, try to vary the pace with questions or discussion.

Sketch the Introduction and Conclusion

The introduction is probably the hardest part of a talk. It is when you first stand up that nerves take over, and that you are most likely to fumble.

To add to their sense of assurance, most speakers carefully think out their introduction.

The classic introduction

1. *focuses the attention of the listener.* People in the audience may have their minds elsewhere, perhaps on work remaining to be done or a phone call that needs to be made. The speaker counters this inclination with a statement or question compelling attention.

2. *gives the audience a reason to listen.* The speaker connects the talk to the concerns of the audience.

3. *orients them to the talk.* The speaker closes the introduction with a summary of the major divisions or points in the talk that will follow.

Such introductions range from the simple to the elaborate. The simplest version would be

1. *focus*—statement of main idea.
2. *reason to listen*—significance of main idea.
3. *orientation*—division of the speech.

This is a standard opening. While it is not dramatic, it is certainly preferable to plunging directly into detail without a statement of the thesis and its significance.

If you are interested in starting with more of a flourish, you might consider

- a representative anecdote illustrating the problem
- a question that you either answer, or involve the audience in answering
- a dramatic example of the problem
- an instance from which you deduce the main point
- a definition from which you proceeded to the thesis

Feynman starts his lecture on probability (Example 20.1) with examples and a definition of chance ("By chance we mean something like a guess"). He asks a question and answers it ("Why do we make guesses?"), gives examples ("Let us consider the flipping of a coin"), and finally settles into the point: "By the 'probability' of a particular outcome of an observation, we mean our estimate for the most likely fraction of a number of repeated observations that will yield that particular outcome."

Examples

"Chance" is a word which is in common use in everyday living. The radio reports speaking of tomorrow's weather may say: "There is a sixty percent chance of rain." You might say: "There is a small chance that I will live to be one hundred years old." Scientists also use the word chance. A seismologist may be interested in the question: "What is the chance that there will be an earthquake of a certain size in Southern California next year?" A physicist might ask the question: "What is the chance that a particular gieger counter will register twenty counts in the next ten seconds?" A politician or statesman might be interested in the question; "What is the chance that there will be a nuclear war within the next ten years?" You may be interested in the chance that you will learn something from this chapter.

Definition and explanation

By *chance*, we mean something like a guess. Why do we make guesses? We make guesses when we wish to make a judgment but have incomplete information or uncertain knowledge. We want to make a guess as to what things are, or what things are likely to happen. Often we wish to make a guess because we have to make a decision. For example: Shall I take my raincoat with me tomorrow? For what earth movement shall I design a new building? Shall I build myself a fallout shelter? Shall I change my stand in international negotiations? Shall I go to class today?

Sometimes we make guesses because we wish, with our limited knowledge, to say as much as we *can* about some situation. Really, any generalization is in the nature of a guess. Any physical theory is a kind of guesswork. There are good guesses and there are bad guesses. The theory of probability is a system for making better guesses. The language of probability allows us to speak quantitively about some situation which may be highly variable, but which does have some consistent average behavior.

Let us consider the flipping of a coin. If the toss—and the coin—are "honest," we have no way of knowing what to expect for the outcome of any particular toss. Yet we would feel that in a large number of tosses there should be about equal numbers of heads and tails. We say: "The probability that a toss will land heads is 0.5."

First main point

We speak of probability only for observations that we contemplate being made in the future. *By the "probability" of a particular outcome of an observation we mean our estimate for the most likely fraction of a number of repeated observations that will yield that particular outcome.* If we imagine repeating an observation—such as looking at a freshly tossed coin—N times, and if we call N_A *our estimate* of the most likely number of our observations that will give some specified result A, say the result "heads," then by $P(A)$, the probability of observing A, we mean

$$P(A) = N_A/N.$$

Example 20.1 Opening of talk on probability. From *The Feynman Lectures on Physics*. Richard P. Feynman, Robert B. Leighton, and Matthew Sands. Addison-Wesley Publishing Company, Reading, Massachusetts, p. 61.

A nice opening, and certainly a more inviting one than

Good afternoon.

By the probability of a particular outcome of an observation we mean our estimate for the most likely fraction of a number of repeated observations that will yield that particular outcome.

The other part of the outline to write out and practice is the conclusion. Again, stage fright is likely to take over, unless the speaker has prepared. "Well, I guess that's all . . . uh . . ." the speaker trails off, suddenly embarrassed to be center stage. That's not the way to end. Instead, a summary, or perhaps a summary and some form of exit line, are a better way to wind up smartly before calling for questions. Send them away with a statement of the key point, not an apology.

Building the Argument

A technical speech, like a technical paper, is built on the scaffolding of logic. The speaker creates a path for the audience to follow, just as a writer does. This path is marked by a clear and unambiguous organization. Your organization may be simple—a question/answer format in which you introduce new lab personnel to equipment and procedures. It may be a step-by-step presentation explaining how to enter data. It may be the distillation of a year of research, in which you present the findings of an investigation.

You will need to shape the argument according to the nature of the material. If you are explaining a procedure, for instance, you will probably use a chronological pattern (time-order) combined with listing. If it is a report of an investigation, you will probably present an abstract and overview first, and then an account of the procedure, results, and conclusions.

You will also need to shape the material according to the audience. For a diversified group, decide which terms to define, rather than presenting an alphabet soup of buzz words and acronyms that people outside your technical specialty cannot follow.

Focus on your topic and its implications, rather than descending into a welter of detail that you find fascinating, but that they may find irrelevant.

Technical background is not the only consideration in audience analysis, although it is the major one. Consider the personal background of those you are addressing. For instance, one polymer scientist who studies compatibility began a technical presentation with a joke defining "incompatibility"—it's the formula for a marriage, he said. He provides the income, she provides the "patability." Most of the people in the audience—both male and female—were offended.

In developing an argument appropriate to your audience, be sure to allow time for devices of coherence and unity:

- *Transitional words and phrases* stitch the argument together and help the audience keep its place.
- *Nutshell summaries*, after you finish each point, wrap up what you've said and connect it to the next point.
- *Explicit coordination with graphics*: Each transparency, slide, or visual you use will take a minimum of 20 seconds, and usually more, to be read by the audience. Allow time for people to see what you are showing them, and weave into the talk explicit summaries of what each visual represents.

WRITING OUT A TALK

Some speakers write out their talks; others do not. Preparation shows in a speech, though, and therefore certain parts are best written in advance:

The Outline A written outline will help you to balance available time against necessary arguments so you do justice to the central idea.

Introduction and Conclusion Writing out the beginning and the end of a talk is also important, because they are the two weakest chains in the argument, the places where the speaker standing before a group is most likely to feel suddenly unsure of a subject which—until that tense moment—seemed completely mastered.

Comments to Accompany Visuals Writing out comments that accompany visuals helps speakers organize a presentation. In the process, the speaker may decide on the best interpretive remarks, and determine how much of the available time will be needed for explanations.

What about the rest of the speech? Some people write out the body word-for-word, others simply practice from a topical outline. The neophyte usually writes out every word, while the more experienced speaker may settle for a dense outline.

STANDARDS OF DELIVERY

Since you will want a conversational quality in the speech, writing sections of the talk is only the beginning. Your talk may be written, but it has to sound spontaneous.

Practice will give you a chance to give the talk a poised, spoken flavor. Go through the talk aloud. Analyze it, revise weak spots, and go through it a second time. Once you have timed it, you can, if you wish, avoid the ordeal of saying it aloud, but silent practice will not work when you are testing how long the talk will run.

How much should you practice? *Until you sound as though you are speaking naturally*. Then reduce your notes to a skeleton outline, including quotations or figures that must be presented verbatim. It is better to read these than to get them wrong. Also include a prompt or cue list for yourself if you are using visuals, so you'll remember the next slide in the sequence, and can coordinate the slide with what you are saying.

As you practice, you will have to consider platform skills. These skills are unique to live performance:

Eye Contact In the movie *Tootsie*, Dorothy Michaels (portrayed by Dustin Hoffman) plays her memorable opening scene in the soap opera opposite the lecherous physician. As they talk, the doctor keeps looking over Dorothy's shoulder, and she naturally keeps swiveling to see why he isn't talking to her. (He is reading the cue cards.)

If you've ever had a conversation with someone who refused to look at you, you know why Dorothy kept checking over her shoulder.

Beginning speakers have trouble maintaining eye contact with their audience. But audiences are disconcerted by speakers who do not look at them, just as Dorothy was surprised by the doctor's looking over her shoulder. If the speaker looks at the ceiling and floor, the audience is going to start by looking there, too.

Speaking is always a struggle to engage the attention of the audience, attention that is easily lost. Looking people in the eye will not prevent their thoughts from being far away, but it is one more way to try to establish contact.

Usually, though, floors and ceilings are not the spots to which shy speakers retreat. Instead, nervous technical speakers tend to talk to their slides, to transparencies, or to the blackboard. It is the perfect way to avoid the audience, as a shy presenter soon learns.

It is understandable that a beginning speaker who is uncertain will prefer to look down at notes or up at slides rather than address the audience directly. However, it is worth the act of will required to look directly at the audience. People respond more favorably when the speaker looks at them, not away.

Try looking from person to person in a slow rotation that includes everyone. Don't look only at the boss—the instructor, for instance, or supervisor. Linger on the occasional friendly face—it will give you courage—but then go on so that you look in turn at all the audience, taking only the occasional reassuring glance at your notes.

Stifle the desire to turn your back on the audience to talk directly to your slides or transparencies. If you are using visuals, position yourself so that you face the audience as you gesture toward the screen.

At a certain point—somewhere around 100 people, according to most accounts—it becomes impractical to look at everyone. You'll have to switch to looking at sections of the crowd, rather than individual faces. Similarly, if there are bright lights focused on you, you may not be able to distinguish individuals beyond the glare.

Dress "He made a very good speech," one student commented after a technical presentation, "but my attention was distracted by the large stain on his tie. I kept wondering if he knew about it." As in an interview, appearance plays a part in a presentation. Adjust your appearance so that it will not be a distraction.

Nervous Mannerisms These blossom when you are standing in front of a group. The sudden and irresistible itch, the quarters in your pants pocket that demand to be jiggled, the push button pen you suddenly start pushing, the pointer transformed into a conductor's baton—these are all fabled nervous tics to avoid. They will distract your audience.

If you have a tendency to tug at a suddenly constricting tie or necklace, force your hands instead to lie quietly at your sides. If you are going to put them in your pockets, empty the loose change first. If it is your feet that decide they want to march in a one-two-three-KICK pattern, force them to stay in one place.

Voice Vocal quality matters to an audience, who tend to become bored when they think the speaker is bored. Unfortunately, many speakers who are not bored sound like they *are*. What happens is that their voices become muted, almost monotonous, because of their nervousness and desire to "get it over with quickly." Enthusiasm, or at least interest, is important in a presentation. *The audience will respond to the intangibles of interest and competence that show through in voice inflection and facial expression.*

EXTRAVERBAL CLUES

Do people respond to a speaker's wit, enthusiasm, authority, poise? Unquestionably they do. Interestingly, the speaker's language and logic, no matter how well chosen and used, are only two of the vehicles that convey authority and poise. There are others.

To illustrate this, consider the instance of a famous physicist who delivered the annual lecture of the scientific research society, Sigma Xi. The lecture was eagerly awaited, for the speaker was known for his brilliant and entertaining essays.

The talk was carefully structured, with an arresting introduction, an elegant development of the thesis, and a clearly focused conclusion. Yet most of the audience failed to follow the argument.

Why was this? He *read* the talk, and did so in a flat, almost uninflected tone.

The speaker's inflection is very important in a talk. To understand why, consider a point made by the physician and author Oliver Sacks. Sacks is describing patients with global aphasia—people who are incapable of understanding words, but who nonetheless understand most of what is said to them. When addressed naturally, they grasp some or most of the meaning. "To demonstrate their aphasia," Sacks comments, "one had to go to extraordinary lengths, as a neurologist, to speak and behave unnaturally, to remove all the extraverbal cues—tone of voice, intonation, suggestive emphasis or inflection, as well as all visual cues (one's expression, one's gestures, one's entire, largely unconscious personal repertoire and posture): one had to remove all of this . . . in order to reduce speech to pure words. . . .

"Why all this? Because speech—natural speech—does *not* consist of words alone. . . . For though the words, the verbal constructions, *per se* might convey nothing, spoken language is normally suffused with 'tone,' embedded in an expressiveness which transcends the verbal—and it is precisely this expressiveness, so deep, so various, so complex, so subtle, which is perfectly preserved in aphasia, though understanding of words be destroyed. Preserved—and often more: preternaturally enhanced. . . ."

Extraverbal clues—tone of voice, intonation, suggestive emphasis—and visual cues—expression, voice, posture—are all a very large part of how people understand. This personal repertoire has more of a chance to shine through if you have worked on your talk until you are comfortable saying it using only a skeleton outline to guide you.

HANDOUTS, VISUALS, AND DEMONSTRATION MODELS

Technical presentations rely heavily on visuals, most often in the form of slides or transparencies. Visuals are suited to technical presentations, but the audience should also have a sense of the speaker, the person whose tone, bearing, and interest can add so much to their understanding of the argument.

Handouts These are invaluable for the person in the audience who wants both to hear and to read a point. A talk goes past word by word, relentlessly serial. A page can be reread. Handouts have disadvantages, though. People given something to read will, naturally, read it, and while they are doing this, they cannot be expected to listen. Many speakers solve this problem by

- allotting time for people to read handouts before talking
- giving out handouts afterwards to those who are interested
- making the handouts so schematic—an outline, a few key words, a drawing—that people can grasp the point with a look, freeing them to pay attention to the lecture

Depending on your subject, it may be helpful to copy the transparencies that will be shown, so that the audience can follow and take notes on, for instance, each chemical reaction.

Slides and Transparencies Beginners tend to use transparencies and slides that are far, far too complex, with three and four times the amount of information on them that can be seen and digested. People simply will not be able to resolve the print in a 14-column table. They won't be able to decipher the 6-point type in a callout.

You also need to allow time for each figure. Do not take a slide or transparency off before people have had a chance to read it. Nothing is more irritating than being on the verge of understanding a visual, only to have it removed. On the other hand, it is ineffective to leave visuals up for many minutes after the speaker begins to talk of something else. The audience naturally continues to look at the figure, which is by now irrelevant.

It is also wise to test the order of slides to make sure they are placed properly in the tray. Transparencies have their own ways of misbehaving during a speech; to keep them from sticking together, try separating them with sheets of paper.

GUIDELINES FOR EVALUATING A TECHNICAL SPEECH

Logic and Organizational Patterns

main topic clearly established

main topic related to audience interest

adequate time allowed for each supporting point

relationship of supporting points to main points established logically

periodic, short summaries given to help listener follow path of argument

conclusion tightly related to body of argument

Language

terms defined

examples, comparison, contrast used effectively to clarify abstractions

transitions provided

Illustrations

adequate time allowed for audience to comprehend technical illustrations

degree of complexity appropriate for intended audience

figures legible and readable, including labels on graphs and callouts

Delivery

eye contact

poise (absence of nervous mannerisms, interest and expertise in subject matter, recovery from mistakes)

voice

audibility

appearance

posture

Handouts (optional)

useful, rather than distracting

legible

Audience Interaction

questions welcomed

responses answered effectively

Box 20.3

Some speakers take questions during the presentation, interacting with the audience. If you want to allow this possibility, consider using transparencies. They are somewhat easier to switch than slides. There can be little last-minute adaptation once you have set up a slide tray.

When possible, turn off the glaring projector lights and talk to the audience. Such looks and interaction are a key part of the talk, not the window dressing.

If it is possible, dim some of the overhead lights, but not all of them. People need to see when they take notes.

COMMON CRITICISMS

Here are some of the common criticisms made by listeners. These comments underscore the importance of structure, language, visual display, and audience analysis to the listener:

- "The talk lacked structure."
- "The talk lacked transitions. I wasn't sure where the speaker was going."
- "Key terms were not defined."
- "The graphics were far too complicated."
- "He only looked at the boss, and at no one else."
- "There was no chance for feedback."
- "She talked to the blackboard, not to us."

- "Several main points were missing."
- "There was no summary."
- "The tables were not shown for long enough. Just as I focused, the transparency was whisked off the screen."
- "He played with the change in his pockets. It was very distracting."

SUMMING UP

- Some of the key ingredients in a successful technical talk are nonverbal: the poise, the energy, and the *brio* that characterize the liveliest speakers cannot be added to a talk like sauce to a steak. These attributes come out in the preparation and effort you put into the talk.
- With the assurance that the details of the talk have been mastered, your interest and competence will have a chance to shine through. There may even be an opportunity for displaying some of the wit and authority that are the prerogatives of any knowledgeable speaker.
- A good talk has a purpose, a desired response, and a main idea. Decide on a goal and main idea, and articulate them in a sentence or two.
- Develop an outline in relation to the amount of time you'll be speaking. Each supporting point must be buttressed with examples and explanations, not simply *announced*. This takes time. The more general the audience background, the more examples, explanations, and contrasts you will want to weave into the talk, and therefore the fewer supporting points you will be able to make. Beginning speakers often try to cram into their presentations too many major points.
- Tight, clear presentation where the speaker makes the point, explains the implications, and then entertains questions are far preferable to lengthy speeches. The successful speaker is one who delivers the message *before* the audience's attention span is depleted.
- Introductions and conclusions are the hardest parts of talks, the two times when nerves take over. To forestall stage fright, skilled speakers prepare these two sections carefully.
- The classic introduction focuses the attention of the listeners, gives them a reason to pay attention, and orients them to the talk that will follow.
- The conclusion should be upbeat, not apologetic. Don't trail off at the end, suddenly embarrassed to be center stage.

• Shape the argument according to the audience. For a diversified group, decide which terms must be defined. Don't present an alphabet soup of buzz words and acronyms that people outside your technical specialty cannot follow.

• Focus on the main idea and its implications, rather than descending into a welter of detail that you find fascinating, but the audience may find irrelevant.

• Use transitional words and phrases to stitch the argument together and help the audience keep its place.

• Use nutshell summaries after you finish each point, to wrap up what you've said and to connect one point to the next.

• Weave into the talk explicit summaries of each slide or transparency you display. Allow time for people to see each visual before you whisk it off the screen.

• Many speakers write out an outline, the introduction and conclusion, and comments on the visuals. Then they practice using a topical outline.

• Practice until you sound as though you are speaking naturally. Reduce your notes to a skeleton outline, including quotations or figures that must be presented verbatim.

• Maintain eye contact. Audiences are disconcerted by speakers who do not look at them. Stifle the desire to turn your back on the group to talk to the slides, transparencies, or blackboard.

• Adjust your appearance so that it will not be a distraction.

• Audiences respond to the intangibles that show through in vocal inflection and facial expression. Extraverbal clues—tone of voice, intonation, suggestive emphasis—and visual clues—expression, voice, posture—are a large part of how people understand us.

EXERCISES

1. Prepare a brief (10-minute) talk explaining a technical product or process with which you are familiar. Some suggestions: how to edit a cell; parts of the dynoptic microscope; path of light through a microscope; memory address switches; how to read the universal product code; how to approach the *New York Times* crossword puzzle; process of desalination; how to read a schematic; how to operate an SLR camera; how to write a 2-part fugue.

2. Prepare a brief (10-minute) talk, defining an important technical term. Examples: price/earning ratio; types of mortgage bonds.

3. Write an analysis of several oral technical presentations. Do it in trip report format, as though you had gone to a short course or conference and heard these speakers. (See Chapter 16 for examples of trip reports.)

4. For discussion: What are the salient differences between an oral presentation of a report and a written presentation?

Appendix: Usage Handbook

1.0 SENTENCES

1.1 Incomplete Sentences

Sentences that are grammatically incomplete are missing a main subject, a main verb, or both. Such sentences are called "fragments."

EXAMPLES OF FRAGMENTS

The yearly Perseid meteor shower.

Appears from the north.

A parachute mounted on an aircraft to be operated in an emergency.

In order to select the one most fit to be used in a mechanical engineering firm.

Note that a sentence may have a subject and verb, but still be a fragment if the subject and verb are subordinate or dependent. Phrases beginning with a subordinate conjunction, for instance, are usually considered incomplete.

> Because it is the 24th state with a mandatory seatbelt law.
>
> Although it is the 24th state with a seatbelt law.

SUBORDINATE CONJUNCTIONS

after	before	when
although	if	whenever
as	since	where
as though	unless	wherever
because	until	while

On occasion, fragments may be used deliberately, for dramatic effect.

> Sales have weakened seriously this year, leading us to a decision to drop the product. But weak sales are not the only reason for our decision.
> fragment
> Is it too soon for a third generation? Probably not.

Incomplete sentences beginning with a coordinate conjunction (*and, but, for, or, nor, so, yet*) are technically fragments. Yet deliberate use of such fragments is becoming increasingly common, particularly when the author seeks a lively, conversational tone, or an extremely informal voice. If you do decide to use a fragment for a transition or for effect, be sure that the audience will understand that the fragment is a deliberate gesture, not a grammatical blunder.

EXERCISES

Underline sentence fragments. Revise incorrect items so that they are complete sentences.

1. He has an important job. Project Engineer for the entire life of the construction contract.

2. He has an essential job. Because he is Project Engineer for the entire life of the construction contract.

3. Press <EXIT> if none of the choices apply.

4. He'll need to contact the vendor. As he is having problems with the program.

5. He'll need to contact the vendor. And try to get his money back, as he is having many problems with the program.

1.2 Comma Splices

English usage prohibits the connecting or splicing of grammatically complete sentences with a comma, unless the sentences are very brief and contain no internal punctuation.

EXAMPLES OF COMMA SPLICES

> Pull the knob on the right-hand side of the console, turn knob clockwise to desired adjustment.
>
> Pull the knob on the right-hand side of the console, then turn the knob clockwise to desired adjustment.
>
> At first the program would not run, however, after repeated tries we finally succeeded.

To correct the comma splice, you may do one of three things, depending on the effect you want:

1. Separate the two sentences with a period.

> Pull the knob on the right-hand side of the console. Turn knob clockwise to desired adjustment.

2. Connect them with a conjunction (and, but, for, or, nor, so, yet).

> Pull the knob on the right-hand side of the console, **and** turn knob clockwise to desired adjustment.

3. Connect them with a semicolon.

> Pull the knob on the right-hand side of the console; turn knob clockwise to desired adjustment.

A common blunder is to use a comma to connect two independent clauses when the second is introduced by a correlative conjunction (*besides, however, hence, moreover, therefore, thus, still, nevertheless, while*).

> We tried running the program, however, the effort was unsuccessful.
>
> **Revision:** We tried running the program; however, the effort was unsuccessful.

EXERCISES

Correct sentences that have comma splices. Explain the reason for the revision.

1. A security lock is on the upper right-hand corner of the keyboard, this lock prevents unauthorized use of the computer.

2. We have been having problems with unauthorized use of the computers, however, security locks should solve some of the difficulties.

3. The computer has been programmed to display messages, furthermore, these prompts appear at each decision point.

1.3 Fused Sentences

In a fused sentence, two sentences are run together and punctuated as though they were one.

EXAMPLE OF FUSED SENTENCES

> The program displays an orderly series of messages the prompts appear at each decision point.

The two independent clauses "The program displays an orderly series of messages" and "the prompts appear at each decision point" have been run together or fused. To correct this error, the writer may separate the sentences with a period:

> The program displays an orderly series of messages The prompts appear at each decision point.

If meaning is better served by connection rather than separation, the writer may use either a semicolon, *and, but, for, or, nor, so,* or *yet*.

The program displays an orderly series of messages; the prompts appear at each decision point.

The program displays an orderly series of messages, for the prompts appear at each decision point.

1.4 Subject-Verb Agreement

Subjects and verbs must agree in number: single subject, single verb; plural subject, plural verb.

EXAMPLES OF SUBJECT-VERB ERROR

Single subject	*Plural verb*
Neither of the instruments	work well.
Each of the hundreds of applicants throughout the United States	are eligible.

To correct this error, the author should change the verbs so that they agree in number with the subjects:

Neither of the instruments works well.

Each of the hundreds of applicants throughout the United States is eligible.

Why are there so many problems with subject-verb agreement in technical writing? Sometimes the problem lies in sentences so long the subject is no longer in shouting distance of the verb. Sometimes, though, the error has to do with the writer's ignorance of the many tricky conventions that govern subject-verb agreement. Here is a brief review:

1. Compound subject. Subjects joined by *and* are usually plural.

His thoughts and actions were centered on success.

The compilation of technical information for the manual and its use with graphics and tabular material are emphasized.

Exceptions: If the compound subject represents a single idea, the verb may be singular.

The long and short of it is that we are leaving.

If the compound subject is modified by *each* or *every*, the verb is singular.

Each computer and printer is available for the students.

2. Collective nouns are usually constructed as singular:

The Board of Directors is pleased to announce a new member.

The laboratory staff has voted.

Collective nouns may be construed as plural if each member of the group is acting individually.

The jury are in strong disagreement, and unable to reach a decision.

A number of chemistry majors are electing medicine as a career. (But: the number of chemistry majors is increasing.)

3. In "either–or" and "neither–nor" constructions, use a singular verb when both subjects are singular. If both are plural, use a plural verb. If they are mixed, the verb agrees with the nearer subject.

Neither the table nor the chair has arrived.

Neither the table nor the chairs have arrived.

Neither the chairs nor the table has arrived.

4. Ignore objects of prepositions or compound prepositions such as "along with," "in addition to," "together with," "accompanied by," when determining the number of the verb.

The CEO, as well as her assistant, visits the floor each week.

The computer, in addition to the printer, is on sale this week.

The problem encountered by the hundreds of users connected by modems has not been addressed.

5. "Either," "neither," and "each" are always singular when they are pronoun subjects.

Each of the hundreds of residents knows what it is like to experience poverty.

Neither of the instruments works well.

Box A.1

ARE PEOPLE WHO SAY 'ONE CRITERIA' INFERIOR?

We retain the spelling of the original language for many words we borrow from other languages, particularly when we are borrowing Greek and Latin terms. Thus we have one spectrum, but two spectra (not spectrums); one alumnus, two alumni; one criterion, two criteria; one phenomenon, two phenomena.

But language changes quickly, and probably no language is more flexible about adaptations than English. For many reasons—from a desire to avoid foreign phrases to a wish to standardize plurals—we've tended to Anglicize foreign plurals—that is, convert their spelling and pronunciation to conform to English language practices. Thus some newer dictionaries now list the preferred plural of appendix as appendixes (as opposed to appendices). *The New York Times Style Guide* prefers curriculums to curricula; many dictionaries list the plural of compendium as compendiums, of encomium as encomiums. If such changes sound odd to you, consider other adaptations that are now institutionalized. Agenda is usually construed as a singular; the plural is agendas. Insignia is now usually used for both singular and plural, although sometimes you may see insigne or insignias.

Does this mean you may now simply add "s" to the plural of any Latin or Greek word? No, it does not. "Curriculums" is acceptable, but "criterions" is not yet on the horizon. Further, even Anglicizations such as curriculums may grate on some people's nerves, however acceptable their use is becoming. If you are conservative in approach, therefore, you may want to stick to traditional Greek and Latin plurals.

If you do preserve the language of the original plural (phenomena, criteria, data), remember that any modifiers must be plural, too. There's no such thing as "one criteria" or "one phenomena."

Wrong: One phenomena remains for analysis.

Revision: One phenomenon remains for analysis.

Wrong: This criteria is the basis for our choice.

Revision: This criterion is the basis for our choice.

6. "None" may be singular or plural, depending on whether you mean "not a single one" (singular) or "all are not" (plural). The trend today is to construe "none" as a plural; therefore, if you mean "not a single one," you may be better off substituting "not one" or "no one" for "none."

None of the applicants were as skilled as their predecessors two years ago.

7. Noun clauses that are the subject of the sentence always take a singular verb.

What the CEO decides is her own business.

What this country needs is a five-cent cigar.

8. Linking verbs (am, is, are, was, were, seem, appear) take their number from the subject. This is confusing when the number of the subject and the complement (that which follows the linking verb) differ:

subject complement

Repeated absences were the reason he was fired.

The reason he was fired was repeated absence.

What was noteworthy was the profit margins.

The profit margins were what was noteworthy.

9. Nouns that end in s, but are singular in meaning, (ethnics, economics, electronics, news, mumps) take a singular verb.

Physics is a difficult subject.

Ten dollars is a lot of money.

EXERCISES: SUBJECT-VERB AGREEMENT

Choose the correct verb. Circle the subject or subjects.

1. Neither the chief chemist nor members of his staff (knows, know) how to analyze the sample.

2. Every one of the hundreds of physicists chosen for the two projects (understands, understand) the criteria for selection.

3. Neither of the instruments (gives, give) proper results.

4. The section head, along with her entire staff, (is, are) planning to attend the conference.

5. Instead of pins popping out and striking the paper, a column of nine ink jets (shoot, shoots) the ink at the paper.

1.5 Pronoun Antecedent

Pronouns should agree in number with their antecedents: singular antecedent, singular pronoun; plural antecedent, plural pronoun. They should also agree in gender (masculine, feminine, neuter) and person (first, second, third).

EXAMPLES OF ERRORS IN PRONOUN ANTECEDENT

Wrong: Each animal in the test group has been in isolation for six weeks before having **their** examination.

Revision: Each animal in the test group has been in isolation for six weeks before having **its** examination.

Wrong: Once we hoisted both her sails and took the yacht out of the bay, we began to move switfly. Soon we realized **it** was the best yacht in the group.

Revision: Once we hoisted both her sails and took the yacht out of the bay, we began to move swiftly. Soon we realized **she** was the best yacht in the group.

Many mistakes come about when the antecedent is "everyone," "anybody," "each," or "either." These antecedents take a singular pronoun.

Each of the men in the battalion had **his** own way of looking at the problem.

When there are women in the battalion, the problem becomes thorny, as English possesses no singular pronoun that includes both sexes. Historically, the solution has been to use "him" or "his" to refer to singular antecedents.

Each staff member should do as he pleases.

The conscientious writer who wants to include females often tries, "Each staff member should do as they please." This creates a new problem. "They" is incorrect, for it is plural and does not agree with the singular antecedent, "each staff member." Should the writer substitute the traditional "his" for "they"? Recently many writers have made an effort to recast such sentences so they are not exclusively male. One solution is to use a plural.

> Staff members should do as they please.

When the relentless plural won't work, the resourceful writer might try either recasting the sentence, or alternating male and female examples.

> A writer should be careful of his antecedents. →
>
> Writers should be careful of their antecedents.
>
> A writer should be careful in the use of antecedents.

Here are some other guidelines for choosing correct pronouns:

1. "Each," "everyone," "everybody," "anybody," and "anyone" use singular pronouns.

> Each of the applicants has an employment form.

2. For "all," "some," and "none," use a singular or plural pronoun, depending on meaning.

> All of the students dislike their assignment.
>
> All of the carbon silicate was emptied from its jar.

3. Antecedents linked by "and" are usually plural. These are called compound antecedents.

> Software and other peripherals have been carefully tested. They are not examples of "vaporware"—programs that exist only in the minds of the manufacturers.

4. When the antecedent is connected by "or" or "nor," the pronoun agrees with the nearer antecedent. When both antecedents are singular, or both plural, this rule presents no problem. When one is singular and the other plural, however, the sentences are often cumbersome. The solution is usually either to place the plural second, or to recast the sentence.

> Neither John nor his brothers have the sense to bring their notebooks.

5. Antecedents of collective nouns take a singular or plural pronoun depending on meaning.

The jury has rendered its verdict.

The jury are unable to make up their minds.

The choir took its seat.

One by one the audience took their seats.

Antecedents should be clear. If the reader must pause to decide on the probable antecedent of a pronoun, recast the sentence so that there is no momentary jolt, ambiguity, or misunderstanding.

Wrong: Smith told Jones his paper was poorly done. (antecedent of "his" may be either Smith or Jones)

Wrong: The tenacity of the American effort suggested their certainty of success. (antecedent of "their" is unstated)

Wrong: NASA engineers, acting on the prodding of the special investigatory committee, presented a series of design changes for the O rings, which pleased the committee. (What pleased the committee—the designs or the engineers' response?)

EXERCISES

Correct any errors in pronoun-antecedent agreement. Circle the antecedent(s) of any pronoun(s) you correct.

1. Most of the time these packages have to be retaped and sent back to its place of origin.

2. Another problem with the letters and packages is that they arrive without a branch name specified. It only says "State Bank."

3. Four speakers from different companies and universities each brought their own expertise to the program.

4. The simple heat and mass transfer models presently composed are inadequate for dealing with the complexities of coal devolatilization; however, it is a productive beginning to the establishment of this process as a way of producing fuel.

5. Amalgamated Corporation distributes their own hiring guidelines.

6. If the printer is capable of more than the minimum criteria, it is a better buy because the added features would make it easier to use; also they probably do a better job of printing.

7. Customers who fail to inform the car service of a change in plans are the biggest problem. The drivers who are dispatched to pick up these customers pay for their short-sightedness.

8. Dr. Korder gave his talk on SIMS imaging which was similar to that presented in Cleveland.

9. Nurses and students will identify with the struggles and stresses depicted; the public will develop a deeper appreciation for what nurses go through—one that I hope is more than the mental one they have now.

1.6 Modification

Introductory phrases should modify the subject of the main sentence. If they don't, they are said to "dangle."

EXAMPLES OF DANGLING MODIFIERS

On arriving at the laboratory, his new staff met him in the reception area. (The sentence says his staff arrived at the laboratory.)

Even as a native of Richmond, it is difficult to adjust to the changes in the city. (The sentence says "it" is a native of Richmond.)

To be considered for a job, a copy editor's test must be taken. (The sentence says a copy editor's test is being considered for a job.)

Some sentences with dangling modifiers may be corrected by simply moving the modifying phrase to its proper position.

Some people hesitate to refer children to the police with drug problems.

Some sentences are best corrected by adding a logical subject.

one finds it

Even as a native of Richmond, ~~it is~~ difficult to adjust to the changes in the city.

an applicant must take

To be considered for a job, a typing test ~~must be taken~~.

Place modifying words and phrases next to the words they explain.

That he finished the job completely amazed the supervisor. ("Completely" may refer either to finishing the job or amazing the supervisor.

They only saw each other at lunch. ("Only" modifies "saw," and suggests that the two people saw each other and nothing else in the room. Only should be moved to modify "at lunch.")

Revision: They only saw each other at lunch.

EXERCISES

Correct any modification errors.

1. If thought about for some time, most people might agree with that statement.
2. No longer in working order, we threw out the equipment.
3. Yet while enjoying the tour of the labs, there were moments that threw the experience out of balance.
4. Cooped up in the tour van of the guide, the countryside shot by as we tore down the two-lane highway.
5. Riding through the squat foliage, dense but low from lack of rain, the road squeezed through the hemp and chickle trees.
6. At the conference, our research team joined the mostly French chemists for dinner.

1.7 Parallel Structure

Use parallel structure when you are joining elements with "and," "but," "not only," "but," and "also."

EXAMPLES OF ERRORS IN PARALLEL STRUCTURE

He likes **swimming**
and
to dance

He likes not only **swimming**
but also **to dance.**

He likes either **swimming**
or **to dance**

The computer is **portable**
and
costs little.

This advertising persuades us

not only

to buy the product

but also

that the advertiser has the highest of morals

Instead, link adjectives with adjectives, prepositional phrases with prepositional phrases, infinitives with infinitives.

He likes not only swimming but dancing.

He likes not only to swim but also to dance.

The computer is portable and inexpensive.

This advertising persuades us not only to buy the product, but also to believe that the advertiser has the highest of morals.

Be sure items in lists, outlines, and tables of contents are parallel. In this excerpt, from a table of contents, the writer combines different syntactic forms.

Wrong:

1. Accessing the System
2. Correcting Errors

    You May Use a Backward Slash(\)
    Using the Backspace
    How to Use Dollar Sign($)

Revision:

1. Accessing the System
2. Correcting Errors

    Backward Slash (\)
    Backspace
    Dollar Sign ($)

EXERCISES

Correct any errors in parallel order.

1. She either is a scientist or an engineer.
2. We needed the new printers because they were faster and they would also reduce the noise level in the office.

3. The manufacturer is not only famous for its reliability but also for its low prices.

4. He is a good manager who not only listens but he works well with others.

5. The committee suggests three changes: upgrading the security on the lot; closing of all corridors to the elevators; we should invest in a closed-circuit monitor system.

2.0 PUNCTUATION

2.1 The Apostrophe

Possessives: Form possessives by adding 's for the singular (the pilot's control) and ' for the plural (the engineers' association).

There are a few exceptions to this rule.

1. When a noun forms the plural without adding "s," use 's for the possessive (children's, men's, women's, mice's).

2. When a singular noun ends in "s," some writers simply add an apostrophe; others follow the rule of 's.

Dickens 's book	Dicken s' book
DeVries 's scholarship	DeVrie s' scholarship

While there is no consensus, there is a trend toward 's for all singular nouns, including those ending in "s."

3. Remember that pronouns that are possessive (hers, its, ours, theirs, yours) require no apostrophe. Indefinite pronouns (anybody, someone else, everyone) do require the apostrophe.

Yours is not a report to be taken lightly.

The cell comes with its own power source. ("It's" would mean "it is.")

It is someone else's laser.

2.2 The Colon

Use a colon to introduce a list, provided you finish the sentence first.

> We had three choices: political, economic, and social.
>
> We had the following three choices: political, economic, and social.
>
> (But not, "Our choices are: political economic, and social.")

Use a colon to link two sentences when the second sentence explains the first. Capitalize the first letter of the second sentence as you would normally.

> His life illustrates an important principle: The written word is powerful.
>
> Only one question remains to be answered: Who will assume responsibility for this project?
>
> Over the piano was printed a notice: Please do not shoot the pianist. He is doing his best. (Oscar Wilde)

Use a colon for double or "hanging" titles.

> *Irascible Genius: A Life of Charles Babbage, Inventor*

2.3 The Comma

1. To connect independent clauses, joined by "and," "but," "for," "or," "nor," "so," "yet."

> It is a carefully thought out program, but I doubt Amalgamated will adopt it.

2. To set off introductory elements such as subordinate clauses or introductory phrases. The comma is used here to prevent an initial misreading of the sentence.

> Although he studied diligently, he failed the entrance examination and had to retake it.
>
> In winter, storms are dangerous.
>
> To succeed, your physical condition must be superlative.
>
> In most cases, we have seen an immediate improvement.

3. To set off parenthetic (nonrestrictive) information.

My wife, Mary, is an engineer.

The difference between restrictive and nonrestrictive information is shown in the following two sentences.

The rules, which we found incomprehensible, were completely revised.

The rules that were incomprehensible were completely revised.

In the first sentence, we are revising all the rules; the expression "which we found incomprehensible" is extra or parenthetic, and is set off by commas. In the second sentence, we are revising only the rules that were incomprehensible. "That were incomprehensible" is essential or restrictive information, and is not set off by commas.

A common blunder is to use the first comma, but omit the second one.

Wrong: A student, Sharon Kelley commented that nursing is a challenge.

Revision: A student, Sharon Kelley, commented that nursing is a challenge.

4. To separate coordinate items in a series, whether the items are words, phrases, or clauses.

He came, he saw, and he conquered.

You have three choices: political, economic, and social.

This text uses the serial comma, the comma before "and." Many writers prefer the sentence without the serial comma, arguing that the "and" does an adequate job of separation.

Other writers cite examples like Robert Frost's line: "The words are lovely, dark and deep." They point out that Frost did not write, "The words are lovely, dark, and deep." Writers who want to preserve the possibility of pairing two items in the series—in this case "dark and deep"—must, they say, use the serial comma consistently in all other cases.

A more prosaic sentence also serves as an example of this point:

The proposal was presented to administrators, engineers and finance officers.

Was the proposal presented to three groups (administrators, engineers, and finance officers) or to one group, administrators, composed of engineers and finance officers?

In general, newspaper style guides omit the serial comma, while academic guidelines use it. While there is no right and wrong in this issue, it's worth noting that the writer who uses the serial comma consistently will avoid some possible misreadings.

5. To separate elements in titles, addresses, and dates, and to set off "not" phrases.

> April 18 , 1948 , was the date the company was founded.
>
> (**Note:** Commas are *not* used after the names of months.)
>
> Dr. Robert Phelan , Section Head , Research , awarded the prize.
>
> Rachel Sinclair, a sophomore at Keuka College in Tennyan , New York , is studying to be a nurse.
>
> We seek facts , not opinions.

2.4 The Dash

The dash is made with two hyphens, with no space before or after the hyphens.

> Integrated circuits--slices of pure silicon embedded with traces of impurities--are used in the new process.

Dashes may function as strong parentheses, as they do in this definition of integrated circuits. They may also be used instead of a colon, if the author seeks a more informal effect:

> The paper is titled "Why Some Professionals Write Articles and Some Do Not--Results of a Survey."
>
> The unit contains three channels--range-sweep generation, range-sweep expansion, and range-sweep limiting.

2.5 The Hyphen

Use hyphens for compound modifiers before a noun:

> He did a 30-hour run on the test machine.
>
> **But not for,** He did a run on the test machine that lasted 30 hours.

They did a step-by-step analysis of the problem.
But not for, They did the analysis step by step.

100-meter dash

22-cent stamp

Use hyphens to avoid confusion:

We need to re-form the committee.

The laboratory will use wire mesh to re-cover the exhaust hoods.

Use suspensive hyphens for a series of hyphenated compound modifiers:

The banjo is a 4- or 5-stringed instrument.

They did the tests at 5-, 10-, and 15-month intervals.

Use hyphens in compound numbers twenty-one through ninety-nine.

Do not use hyphens with **ly** adverbs:

The newly refurbished laboratory was opened last week.

2.6 Parentheses

Use parentheses to set off technical definitions, acronyms, figure citations, abbreviations, and asides.

Nuclear Magnetic Resonance (NMR) is an increasingly popular diagnostic tool.

2.7 Quotation Marks

Use commas and periods inside quotation marks; use colons and semicolons outside.

"The feasibility study was completed in May," the chairman said.

The secretary spoke about "misspellings in the minutes"; however, I see none.

Question marks and exclamation points go within the quotation only if the punctuation is part of the quotation.

"Danger!" was printed in orange letters on the caution.

Is the name of the article "Revisions to the Tax Code"?

Use quotation marks to enclose titles of articles from periodical literature.

We found the source of the quotation in a recent article, "Bits or Bytes?"

2.8 The Semicolon

Use semicolons to connect independent clauses when you seek a closer link than that provided by a coordinating conjunction (see Section 1.1).

The homogeneal light and rays which appear red, or rather make objects appear so, I call Rubrifick or red-making; those which make objects appear yellow, green, blue, and violet, I call yellow-making, green-making, blue-making, violet-making, and so of the rest. (Isaac Newton)

Use semicolons to connect independent clauses joined by a conjunctive adverb (*however, therefore, thus, then, still, hence, indeed, instead, nonetheless, otherwise*).

Mr. Sinclair had difficulty scheduling the examination; however, he appeared promptly at the inspection.

The heat and mass transfer models presently composed are inadequate for dealing with the complexities of coal devolatilization; however, they are a productive beginning.

Use semicolons to separate items in a list when the items themselves are punctuated.

The modules are 300 mm deep; 200, 300, or 400 mm long; and 250 mm high.

Do not use a semicolon to introduce a list. (The proper mark of punctuation is a colon.)

Wrong: We have three choices in the matter; political, economic, and social.

Revision: We have three choices in the matter: political, economic, and social.

Do not use a semicolon to link a subordinate clause with an independent clause. Remember, the semicolon links complete sentences.

Wrong: Although Mr. Sinclair had a serious cold; he appeared promptly at the inspection.

Revision: Although Mr. Sinclair had a serious cold, he appeared promptly at the inspection.

2.9 Underlining

Use underlining to represent italics in

- names of books, newspapers, and periodical literature

The study appeared in <u>Science.</u>

- foreign words and phrases

We did an <u>in vitro</u> sample.

- letters

Delete item <u>c.</u>

3.0 NUMBERS

If a number begins a sentence, write it out. If you do not want to write out the number, recast the sentence.

One hundred samples were received.

Received were 100 samples.

In general, write out numbers, both ordinal and cardinal, below ten, unless the numbers are accompanied by units of measurement.

Only five people appeared for the committee meeting.

We are doing the fifth trial this week.

When numbers are accompanied by units of measurement, they are rarely spelled out.

We added 3 ml.

The graph showed a 3-point rise.

We measured 3.7 inches of rain.

3-year-old data.

In narrative text, numbers in the millions and above may be rendered by a numeral, followed by the word "million," "billion," etc.

1.3 million tons

555 billion gallons

100 billion times fewer atoms than in the earth's normal atmosphere.

CUMULATIVE EXERCISES

Correct errors in sentence structure, subject-verb agreement, pronoun-antecedent agreement, modification, parallel order, or punctuation.

1. Those honored are also leaders in academia, but it is not the deciding factor in their choice.
2. The nurses' environment can contribute to the stress she endures on her job.
3. Unlike the beginning of her career, the notion of quitting is becoming increasingly feasible.
4. Following the past research of Louis Pasteur in the area of sepsis, carbolic acid was introduced to different traumas.
5. Injuries such as: compound fractures, systemic wounds, and abscesses were treated.

6. At the end of the study period, each woman was classified as belonging to one of four categories: success, noted by the passing of an immunologic pregnancy test, lost to follow-up, if the result of the last cycle was not known, open case, if the last cycle had not been performed, or dropout, if the women discontinued treatment.

7. Oils do influence combustion chamber deposits, however, the mechanism is somewhat obscure at this time.

8. Piston land deposits, while poor for the GM engine, was not affected by high ash oils. The general rating for land deposits were somewhat better for the Chrysler engine.

9. The rates at which hydrogen reduction occurs with pyrite samples from different sources was found to depend upon the samples' impurities and the extent to which various crystallographic faces were exposed.

10. Room 341B has its own intrinsic features that separates it from other rooms at the Polytechnic.

11. The brick wall has two large openings which accommodates the windows.

12. The director, as well as his assistants, were pleased with the results.

13. Each of the hundreds of students studying physics in the elementary sections have a different problem.

14. At the end of each month in which a staff member incurs reimbursable expenses, he should submit a completed travel voucher with the required receipts attached to his department head.

15. A set trap is very dangerous and should be placed where children and pets cannot reach them.

16. All traps incorporate the highest quality workmanship and material which guarantees ease of use.

17. To the right was installed the computer, printer and monitor.

18. There are three electrical outlets, each about a foot above the floor. One behind the bed, one next to the book case, and the third under the desk.

19. Two window frames, each wooden are located in the south wall.

20. The window, facing east receives great amounts of sunshine.

21. The floor is covered with a worn shag carpet of mixed, interwoven colors; khaki, green, brown, black, and white.

22. The moldings of the ceiling are painted dark brown, which blends in with the color of the walls.

23. The research for this report was conducted among 2,193 women whose husbands were totally sterile and as a result conceived with donor semen through a process known as artificial insemination.

Index

D

E

F

G

Q

R

S

T

U

V

W

ABOUT THE AUTHOR

Anne Eisenberg is an Associate Professor at Polytechnic University, where she directs the graduate program in technical and scientific writing. The author of two other textbooks and many articles, she also wrote the American Chemical Society film *Technical Writing*, and gives a course in writing for that society. She has wide experience teaching technical communication for American as well as Japanese corporations.

#435